CAMBRIDGE STUDIES IN ADVANCED MATHEMATICS 109

AN INTRODUCTION TO CONTACT TOPOLOGY

This text on contact topology is the first comprehensive introduction to the subject, including recent striking applications in geometric and differential topology: Eliashberg's proof of Cerf's theorem $\Gamma_4 = 0$ via the classification of tight contact structures on the 3-sphere, and the Kronheimer–Mrowka proof of Property P for knots via symplectic fillings of contact 3-manifolds.

Starting with the basic differential topology of contact manifolds, all aspects of 3-dimensional contact manifolds are treated in this book. One notable feature is a detailed exposition of Eliashberg's classification of overtwisted contact structures. Later chapters also deal with higher-dimensional contact topology. Here the focus is on contact surgery, but other constructions of contact manifolds are described, such as open books or fibre connected sums.

This book serves both as a self-contained introduction to the subject for advanced graduate students and as a reference for researchers.

HANSJÖRG GEIGES is Professor of Mathematics at the Universität zu Köln.

CAMBRIDGE STUDIES IN ADVANCED MATHEMATICS

All the titles listed below can be obtained from good booksellers or from Cambridge University Press. For a complete series listing visit: http://www.cambridge.org/series/sSeries.asp?code=CSAM

Already published

60 M. P. Brodmann & R. Y. Sharp *Local cohomology*
61 J. D. Dixon *et al. Analytic pro-p groups*
62 R. Stanley *Enumerative combinatorics II*
63 R. M. Dudley *Uniform central limit theorems*
64 J. Jost & X. Li-Jost *Calculus of variations*
65 A. J. Berrick & M. E. Keating *An introduction to rings and modules*
66 S. Morosawa *Holomorphic dynamics*
67 A. J. Berrick & M. E. Keating *Categories and modules with K-theory in view*
68 K. Sato *Levy processes and infinitely divisible distributions*
69 H. Hida *Modular forms and Galois cohomology*
70 R. Iorio & V. Iorio *Fourier analysis and partial differential equations*
71 R. Blei *Analysis in integer and fractional dimensions*
72 F. Borceaux & G. Janelidze *Galois theories*
73 B. Bollobas *Random graphs*
74 R. M. Dudley *Real analysis and probability*
75 T. Sheil-Small *Complex polynomials*
76 C. Voisin *Hodge theory and complex algebraic geometry, I*
77 C. Voisin *Hodge theory and complex algebraic geometry, II*
78 V. Paulsen *Completely bounded maps and operator algebras*
79 F. Gesztesy & H. Holden *Soliton Equations and Their Algebro-Geometric Solutions, I*
81 S. Mukai *An Introduction to Invariants and Moduli*
82 G. Tourlakis *Lectures in Logic and Set Theory, I*
83 G. Tourlakis *Lectures in Logic and Set Theory, II*
84 R. A. Bailey *Association Schemes*
85 J. Carlson, S. Müller-Stach & C. Peters *Period Mappings and Period Domains*
86 J. J. Duistermaat & J. A. C. Kolk *Multidimensional Real Analysis I*
87 J. J. Duistermaat & J. A. C. Kolk *Multidimensional Real Analysis II*
89 M. Golumbic & A. N. Trenk *Tolerance Graphs*
90 L. Harper *Global Methods for Combinatorial Isoperimetric Problems*
91 I. Moerdijk & J. Mrcun *Introduction to Foliations and Lie Groupoids*
92 J. Kollár, K. E. Smith & A. Corti *Rational and Nearly Rational Varieties*
93 D. Applebaum *Levy Processes and Stochastic Calculus*
94 B. Conrad *Modular Forms and the Ramanujan Conjecture*
95 M. Schechter *An Introduction to Nonlinear Analysis*
96 R. Carter *Lie Algebras of Finite and Affine Type*
97 H. L. Montgomery, R. C. Vaughan & M. Schechter *Multiplicative Number Theory I*
98 I. Chavel *Riemannian Geometry*
99 D. Goldfeld *Automorphic Forms and L-Functions for the Group GL(n,R)*
100 M. Marcus & J. Rosen *Markov Processes, Gaussian Processes, and Local Times*
101 P. Gille & T. Szamuely *Central Simple Algebras and Galois Cohomology*
102 J. Bertoin *Random Fragmentation and Coagulation Processes*
103 E. Frenkel *Langlands Correspondence for Loop Groups*
104 A. Ambrosetti & A. Malchiodi *Nonlinear Analysis and Semilinear Elliptic Problems*
105 T. Tao & V. H. Vu *Additive Combinatorics*
106 E. B. Davies *Linear Operators and their Spectra*
107 K. Kodaira *Complex Analysis*
108 T. Ceccherini-Silberstein, F. Scarabotti, F. Tolli *Harmonic Analysis on Finite Groups*

An Introduction to Contact Topology

HANSJÖRG GEIGES

Universität zu Köln

CAMBRIDGE UNIVERSITY PRESS
Cambridge, New York, Melbourne, Madrid, Cape Town,
Singapore, São Paulo, Delhi, Mexico City

Cambridge University Press
The Edinburgh Building, Cambridge CB2 8RU, UK

Published in the United States of America by Cambridge University Press, New York

www.cambridge.org
Information on this title: www.cambridge.org/9780521865852

First published 2008

A catalogue record for this publication is available from the British Library

ISBN 978-0-521-86585-2 Hardback

Dedicated to the memory of
Charles B. Thomas (1938–2005)

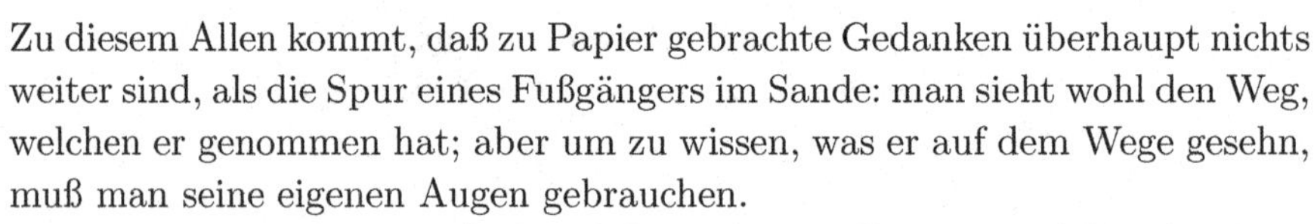

Zu diesem Allen kommt, daß zu Papier gebrachte Gedanken überhaupt nichts weiter sind, als die Spur eines Fußgängers im Sande: man sieht wohl den Weg, welchen er genommen hat; aber um zu wissen, was er auf dem Wege gesehn, muß man seine eigenen Augen gebrauchen.

Arthur Schopenhauer, *Parerga und Paralipomena*

Contents

Preface

> 'We are all familiar with the after-the-fact tone – weary, self-justificatory, aggrieved, apologetic – shared by ship's captains appearing before boards of inquiry to explain how they came to run their vessels aground and by authors composing Forewords.'
>
> John Lanchester,
> *The Debt to Pleasure*

Contact geometry, as a subject in its own right, was born in 1896 in the monumental work of Sophus Lie on *Berührungstransformationen* (contact transformations). Lie traces the pedigree of contact geometric notions back to the work of Christiaan Huygens on geometric optics in the *Traité de la Lumière* of 1690 – or even Apollonius of Perga's *Conica* from the third century BC – and to practically all the famous mathematicians of the eighteenth and nineteenth century.

But as late as 1990, when I began my journey into contact geometry, the field still seemed rather arcane. To the prescience of Charles Thomas I owe the privilege of starting graduate work in an area that was only just beginning to flourish. This, of course, had its drawbacks – there were no texts from which to learn the essentials. Even contact geometry's elder sibling, symplectic geometry – firmly established as the natural language for classical mechanics, and brought into prominence by Gromov's influential 1985 paper on pseudoholomorphic curves – suffered from a similar dearth.

One of the most eloquent of modern panegyrists of contact geometry is Vladimir Arnold, who proclaimed on several occasions since 1989 that 'contact geometry is all geometry'. His *Mathematical Methods of Classical Mechanics*, first published (in Russian) in 1974, was then the only textbook covering the basic notions of contact geometry, albeit in the relative obscurity of an appendix.

A *contact structure* is a maximally non-integrable hyperplane field on an odd-dimensional manifold (Defn. 1.1.3). This sounds like a pretty abstruse object, but, as I hope to convince you in Chapter 1, contact structures appear virtually everywhere in nature.*

There are various ways in which contact manifolds may be regarded as the odd-dimensional analogue of symplectic manifolds. They share the property that there is a *Darboux theorem* providing a local model for such structures. Thus, roughly speaking, there are no *local* invariants, but only *global* phenomena of interest.

The investigation of contact manifolds from a more topological perspective was taken up in the late 1950s by John Gray and Boothby–Wang. Gray proved the stability of contact structures on closed manifolds, i.e. the fact that there are no non-trivial deformations of such structures. This is analogous to the stability of symplectic forms proved by Jürgen Moser, and indeed the 'modern' proof of Gray stability uses the famous 'Moser trick'. Boothby and Wang constructed contact structures on certain principal S^1–bundles; these were the first non-trivial examples of contact manifolds.

The classical period† of contact topology began with the existence results for contact structures on 3–manifolds due to Jean Martinet and Robert Lutz in the early 1970s. Another landmark in this period was Daniel Bennequin's surprising discovery of exotic contact structures on standard 3–space in 1983. The topological flavour of global contact geometry was affirmed in 1989 by Yasha Eliashberg's classification of what he called *overtwisted* contact structures on 3–manifolds; this seminal result marks the culmination of the classical age.

Eliashberg's proof hinted at the importance of using surfaces in contact 3–manifolds as a tool for understanding the latter's global structure. A contact structure induces on such a surface a singular 1–dimensional foliation, the so-called *characteristic foliation*. Control of this characteristic foliation permits various cut-and-paste constructions. Then, in his 1991 thesis, Emmanuel Giroux introduced the notion of a *convex surface*, that is, a surface transverse to a flow preserving the contact structure. It turns out that the characteristic foliation on a convex surface is, in essence, determined by a set of simple closed curves, the *dividing set*, and therefore can be controlled much more effectively.

It seems fair to say that Giroux's thesis inaugurated the modern era of contact topology. Convex surface theory lies at the heart of many of the

* Admittedly, my notion of 'nature' suffers from a certain *déformation professionnelle.*
† On the correct usage of 'classical' see the footnote on page 338.

recent developments, be it the classification of *tight* (i.e. non-overtwisted) contact structures on 3–manifolds, or the classification of various types of knots in contact 3–manifolds.

In 1992, Eliashberg's contact geometric proof of Jean Cerf's celebrated result $\Gamma_4 = 0$ gave the first inkling that contact topology might develop into a powerful tool for tackling pure topological questions. In 2004, this expectation was confirmed in a spectacular manner by Peter Kronheimer and Tom Mrowka's establishing Property P for non-trivial knots. (One consequence of this result is a new proof of the Gordon–Luecke theorem that knots in the 3–sphere are determined by their complement.)

The Kronheimer–Mrowka argument is based in an essential way on a result about 'capping off' symplectic fillings of contact 3–manifolds, proved independently by Eliashberg and John Etnyre. This result, in turn, relies crucially on a correspondence between contact structures and open book decompositions, established by Giroux. This correspondence is another instance of the topological nature of much of contact geometry.

One of the aims of this book is to explain these exciting developments, starting from an exposition of first principles. As yet, there is no textbook or monograph giving a comprehensive introduction to contact geometry. Some basic aspects are treated in Chapter 8 of [7] and in Section 3.4 of [177], but these texts do not prepare the reader for any of the modern topological techniques in contact geometry. The introductory lectures by Etnyre [80], though brief, give a better impression of the state of the art. In the *Handbook* article [97] I was more thorough, but my focus lay on the basic differential topological aspects – many of which had remained folklore until then – and no attempt was made to cover more than the classical period.

Fortunately, that article could serve as a germ for the present monograph and now constitutes part of Chapters 2, 3, and Sections 4.1 to 4.5. Otherwise the task of writing might have seemed too daunting.

Previous versions of Sections 4.1 to 4.3 and Chapter 8 were written as early as 1997 for a series of lectures I gave in Les Diablerets as part of the IIIe Cycle Romand, but contact topology simply developed too rapidly for an exhaustive account to be feasible even then. That a tortoise cannot catch up with Achilles is hardly paradoxical. So when I started writing this book in earnest, I had to be modest and realistic about its aims. The reader I had primarily in mind is an advanced graduate student educated in differential topology and the geometric aspects of algebraic topology, say on the level of the textbooks by Bröcker–Jänich [38] or Hirsch [132], and that by Bredon [35], respectively. A good dose of geometric topology, e.g. from Rolfsen [215] or Prasolov–Sossinsky [209], would be helpful, but most geometric topological concepts will be explained at length when we need them. The present book should serve as a preparation for studying a substantial body of research literature; ample references for further reading will be provided.

Here is a brief summary of the contents of this book; more details can be found at the beginning of each chapter. Chapter 1 is something of an advertisement, aimed at readers not necessarily inclined to devote the rest of their life to contact geometry. Chapter 2 deals with the basic differential topology of contact manifolds. Sections 2.1 to 2.4 are foundational for everything that is to follow. The neighbourhood and isotopy extension theorems in Sections 2.5 and 2.6 are likewise indispensable, but their proofs – though useful for learning the nuts and bolts of the Moser trick – should probably be skimmed on a first reading. The contact disc theorem in Section 2.6.2 deserves special mention. This theorem has been assumed at various places in the contact topological literature. The proof is analogous to that of the usual disc theorem in differential topology, but I thought it opportune to be explicit about the 'contact Alexander trick'.

Chapter 3 provides a basic introduction to the theory of knots in contact 3–manifolds. There are two types of knots to investigate: those tangent to the contact structure (called *Legendrian knots*), and those transverse to the contact structure (called, rather imaginatively, *transverse knots*). That chapter constitutes the basis for the surgical constructions of contact manifolds along transverse knots (Sections 4.1 to 4.3) and Legendrian knots (Section 6.4), respectively.

My initial plan for Chapter 4 on contact 3–manifolds was to heed Polonius's advice that 'brevity is the soul of wit' [220]. If that chapter has grown beyond proportion, it is less owing to the 'outward flourishes', but to my realising that this monograph would be inadequate if it failed to introduce the reader to some modern techniques like convex surfaces (Sections 4.6.2 and 4.8) and Giroux's tomography (Section 4.9). Even so, there are some regrettable omissions. I say virtually nothing about Ko Honda's important contributions to convex surface theory – Honda's approach is parallel to, but independent of, much of Giroux's work on tomography. My impression is that tomography *à la* Giroux is more amenable to an elementary treatment, in particular for the classification of tight contact structures on some simple 3–manifolds (Section 4.10).

Concerning the relation between open books and contact structures, I only discuss the *existence* of contact structures on open books, both in dimension 3 (Section 4.4.2) and in higher dimensions (Section 7.3). For the converse, how to find an open book decomposition adapted to a given contact structure, there are some useful lecture notes by Etnyre [83], at least for the 3–dimensional case.

Other lacunæ in the chapter on 3–manifolds stem from my failure to mention anything related to the Heegaard Floer theory of Peter Ozsváth and Zoltán Szabó, or the analytical side of contact geometry (dynamics of the Reeb flow, including the Weinstein conjecture, and techniques involving pseudoholomorphic curves), initiated by Helmut Hofer. Here are some pertinent references: [165], [200], [201], [202], [226]; [135], [136], [228].

Section 4.7 contains a detailed proof of Eliashberg's classification of overtwisted contact structures, following the lines of the original argument. Eliashberg's paper – together with his 1992 paper on Cerf's theorem – is probably one of the two most-quoted papers in all of contact geometry, but one which is notoriously hard to read. I hope that my exposition goes some way towards making Eliashberg's ideas more accessible.‡

Chapter 4 also includes Eliashberg's proof of Cerf's theorem (Section 4.11), and the analogue for tight contact 3–manifolds of John Milnor's unique decomposition theorem for 3–manifolds (Section 4.12, based on joint work with Fan Ding).

The remainder of the book is concerned mostly with higher-dimensional contact topology. Chapter 5 introduces the notions of symplectic fillings and convexity. There is an intrinsic interest in relating contact structures to notions of convexity in complex geometry, but the main purpose of that chapter is to lead into Chapter 6 on contact surgery. Being the mathematical son of a mathematical surgeon and the real grandson of a real surgeon, I regard this chapter (and its companion Chapter 8, where the theory is applied to 5–manifolds) as my earliest motivation for writing this book. Sections 6.1 to 6.3 contain an exposition of the Eliashberg–Weinstein contact surgery. Some ideas about homotopy principles *à la* Gromov are explained along the way. I also allow myself a small detour to give a contact geometric proof of the Whitney–Graustein theorem on the classification of immersions of the circle in the plane.

With Section 6.4 we return to the study of contact 3–manifolds. In the 3–dimensional setting one has a more general concept of contact *Dehn* surgery, introduced by Fan Ding and myself. This type of surgery is then used to prove the Eliashberg–Etnyre theorem about symplectic caps – in contrast with the proofs of Eliashberg and Etnyre, the one presented here does not rely on the Giroux correspondence between contact structures and open books. That section completes the account of the contact topological aspects of the Kronheimer–Mrowka proof of Property P.

‡ Charles Thomas might have referred to this exposition – with apologies to Moses Maimonides – as a 'guide for the perplexed'.

In Section 6.5, contact surgery of 3–manifolds is used to show, modulo a result from Seiberg–Witten theory that I quote only, that symplectically fillable contact structures are tight.

As mentioned before, Chapter 8 may be regarded as a rather long example how contact surgery can be applied. Section 8.2 is an exercise in the classification of manifolds; Section 8.1 may be instructive as an application of obstruction theory, and it contains some neat geometry of complex projective 3–space.

Chapter 7, which can be read right after Chapter 2, gives further justification – if such be required – for calling this a book on contact *topology*. That chapter explains a variety of topological constructions that can be made compatible with contact structures. Besides the open books referred to earlier, these are fibre connected sums, branched covers, and plumbing. Applications of these techniques include the construction of contact structures on all odd-dimensional tori, due to Frédéric Bourgeois, and that of exotic contact structures on all odd-dimensional spheres. Here I have to confess another glaring sin: contact homology, not to mention symplectic field theory, gets only a passing reference as a tool for distinguishing such exotic structures. Slightly more geometric constructions discussed in Chapter 7 are Brieskorn manifolds, the Boothby–Wang construction, and contact reduction.

The two brief appendices contain some minor technicalities concerning the generalised Poincaré lemma and time-dependent vector fields.

Exercises are conspicuous by their absence. I hope that the numerous explicit examples make up for this defect.

As pointed out earlier, this book had its beginnings in a series of lectures in the IIIe Cycle Romand in 1997. The book began to take shape during a sabbatical at the Forschungsinstitut für Mathematik of the ETH Zürich in 2004. I lectured on parts of this book in various seminars, and in 2005 at a Frühlingsschule in Waren (Müritz) organised by Christian Becker, Hartmut Weiss and Anna Wienhard. In the winter term 2006/07 I gave a lecture course on contact geometry at the Universität zu Köln. In June 2007 I presented the Lisbon Summer Lectures in Geometry based on this book, at the invitation of Miguel Abreu and Sílvia Anjos. The questions and comments of these various audiences have helped to improve the exposition.

Many people read parts of this book and found numerous smaller or bigger misprints, obscure statements, or plain inaccuracies. Countless stimulating conversations have left their traces in this book. For their assistance I thank Frédéric Bourgeois, Yuri Chekanov, Yvonne Deuster, Thomas Eckl, John Etnyre, Otto van Koert, Paolo Lisca, Klaus Niederkrüger, Federica

Pasquotto, Bijan Sahamie, Felix Schlenk and Matthias Zessin. Fabian Meier sent me a copy of his Cambridge Part III essay on Eliashberg's classification of overtwisted contact structures; this helped with working out some of the details in Section 4.7. Stephan Schönenberger allowed me to use his Figures 1.1 and 2.1; John Etnyre contributed Figures 2.2 and 6.8. John Etnyre's encouraging me to go ahead with this book project is invaluable. I am grateful to my collaborators Fan Ding, Jesús Gonzalo and András Stipsicz; talking mathematics with them has been a wonderful experience over many years, and the fruits of our joint projects are visible throughout this book.

Yasha Eliashberg has been a source of inspiration ever since he commented encouragingly on my first mathematical publication. His influence on this book is plainly evident. Especially Section 4.7 would never have been completed without his spending many hours patiently explaining to me his classification of overtwisted contact structures. In passing we noticed little gems such as the contact geometric proof of the Whitney–Graustein theorem.

Much of my intellectual education, mathematical and otherwise, I owe to my mentor and friend Charles Thomas. I should have liked to present this book to him as a small thank-you for introducing me to the fascinating world of contact geometry when I was his graduate student and – perhaps even more important – for showing me by his example what it means to lead a scholarly life. Now, sadly, all I can do is dedicate this book to his memory.

Köln, July 2007 Hansjörg Geiges

1
Facets of contact geometry

'After a while the style settles down a bit and it begins to tell you things you really need to know.'

Douglas Adams,
The Hitch Hiker's Guide to the Galaxy

This opening chapter is not meant as an introduction in the conventional sense (if only because it is much too long for that). Perhaps it would be appropriate to call it a *proem*, defined by the *Oxford English Dictionary* as 'an introductory discourse at the beginning of a book'. Although some basic concepts are introduced along the way, the later chapters are largely independent of the present one. Primarily this chapter gives a somewhat rambling tour of contact geometry.

Specifically, we consider polarities in projective geometry, the Hamiltonian flow of a mechanical system, the geodesic flow of a Riemannian manifold, and Huygens' principle in geometric optics. Contact geometry is the theme that connects these diverse topics. This may serve to indicate that Arnold's claim that 'contact geometry is all geometry' ([15], [17]) is not entirely facetious. I also present two remarkable applications of contact geometry to questions in differential and geometric topology: Eliashberg's proof of Cerf's theorem $\Gamma_4 = 0$ via the classification of contact structures on the 3–sphere, and the proof of Property P for non-trivial knots by Kronheimer and Mrowka, where symplectic fillings of contact manifolds play a key role. One of the major objectives of this book will be to develop the contact topological methods necessary to understand these results.

The subsequent chapters will not rely substantively on the material presented here. Therefore, readers interested in a more formal treatment of contact geometry and topology may skim this proem and refer back to it for some basic definitions only. But I hope that the present chapter, while serving as an invitation to contact geometry for the novice, contains material that will be of interest even to readers with prior exposure to the subject. In the survey article [94] I have touched upon some of the themes of this introduction, but without the detailed proofs given here; on the other hand, that article contains more on the historical development of contact geometry. On that topic, see also the historical survey by Lutz [170]. As an *amuse-gueule* you may enjoy [96].

1.1 Contact structures and Reeb vector fields

Let M be a differential manifold, TM its tangent bundle, and $\xi \subset TM$ a field of hyperplanes on M, that is, a smooth† sub-bundle of codimension 1. The term codimension 1 **distribution** is quite common for such a tangent hyperplane field (and not to be confused with distributions in the analysts' sense, of course). In order to describe special types of hyperplane fields, it is useful to present them as the kernel of a differential 1–form.

Lemma 1.1.1 *Locally, ξ can be written as the kernel of a differential 1–form α. It is possible to write $\xi = \ker\alpha$ with a 1–form α defined globally on all of M if and only if ξ is* coorientable, *which by definition means that the quotient line bundle TM/ξ is trivial.*

Proof Choose an auxiliary Riemannian metric g on M and define the line bundle $\xi^\perp$ as the orthogonal complement of ξ in TM with respect to that metric. Then $TM \cong \xi \oplus \xi^\perp$ and $TM/\xi \cong \xi^\perp$. Around any given point p of M, there is a neighbourhood $U = U_p$ over which the line bundle $\xi^\perp$ is trivial. Let X be a non-zero section of $\xi^\perp|_U$ and define a 1–form α_U on U by $\alpha_U = g(X, -)$. Then clearly $\xi|_U = \ker \alpha_U$.

Saying that ξ is coorientable is the same as saying that $\xi^\perp$ is orientable and hence (being a line bundle) trivial. In that case, X and thus also α exist globally. Conversely, if $\xi = \ker\alpha$ with a globally defined 1–form α, one can define a global section of $\xi^\perp$ by the conditions $g(X, X) \equiv 1$ and $\alpha(X) > 0$, hence ξ is coorientable. □

Remark 1.1.2 As a student at Cambridge, I was taught that in Linear Algebra 'a gentleman should never use a basis unless he really has to'. By the same token, one should not use auxiliary Riemannian metrics if one can do without them. So here is the alternative argument, which may well serve

† The terms *smooth* and *differentiable* are used synonymously with C^∞; a *differential* manifold is a manifold with a choice of differentiable structure. Throughout this book, manifolds, bundles, vector fields, and related objects, are assumed to be smooth.

as a warm-up for (slightly less pedestrian) bundle-theoretic considerations later on.

The quotient bundle TM/ξ and its dual bundle are locally trivial. Thus, over small neighbourhoods U, one can define a differential 1–form α_U as the pull-back to the cotangent bundle T^*M of a non-zero section of $(TM/\xi)^*|_U$ under the bundle projection $TM \to TM/\xi$. This 1–form clearly satisfies $\ker \alpha_U = \xi|_U$.

If ξ is coorientable, $(TM/\xi)^*$ admits a global section, and the above construction yields a global 1–form α defining ξ. Conversely, a global 1–form α defining $\xi = \ker \alpha$ induces a global non-zero section of $(TM/\xi)^*$.

Except in certain isolated examples below, we shall always assume our hyperplane fields ξ to be coorientable.

One class of hyperplane fields that has received a great deal of attention are the **integrable** ones. This term denotes hyperplane fields with the property that through any point $p \in M$ one can find a codimension 1 submanifold N whose tangent spaces coincide with the hyperplane field, i.e. such that $T_qN = \xi_q$ for all $q \in N$. Such an N is called an **integral submanifold** of ξ. It turns out that $\xi = \ker \alpha$ is integrable precisely if α satisfies the *Frobenius integrability condition*

$$\alpha \wedge d\alpha \equiv 0.$$

In terms of Lie brackets of vector fields, this integrability condition can be written as

$$[X, Y] \in \xi \text{ for all } X, Y \in \xi;$$

here $X \in \xi$ means that X is a smooth section of TM with $X_p \in \xi_p$ for all $p \in M$. A third equivalent formulation of integrability is that ξ is locally of the form $dz = 0$, where z is a local coordinate function on M. A good textbook reference for these facts is Warner [238]. The collection of integral submanifolds of an integrable hyperplane field constitutes what is called a codimension 1 *foliation*. For the global topology of foliations (of arbitrary codimension) see Tamura [227].

Contact structures are in a certain sense the exact opposite of integrable hyperplane fields.

Definition 1.1.3 Let M be a manifold of odd dimension $2n+1$. A **contact structure** is a maximally non-integrable hyperplane field $\xi = \ker \alpha \subset TM$, that is, the defining differential 1–form α is required to satisfy

$$\alpha \wedge (d\alpha)^n \neq 0$$

(meaning that it vanishes nowhere). Such a 1–form α is called a **contact form**. The pair (M, ξ) is called a **contact manifold**.

Remark 1.1.4 As a somewhat degenerate case, this definition includes 1–dimensional manifolds with a non-vanishing 1–form α. The corresponding contact structure $\xi = \ker \alpha$ is the zero section of the tangent bundle.

Example 1.1.5 On $\mathbb{R}^{2n+1}$ with Cartesian coordinates

$$(x_1, y_1, \ldots, x_n, y_n, z),$$

the 1–form

$$\alpha_1 = dz + \sum_{j=1}^{n} x_j \, dy_j$$

is a contact form. The contact structure $\xi_1 = \ker \alpha_1$ is called the **standard contact structure** on $\mathbb{R}^{2n+1}$. See Figure 1.1 for the 3–dimensional case.

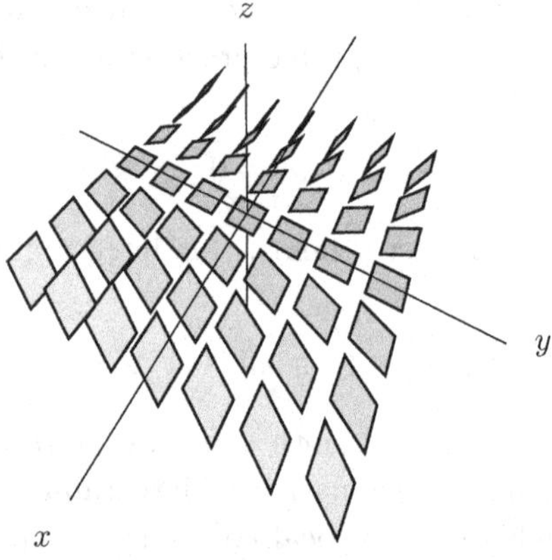

Fig. 1.1. The contact structure $\ker(dz + x\, dy)$.

Remark 1.1.6 Observe that α is a contact form precisely if $\alpha \wedge (d\alpha)^n$ is a **volume form** on M (i.e. a nowhere vanishing top-dimensional differential form); in particular, M needs to be orientable. The condition $\alpha \wedge (d\alpha)^n \neq 0$ is independent of the specific choice of α and thus is indeed a property of $\xi = \ker \alpha$: any other 1–form defining the same hyperplane field must be of the form $\lambda\alpha$ for some smooth function $\lambda\colon M \to \mathbb{R} \setminus \{0\}$, and we have

$$(\lambda\alpha) \wedge (d(\lambda\alpha))^n = \lambda\alpha \wedge (\lambda \, d\alpha + d\lambda \wedge \alpha)^n = \lambda^{n+1} \alpha \wedge (d\alpha)^n \neq 0.$$

We see that if n is odd, the sign of this volume form depends only on ξ, not the choice of α, so the contact structure ξ induces a natural orientation of M. If M comes equipped with a specific orientation, one can speak of *positive* and *negative* contact structures.

Lemma 1.1.7 *In the 3–dimensional case the contact condition can also be formulated as*

$$[X,Y]_p \notin \xi_p \quad \text{at every } p \in M, \text{ for all pointwise linearly independent vector fields } X, Y \in \xi.$$

Proof The equation

$$d\alpha(X,Y) = X(\alpha(Y)) - Y(\alpha(X)) - \alpha([X,Y]),$$

which holds for arbitrary 1–forms α and vector fields X, Y on M, see [238, p. 70], implies

$$d\alpha(X,Y) = -\alpha([X,Y]) \text{ for all } X, Y \in \xi = \ker \alpha.$$

The contact condition $\alpha \wedge d\alpha \neq 0$ in dimension 3 is equivalent to $d\alpha|_\xi \neq 0$. This implies the claim. □

Example 1.1.8 The standard contact structure ξ_1 on $\mathbb{R}^3$ is given by $dz + x\,dy = 0$, hence ξ_1 is spanned by the vector fields ∂_x and $\partial_y - x\,\partial_z$, with $[\partial_x, \partial_y - x\,\partial_z] = -\partial_z \notin \xi_1$.

Here is another fundamental concept of contact geometry.

Lemma/Definition 1.1.9 *Associated with a contact form α one has the so-called* **Reeb vector field** R_α, *uniquely defined by the equations*

(i) $d\alpha(R_\alpha, -) \equiv 0$,
(ii) $\alpha(R_\alpha) \equiv 1$.

Proof This is essentially a matter of linear algebra. For each point $p \in M$, the form $d\alpha|_{T_pM}$ is, by the contact condition $\alpha \wedge (d\alpha)^n \neq 0$, a skew-symmetric form of maximal rank $2n$ (for M of dimension $2n+1$). Hence $d\alpha|_{T_pM}$ has a 1–dimensional kernel (see the section on symplectic linear algebra below) and equation (i) defines R_α uniquely up to scaling, in other words, a unique line field $\langle R_\alpha \rangle \subset TM$. (The smoothness of this line field follows from the smoothness of α.) Again by the contact condition, α is non-trivial on that line field, so the normalisation condition (ii) specifies a non-vanishing section of it. □

Example 1.1.10 The Reeb vector field R_{α_1} of the standard contact form α_1 on $\mathbb{R}^{2n+1}$ equals ∂_z.

Notice that one cannot reasonably speak of the Reeb vector field of a contact *structure*. The consequences of the fact that different contact forms defining the same contact structure may have Reeb vector fields with wildly differing dynamics will be addressed in Example 2.2.5 below.

The name 'contact structure' has its origins in the fact that one of the first historical sources of contact manifolds are the so-called *spaces of contact elements*. We shall presently discuss these and other classical examples of contact manifolds.

1.2 The space of contact elements

In 1872, Lie [159] (see also [160], [161]) introduced the notion of *contact transformation* (Berührungstransformation) as a geometric tool for studying systems of differential equations. This may be regarded as the earliest precursor of modern contact geometry.

Contact transformations constitute a particular case of a local transformation group defined by the integrals of a system of differential equations. These transformations were studied extensively during the later part of the nineteenth century and the beginning of the twentieth century by, amongst others, Engel, Poincaré, Goursat, and Cartan.

In the present section we phrase in modern language some of the contact geometric notions that can be traced back to the work of Lie.

Definition 1.2.1 Let B be a smooth n–dimensional manifold. A **contact element** is a hyperplane in a tangent space to B. The **space of contact elements** of B is the collection of pairs (b, V) consisting of a point $b \in B$ and a contact element $V \subset T_bB$.

Lemma 1.2.2 *The space of contact elements of B can be naturally identified with the projectivised cotangent bundle $\mathbb{P}T^*B$, which is a manifold of dimension $2n-1$.*

Proof A hyperplane V in the tangent space T_bB is defined as the kernel of a non-trivial linear map $u_V : T_bB \to \mathbb{R}$, and u_V is determined by V up to multiplication by a non-zero scalar. So the space of contact elements at $b \in B$ may be thought of as the projectivisation of the dual space T_b^*B. It is standard bundle theory that this fibrewise projectivisation yields a smooth bundle, see [38]. □

Next we want to see that the space of contact elements comes equipped with a contact structure.

Lemma 1.2.3 *Write π for the bundle projection $\mathbb{P}T^*B \to B$. For $u = u_V \in \mathbb{P}T_b^*B$, let ξ_u be the hyperplane in $T_u(\mathbb{P}T^*B)$ such that $T\pi(\xi_u)$ is the hyperplane V in $T_{\pi(u)}B = T_bB$ defined by u. Then ξ defines a contact structure on $\mathbb{P}T^*B$.*

We call this the **natural contact structure** on the space of contact elements. Figure 1.2 illustrates the construction for $B = \mathbb{R}^2$. Here $\mathbb{P}T^*B = \mathbb{R}^2 \times \mathbb{R}P^1$.

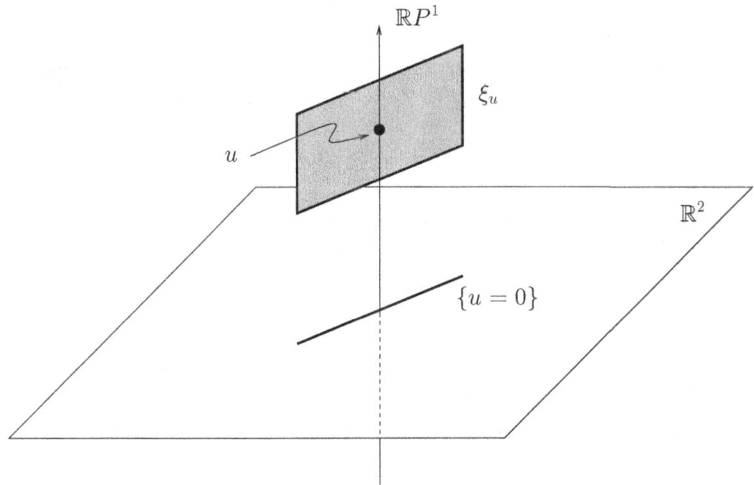

Fig. 1.2. The space of contact elements.

Proof of Lemma 1.2.3 Let $q_1, \ldots, q_n$ be local coordinates on B, and denote the corresponding dual coordinates in the fibres of the cotangent bundle T^*B by $p_1, \ldots, p_n$. This means that the coordinate description of covectors is given by

$$(q_1, \ldots, q_n, p_1, \ldots, p_n) = \Big(\sum_{j=1}^n p_j \, dq_j\Big)_{(q_1, \ldots, q_n)}.$$

Thus, a point

$$(q_1, \ldots, q_n, (p_1 : \ldots : p_n))$$

in the projectivised cotangent bundle $\mathbb{P}T^*B$ defines the hyperplane

$$\sum_{j=1}^{n} p_j \, dq_j = 0$$

in T_bB, where $b = (q_1, \ldots, q_n)$. By construction, the natural contact structure ξ on $\mathbb{P}T^*B$ is defined by

$$\xi = \ker\Bigl(\sum_{j=1}^{n} p_j \, dq_j\Bigr);$$

notice that this kernel is indeed well defined in terms of the coordinates on $\mathbb{P}T^*B$, although the 1–form $\sum p_j \, dq_j$ is not.

In order to verify the contact condition for ξ, we restrict to affine subspaces of the fibre. Over the open set $\{p_1 \neq 0\}$,† for instance, ξ is defined in terms of affine coordinates $p'_j = p_j/p_1$, $j = 2, \ldots, n$, by the equation

$$dq_1 + p'_2 \, dq_2 + \cdots + p'_n \, dq_n = 0,$$

which is exactly the description of the standard contact structure on $\mathbb{R}^{2n-1}$ from Example 1.1.5. □

Example 1.2.4 Consider the 2–torus $B = T^2 = S^1 \times S^1$ and let x, y be S^1–valued coordinates on B. Since B has trivial (co-)tangent bundle, the space of contact elements of B is $B \times \mathbb{R}P^1$. When identifying $\mathbb{R}P^1$ with $\mathbb{R}/\pi\mathbb{Z}$ with coordinate θ, the natural contact structure can be written as

$$\sin\theta \, dx - \cos\theta \, dy = 0.$$

This is an example of a contact structure that is not coorientable. It lifts to a coorientable contact structure, given by the same equation, on $B \times S^1$, with $S^1 := \mathbb{R}/2\pi\mathbb{Z}$.

This standard contact structure on the space of contact elements also plays a role in the Hamiltonian formalism of classical mechanics, a point to which we shall return in Section 1.4.

Definition 1.2.5 A **contact transformation** is a diffeomorphism of a space of contact elements that preserves the natural contact structure on that space.

† By this notation I mean the set of points (here: in the projectivised cotangent bundle over a coordinate neighbourhood) that satisfy the inequality in braces. Similar shorthand notation will be used throughout this text.

We briefly elaborate on the relevance of contact transformations for the theory of differential equations. Let $B = \mathbb{R}^2$. In this case, contact elements are also called **line elements**. Following Lie, we write (x, z) for the Cartesian coordinates on $\mathbb{R}^2$, and p for the slope of a line passing through a given point. The space of line elements whose slope p is finite can be identified with $\mathbb{R}^3$ with coordinates (x, z, p). The equation for lines of slope p is given by

$$dz - p\,dx = 0,$$

and when regarded as an equation on the space of line elements, it defines the natural contact structure. A solution $z = z(x)$ of a differential equation $F(x, z, z') = 0$ corresponds to an integral curve

$$x \longmapsto (x, z(x), z'(x))$$

of that contact structure.

Observe that a diffeomorphism

$$f\colon\ (x, z, p) \longmapsto (x_1, z_1, p_1)$$

of $\mathbb{R}^3$ is a contact transformation if and only if

$$dz_1 - p_1\,dx_1 = \rho(dz - p\,dx)$$

for some nowhere zero function $\rho\colon \mathbb{R}^3 \to \mathbb{R}$, that is,

$$f^*(dz - p\,dx) = \rho(dz - p\,dx).$$

Equivalently, this is saying that f maps all integral curves of the contact structure $dz - p\,dx = 0$ to integral curves of $dz_1 - p_1\,dx_1 = 0$.

Define a function F_1 of the transformed coordinates (x_1, z_1, p_1) by

$$F_1(x_1, z_1, p_1) = F(x, z, p),$$

i.e. $F = F_1 \circ f$.

Lemma 1.2.6 *Let f be a contact transformation as above. Suppose the curve $x \mapsto z(x)$ is a local solution of the differential equation*

$$F(x, z, z') = 0.$$

Define a curve in the transformed variables by

$$(x_1(x), z_1(x), p_1(x)) := f(x, z(x), z'(x)).$$

If the curve $x \mapsto (x_1(x), z_1(x))$ is regular, i.e. $(x_1'(x), z_1'(x)) \neq (0, 0)$ for all x, *then $z_1(x)$ may be regarded as a function of the transformed variable $x_1(x)$, and the curve $x_1 \mapsto z_1(x_1)$ is a local solution of the transformed equation*

$$F_1\Big(x_1, z_1, \frac{dz_1}{dx_1}\Big) = 0.$$

Proof The curve $x \mapsto (x, z(x), z'(x))$ is an integral curve of the natural contact structure $dz - p\,dx = 0$. Since f is a contact transformation, the curve

$$x \longmapsto (x_1(x), z_1(x), p_1(x)) := f(x, z(x), z'(x))$$

is an integral curve of the contact structure $dz_1 - p_1\,dx_1 = 0$. It follows that $z_1'(x) - p_1(x)x_1'(x) = 0$. With the regularity condition on the curve $x \mapsto (x_1(x), z_1(x))$ this forces $x_1'(x) \neq 0$. We may therefore write x locally as a function of x_1. Hence

$$\frac{dz_1}{dx_1}(x_1) = \frac{dz_1}{dx}(x(x_1)) \cdot \frac{dx}{dx_1}(x_1) = \frac{z_1'(x(x_1))}{x_1'(x(x_1))} = p_1(x(x_1))$$

and

$$\begin{aligned} F_1\Big(x_1, z_1, \frac{dz_1}{dx_1}\Big) &= F_1(x_1(x), z_1(x), p_1(x)) \\ &= F_1 \circ f(x, z(x), z'(x)) \\ &= F(x, z(x), z'(x)) = 0. \end{aligned}$$

This proves the lemma. □

Example 1.2.7 Let $z\colon \mathbb{R} \to \mathbb{R}$, $x \mapsto z(x)$ be a strictly convex function (i.e. $z'' > 0$), and assume that $x \mapsto z'(x)$ defines a diffeomorphism of the real line onto itself. Then, for any $p \in \mathbb{R}$, the function $Z_p\colon \mathbb{R} \to \mathbb{R}$ defined by

$$Z_p(x) := px - z(x)$$

has a unique maximum at the point $x = x(p)$ given by

$$\frac{dz}{dx}(x(p)) = p.$$

Define a new function z_1 of the variable p by

$$z_1(p) = Z_p(x(p)) = p \cdot x(p) - z(x(p)).$$

Then

$$\frac{dz_1}{dp}(p) = x(p) + p \cdot \frac{dx}{dp}(p) - \frac{dz}{dx}(x(p)) \cdot \frac{dx}{dp}(p) = x(p).$$

The transformation

$$f\colon (x, z, p) \longmapsto (x_1 := p, z_1 := px - z, p_1 := x)$$

is a contact transformation of $\mathbb{R}^3$ (with $\rho \equiv 1$); it is called the **Legendre transformation** of the function z with respect to the variable x, see [14].

Spaces of contact elements appear implicitly or explicitly in a multitude of geometric situations. Here is an example of a contact geometric interpretation of polarities in projective geometry. This may justify calling Apollonius of Perga [12] the earliest contact geometer *avant la lettre.*

We take $B = \mathbb{R}P^2$, the projective 2–plane. Represent points of $\mathbb{R}P^2$ in homogeneous coordinates, written as column vectors. Write $(\mathbb{R}P^2)^*$ for a dual copy of $\mathbb{R}P^2$, that is, the space of projective lines in $\mathbb{R}P^2$. In $(\mathbb{R}P^2)^*$ we write points as row vectors. Thus, the point

$$\mathbf{p} = (p^0 : p^1 : p^2) \in (\mathbb{R}P^2)^*$$

defines the line

$$\left\{\mathbf{q} = \begin{pmatrix} q^0 \\ q^1 \\ q^2 \end{pmatrix} \in \mathbb{R}P^2 \colon \ \mathbf{pq} := p^0q^0 + p^1q^1 + p^2q^2 = 0\right\}.\dagger$$

By slight abuse of terminology, we call this subset of $\mathbb{R}P^2$ 'the line $\mathbf{p}$'. Observe that the contact elements at a point $\mathbf{q} \in \mathbb{R}P^2$ correspond in a one-to-one fashion to projective lines through $\mathbf{q}$, so we can give a global coordinate description of the space of contact elements of $\mathbb{R}P^2$ as

$$M = \big\{(\mathbf{q}, \mathbf{p}) \in \mathbb{R}P^2 \times (\mathbb{R}P^2)^* \colon \ \mathbf{pq} = 0\big\}.$$

There are two distinguished projective lines in M passing through a given point $(\mathbf{q}_0, \mathbf{p}_0) \in M$, to wit,

$$L_{(\mathbf{q}_0, \mathbf{p}_0)} = \{(\mathbf{q}, \mathbf{p}_0) \in M\}$$

and

$$L^*_{(\mathbf{q}_0, \mathbf{p}_0)} = \{(\mathbf{q}_0, \mathbf{p}) \in M\}.$$

The tangent lines to these projective lines at the point $(\mathbf{q}_0, \mathbf{p}_0)$ determine the contact plane $\xi_{(\mathbf{q}_0, \mathbf{p}_0)}$ of the natural contact structure on M.

Now consider a non-singular projective conic $\mathfrak{C}$ in $\mathbb{R}P^2$. This is given by an equation $\mathbf{q}^t A\mathbf{q} = 0$, where A is a symmetric 3×3 matrix of rank 3 and index 2. Define maps $\varphi\colon \mathbb{R}P^2 \to (\mathbb{R}P^2)^*$ and $\varphi^*\colon (\mathbb{R}P^2)^* \to \mathbb{R}P^2$ by $\varphi(\mathbf{q}) = \mathbf{q}^t A$ and $\varphi^*(\mathbf{p}) = A^{-1}\mathbf{p}^t$. This defines a **correlation**: the image of a point $\mathbf{q}$ on a line $\mathbf{p}$ is a line $\varphi(\mathbf{q})$ containing the point $\varphi^*(\mathbf{p})$. Indeed, if $\mathbf{pq} = 0$, then

$$\varphi(\mathbf{q})\varphi^*(\mathbf{p}) = \mathbf{q}^t AA^{-1}\mathbf{p}^t = \mathbf{q}^t\mathbf{p}^t = \mathbf{pq} = 0.$$

† Products of vector-valued quantities will always be read as scalar products; likewise: $\mathbf{p}\,d\mathbf{q} = \sum_j p_j\,dq_j$, $\mathbf{q}\,\partial_{\mathbf{q}} = \sum_j q_j\,\partial_{q_j}$, etc.

Obviously $\varphi^* \circ \varphi = \mathrm{id}_{\mathbb{R}P^2}$ and $\varphi \circ \varphi^* = \mathrm{id}_{(\mathbb{R}P^2)^*}$. A correlation with this property is called a **polarity**.

Since we defined φ and φ^* in terms of a coordinate description of the conic $\mathfrak{C}$, it is not clear, *a priori*, that these maps have any geometric meaning. The following proposition shows that they do.

Proposition 1.2.8 *(a) The maps φ and φ^* depend only on the conic $\mathfrak{C}$, not on the specific coordinate representation.*

(b) If $\mathfrak{C}$ is realised as an ellipse in an affine subspace $\mathbb{R}^2 \subset \mathbb{R}P^2$, the geometric meaning of φ is as follows.

(i) If $\mathbf{q} \in \mathfrak{C}$, then $\varphi(\mathbf{q})$ is the tangent line to $\mathfrak{C}$ at $\mathbf{q}$.

(ii) If $\mathbf{q}$ is in the exterior of $\mathfrak{C}$, let $\mathbf{q}_1, \mathbf{q}_2$ be the points of $\mathfrak{C}$ where the corresponding tangent lines to $\mathfrak{C}$ pass through $\mathbf{q}$. Then $\varphi(\mathbf{q})$ is the line through $\mathbf{q}_1$ and $\mathbf{q}_2$.

(iii) If $\mathbf{q}$ is in the interior of $\mathfrak{C}$, then $\varphi(\mathbf{q})$ is constructed as illustrated in Figure 1.3.

(c) The map

$$\begin{array}{ccccccccc} \Phi: & \mathbb{R}P^2 & \times & (\mathbb{R}P^2)^* & \longrightarrow & \mathbb{R}P^2 & \times & (\mathbb{R}P^2)^* \\ & (\mathbf{q} & , & \mathbf{p}) & \longmapsto & (\varphi^*(\mathbf{p}) & , & \varphi(\mathbf{q})) \end{array}$$

restricts to a contact transformation of M.

Proof (a) Consider a coordinate transformation

$$\mathbf{q} = B\widetilde{\mathbf{q}}$$

of $\mathbb{R}P^2$, with B an invertible 3×3 matrix. Then

$$\mathbf{q}^t A\mathbf{q} = \widetilde{\mathbf{q}}^t B^t AB\widetilde{\mathbf{q}},$$

i.e. with respect to the new coordinates (with tildes) the conic is described by the matrix $\widetilde{A} = B^t AB$. Moreover, we have $\mathbf{pq} = 0$ if and only if $\mathbf{p}B\widetilde{\mathbf{q}} = 0$, which implies that the corresponding coordinate transformation of $(\mathbb{R}P^2)^*$ is given by

$$\widetilde{\mathbf{p}} = \mathbf{p}B.$$

With respect to the old coordinates, we had

$$\varphi(\mathbf{q}) = \mathbf{q}^t A = \widetilde{\mathbf{q}}^t B^t A;$$

with respect to the new coordinates we have

$$\widetilde{\varphi}(\widetilde{\mathbf{q}}) = \widetilde{\mathbf{q}}^t \widetilde{A} = \widetilde{\mathbf{q}}^t B^t AB = \varphi(\mathbf{q})B.$$

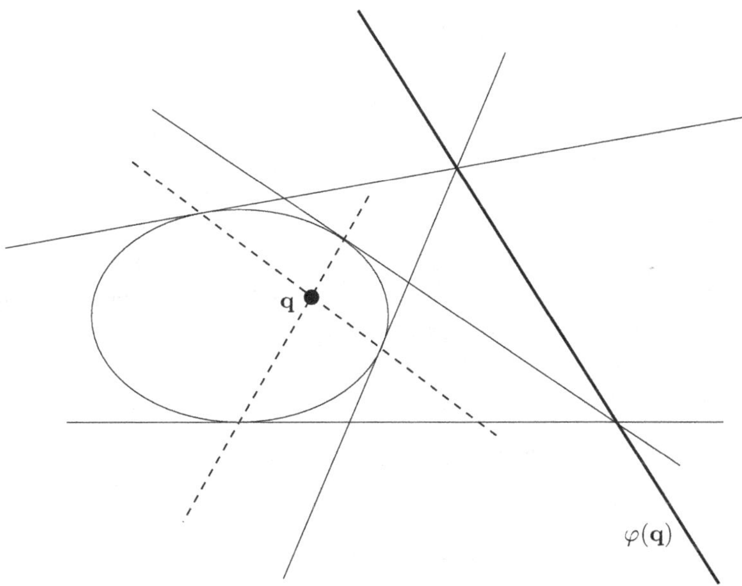

Fig. 1.3. Polarity in affine space.

This transformation behaviour shows that $\varphi(\mathbf{q})$ in the old coordinates and $\widetilde{\varphi}(\widetilde{\mathbf{q}})$ in the new coordinates represent the same line. Similarly, we have

$$\varphi^*(\mathbf{p}) = A^{-1}\mathbf{p}^t = A^{-1}(B^{-1})^t\widetilde{\mathbf{p}}^t$$

and

$$\widetilde{\varphi}^*(\widetilde{\mathbf{p}}) = (\widetilde{A})^{-1}\widetilde{\mathbf{p}}^t = B^{-1}A^{-1}(B^{-1})^t B^t\mathbf{p}^t = B^{-1}\varphi^*(\mathbf{p}),$$

so $\varphi^*(\mathbf{p})$ in the old coordinates and $\widetilde{\varphi}^*(\widetilde{\mathbf{p}})$ in the new coordinates represent the same point.

(b) Let $\mathbf{x}$ be a point on $\mathfrak{C}$. Then $\varphi(\mathbf{x})\mathbf{x} = \mathbf{x}^t A\mathbf{x} = 0$, so $\mathbf{x}$ also lies on its polar $\varphi(\mathbf{x})$. We want to show that $\varphi(\mathbf{x})$ is actually the tangent to $\mathfrak{C}$ at $\mathbf{x}$. To that end, assume that $\mathbf{y}$ is another point that lies both on $\mathfrak{C}$ and on the polar $\varphi(\mathbf{x})$ of $\mathbf{x}$, that is, $\mathbf{y}^t A\mathbf{y} = 0$ and $\mathbf{x}^t A\mathbf{y} = 0$. By (a) we may assume without loss of generality that

$$A = \begin{pmatrix} -1 & 0 & 0 \\ 0 & 1 & 0 \\ 0 & 0 & 1 \end{pmatrix},$$

and we may then choose homogeneous coordinates for $\mathbf{x}$ and $\mathbf{y}$ in the form

$$\mathbf{x} = \begin{pmatrix} 1 \\ x_1 \\ x_2 \end{pmatrix}, \quad \mathbf{y} = \begin{pmatrix} 1 \\ y_1 \\ y_2 \end{pmatrix}.$$

From $(\mathbf{x}-\mathbf{y})^t A(\mathbf{x}-\mathbf{y}) = 0$ we obtain $(x_1-y_1)^2 + (x_2-y_2)^2 = 0$, i.e. $\mathbf{x} = \mathbf{y}$. This proves (i).

Statements (ii) and (iii) then follow from the general properties of a polarity: the tangent line to $\mathfrak{C}$ at $\mathbf{q}_j$, $j = 1, 2$, is $\varphi(\mathbf{q}_j)$ by (i). Hence $\mathbf{q} \in \varphi(\mathbf{q}_j)$ and therefore $\varphi(\mathbf{q}) \ni \varphi^*(\varphi(\mathbf{q}_j)) = \mathbf{q}_j$. This proves (ii). Statement (iii) follows by a further iteration of this argument.

(c) One easily verifies

$$\Phi(L_{(\mathbf{q}_0,\mathbf{p}_0)}) = L^*_{\Phi(\mathbf{q}_0,\mathbf{p}_0)}$$

and

$$\Phi(L^*_{(\mathbf{q}_0,\mathbf{p}_0)}) = L_{\Phi(\mathbf{q}_0,\mathbf{p}_0)}.$$

Since the natural contact structure ξ is determined by the tangent lines to these projective lines, we have

$$T\Phi_{(\mathbf{q}_0,\mathbf{p}_0)}(\xi_{(\mathbf{q}_0,\mathbf{p}_0)}) = \xi_{\Phi(\mathbf{q}_0,\mathbf{p}_0)}.$$

This proves that Φ restricts to a contact transformation of M. □

1.3 Interlude: symplectic linear algebra

Before we proceed with the contact geometric interpretation of various phenomena in geometry and physics, we pause to recall some simple symplectic linear algebra. For a complete treatment of these issues the reader should refer to the textbooks by Cannas da Silva [40] or McDuff–Salamon [177]; here I restrict myself to those aspects that will be immediately relevant throughout this book.

Let V be a finite-dimensional real vector space and ω a **symplectic linear form** on V, that is, a skew-symmetric bilinear form with the property that the homomorphism

$$\begin{array}{rccc} \phi_\omega: & V & \longrightarrow & V^* \\ & \mathbf{v} & \longmapsto & \omega(\mathbf{v}, -) \end{array}$$

is an isomorphism. The pair (V, ω) is called a **symplectic vector space**. For a subspace $U \subset V$, define

$$U^\perp := \{\mathbf{v} \in V\colon\ \omega(\mathbf{u}, \mathbf{v}) = 0 \text{ for all } \mathbf{u} \in U\},$$

the **symplectic orthogonal complement** of U.

Lemma 1.3.1 *For any subspace U of a symplectic vector space (V, ω) we have*

$$\dim U + \dim U^{\perp} = \dim V$$

and $(U^{\perp})^{\perp} = U$.

Proof Consider the homomorphism

$$\begin{array}{rccc} \phi_U : & V & \longrightarrow & U^* \\ & \mathbf{v} & \longmapsto & \omega(\mathbf{v}, -)|_U . \end{array}$$

Then $\ker \phi_U = U^{\perp}$ by the definition of $U^{\perp}$.

Given any $\varphi \in U^*$, extend it to a linear form $\widetilde{\varphi} \in V^*$ (that is, $\widetilde{\varphi}|_U = \varphi$). Since ϕ_ω is surjective, we find a vector $\mathbf{v} \in V$ with $\phi_\omega(\mathbf{v}) = \widetilde{\varphi}$. It follows

$$\phi_U(\mathbf{v}) = \phi_\omega(\mathbf{v})|_U = \widetilde{\varphi}|_U = \varphi.$$

Thus, ϕ_U is surjective. The dimension formula for vector space homomorphisms gives

$$\dim V = \dim \operatorname{im} \phi_U + \dim \ker \phi_U = \dim U + \dim U^{\perp}.$$

The inclusion $U \subset (U^{\perp})^{\perp}$ follows directly from the definition of the symplectic orthogonal complement. The dimension formula yields $\dim U = \dim (U^{\perp})^{\perp}$. This implies $(U^{\perp})^{\perp} = U$. □

Definition 1.3.2 *A subspace U of a symplectic vector space (V, ω) is called*

- **symplectic** *if $U \cap U^{\perp} = \{\mathbf{0}\}$,*
- **isotropic** *if $U \subset U^{\perp}$,*
- **coisotropic** *if $U \supset U^{\perp}$,*
- **Lagrangian** *if $U = U^{\perp}$.*

From $U = (U^{\perp})^{\perp}$ we immediately deduce that U is isotropic if and only if $U^{\perp}$ is coisotropic.

Once we pass from symplectic vector spaces to symplectic manifolds, we shall encounter the corresponding notions for submanifolds. There will also be related contact geometric concepts. For instance, the next lemma will play a role in the discussion of isotropic submanifolds of contact manifolds, see Definition 1.5.11 and Section 2.5.2.

Lemma 1.3.3 *Let U be an isotropic subspace of a symplectic vector space (V, ω). Then ω induces a well-defined symplectic structure on the quotient space $U^{\perp}/U$.*

Proof For $\mathbf{u}, \mathbf{u}' \in U$ and $\mathbf{v}, \mathbf{v}' \in U^\perp$ we have from $U \subset U^\perp$ that

$$\omega(\mathbf{v} + \mathbf{u}, \mathbf{v}' + \mathbf{u}') = \omega(\mathbf{v}, \mathbf{v}'),$$

so ω induces a well-defined bilinear form on $U^\perp/U$.

If $\mathbf{v}_0 \in U^\perp$ has the property that $\omega(\mathbf{v}_0, \mathbf{v}) = 0$ for all $\mathbf{v} \in U^\perp$, then $\mathbf{v}_0 \in (U^\perp)^\perp = U$, so $[\mathbf{v}_0] = \mathbf{0}$ in $U^\perp/U$. This means that ω is symplectic on $U^\perp/U$. □

The following observation justifies the name 'symplectic' subspace.

Lemma 1.3.4 *Let U be a subspace of a symplectic vector space (V, ω). Then $U \cap U^\perp = \{\mathbf{0}\}$ if and only if $\omega|_U$ is symplectic, i.e. if $\phi_U|_U : U \to U^*$ is an isomorphism. Furthermore, U has this property if and only if so does $U^\perp$.*

Proof If $U \cap U^\perp = \{\mathbf{0}\}$, then $V = U \oplus U^\perp$ by the dimension formula of the preceding lemma. Given $\mathbf{0} \neq \mathbf{u}_0 \in U$, there is a $\mathbf{v} \in V$ such that $\omega(\mathbf{u}_0, \mathbf{v}) \neq 0$. Write $\mathbf{v} = \mathbf{u} + \mathbf{u}^\perp \in U \oplus U^\perp$. Then $0 \neq \omega(\mathbf{u}_0, \mathbf{v}) = \omega(\mathbf{u}_0, \mathbf{u})$. This means that the homomorphism $\phi_U|_U : U \to U^*$ is injective, and hence an isomorphism, since $\dim U = \dim U^*$.

Conversely, suppose we know this homomorphism to be injective. Let $\mathbf{u} \in U \cap U^\perp$. Then $\omega(\mathbf{u}, \mathbf{u}') = 0$ for all $\mathbf{u}' \in U$. This means that $\phi_U(\mathbf{u}) = \mathbf{0}$, and hence $\mathbf{u} = \mathbf{0}$.

Because of $(U^\perp)^\perp = U$, the situation is symmetric in U and $U^\perp$. □

Proposition 1.3.5 *A symplectic vector space (V, ω) has even real dimension. If $\dim V = 2n$, there is a basis $\{\mathbf{e}_1^*, \mathbf{f}_1^*, \ldots, \mathbf{e}_n^*, \mathbf{f}_n^*\}$ for V^* such that*

$$\omega = \mathbf{e}_1^* \wedge \mathbf{f}_1^* + \cdots + \mathbf{e}_n^* \wedge \mathbf{f}_n^*.$$

Hence $\omega^n = n!\, \mathbf{e}_1^* \wedge \mathbf{f}_1^* \wedge \ldots \wedge \mathbf{e}_n^* \wedge \mathbf{f}_n^* \neq 0$.

Conversely, a skew-symmetric bilinear form ω on a $2n$–dimensional vector space V with the property that $\omega^n \neq 0$ is a symplectic linear form.

Proof Let (V, ω) be a symplectic vector space. Choose an arbitrary non-zero vector $\mathbf{e}_1 \in V$ (assuming V not to be the zero vector space). By the definition of a symplectic linear form we find $\mathbf{f}_1 \in V$ with $\omega(\mathbf{e}_1, \mathbf{f}_1) = 1$. By the skew-symmetry of ω, the vector space U_1 spanned by $\mathbf{e}_1$ and $\mathbf{f}_1$ is 2–dimensional. Moreover, $\omega|_{U_1}$ is symplectic, hence $V = U_1 \oplus U_1^\perp$ by the preceding lemma, and $\omega|_{U_1^\perp}$ is also symplectic.

By induction, we find a basis $\{\mathbf{e}_1, \mathbf{f}_1, \ldots, \mathbf{e}_n, \mathbf{f}_n\}$ for V with

$$\omega(\mathbf{e}_i, \mathbf{e}_j) = \omega(\mathbf{f}_i, \mathbf{f}_j) = 0, \quad \omega(\mathbf{e}_i, \mathbf{f}_j) = \delta_{ij}, \quad i, j = 1, \ldots, n.$$

Hence $\dim V = 2n$ is even. The dual basis $\{\mathbf{e}_1^*, \mathbf{f}_1^*, \ldots, \mathbf{e}_n^*, \mathbf{f}_n^*\}$ for V^* yields the desired description of ω.

Conversely, assume that ω is a skew-symmetric bilinear form on a $2n$–dimensional vector space V with $\omega^n \neq 0$. Given a non-zero vector $\mathbf{v} \in V$, there must be a vector $\mathbf{v}' \in V$ with $\omega(\mathbf{v}, \mathbf{v}') \neq 0$, otherwise $\omega^n(\mathbf{v}, \ldots) = 0$. This means that $\phi_\omega(\mathbf{v}) \neq \mathbf{0} \in V^*$. In other words, ϕ_ω is injective, and hence ω symplectic. □

Remark 1.3.6 From the dimension formula (Lemma 1.3.1) it follows that an isotropic subspace of a $2n$–dimensional symplectic vector space has dimension less than or equal to n; Lagrangian subspaces are precisely the isotropic subspaces of dimension n.

Remark 1.3.7 Observe that an equivalent condition for $\xi \subset TM$ to be a contact structure on the $(2n+1)$–dimensional manifold M is that for any (local) 1–form α with $\xi = \ker \alpha$ we have that $(d\alpha)^n|_\xi \neq 0$, in other words, that $(\xi_p, d\alpha|_{\xi_p})$ be a symplectic vector space for all points $p \in M$.

In that last remark, we actually have an example of a *symplectic vector bundle* $(\xi, d\alpha)$, i.e. a vector bundle with a symplectic form on every fibre, varying smoothly (in a sense that will be made precise) with the base point. When working with such bundles, certain constructions require one to find complementary bundles of a given sub-bundle, e.g. a coisotropic complement of an isotropic sub-bundle. There is no canonical way of doing this, but a specific choice can be made once a compatible complex bundle structure has been fixed. Here we develop this concept on the vector space level.

Definition 1.3.8 A **complex structure** on a real vector space V is an automorphism $J\colon V \to V$ satisfying $J^2 = -\mathrm{id}_V$. If (V, ω) is a symplectic vector space, then a complex structure J on V is called ω–**compatible** if it satisfies

(i) $\omega(J\mathbf{u}, J\mathbf{v}) = \omega(\mathbf{u}, \mathbf{v})$ for all $\mathbf{u}, \mathbf{v} \in V$,
(ii) $\omega(\mathbf{v}, J\mathbf{v}) > 0$ for all non-zero $\mathbf{v} \in V$.

Remarks 1.3.9 (1) If J is a complex structure on a vector space V, one can define a complex scalar multiplication by $(a + ib)\mathbf{v} = a\mathbf{v} + bJ\mathbf{v}$ for $a, b \in \mathbb{R}$ and $\mathbf{v} \in V$; here $\mathrm{i} := \sqrt{-1}$. This turns V into a vector space over $\mathbb{C}$, and in particular the real dimension of V has to be even.

(2) An inner product g on V is called J–**compatible** if

$$g(J\mathbf{u}, J\mathbf{v}) = g(\mathbf{u}, \mathbf{v}) \quad \text{for all } \mathbf{u}, \mathbf{v} \in V.$$

Starting from an arbitrary inner product $\widetilde{g}$ on V, one can define a J–compatible inner product g by setting

$$g(\mathbf{u},\mathbf{v}) = \widetilde{g}(\mathbf{u},\mathbf{v}) + \widetilde{g}(J\mathbf{u},J\mathbf{v}).$$

(3) Given a J–compatible inner product g on V, a Hermitian inner product h on V (regarded as a complex vector space as in (1)) can be defined by

$$h(\mathbf{u},\mathbf{v}) = g(\mathbf{u},\mathbf{v}) + ig(\mathbf{u},J\mathbf{v}).$$

(4) If J is a complex structure compatible with a symplectic form ω, then $g_J(\mathbf{u},\mathbf{v}) := \omega(\mathbf{u},J\mathbf{v})$ defines a J–compatible inner product on V.

Proposition 1.3.10 *The space $\mathcal{J}(\omega)$ of ω–compatible complex structures on the symplectic vector space (V,ω) is non-empty and contractible (with the topology $\mathcal{J}(\omega)$ inherits as a subset of the vector space of endomorphisms of V).*

Proof Choose an arbitrary inner product g on V and define an automorphism A of V by $\omega(\mathbf{u},\mathbf{v}) = g(A\mathbf{u},\mathbf{v})$. Formally, if ϕ_g denotes the isomorphism $V \to V^*$ given by $\mathbf{v} \mapsto g(\mathbf{v},-)$, then $A\mathbf{u} = \phi_g^{-1}(\phi_\omega(\mathbf{u}))$. Then

$$g(A\mathbf{u},\mathbf{v}) = \omega(\mathbf{u},\mathbf{v}) = -\omega(\mathbf{v},\mathbf{u}) = -g(A\mathbf{v},\mathbf{u}) = g(\mathbf{u},-A\mathbf{v}),$$

so the adjoint A^* of A with respect to g is $A^* = -A$. It follows that $-A^2 = AA^*$ is symmetric and positive definite with respect to g.

Let Q be the uniquely defined positive square root of $-A^2$. This automorphism is determined geometrically and without reference to any basis by the property that on any eigenspace of $-A^2$ it acts as multiplication by the positive square root of the corresponding eigenvalue.

Define $J = AQ^{-1}$. One checks easily that A preserves the eigenspaces of $-A^2$ and hence commutes with Q. We conclude

$$J^2 = AQ^{-1}AQ^{-1} = A^2(Q^2)^{-1} = -\mathrm{id}_V,$$

so J is a complex structure on V.

The automorphism Q is self-adjoint, and A is skew-adjoint, hence J is skew-adjoint (and commutes with A). It follows that

$$\omega(J\mathbf{u},J\mathbf{v}) = \omega(\mathbf{u},-J^2\mathbf{v}) = \omega(\mathbf{u},\mathbf{v}).$$

Similarly one finds

$$\begin{aligned}\omega(\mathbf{u},J\mathbf{v}) &= g(A\mathbf{u},J\mathbf{v}) = g(\mathbf{u},-AJ\mathbf{v}) \\ &= g(\mathbf{u},-A^2Q^{-1}\mathbf{v}) = g(\mathbf{u},Q\mathbf{v}),\end{aligned}$$

and hence $\omega(\mathbf{v}, J\mathbf{v}) > 0$ for $\mathbf{v} \neq \mathbf{0}$. This concludes the proof that the space $\mathcal{J}(\omega)$ is non-empty.

In order to show the contractibility of that space, we introduce the space $\mathcal{G}$ of inner products on V. (The topology on that space can most easily be defined by choosing a basis for V – violating my own dictum – and identifying $\mathcal{G}$ with the space of symmetric and positive definite matrices.) The construction above gives a smooth map

$$\begin{array}{rccc} j\colon & \mathcal{G} & \longrightarrow & \mathcal{J}(\omega) \\ & g & \longmapsto & J = j(g). \end{array}$$

If we start with $J \in \mathcal{J}(\omega)$ and define the J–compatible inner product g_J by $g_J(\mathbf{u}, \mathbf{v}) = \omega(\mathbf{u}, J\mathbf{v})$, then $\omega(\mathbf{u}, \mathbf{v}) = g_J(J\mathbf{u}, \mathbf{v})$. So the above construction applied to g_J yields $A = J$, $Q = \mathrm{id}_V$, and hence $j(g_J) = J$.

Fix an element $J_0 \in \mathcal{J}(\omega)$. Define a family of continuous maps

$$f_t\colon \mathcal{J}(\omega) \to \mathcal{J}(\omega), \quad t \in [0, 1],$$

by

$$f_t(J) = j((1-t)g_{J_0} + t g_J).$$

This defines a homotopy between the constant map $f_0 \equiv J_0$ and the identity map $f_1 = \mathrm{id}_{\mathcal{J}(\omega)}$, which means that the space $\mathcal{J}(\omega)$ is contractible. □

1.4 Classical mechanics

In the Hamiltonian formalism, a mechanical system is described by a configuration space B, which we shall assume to be a smooth manifold, with corresponding phase space (i.e. space of positions and momenta) described by the cotangent bundle T^*B of B. For a comprehensive introduction to the mathematical treatment of classical mechanics see the textbook by Arnold [13]. Here I only want to highlight one aspect of this mathematical set-up where contact geometry comes into play.

Consider the commutative diagram

$$\begin{array}{ccc} TT^*B & \longrightarrow & T^*B \\ {\scriptstyle T\pi}\downarrow & & \downarrow{\scriptstyle \pi} \\ TB & \longrightarrow & B \end{array}$$

and define a differential 1–form λ on T^*B by $\lambda_u = u \circ T\pi$ for $u \in T^*B$. This 1–form is called the **Liouville form** on T^*B.

Lemma 1.4.1 *In local coordinates $\mathbf{q} = (q_1, \ldots, q_n)$ on the manifold B and* *dual coordinates* $\mathbf{p} = (p_1, \ldots, p_n)$ *on the fibres of* T^*B*, the Liouville form* λ *is equal to*

$$\lambda = \sum_{j=1}^{n} p_j \, dq_j =: \mathbf{p} \, d\mathbf{q}.$$

This form can also be characterised by the property that for any differential 1–form τ *on* B *(i.e. section of the bundle* $T^*B \to B$*) one has* $\tau = \tau^*\lambda$*.*

Proof Denote the fibre coordinates on TT^*B corresponding to the local coordinates $(\mathbf{q}, \mathbf{p})$ on T^*B by $(\dot{\mathbf{q}}, \dot{\mathbf{p}})$. This means that tangent vectors to T^*B are written in the form $\dot{\mathbf{q}} \, \partial_{\mathbf{q}} + \dot{\mathbf{p}} \, \partial_{\mathbf{p}}$. Recall that $(\mathbf{q}, \mathbf{p})$ is the coordinate representation of $(\mathbf{p} \, d\mathbf{q})_{\mathbf{q}}$. Thus

$$\begin{aligned} \lambda_{(\mathbf{q},\mathbf{p})}(\dot{\mathbf{q}} \, \partial_{\mathbf{q}} + \dot{\mathbf{p}} \, \partial_{\mathbf{p}}) &= (\mathbf{p} \, d\mathbf{q})_{\mathbf{q}}(T\pi(\dot{\mathbf{q}} \, \partial_{\mathbf{q}} + \dot{\mathbf{p}} \, \partial_{\mathbf{p}})) \\ &= (\mathbf{p} \, d\mathbf{q})_{\mathbf{q}}(\dot{\mathbf{q}} \, \partial_{\mathbf{q}}) \\ &= \mathbf{p}\dot{\mathbf{q}} \\ &= (\mathbf{p} \, d\mathbf{q})_{(\mathbf{q},\mathbf{p})}(\dot{\mathbf{q}} \, \partial_{\mathbf{q}} + \dot{\mathbf{p}} \, \partial_{\mathbf{p}}), \end{aligned}$$

i.e. $\lambda = \mathbf{p} \, d\mathbf{q}$.

For the alternative characterisation of λ, let τ be any differential 1–form on B, and b a point of B. Then

$$\begin{aligned} (\tau^*\lambda)_b &= \lambda_{\tau_b} \circ T\tau \\ &= \tau_b \circ T\pi \circ T\tau \\ &= \tau_b, \end{aligned}$$

since $\pi \circ \tau = \mathrm{id}_B$. Conversely, if μ is another 1–form on T^*B with the property that $\tau^*\mu = \tau$ for all 1–forms τ on B, then $\tau^*(\lambda - \mu) = 0$ for all τ. By choosing suitable τ, one concludes that $\lambda - \mu$ vanishes identically. □

Observe that $\ker \lambda$ defines the natural contact structure on the space of contact elements $\mathbb{P}T^*B$ described in the preceding section.

Definition 1.4.2 A **symplectic form** on a differential manifold W of even dimension $2n$ is a differential 2–form that is closed ($d\omega = 0$) and non-degenerate, i.e. the $2n$–form ω^n is required to be nowhere zero. The pair (W, ω) is then called a **symplectic manifold**.

Notice that if (W, ω) is a symplectic manifold, then each tangent space T_xW, $x \in W$, is a symplectic vector space with symplectic linear form ω_x.

The natural 2–form $\omega := d\lambda$ on T^*B is clearly a symplectic form: ω is exact, and so in particular closed; the non-degeneracy of ω can best be seen in local coordinates, here

$$\omega = d\mathbf{p} \wedge d\mathbf{q} = \sum_{j=1}^{n} dp_j \wedge dq_j ,$$

from which we compute $\omega^n = n!\, dp_1 \wedge dq_1 \wedge \ldots \wedge dp_n \wedge dq_n \neq 0$.

Let $L = L(t, \mathbf{q}, \mathbf{v})$ be a C^2-function of the variables $t \in \mathbb{R}$ and $\mathbf{q}, \mathbf{v} \in \mathbb{R}^n$. From elementary variational calculus, see [143] or [177, p. 13], we know that critical points $\mathbf{q} \in C^1([t_0, t_1], \mathbb{R}^n)$ of the **action functional**

$$\mathbf{q} \mapsto \int_{t_0}^{t_1} L(t, \mathbf{q}(t), \dot{\mathbf{q}}(t))\, dt$$

have to satisfy the **Euler–Lagrange equations**

$$\frac{d}{dt}\Big(\frac{\partial L}{\partial \dot{\mathbf{q}}}\Big) - \frac{\partial L}{\partial \mathbf{q}} = 0,$$

that is,

$$\frac{d}{dt}\Big(\frac{\partial L}{\partial v_j}(t, \mathbf{q}(t), \dot{\mathbf{q}}(t))\Big) - \frac{\partial L}{\partial q_j}(t, \mathbf{q}(t), \dot{\mathbf{q}}(t)) = 0, \ j = 1, \ldots, n.$$

We call L the **Lagrangian function** of the variational problem.

Many mechanical systems can be described by such a variational principle of least (strictly speaking: critical) action.† For instance, the Newtonian equations of dynamics describing the motion of particles in a potential $U = U(\mathbf{q})$ are simply the Euler–Lagrange equations for the Lagrangian $L := T - U$, i.e. the difference between the kinetic energy $T = T(\mathbf{v})$ and the potential U, see [13, Section 13].

Now assume that L satisfies the **Legendre condition**

$$\det\Big(\frac{\partial^2 L}{\partial v_i\, \partial v_j}\Big) \neq 0.$$

Consider the equation

$$\mathbf{p} - \frac{\partial L}{\partial \mathbf{v}}(t, \mathbf{q}, \mathbf{v}) = 0.$$

By the implicit function theorem, the Legendre condition is precisely what is needed to solve this equation for $\mathbf{v}$, i.e. to write $\mathbf{v} = \mathbf{v}(\mathbf{q}, \mathbf{p}, t)$.

The **Hamiltonian function** can then be defined as the Legendre transform of L with respect to $\mathbf{v}$, see Example 1.2.7, i.e.

$$H(t, \mathbf{q}, \mathbf{p}) := \mathbf{p}\mathbf{v} - L(t, \mathbf{q}, \mathbf{v}),$$

† 'Nach Leibniz ist unsere Welt die bestmögliche aller Welten, und daher lassen sich die Naturgesetze durch Extremalprinzipien beschreiben.' [221]

and it is a simple matter to show that the Euler–Lagrange equations for L transform into the **Hamiltonian equations** for H:

$$\dot{\mathbf{q}} = \frac{\partial H}{\partial \mathbf{p}}, \quad \dot{\mathbf{p}} = -\frac{\partial H}{\partial \mathbf{q}};$$

these have to be read as $\dot{q_j} = \partial H/\partial p_j$, $j = 1, \ldots, n$; likewise for the second system of n equations. Notice that we have reduced a system of n second-order equations to a system of $2n$ first-order equations.

If $L = T - U$ with $T = T(\mathbf{v})$ homogeneous of degree 2 in $\mathbf{v}$, and $U = U(\mathbf{q})$ (as will be the case in a typical mechanical system), then

$$H(\mathbf{q}, \mathbf{p}) = \mathbf{pv} - L = \frac{\partial L}{\partial \mathbf{v}}\mathbf{v} - T + U = T + U$$

is the total energy.

The preceding discussion can be generalised to the consideration of Hamiltonian functions on cotangent bundles T^*B. With the following lemma we return to a more mathematical mode.

Lemma/Definition 1.4.3 *Let $H\colon W \to \mathbb{R}$ be a differentiable function on a symplectic manifold (W, ω). Then there is a unique vector field X_H on W defined by*

$$\omega(X_H, -) = -dH.$$

This X_H is called the **Hamiltonian vector field** *corresponding to the Hamiltonian function H.*

The Hamiltonian vector field X_H satisfies the condition $dH(X_H) = 0$, so its flow preserves the level sets of H.

Proof On each tangent space T_xW, $x \in W$, the 2–form ω_x is a linear symplectic form. Hence $X_H(x)$ is uniquely defined as $\phi_{\omega_x}^{-1}(-dH_x)$.

Further, we compute $dH(X_H) = -\omega(X_H, X_H) = 0$. □

It is not difficult to see (for instance in a local coordinate representation) that X_H is a smooth vector field.

The next proposition gives a succinct way of writing the Hamiltonian equations. The statement $dH(X_H) = 0$ from the preceding lemma can then be read as a mathematical formulation of the preservation of energy.

Proposition 1.4.4 *The Hamiltonian equations for a smooth function H on the symplectic manifold $(T^*B, \omega = d\lambda)$ can be written in a coordinate-free manner as*

$$\dot{x} = X_H(x).$$

Proof In local coordinates $x = (\mathbf{q}, \mathbf{p})$ we have

$$dH = \frac{\partial H}{\partial \mathbf{q}}\, d\mathbf{q} + \frac{\partial H}{\partial \mathbf{p}}\, d\mathbf{p}$$

and $\omega = d\mathbf{p} \wedge d\mathbf{q}$. This implies

$$X_H = \frac{\partial H}{\partial \mathbf{p}}\, \partial_{\mathbf{q}} - \frac{\partial H}{\partial \mathbf{q}}\, \partial_{\mathbf{p}}.$$

So the equation $\dot{x} = X_H(x)$ is equivalent to the Hamiltonian equations

$$(\dot{\mathbf{q}}, \dot{\mathbf{p}}) = \left(\frac{\partial H}{\partial \mathbf{p}}, -\frac{\partial H}{\partial \mathbf{q}}\right),$$

as was to be shown. □

We now want to relate this Hamiltonian flow to the Reeb flow of a suitable contact form. The connection between the two is given by the fact that certain hypersurfaces in symplectic manifolds carry contact forms.

Lemma/Definition 1.4.5 *A* **Liouville vector field** *Y on a symplectic manifold (W, ω) is a vector field satisfying the equation $\mathcal{L}_Y \omega = \omega$, where $\mathcal{L}$ denotes the Lie derivative. In this case, the 1–form $\alpha := i_Y \omega := \omega(Y, -)$ is a contact form on any hypersurface M transverse to Y (that is, with Y nowhere tangent to M). Such hypersurfaces are said to be of* **contact type**.

Proof The *Cartan formula*

$$\mathcal{L}_Y = d \circ i_Y + i_Y \circ d$$

for the Lie derivative (see [238, p. 70]) and the fact that ω is closed allow us to write the Liouville condition on Y as $d(i_Y \omega) = \omega$. Assuming W to be of dimension $2n$, we compute

$$\begin{aligned} \alpha \wedge (d\alpha)^{n-1} &= i_Y \omega \wedge (d(i_Y \omega))^{n-1} \\ &= i_Y \omega \wedge \omega^{n-1} \\ &= \frac{1}{n} i_Y (\omega^n). \end{aligned}$$

Since ω^n is a volume form on W, it follows that $\alpha \wedge (d\alpha)^{n-1}$ is a volume form when restricted to the tangent bundle of any hypersurface transverse to Y. □

Example 1.4.6 A Liouville vector field Y on $(T^*B, \omega = d\lambda)$ (or any manifold with an exact symplectic form) can be defined by $i_Y \omega = \lambda$. In local coordinates $(\mathbf{q}, \mathbf{p})$ on T^*B we have $Y = \mathbf{p}\, \partial_{\mathbf{p}}$, that is, the radial vector field in fibre direction.

Example 1.4.7 Given a contact manifold $(M, \xi = \ker \alpha)$, one can form the symplectic manifold $(\mathbb{R} \times M, \omega = d(e^t \alpha))$, with t denoting the $\mathbb{R}$–coordinate. (Here α is interpreted as a 1–form on $\mathbb{R} \times M$, i.e. we identify α with its pull-back under the projection $\mathbb{R} \times M \to M$.) Indeed, if M has dimension $2n-1$, then

$$\omega^n = (e^t (dt \wedge \alpha + d\alpha))^n = ne^{nt} dt \wedge \alpha \wedge (d\alpha)^{n-1} \neq 0.$$

One easily sees that the vector field ∂_t is a Liouville vector field for this symplectic form ω. The manifold $(\mathbb{R} \times M, \omega)$ is called the **symplectisation** of (M, α). The orientation of $\mathbb{R} \times M$ induced by the volume form ω^n coincides with the product orientation on $\mathbb{R} \times M$, where M is oriented by $\alpha \wedge (d\alpha)^{n-1}$.

Example 1.4.8 Consider $W = \mathbb{R}^4$ with its standard symplectic form

$$\omega = dx_1 \wedge dy_1 + dx_2 \wedge dy_2.$$

The radial vector field

$$Y = \frac{1}{2} r\, \partial_r = \frac{1}{2}(x_1\, \partial_{x_1} + y_1\, \partial_{y_1} + x_2\, \partial_{x_2} + y_2\, \partial_{y_2})$$

(where $r^2 = x_1^2 + y_1^2 + x_2^2 + y_2^2$) is a Liouville vector field for ω. On the unit sphere $S^3 \subset \mathbb{R}^4$ it induces the **standard contact form**

$$\alpha := i_Y \omega = \frac{1}{2}(x_1\, dy_1 - y_1\, dx_1 + x_2\, dy_2 - y_2\, dx_2).$$

Lemma 1.4.9 *The Reeb vector field of the standard contact form α on S^3 of the preceding example is*

$$R_\alpha = 2(x_1\, \partial_{y_1} - y_1\, \partial_{x_1} + x_2\, \partial_{y_2} - y_2\, \partial_{x_2}).$$

The orbits of the flow of R_α define the fibres of the **Hopf fibration**

$$\begin{array}{ccc} \mathbb{C}^2 \supset S^3 & \longrightarrow & S^2 = \mathbb{C}P^1 \\ (z_1, z_2) & \longmapsto & (z_1 : z_2), \end{array}$$

in particular, all the orbits of R_α are closed.

Proof Write $Z = 2(x_1\, \partial_{y_1} - y_1\, \partial_{x_1} + x_2\, \partial_{y_2} - y_2\, \partial_{x_2})$. We have

$$\alpha(Z) = x_1^2 + y_1^2 + x_2^2 + y_2^2 = 1$$

along S^3 and

$$\begin{aligned} i_Z d\alpha &= i_Z \omega \\ &= -2(x_1\, dx_1 + y_1\, dy_1 + x_2\, dx_2 + y_2\, dy_2) \\ &= -2r\, dr, \end{aligned}$$

which is identically zero when restricted to the tangent spaces of S^3, since S^3 is given by the equation $r^2 = 1$. Moreover, $r\,dr(Z) = 0$, so we may regard Z (more precisely: its restriction to S^3) as a vector field on S^3. This proves that $Z = R_\alpha$.

The fibre of the Hopf fibration through a point

$$(z_1, z_2) = (x_1 + \mathrm{i}y_1, x_2 + \mathrm{i}y_2) \in S^3 \subset \mathbb{C}^2$$

can be parametrised as

$$\gamma(t) = (\mathrm{e}^{\mathrm{i}t} z_1, \mathrm{e}^{\mathrm{i}t} z_2), \quad t \in \mathbb{R}.$$

We compute

$$\dot{\gamma}(0) = (\mathrm{i}z_1, \mathrm{i}z_2) = (\mathrm{i}x_1 - y_1, \mathrm{i}x_2 - y_2).$$

Expressed as a real vector field on $\mathbb{R}^4$ (tangent to S^3), this is equal to

$$\dot{\gamma}(0) = x_1\,\partial_{y_1} - y_1\,\partial_{x_1} + x_2\,\partial_{y_2} - y_2\,\partial_{x_2} = \frac{1}{2} R_\alpha,$$

so the Reeb flow does indeed define the Hopf fibration. □

Here is the promised result relating the Hamiltonian to the Reeb flow.

Lemma 1.4.10 *If a codimension* 1 *submanifold $M \subset T^*B$ is both a hypersurface of contact type (with contact form $\alpha = i_Y\omega$ for some Liouville vector field Y) and the level set of a Hamiltonian function $H\colon T^*B \to \mathbb{R}$, then the Reeb flow of α is a reparametrisation of the Hamiltonian flow.*

Proof This follows from the observation that the 2–form $d\alpha = \omega$ has a 1–dimensional kernel on (each fibre of) TM – again this is fibrewise a matter of linear algebra. This kernel is defined both by the Reeb vector field R_α and the Hamiltonian vector field X_H, since

$$i_{R_\alpha} d\alpha|_{TM} = 0$$

and

$$i_{X_H}\omega|_{TM} = -dH|_{TM} = 0.$$

So R_α and X_H agree up to scaling. □

1.5 The geodesic flow and Huygens' principle

We now make a brief excursion into Riemannian geometry and give a contact geometric interpretation of the so-called geodesic flow on the tangent bundle of a Riemannian manifold. (For a comprehensive treatment of geodesic flows in the context of the theory of dynamical systems see the monograph by Paternain [204]. However, the reader should beware the possible confusion arising from the different definitions used there, which seem to render the main result of the present section tautological.) We also consider the dual flow on the cotangent bundle; here the discussion ties up with the space of contact elements and yields a simple contact geometric proof of Huygens' principle concerning the propagation of wave fronts.

Proposition/Definition 1.5.1 *Let B be a manifold with a Riemannian metric g. There is a unique vector field G on the tangent bundle TB whose trajectories are of the form $t \mapsto \dot{\gamma}(t) \in T_{\gamma(t)}B \subset TB$, where γ is a geodesic on B (not necessarily of unit speed). This vector field G is called the* **geodesic field**, *and its (local) flow the* **geodesic flow**.

Note that the geodesic flow being defined globally (i.e. for all times) is equivalent to saying that (B, g) is a complete Riemannian manifold.

Proof In local coordinates $(q_1, \ldots, q_n)$ on B, geodesics are found as solutions of the system of second-order differential equations

$$\ddot{q}_k + \sum_{i,j} \Gamma^k_{ij} \dot{q}_i \dot{q}_j = 0, \quad k = 1, \ldots, n,$$

where the Γ^k_{ij} are the Christoffel symbols of the Riemannian metric g, see any book on Riemannian geometry, e.g. the one by do Carmo [42]. Choose local coordinates on the tangent bundle TB such that

$$(q_1, \ldots, q_n, v_1, \ldots, v_n) = \Big(\sum_{j=1}^{n} v_j\, \partial_{q_j}\Big)_{(q_1, \ldots, q_n)}.$$

Then the second-order system above is equivalent to the following system of first-order differential equations on TB:

$$\left.\begin{array}{rcl} \dot{q}_k & = & v_k \\ \dot{v}_k & = & -\sum_{i,j} \Gamma^k_{ij} v_i v_j \end{array}\right\} k = 1, \ldots, n.$$

The solutions of this first-order system are precisely the integral curves of the vector field

$$G(q_1, \ldots, q_n, v_1, \ldots, v_n) = \Big(v_1, \ldots, v_n, -\sum_{i,j} \Gamma^1_{ij} v_i v_j, \ldots, -\sum_{i,j} \Gamma^n_{ij} v_i v_j\Big).$$

This proves the existence of the desired geodesic field G in a local coordinate neighbourhood. Thus, if we can prove uniqueness of G, this will automatically entail its global existence.

By definition, the integral curves of the (locally defined) vector field G are of the form $t \mapsto \dot{\gamma}(t) \in T_{\gamma(t)}B$ with γ a geodesic on B. But geodesics can be defined without any reference to local coordinates as autoparallel curves, i.e. curves for which the covariant derivative $\nabla_{\dot{\gamma}(t)}\dot{\gamma}(t)$ vanishes identically. This implies that the definition of G is likewise independent of the choice of local coordinates. □

The Riemannian metric g on B allows us to define a bundle isomorphism Ψ from the tangent bundle TB to the cotangent bundle T^*B, which is fibrewise given by

$$\begin{array}{rccc} \Psi_b\colon & T_bB & \longrightarrow & T_b^*B \\ & X & \longmapsto & g_b(X,-). \end{array}$$

This induces a bundle metric g^* on T^*B, defined by

$$g_b^*(\mathbf{u}_1,\mathbf{u}_2) = g_b(\Psi_b^{-1}(\mathbf{u}_1),\Psi_b^{-1}(\mathbf{u}_2)) \text{ for } \mathbf{u}_1,\mathbf{u}_2 \in T_b^*B.$$

The **unit tangent bundle** STB is defined fibrewise by

$$ST_bB = \{X \in T_bB\colon\ g_b(X,X) = 1\};$$

likewise the **unit cotangent bundle** ST^*B is defined in terms of g^*.

For a geodesic γ on B, the length of its tangent vector $\dot{\gamma}$ is constant:

$$\frac{d}{dt}\, g(\dot{\gamma},\dot{\gamma}) = 2g(\nabla_{\dot{\gamma}}\dot{\gamma},\dot{\gamma}) = 0.$$

This implies that the flow of the geodesic vector field G preserves the unit tangent bundle STB, and hence along STB the vector field G is actually tangent to STB. It is therefore reasonable to speak of the geodesic flow on STB.

Theorem 1.5.2 *Let (B,g) be a Riemannian manifold.*

(a) The Liouville form λ on the cotangent bundle T^*B induces a contact form on the unit cotangent bundle ST^*B. The Reeb vector field R_λ of this contact form is dual to the geodesic vector field G in the sense that

$$T\Psi(G) = R_\lambda.$$

(b) Let $H\colon T^*B \to \mathbb{R}$ be the Hamiltonian function defined by $H(\mathbf{u}) = \frac{1}{2}g^*(\mathbf{u},\mathbf{u})$. Then along $ST^*B = H^{-1}(\frac{1}{2})$, the Reeb vector field R_λ equals the Hamiltonian vector field X_H (with respect to the symplectic form $\omega = d\lambda$ on T^*B).

The flow of $T\Psi(G)$ on ST^*B is called the **cogeodesic flow**. The theorem says that the cogeodesic flow is equivalent both to the *Reeb flow* of λ (i.e. the flow of R_λ) and the *Hamiltonian flow* of H on ST^*B (i.e. the flow of X_H).

In principle, the theorem can be proved by a relentless calculation in terms of arbitrary local coordinates. These calculations, however, simplify greatly when performed in special coordinates, so we first recall some Riemannian geometry.

Definition 1.5.3 Let $b_0 \in B$ be given. Pick a g_{b_0}–orthonormal basis $\mathbf{e}_1, \ldots, \mathbf{e}_n$ for $T_{b_0}B$. Take $\mathbf{q} = (q_1, \ldots, q_n)$ as the coordinates of the point $\exp_{b_0}(\sum_{j=1}^n q_j \mathbf{e}_j)$, where exp denotes the exponential map $TB \to B$ of the metric g. These coordinates are called **normal coordinates** centred at the point b_0.

Of course, these coordinates are only defined in a neighbourhood of b_0 chosen sufficiently small so as to be the diffeomorphic image under the exponential map $\exp_{b_0}$ of a neighbourhood of $\mathbf{0} \in T_{b_0}B$.

Write $g_{ij} = g(\partial_{q_i}, \partial_{q_j})$ for the metric coefficients in terms of these local coordinates. More formally, identify $T_{b_0}B$ with $\mathbb{R}^n$ via the basis $\mathbf{e}_1, \ldots, \mathbf{e}_n$; then

$$g_{ij}(\mathbf{q}) := g_{\exp_{b_0}(\mathbf{q})}\big(T_{\mathbf{q}} \exp_{b_0}(\partial_{q_i}), T_{\mathbf{q}} \exp_{b_0}(\partial_{q_j})\big).$$

Lemma 1.5.4 *In terms of normal coordinates centred at $b_0 \in B$ we have*

(i) $b_0 = (0, \ldots, 0) =: \mathbf{0}$;
(ii) $g_{ij}(\mathbf{0}) = \delta_{ij}$, for $i, j = 1, \ldots, n$;
(iii) $\Gamma^k_{ij}(\mathbf{0}) = 0$, for $i, j, k = 1, \ldots, n$;
(iv) $g_{ij,k}(\mathbf{0}) := \dfrac{\partial g_{ij}}{\partial q_k}(\mathbf{0}) = 0$, for $i, j, k = 1, \ldots, n$.

Proof (i) This simply expresses the fact that $b_0 = \exp_{b_0}(\mathbf{0})$.

(ii) Since $T_0 \exp_{b_0} : T_0(T_{b_0}B) \to T_{b_0}B$ is the identity map under the natural identification of $T_0(T_{b_0}B)$ with $T_{b_0}B$, we have $T_0 \exp_{b_0}(\partial_{q_i}) = \mathbf{e}_i$ and hence $g_{ij}(\mathbf{0}) = g_{b_0}(\mathbf{e}_i, \mathbf{e}_j) = \delta_{ij}$.

(iii) By definition of the exponential map, the geodesics through b_0 are in normal coordinates given as linear maps

$$t \longmapsto \gamma(t) = (ta_1, \ldots, ta_n) =: t\mathbf{a}.$$

From $\ddot{\gamma}(t) = \mathbf{0}$ and the second-order differential equation for geodesics in terms of local coordinates it follows that $\Gamma^k_{ij}(\mathbf{0}) = 0$.

(iv) From

$$\Gamma^k_{ij} = \frac{1}{2} g^{kl}(g_{il,j} + g_{jl,i} - g_{ij,l})$$

and (ii) one finds

$$g_{ij,k}(\mathbf{0}) = \Gamma^i_{jk}(\mathbf{0}) + \Gamma^j_{ik}(\mathbf{0}) = 0,$$

as claimed. □

Lemma 1.5.5 *Let* $\mathbf{q}$ *be normal coordinates centred at* $b_0 \in B$*. Let* $(\mathbf{q}, \mathbf{v})$ *and* $(\mathbf{q}, \mathbf{p})$ *be the corresponding local coordinates on* TB *and* T^*B*, respectively. In these coordinates, the isomorphism*

$$\begin{array}{rccc} \Psi_{b_0}: & T_{b_0}B & \longrightarrow & T^*_{b_0}B \\ & (\mathbf{0}, \mathbf{v}) & \longmapsto & (\mathbf{0}, \mathbf{p}) \end{array}$$

is given by the identity map; the same is true – for $X = (\mathbf{0}, \mathbf{v}) \in T_{b_0}B$ *– for the differential*

$$\begin{array}{rccc} T\Psi|_{T_X(TB)}: & T_X(TB) & \longrightarrow & T_{\Psi(X)}(T^*B) \\ & (\dot{\mathbf{q}}, \dot{\mathbf{v}})_{(\mathbf{0},\mathbf{v})} & \longmapsto & (\dot{\mathbf{q}}, \dot{\mathbf{p}})_{\Psi(\mathbf{0},\mathbf{v})}. \end{array}$$

The unit spheres in $T_{b_0}B$ *and* $T^*_{b_0}B$ *are given by* $v_1^2 + \cdots + v_n^2 = 1$ *and* $p_1^2 + \cdots + p_n^2 = 1$*, respectively.*

Proof In arbitrary local coordinates $\mathbf{q}$, write a vector $X \in T_bB$ in the form $X = \sum_{i=1}^n v_i\, \partial_{q_i}$. Then

$$\begin{aligned} \Psi_b(X)(\partial_{q_j}) &= g_b(X, \partial_{q_j}) \\ &= g_b\Big(\sum_i v_i\, \partial_{q_i}, \partial_{q_j}\Big) \\ &= \sum_i g_{ij}(\mathbf{q}) v_i. \end{aligned}$$

Thus, in terms of the basis $\partial_{q_1}, \ldots, \partial_{q_n}$ for T_bB and $dq_1, \ldots, dq_n$ for T^*_bB, the isomorphism Ψ_b is given by†

$$\begin{pmatrix} p_1 \\ \vdots \\ p_n \end{pmatrix} = (g_{ij}(\mathbf{q})) \begin{pmatrix} v_1 \\ \vdots \\ v_n \end{pmatrix}.$$

Then ψ_b^{-1} is given by the inverse matrix $(g^{kl}(\mathbf{q}))$, and a straightforward computation shows that the metric g^* on T^*B is also given by $(g^{kl}(\mathbf{q}))$, that is,

$$g^*_b(\mathbf{p}, \mathbf{p}') = \mathbf{p}^t (g^{kl}(\mathbf{q}))\mathbf{p}',$$

where $\mathbf{p}$ and $\mathbf{p}'$ are read as column vectors.

Now let $\mathbf{q}$ be *normal* coordinates centred at $b_0 \in B$. The statements about the unit spheres in $T_{b_0}B$ and $T^*_{b_0}B$, as well as the coordinate description of Ψ_{b_0}, follow immediately from the preceding discussion and Lemma 1.5.4 (ii).

† Here we use that the matrix $(g_{ij}(\mathbf{q}))$ is symmetric.

The statement about $T\Psi|_{T_X(TB)}$ for $X \in T_{b_0}B$ follows by additionally taking Lemma 1.5.4 (iv) into account: we have

$$\Psi(\mathbf{q}, \mathbf{v}) = (\mathbf{q}, (g_{ij}(\mathbf{q}))\mathbf{v}),$$

hence

$$\begin{aligned} T_{(\mathbf{0},\mathbf{v})}\Psi(\dot{\mathbf{q}}, \dot{\mathbf{v}}) &= \Big(\dot{\mathbf{q}}, (g_{ij}(\mathbf{0}))\dot{\mathbf{v}} + \Big(\sum_k g_{ij,k}(\mathbf{0})\dot{\mathbf{q}}_k\Big)\mathbf{v}\Big) \\ &= (\dot{\mathbf{q}}, \dot{\mathbf{v}}), \end{aligned}$$

as was to be shown. □

Proof of Theorem 1.5.2 (a) We have seen that in local coordinates $\mathbf{q} = (q_1, \ldots, q_n)$ on B and dual coordinates $\mathbf{p} = (p_1, \ldots, p_n)$ on the fibres of T^*B we have

$$\lambda = \mathbf{p}\, d\mathbf{q} \quad \text{and} \quad \omega := d\lambda = d\mathbf{p} \wedge d\mathbf{q}.$$

Since ω is a symplectic form on T^*B, we can define a unique vector field Y on T^*B by $i_Y\omega = \lambda$. Then $d(i_Y\omega) = d\lambda = \omega$, i.e. this Y is a Liouville vector field. In the chosen local coordinates we have $Y = \mathbf{p}\,\partial_{\mathbf{p}}$, which is the radial vector field in the fibre direction of T^*B.

It is easy to see that Y is transverse to the unit cotangent bundle ST^*B. Indeed, the integral curves of Y are tangent to the fibres of T^*B and can be written in the form

$$\mathbf{u}(t) = (b, \mathrm{e}^t\mathbf{u}_0) \in T_b^*B, \; t \in \mathbb{R}.$$

Hence $g_b^*(\mathbf{u}(t), \mathbf{u}(t)) = \mathrm{e}^{2t}g_b^*(\mathbf{u}_0, \mathbf{u}_0)$. For $\mathbf{u}_0 \neq 0$ we obtain

$$\frac{d}{dt}g_b^*(\mathbf{u}(t), \mathbf{u}(t)) > 0,$$

which means that Y is transverse to the positive level sets of g_b^*. (Along the zero section of T^*B, the vector field Y vanishes identically.) By Lemma 1.4.5 we conclude that $\lambda = i_Y\omega$ induces a contact form on ST^*B (which we continue to denote by λ).

In order to prove that $T\Psi(G) = R_\lambda$, let a point $b_0 \in B$ be given. Let $\mathbf{q} = (q_1, \ldots, q_n)$ be normal coordinates centred at b_0. Let $(\mathbf{q}, \mathbf{v})$ and $(\mathbf{q}, \mathbf{p})$ be the corresponding local coordinates on TB and T^*B, respectively.

From the preceding lemmata it follows that the coordinate description of the integral curves of the geodesic vector field G passing through the fibre $ST_{b_0}B$ at time $t = 0$ is of the form

$$t \longmapsto (\gamma(t), \dot{\gamma}(t)) = (t\mathbf{a}, \mathbf{a})$$

with $\sum_j a_j^2 = 1$. Hence at the point $(\mathbf{q}, \mathbf{v}) = (\mathbf{0}, \mathbf{a}) \in ST_{b_0}B$ we have

$$G(\mathbf{0}, \mathbf{a}) = (\dot{\gamma}(0), \ddot{\gamma}(0)) = (\mathbf{a}, \mathbf{0}) \in T_{(\mathbf{0},\mathbf{a})}(ST_{b_0}B).$$

Since $T\Psi$ is the identity map at points $(\mathbf{0}, \mathbf{a})$, it only remains to check that

$$(\mathbf{a}, \mathbf{0}) \in T_{(\mathbf{0},\mathbf{a})}(ST^*_{b_0}B)$$

is the coordinate description of $R_\lambda(\mathbf{0}, \mathbf{a})$. To that end, we compute

$$\lambda_{(\mathbf{0},\mathbf{a})}(\mathbf{a}, \mathbf{0}) = \mathbf{a}\, d\mathbf{q}(\mathbf{a}, \mathbf{0}) = \sum_{j=1}^{n} a_j^2 = 1$$

and

$$\begin{aligned} d\lambda_{(\mathbf{0},\mathbf{a})}\big((\mathbf{a}, \mathbf{0}), -\big) &= (d\mathbf{p} \wedge d\mathbf{q})_{(\mathbf{0},\mathbf{a})}\big((\mathbf{a}, \mathbf{0}), -\big) \\ &= -\mathbf{a}\, d\mathbf{p}_{(\mathbf{0},\mathbf{a})} \\ &= -\mathbf{p}\, d\mathbf{p}_{(\mathbf{0},\mathbf{a})}. \end{aligned}$$

Thus, in order to establish $R_\lambda(\mathbf{0}, \mathbf{a}) = (\mathbf{a}, \mathbf{0})$, we need to show that last term to be equal to zero on $T(ST^*_{b_0}B)$. That this is indeed the case follows from the fact that ST^*B is defined locally by the condition

$$\mathbf{p}^t(g^{kl}(\mathbf{q}))\mathbf{p} = 1,$$

which on taking the differential at the point $(\mathbf{q}, \mathbf{p}) = (\mathbf{0}, \mathbf{a})$ yields, by Lemma 1.5.4 (ii) and (iv), the vanishing of $\mathbf{p}\, d\mathbf{p}$ on $T(ST^*_{b_0}B)$.

Since the point $b_0 \in B$ and the choice of $(\mathbf{q}, \mathbf{v}) = (\mathbf{0}, \mathbf{a}) \in T_{b_0}B$ were arbitrary, this proves part (a) of the theorem.

(b) This statement follows by similar considerations. Again, we need to show that in normal coordinates $\mathbf{q}$ and dual coordinates $\mathbf{p}$ we have $X_H(\mathbf{0}, \mathbf{a}) = (\mathbf{a}, \mathbf{0})$.

In (arbitrary) local coordinates we have

$$H(\mathbf{q}, \mathbf{p}) = \frac{1}{2}\mathbf{p}^t(g^{kl}(\mathbf{q}))\mathbf{p}.$$

Hence, if $\mathbf{q}$ are normal coordinates, we find

$$dH|_{(\mathbf{0},\mathbf{a})} = \mathbf{p}\, d\mathbf{p}_{(\mathbf{0},\mathbf{a})} = \mathbf{a}\, d\mathbf{p}.$$

From

$$d\mathbf{p} \wedge d\mathbf{q}(X_H(\mathbf{0}, \mathbf{a}), -) = -dH|_{(\mathbf{0},\mathbf{a})} = -\mathbf{a}\, d\mathbf{p}$$

we obtain $X_H(\mathbf{0}, \mathbf{a}) = (\mathbf{a}, \mathbf{0})$, as desired. □

We now want to show that the result just proved allows us to give a mathematical formulation of Huygens' principle in geometric optics. In order to do so, we introduce the space of *cooriented* contact elements of a manifold B, made up of the collection of pairs (b, V) consisting of a point $b \in B$ and a cooriented hyperplane $V \subset T_bB$. If B comes equipped with a Riemannian metric g, the space of cooriented contact elements can be naturally identified with either of STB and ST^*B: the identification with STB is given by associating to (b, V) the vector $X \in T_bB$ positively orthonormal to the cooriented hyperplane $V \subset T_bB$ with respect to the inner product g_b; the corresponding element in ST^*B is $\Psi_b(X) = g_b(X, -)$. Under this identification, the Liouville form λ on ST^*B defines the natural contact structure on the space of cooriented contact elements.

For simplicity we assume that (B, g) is a complete Riemannian manifold. Then the time-t-map Φ_t of the geodesic flow is, for every $t \in \mathbb{R}$, a diffeomorphism of STB.

Corollary 1.5.6 *Under the identification of the space M of cooriented contact elements of the complete Riemannian manifold (B, g) with the unit tangent bundle STB, the time-t-map Φ_t of the geodesic flow is given by*

$$\Phi_t(b_0, V_0) = (\gamma(t), V(t)),$$

where γ is the unique geodesic on B with $\gamma(0) = b_0$ and $\dot{\gamma}(0)$ positively orthonormal to V_0, and $V(t)$ the hyperplane in $T_{\gamma(t)}B$ orthogonal to and cooriented by $\dot{\gamma}(t)$. This Φ_t is a contact transformation of M.

The proof of the last statement in this corollary requires an understanding of how a contact structure behaves under the flow of a given vector field. Thus, before we turn to the proof of the corollary, we briefly address that issue. In Section 2.3 this discussion will be extended to time-dependent vector fields.

Definition 1.5.7 Let X be a vector field on an arbitrary contact manifold $(M, \xi = \ker \alpha)$. Write, by slight abuse of notation, ψ_t for the local flow of X; if M is not closed, the map ψ_t (for a fixed $t \neq 0$) will not, in general, be defined on all of M. The vector field X will be called an **infinitesimal automorphism of** ξ (or a **contact vector field**) if $T\psi_t(\xi) = \xi$ for all $t \in \mathbb{R}$ (and for all points on M where the relevant terms are defined). It will be called an **infinitesimal automorphism of** α (or a **strict contact vector field**) if $\psi_t^*\alpha = \alpha$ for all t.

Recall that the relation between the flow ψ_t and the Lie derivative $\mathcal{L}_X$ is given by $\mathcal{L}_X = \frac{d}{dt}|_{t=0}\psi_t^*$.

Lemma 1.5.8 *(a) The vector field X is an infinitesimal automorphism of α if and only if $\mathcal{L}_X\alpha \equiv 0$.*

(b) The vector field X is an infinitesimal automorphism of ξ if and only if $\mathcal{L}_X\alpha = \mu\alpha$ for some function $\mu\colon M \to \mathbb{R}$ (and this condition is independent of the choice of contact form α defining a given ξ).

Proof We have $\psi_{t+t_0} = \psi_{t_0} \circ \psi_t$, hence

$$\frac{d}{dt}|_{t=t_0}\left(\psi_t^*\alpha\right) = \psi_{t_0}^*\left(\frac{d}{dt}|_{t=0}\psi_t^*\alpha\right) = \psi_{t_0}^*\left(\mathcal{L}_X\alpha\right).$$

(a) If X is an infinitesimal automorphism of α, then

$$\psi_t^*\left(\mathcal{L}_X\alpha\right) = \frac{d}{dt}\left(\psi_t^*\alpha\right) \equiv 0 \text{ for all } t \in \mathbb{R},$$

and hence $\mathcal{L}_X\alpha \equiv 0$.

Conversely, if $\mathcal{L}_X\alpha \equiv 0$, the computation above shows that $\frac{d}{dt}(\psi_t^*\alpha) \equiv 0$, hence $\psi_t^*\alpha \equiv \psi_0^*\alpha = \alpha$.

(b) First of all we notice that if X satisfies $\mathcal{L}_X\alpha = \mu\alpha$, and λ is a nowhere zero function on M, then

$$\mathcal{L}_X(\lambda\alpha) = (\mathcal{L}_X\lambda)\alpha + \lambda(\mathcal{L}_X\alpha) = (X(\lambda) + \lambda\mu)\alpha. \tag{1.1}$$

This shows that the condition $\mathcal{L}_X\alpha = \mu\alpha$ for some function μ on M is a condition on ξ, independent of the specific choice of contact form α.

Now suppose that X is an infinitesimal automorphism of ξ. The condition $T\psi_t(\xi) = \xi$ can be rewritten as $\psi_t^*\alpha = \lambda_t\alpha$ with† $\lambda_t\colon M \to \mathbb{R}^+$. Then

$$\mathcal{L}_X\alpha = \frac{d}{dt}|_{t=0}\psi_t^*\alpha = \frac{d}{dt}|_{t=0}(\lambda_t\alpha) = \mu\alpha$$

with $\mu := \frac{d}{dt}|_{t=0}\lambda$.

Conversely, if $\mathcal{L}_X\alpha = \mu\alpha$, then the local flow ψ_t of X satisfies

$$\frac{d}{dt}\psi_t^*\alpha = \psi_t^*(\mu\alpha) = (\mu \circ \psi_t)\psi_t^*\alpha,$$

whence

$$\psi_t^*\alpha = \lambda_t\alpha \text{ with } \lambda_t = \exp\Big(\int_0^t (\mu \circ \psi_s)\, ds\Big),$$

so X is an infinitesimal automorphism of ξ. □

† The nowhere zero functions λ_t constitute a smooth family with $\lambda_0 \equiv 1$, hence all the λ_t take values in $\mathbb{R}^+$.

Example 1.5.9 The Reeb vector field R_α is a strict contact vector field for α, since

$$\mathcal{L}_{R_\alpha}\alpha = d(i_{R_\alpha}\alpha) + i_{R_\alpha} d\alpha \equiv 0.$$

Proof of Corollary 1.5.6 The geodesic flow Φ_t is characterised by the property that for a geodesic γ on B we have

$$\Phi_t(\gamma(0), \dot{\gamma}(0)) = (\gamma(t), \dot{\gamma}(t)).$$

This restricts to a flow on STB by considering unit speed geodesics only. The description of the flow on M given in the corollary is an immediate consequence of the identification of STB with M, given by associating to $(b, X) \in STB$ the cooriented contact element (b, V), where $V \subset T_bB$ is the hyperplane orthogonal to and cooriented by X.

By the foregoing example, the Reeb flow of λ on ST^*B consists of (strict) contact transformations. So Φ_t is a contact transformation by Theorem 1.5.2. □

We now interpret the Riemannian manifold (B, g) as a model for an optical medium, so that geodesics with respect to the metric g correspond to light rays.

Definition 1.5.10 The **wave front** $F_{b_0}(t)$ of a point $b_0 \in B$ at time t is the set of points $b \in B$ with the property that there exists a unit speed geodesic γ in B with $\gamma(0) = b_0$ and $\gamma(t) = b$.

For $|t|$ small, the wave front $F_{b_0}(t)$ will be a smooth hypersurface in B, diffeomorphic to a sphere. As $|t|$ increases, $F_{b_0}(t)$ will, in general, develop singularities. In order to formulate Huygens' principle, we also need the following concept.

Definition 1.5.11 Let (M, ξ) be a contact manifold. A submanifold L of (M, ξ) is called an **isotropic submanifold** if $T_pL \subset \xi_p$ for all points $p \in L$.

The contact condition $\alpha \wedge (d\alpha)^n \neq 0$ (for $\dim M = 2n + 1$) says that a contact structure $\xi = \ker \alpha$ is a hyperplane field that is, in some sense, as far from satisfying the Frobenius integrability condition $\alpha \wedge d\alpha \equiv 0$ as possible. The following proposition gives a more concrete geometric interpretation of this statement.

Proposition 1.5.12 *Let (M, ξ) be a contact manifold of dimension $2n + 1$ and $L \subset (M, \xi)$ an isotropic submanifold. Then $\dim L \leq n$.*

Proof Write i for the inclusion of L in M and let α be an (at least locally defined) contact form defining ξ. Then the condition for L to be isotropic becomes $i^*\alpha \equiv 0$. It follows that $i^*d\alpha \equiv 0$. In particular, $T_pL \subset \xi_p$ is an isotropic subspace of the $2n$–dimensional symplectic vector space $(\xi_p, d\alpha|_{\xi_p})$. By Remark 1.3.6 we have $\dim T_pL \leq (\dim \xi_p)/2 = n$. □

Definition 1.5.13 An isotropic submanifold $L \subset (M^{2n+1}, \xi)$ of maximal possible dimension n is called a **Legendrian submanifold**.

Examples 1.5.14 (1) Regard STB as the space of cooriented contact elements with its natural contact structure, and write $\pi\colon STB \to B$ for the bundle projection. This particular bundle projection is called the **(wave) front projection**. Then every fibre $\pi^{-1}(b)$, $b \in B$, is a Legendrian submanifold of STB. This is immediate from the geometric definition of the natural contact structure on the space of contact elements.

(2) The Legendrian fibre $\pi^{-1}(b)$ of the bundle $\pi\colon STB \to B$ consists of the pairs $(b, \mathbf{v})$ with $\mathbf{v}$ a unit tangent vector at b. With Φ_t denoting the geodesic flow on STB, the front $F_b(t)$ can be described as

$$F_b(t) = \pi\big(\Phi_t(\pi^{-1}(b))\big).$$

Since Φ_t is a contact transformation of STB for each t, the submanifold $\Phi_t(\pi^{-1}(b)) \subset STB$ is a Legendrian submanifold.

Remarks 1.5.15 (1) Let L_0, L_1 be Legendrian submanifolds of STB that intersect in a point $\mathbf{v} \in ST_bB$. If the front projections $\pi(L_0), \pi(L_1)$ are smooth hypersurfaces near the point $b \in \pi(L_0) \cap \pi(L_1)$, then these front projections are tangent to each other at that point. Again this follows immediately from the geometric definition of the natural contact structure on STB.

(2) Notice that $\pi^{-1}(b)$ consists of the unit tangent vectors in T_bB, i.e. the set of possible initial velocities of unit speed geodesics starting at b. Assume that $b_1 \in F_b(t)$ is a smooth point of the wave front $F_b(t)$, that is, this wave front is a smooth hypersurface near b_1. Then, in particular, there is a unique unit speed geodesic γ with $\gamma(0) = b$ and $\gamma(t) = b_1$. Write $\mathbf{v}_1 = \dot\gamma(t) = \Phi_t(\dot\gamma(0))$. Since $\Phi_t(\pi^{-1}(b)) \subset STB$ is a Legendrian submanifold, its tangent space at $\mathbf{v}_1 \in ST_{b_1}B$ is contained in $\xi_{\mathbf{v}_1}$, with ξ denoting the natural contact structure on STB. By the definition of that natural contact structure, the tangent plane to the front $F_b(t) = \pi\big(\Phi_t(\pi^{-1}(b))\big)$ at the point b_1 is the hyperplane in $T_{b_1}B$ defined by (i.e. orthogonal to) $\mathbf{v}_1$.

It is not clear, *a priori*, that Legendrian submanifolds exist in any given contact manifold. It turns out, however, that such submanifolds even satisfy what is called an h–principle† (homotopy principle) and consequently exist in abundance. In the case of Legendrian knots in contact 3–manifolds we shall meet a proof of this fact later.

Huygens' principle says that the wave front $F_{q_0}(t_0 + t)$ is the envelope of the fronts $F_q(t)$ with $q \in F_{q_0}(t_0)$ (see Figure 1.4). Some care has to be taken in making this statement precise, in particular when the wave fronts develop singularities. For simplicity, we only prove Huygens' principle in the following form, where Φ_t denotes the geodesic flow on STB and π, as above, the bundle projection $STB \to B$.

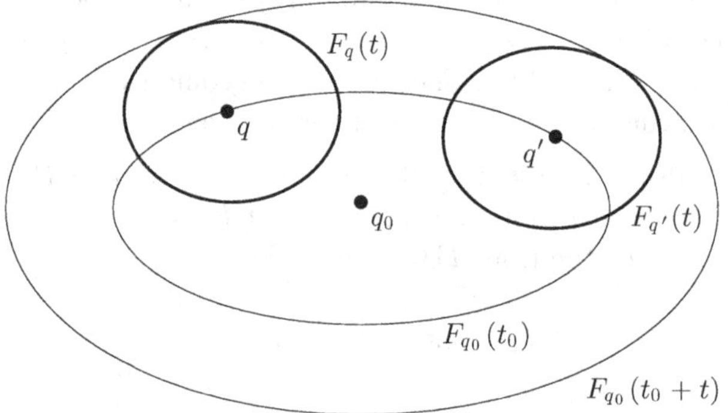

Fig. 1.4. Huygens' principle.

Corollary 1.5.16 *Let γ be a unit speed geodesic in B. Suppose that $t_1, t \in \mathbb{R}^+$ are such that – with $b_0 := \gamma(0)$ and $b_1 := \gamma(t_1)$ – the wave fronts $F_{b_0}(t_1 + t)$ and $F_{b_1}(t)$ are smooth at the point*

$$b := \gamma(t_1 + t) \in F_{b_0}(t_1 + t) \cap F_{b_1}(t).$$

Then $F_{b_0}(t_1 + t)$ and $F_{b_1}(t)$ are tangent at b.

Proof The sets $L_0 := \Phi_{t_1}(\pi^{-1}(b_0))$ and $L_1 := \pi^{-1}(b_1)$ are Legendrian submanifolds of STB that intersect in the point $\dot{\gamma}(t_1) \in ST_{b_1}B$. Hence $L_0' := \Phi_{t_1+t}(\pi^{-1}(b_0)) = \Phi_t(L_0)$ and $L_1' := \Phi_t(L_1)$ are Legendrian submanifolds that intersect in $\dot{\gamma}(t_1 + t) \in ST_bB$. By assumption, the wave fronts $F_{t_1+t}(b_0) = \pi(L_0')$ and $F_t(b_1) = \pi(L_1')$ are smooth hypersurfaces near the

† The bible on h–principles was written by Gromov [119]; a more recent monograph on the subject is the one by Eliashberg–Mishachev [77]; my own modest contribution to the literature is [95].

point b. By the geometric properties of the natural contact structure on STB, see Remarks 1.5.15, these hypersurfaces are tangent to each other at the point b. □

A related discussion of Huygens' principle as a contact geometric theorem in the context of partial differential equations can be found in Lecture 3 by Arnold [18]. For further contact geometric aspects of geometric optics and wave propagation see [15].

Yet another physical theory admitting a contact geometric formulation is thermodynamics, see [16]. It is worth quoting the opening lines of that paper: 'Every mathematician knows that it is impossible to understand any elementary course in thermodynamics. The reason is that thermodynamics is based – as Gibbs has explicitly proclaimed – on a rather complicated mathematical theory, on contact geometry.'

1.6 Order of contact

In Proposition 1.5.12 we saw that an isotropic submanifold in a $(2n+1)$–dimensional contact manifold has dimension at most equal to n. Thus, in a contact 3–manifold (M, ξ) we can find curves tangent to ξ, but no surfaces.

The aim of the present section is to analyse this situation for 3–manifolds a little more carefully. This will allow us to give an entirely geometric definition of the notion of contact structure, without reference to the exterior derivative of a differential 1–form. I learned this characterisation of contact structures from Jesús Gonzalo.

Consider a 2–plane field ξ on a 3–manifold M and an embedded surface $\Sigma \subset M$. Let (u, v) be local coordinates on Σ near some point $p = (0,0) \in \Sigma$. Define $\theta(u, v)$ as the angle between the tangent plane $T_{(u,v)}\Sigma$ and the plane $\xi_{(u,v)}$.

Definition 1.6.1 We say that ξ has **contact of order at least equal to** k with Σ at $p = (0,0)$ if $\theta(u, v)$ is a function of type $O(\|(u, v)\|^k)$ for $(u, v) \to (0,0)$, where O denotes the Landau symbol. This means that $\theta(u, v)/\|(u, v)\|^k$ is bounded above by a constant as $(u, v) \to (0,0)$.†

Thus, ξ having contact of order at least 1 with Σ at p is simply saying that $\xi_p = T_p\Sigma$. Obviously, contact of order *equal* to k is going to mean that $\theta(u, v)$ is of type $O(\|(u, v)\|^k)$, but not of type $O(\|(u, v)\|^{k+1})$. You may

† For a differentiable function $f(u, v)$ to be of type $O(\|(u, v)\|^k)$ is equivalent to saying that the ith order partial derivatives of f vanish for $i = 0, 1, \ldots, k-1$ (one says that 'f vanishes to kth order'). Unfortunately, $\theta(u, v)$ may not be differentiable in the points where $\theta(u, v) = 0$.

want to convince yourself that the order of contact does not depend on the choice of local coordinates on Σ.

The following theorem gives a characterisation of contact structures on 3–manifolds in terms of this notion of contact. I should enter the caveat that the name 'contact structure' does *not* derive from this characterisation, but rather – as mentioned before – from the space of contact elements, which (in the 3–dimensional case) has to do with tangencies of curves.

Theorem 1.6.2 *Let ξ be a tangent 2–plane field on a 3–manifold M. Then ξ is a contact structure if and only if, for every embedded surface $\Sigma \subset M$ and every point $p \in \Sigma$, the plane field ξ has order of contact at most equal to 1 with Σ at p.*

Proof Since all considerations are local in nature, we may assume that $M = \mathbb{R}^3$ and that $p = \mathbf{0}$, the origin in $\mathbb{R}^3$. Moreover, we may assume without loss of generality that ξ_0 coincides with the (x, y)–plane. Then we can write ξ near $\mathbf{0}$ as the kernel of a 1–form

$$dz + a(x, y, z)\, dx + b(x, y, z)\, dy$$

with

$$a(\mathbf{0}) = b(\mathbf{0}) = 0. \tag{1.2}$$

The contact condition at $\mathbf{0}$ becomes

$$a_y(\mathbf{0}) \neq b_x(\mathbf{0}), \tag{1.3}$$

with subscripts denoting partial derivatives. Thus, we have to show that the condition $a_y(\mathbf{0}) = b_x(\mathbf{0})$ is equivalent to the existence of a surface Σ whose order of contact with ξ at $\mathbf{0} \in \Sigma$ is at least 2.

Any surface Σ having order of contact at least 1 with ξ at $\mathbf{0}$ can be described near $\mathbf{0}$ as a graph

$$\Sigma = \{(x, y, z) \colon\; z = f(x, y),\; (x, y) \in U\}$$

with

$$f(0, 0) = f_x(0, 0) = f_y(0, 0) = 0. \tag{1.4}$$

Here U is a neighbourhood of $(0, 0)$ in the (x, y)–plane. We can then take x and y as local coordinates on Σ.

The 2–plane field ξ is spanned by the vector fields

$$\partial_x - a\,\partial_z \quad \text{and} \quad \partial_y - b\,\partial_z,$$

so a vector field orthogonal to ξ (along Σ) is given by

$$\mathbf{n}_\xi(x,y) := a(x,y,f(x,y))\,\partial_x + b(x,y,f(x,y))\,\partial_y + \partial_z.$$

Likewise, a vector field orthogonal to Σ is given by

$$\mathbf{n}_\Sigma(x,y) := -f_x(x,y)\,\partial_x - f_y(x,y)\,\partial_y + \partial_z.$$

Write $\theta(x,y)$ for the angle between $\mathbf{n}_\xi(x,y)$ and $\mathbf{n}_\Sigma(x,y)$, and set

$$\begin{aligned} h(x,y) &:= \sin^2\theta(x,y) \\ &= \frac{\|\mathbf{n}_\xi(x,y)\times\mathbf{n}_\Sigma(x,y)\|^2}{\|\mathbf{n}_\xi(x,y)\|^2\cdot\|\mathbf{n}_\Sigma(x,y)\|^2} \\ &= \frac{\|(b+f_y,-a-f_x,bf_x-af_y)\|^2}{(1+a^2+b^2)\cdot(1+f_x^2+f_y^2)} \\ &= \frac{(b+f_y)^2+(a+f_x)^2+(bf_x-af_y)^2}{(1+a^2+b^2)\cdot(1+f_x^2+f_y^2)}. \end{aligned}$$

Notice that this is a differentiable function, although $\theta(x,y)$ need not be. Since $\lim_{\theta\to 0}(\sin\theta)/\theta = 1$, the condition for ξ to have contact with Σ of order at least equal to 2 translates into the requirement that $h(x,y)$ vanish to fourth order at $(0,0)$. Given Equations (1.2) and (1.4), this in turn is equivalent to the vanishing to second order of the functions $a(x,y,f(x,y)) + f_x(x,y)$ and $b(x,y,f(x,y)) + f_y(x,y)$, that is,

$$\begin{aligned} a_x(\mathbf{0}) + f_{xx}(0,0) &= 0, \\ a_y(\mathbf{0}) + f_{xy}(0,0) &= 0, \\ b_x(\mathbf{0}) + f_{xy}(0,0) &= 0, \\ b_y(\mathbf{0}) + f_{yy}(0,0) &= 0. \end{aligned} \tag{1.5}$$

One direction of the proof is now simple. Suppose we can find a surface $\Sigma = \{z = f(x,y)\}$ containing the point $\mathbf{0}$ such that ξ has contact of order at least 2 with Σ at $\mathbf{0}$. Then in particular

$$a_y(\mathbf{0}) + f_{xy}(0,0) = 0 = b_x(\mathbf{0}) + f_{xy}(0,0),$$

and hence $a_y(\mathbf{0}) = b_x(\mathbf{0})$. This means that ξ violates the contact condition (1.3) at $\mathbf{0}$.

Conversely, suppose that $a_y(\mathbf{0}) = b_x(\mathbf{0})$. Define

$$f(x,y) := -\int_0^1 \big[a(tx,ty,0)x + b(tx,ty,0)y\big]\,dt.$$

This definition is modelled on the usual definition of the antiderivative of a closed 1–form in the proof of the Poincaré Lemma. We then compute

$$\begin{aligned}
f_x(x,y) &= -\int_0^1 [a_x tx + a + b_x ty]\,dt,\\
f_y(x,y) &= -\int_0^1 [a_y tx + b_y ty + b]\,dt,\\
f_{xx}(x,y) &= -\int_0^1 [a_{xx} t^2 x + 2a_x t + b_{xx} t^2 y]\,dt,\\
f_{xy}(x,y) &= -\int_0^1 [a_{xy} t^2 x + a_y t + b_{xy} t^2 y + b_x t]\,dt,\\
f_{yy}(x,y) &= -\int_0^1 [a_{yy} t^2 x + 2b_y t + b_{yy} t^2 y]\,dt,
\end{aligned}$$

where all the integrands on the right are to be evaluated at $(tx, ty, 0)$. Hence

$$f_{xx}(0,0) = -a_x(\mathbf{0}),\ f_{xy}(0,0) = -\frac{1}{2}\big(a_y(\mathbf{0}) + b_x(\mathbf{0})\big),\ f_{yy}(0,0) = -b_y(\mathbf{0}).$$

Thus, if $a_y(\mathbf{0}) = b_x(\mathbf{0})$, then $f_{xy}(0,0) = -a_y(\mathbf{0}) = -b_x(\mathbf{0})$, and it follows that Equations (1.5) are satisfied, i.e. we have found a surface $\Sigma := \{z = f(x,y)\}$ with second-order contact to ξ. □

1.7 Applications of contact geometry to topology

When attempting to prove topological results about differentiable manifolds, it is often useful to endow these manifolds with additional geometric structures. Thurston's geometrisation programme for 3–manifolds – see [232], [219] – is arguably the most striking instance of this philosophy; here the additional structure comes from Riemannian geometry. In the present section I want to present two topological results where the supporting role is played by contact geometry. I include some background material that most readers will have met in a first course on differential and geometric topology, respectively.

1.7.1 Cerf's theorem

Write Diff (M) for the group of *orientation-preserving* diffeomorphisms of an orientable differential manifold M (the group multiplication being given by composition of diffeomorphisms). Let D^n be the n–dimensional unit disc in $\mathbb{R}^n$, and $S^{n-1} = \partial D^n$ its boundary, the standard $(n-1)$–dimensional unit sphere. Since diffeomorphisms of a manifold with boundary preserve that boundary, we have a natural restriction homomorphism

$$\begin{array}{rccc} \rho_n: & \mathrm{Diff}\,(D^n) & \longrightarrow & \mathrm{Diff}\,(S^{n-1}) \\ & f & \longmapsto & f|_{S^{n-1}}. \end{array}$$

The group Γ_n is defined as

$$\Gamma_n = \mathrm{Diff}\,(S^{n-1})/\mathrm{im}\,\rho_n.$$

In order to show that this is indeed a group, we need to prove that $\mathrm{im}\,\rho_n$ is a normal subgroup in $\mathrm{Diff}\,(S^{n-1})$.

We begin with two lemmata. Write $\mathrm{Diff}_0(S^{n-1})$ for the group of diffeomorphisms of S^{n-1} that are isotopic to the identity.

Lemma 1.7.1 *With the commutator* $[f,g]$ *of two diffeomorphisms* f,g *defined by* $[f,g] = f \circ g \circ f^{-1} \circ g^{-1}$ *we have*

$$[\mathrm{Diff}\,(S^{n-1}), \mathrm{Diff}\,(S^{n-1})] \subset \mathrm{Diff}_0(S^{n-1}).$$

Proof Let $f,g \in \mathrm{Diff}\,(S^{n-1})$ be given. It follows from the disc theorem [132, Thm. 8.3.1] that f is isotopic to a diffeomorphism f_+ that restricts to the identity on the (closed) upper hemisphere of S^{n-1}. Similarly, g is isotopic to a diffeomorphism g_- that restricts to the identity on the lower hemisphere. It follows that $[f,g]$ is isotopic to $[f_+, g_-] = \mathrm{id}_{S^{n-1}}$. □

Lemma 1.7.2 $\mathrm{Diff}_0(S^{n-1}) \subset \mathrm{im}\,\rho_n$.

Proof Let $f \in \mathrm{Diff}_0(S^{n-1})$. Choose a diffeotopy F_t (smooth in $t \in [0,1]$) of S^{n-1} with $F_t = \mathrm{id}_{S^{n-1}}$ for t near 0 and $F_1 = f$. Define $h \in \mathrm{Diff}\,(D^n)$ by

$$h(\mathbf{x}) = \begin{cases} \|\mathbf{x}\| \cdot F_{\|\mathbf{x}\|}\left(\frac{\mathbf{x}}{\|\mathbf{x}\|}\right) & \text{for } 0 < \|\mathbf{x}\| \leq 1, \\ \mathbf{0} & \text{for } \mathbf{x} = \mathbf{0}. \end{cases}$$

Then $\rho_n(h) = f$. □

Proposition 1.7.3 Γ_n *is an abelian group.*

Proof Let $f \in \mathrm{Diff}\,(S^{n-1})$ and $h \in \mathrm{im}\,\rho_n$. By the preceding lemmata we have

$$[f,h] \in \mathrm{Diff}_0(S^{n-1}) \subset \mathrm{im}\,\rho_n,$$

hence

$$f \circ h \circ f^{-1} \in \mathrm{im}\,\rho_n.$$

This means that $\operatorname{im} \rho_n$ is a normal subgroup of $\operatorname{Diff}(S^{n-1})$, so that Γ_n is a group.

In fact, those lemmata show that $[f, g] \in \operatorname{im} \rho_n$ for arbitrary $f, g \in \operatorname{Diff}(S^{n-1})$, which implies that Γ_n is abelian. □

Using once again the disc theorem, it is not difficult to show that any orientation-preserving diffeomorphism of S^1 is isotopic to the identity [132, Thm. 8.3.3], and hence extends to a diffeomorphism of D^2, that is, $\Gamma_2 = 0$. The result $\Gamma_3 = 0$ was shown by Smale [222] and Munkres [195]. Both use essentially elementary arguments.†

The statement $\Gamma_4 = 0$ is known as Cerf's theorem [44]. Its proof covers some 130 pages. A conceptually completely different and – once the contact geometric background has been established – extremely short proof of this theorem was given by Eliashberg [69]. Here is a very rough sketch of Eliashberg's argument; details will be given in Section 4.11.

The key point is to establish a classification of contact structures on S^3. Eliashberg has introduced a dichotomy of so-called 'tight' and 'overtwisted' contact structures on 3–manifolds. Whereas any homotopy class of 2–plane fields on an arbitrary closed, orientable 3–manifold contains an essentially unique overtwisted contact structure (Section 4.7), the standard contact structure ξ_{st} on the 3–sphere (Example 2.1.7) is unique up to isotopy (Section 4.10). From the definition of tightness it is easy to see that this property is preserved under diffeomorphisms.

Now, let f be any diffeomorphism of S^3. The fact that the image of ξ_{st} under f is isotopic to ξ_{st} implies that f is isotopic to a diffeomorphism f_0 that preserves ξ_{st}. Such a diffeomorphism, however, is geometrically much more rigid, and Eliashberg [69] showed that the technique of filling by holomorphic discs ([25],[118], [66]) allows one to construct an extension of f_0 to a diffeomorphism over D^4. By sweeping out the isotopy between f and f_0 over a collar neighbourhood of S^3 in D^4 as in the proof of Lemma 1.7.2, one obtains an extension of f to a diffeomorphism of D^4.

Remark 1.7.4 For the relevance of the groups Γ_n to questions in differential topology, in particular the classification of exotic differentiable structures on spheres, see [153], [145], and the preface to [44] by Kuiper.

† Since S^{n-1} admits an orientation-reversing diffeomorphism that extends to a diffeomorphism of D^n, e.g. a reflection in a hyperplane of $\mathbb{R}^n$ through the origin, it follows from $\Gamma_n = 0$ that every diffeomorphism of S^{n-1} extends.

1.7.2 Property P for knots

We begin by recalling a few facts about Dehn surgery on knots in 3–manifolds, mostly to set up notation. For a textbook reference on this topic see [209] or [215].

Let K be a knot in the 3–sphere S^3 (or, more generally, in some oriented 3–manifold M), by which we mean a smoothly embedded copy of the circle S^1. Write νK for a (closed) tubular neighbourhood of K. The neighbourhood νK is diffeomorphic to a solid torus $S^1 \times D^2$, since this is the only orientable D^2–bundle over S^1. Let C be the closure of the complement $S^3 \setminus \nu K$ of νK in S^3. (Part of) the Mayer–Vietoris sequence† for $S^3 = \nu K \cup C$ with $\nu K \cap C = T^2$ reads

$$\begin{array}{ccccccccc} H_2(S^3) & \rightarrow & H_1(T^2) & \rightarrow & H_1(\nu K) & \oplus & H_1(C) & \rightarrow & H_1(S^3) \\ 0 & \rightarrow & \mathbb{Z}\oplus\mathbb{Z} & \rightarrow & \mathbb{Z} & \oplus & H_1(C) & \rightarrow & 0. \end{array}$$

We conclude that $H_1(C) \cong \mathbb{Z}$. We also see that on $T^2 = \partial(\nu K)$ there are two distinguished curves, unique up to isotopy.

(1) The **meridian** μ, defined as a simple closed curve that generates the kernel of the homomorphism $H_1(T^2) \rightarrow H_1(\nu K)$.
(2) The **preferred longitude** λ, a simple closed curve that generates the kernel of the homomorphism $H_1(T^2) \rightarrow H_1(C)$.

We assume that S^3 is equipped with its standard orientation as the boundary of D^4.‡ We give $T^2 = \partial(\nu K)$ the boundary orientation. We also assume K to be oriented. Then λ can be oriented by requiring it to be isotopic to K in νK as oriented curve; the orientation we choose for μ is the one that turns μ, λ into a positive basis for that homology group. (Occasionally we allow ourselves to denote a simple closed curve on T^2 by the same symbol as the class it represents in $H_1(T^2)$, since that class determines the curve up to isotopy.) This is illustrated in Figure 1.5, with the standard (right-hand) orientation of ambient 3–space.

With an appropriate choice of generator for $H_1(C) \cong \mathbb{Z}$, the homomorphism $H_1(T^2) \rightarrow H_1(\nu K) \oplus H_1(C)$ is then characterised by $\mu \mapsto (0,1)$ and $\lambda \mapsto (1,0)$.

Dehn surgery along K means that one removes the tubular neighbourhood νK and glues in a solid torus $S^1 \times D^2$, using some diffeomorphism

$$\partial(S^1 \times D^2) \longrightarrow \partial(\nu K)$$

† Strictly speaking, in this Mayer–Vietoris sequence one should be working with open neighbourhoods of νK and C which retract to those sets, and whose intersection retracts to T^2.

‡ The rule for *boundary orientations* is: outward normal first, followed by a positively oriented basis for a tangent space to the boundary, gives a positively oriented basis for the tangent space of the manifold with boundary.

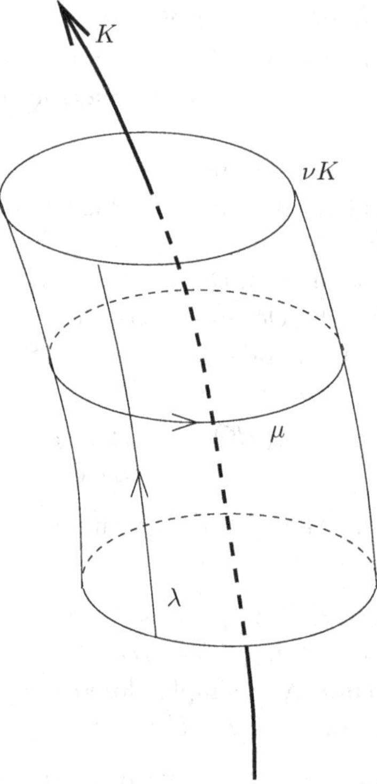

Fig. 1.5. Meridian μ and longitude λ.

to identify the boundaries.

Write μ_0 for the meridian $* \times \partial D^2$ of $S^1 \times D^2$, and λ_0 for the longitude $S^1 \times *$, with $*$ denoting a point in S^1 or ∂D^2, respectively. Then the gluing diffeomorphism can be described by

$$\mu_0 \longmapsto p\mu + q\lambda, \quad \lambda_0 \longmapsto m\mu + n\lambda, \tag{1.6}$$

with $\left(\begin{smallmatrix} p & m \\ q & n \end{smallmatrix}\right) \in \mathrm{GL}(2, \mathbb{Z})$. Actually, the effect of the Dehn surgery along K (up to an isotopy of the resulting manifold fixed outside νK) is completely determined by the image $p\mu + q\lambda$ of μ_0, i.e. the curve on $\partial(\nu K)$ that becomes homotopically trivial in the surgered manifold. This follows by observing that the surgered manifold can be obtained from C by first attaching a 2–disc along the curve $p\mu + q\lambda$, and then gluing in a 3–cell. That latter gluing is unique up to isotopy (recall $\Gamma_3 = 0$ from the preceding section).

In fact, the Dehn surgery is completely determined by the **surgery coefficient** $p/q \in \mathbb{Q} \cup \{\infty\}$, since the diffeomorphism of $\partial(S^1 \times D^2)$ given by $(\lambda_0, \mu_0) \mapsto (\lambda_0, -\mu_0)$, which has the effect of changing the sign of both p

and q in the gluing map, extends to a diffeomorphism of the solid torus that we glue back. We therefore speak unambiguously of (p/q)–surgery. Notice that ∞–surgery (that is, $q = 0$, $p = \pm 1$) is topologically trivial.

Observe that fixing some longitude on $\partial(\nu K)$, that is, a curve λ that together with μ defines a basis for $H_1(\partial(\nu K))$, amounts to choosing a trivialisation of the normal bundle of K in S^3, or what is called a **framing** of K. The *preferred* longitude λ we have chosen above can be characterised by the property that it bounds a surface in C, or equivalently by the fact that its linking number† $\mathtt{lk}(\lambda, K)$ with K is equal to zero.

If instead we choose the longitude $\lambda' = \lambda + k\mu$ for some $k \in \mathbb{Z}$, then the gluing map changes to

$$\mu_0 \longmapsto (p - kq)\mu + q\lambda',$$

so that the surgery coefficient changes to $p'/q' = (p/q) - k$.

It follows that Dehn surgeries with integer surgery coefficient are exactly the Dehn surgeries that can, by a different choice of longitude, be turned into Dehn surgeries with surgery coefficient 0. This corresponds to choosing some embedding

$$S^1 \times D^2 \xrightarrow{\cong} \nu K \subset S^3$$

(which we use to identify νK with a solid torus $S^1 \times D^2$), then cutting out this solid torus $S^1 \times D^2$ and gluing back a solid torus $D^2 \times S^1$ with the identity map

$$\partial(D^2 \times S^1) \longrightarrow \partial(S^1 \times D^2);$$

see Figure 1.6, which obviously suffers from lack of dimensions. Such a surgery is called **handle surgery** or simply **surgery**. Figure 1.6 illustrates that attaching a 'handle' $D^2 \times D^2$ along $S^1 \times D^2$ to the boundary M of a 4–manifold W produces – after the 'smoothing of corners' or 'straightening the angle', see [38] – a 4–manifold with new boundary being the result of a surgery on M. (A more formal discussion of this point will be given in Chapter 6.)

The following is perhaps the most fundamental theorem about the structure of 3–manifolds. Recall that a manifold being **closed** means that it is compact and without boundary; a **link** is a finite collection of disjoint knots. Implicitly we shall always assume our manifolds to be connected, the necessary modifications for manifolds with several components being obvious.

† See Section 3.4 for a formal discussion of linking numbers and the realisation of homology cycles by submanifolds.

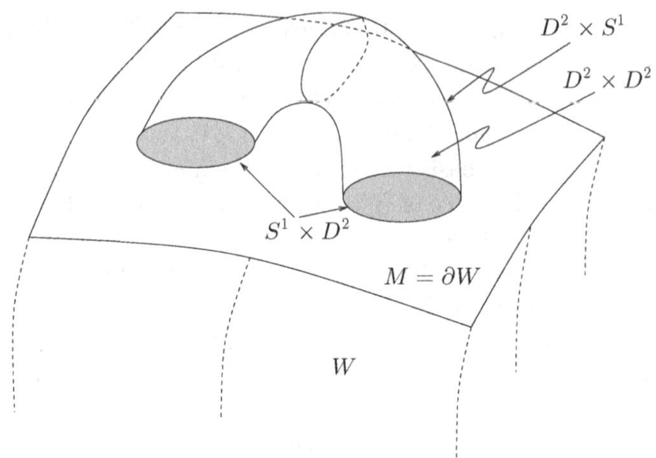

Fig. 1.6. Attaching a handle.

Theorem 1.7.5 (Lickorish [158], Wallace [237]) *Any closed, connected, orientable 3–manifold can be obtained from S^3 by surgery along a link.*

The recent work of Perelman [208] has closed the coffin on the 3–dimensional Poincaré conjecture (saying that a closed, simply connected 3–manifold is necessarily the 3–sphere), see also [41]. Previous to that work, and given the Lickorish–Wallace theorem, it was a natural question for geometric topologists whether one might be able to produce a counterexample to the Poincaré conjecture by a single Dehn surgery. This led to the definition of the following property, whose name is generally regarded as a little unfortunate.

Definition 1.7.6 A knot K in S^3 has **Property P** if every non-trivial surgery along K yields a 3–manifold that is not simply connected.

In the early 1970s, Bing and Martin [27] and González-Acuña [113] conjectured that every non-trivial knot has Property P. By work of Kronheimer and Mrowka [154], this is now a theorem.

Theorem 1.7.7 (Kronheimer–Mrowka) *Every non-trivial knot in S^3 has Property P.*

Before describing the role that contact geometry has played in the proof of this theorem, I want to indicate the importance of this theorem beyond the negative statement – now superseded by Perelman's work – that counterexamples to the Poincaré conjecture cannot result from a single surgery.

Write $K_{p/q}$ for the 3–manifold obtained from S^3 by (p/q)–surgery along the knot K.

Lemma 1.7.8 *The first integral homology group of the surgered manifold* $K_{p/q}$ *is* $H_1(K_{p/q}) \cong \mathbb{Z}_{|p|}$.

Proof Write $\mathbb{Z}^{\mu_0}$, $\mathbb{Z}^{\lambda_0}$ for a copy of $\mathbb{Z}$ generated by the class μ_0 or λ_0, respectively. Think of T^2 as the boundary of $S^1 \times D^2$. Then the relevant part of the Mayer–Vietoris sequence of $K_{p/q} = C \cup (S^1 \times D^2)$ is as follows. Recall that $H_1(C)$ is generated by the class μ, and that the gluing map is described by (1.6).

$$\begin{array}{ccccccccc} H_1(T^2) & \to & H_1(S^1 \times D^2) & \oplus & H_1(C) & \to & H_1(K_{p/q}) & \to & 0 \\ \mathbb{Z}^{\mu_0} \oplus \mathbb{Z}^{\lambda_0} & \to & \mathbb{Z} & \oplus & \mathbb{Z} & \to & H_1(K_{p/q}) & \to & 0 \\ (1,0) & \mapsto & (\,0 & , & p\,) & & & & \\ (0,1) & \mapsto & (\,1 & , & m\,) & & & & \end{array}$$

From the exactness of this sequence the result is immediate. □

Consequently, for a knot K to have Property P is equivalent to the condition that the fundamental group $\pi_1(K_{1/q})$ is trivial only for $q = 0$.

Example 1.7.9 The unknot does *not* have Property P. Indeed, every $(1/q)$–surgery on the unknot yields S^3, which is seen as follows. If K is the unknot, then the closure C of $S^3 \setminus \nu K$ is also a solid torus. Write μ_C and λ_C for meridian and preferred longitude on ∂C. We may assume $\mu = \lambda_C$ and $\lambda = \mu_C$. When performing $(1/q)$–surgery on K, a solid torus is glued to C by sending its meridian μ_0 to $\mu + q\lambda = \lambda_C + q\mu_C$. Now, there clearly is a diffeomorphism of C that sends μ_C to itself and λ_C to $\lambda_C + q\mu_C$. It follows that the described surgery is equivalent to the one where we send μ_0 to $\lambda_C = \mu$, which is a trivial ∞–surgery.

The next proposition gives an alternative characterisation of Property P. As always, C denotes the closure of the complement of a tubular neighbourhood νK.

Proposition 1.7.10 *A knot K in S^3 has Property P if and only if both of the following conditions hold.*

(i) One cannot produce a counterexample to the Poincaré conjecture by surgery along K.
(ii) Any embedding of $C \subset S^3$ into S^3 extends to a diffeomorphism of S^3.

Proof Suppose that K has Property P; in particular K is not the unknot by the preceding example. Then (i) is clear, since non-trivial surgery along K does not produce any simply connected 3–manifold. Moreover, C is not a solid torus (that would only be the case for the unknot [215, p. 103]). Now let $\varphi\colon C \to S^3$ be an embedding. Then $\varphi(\partial C)$ is an embedded 2–torus, and the solid torus theorem [215, p. 107] states that the closure of one component of $S^3 \setminus \varphi(\partial C)$ is a solid torus, so this must be the component $\overline{S^3 \setminus \varphi(C)}$. This implies that by attaching a solid torus to $\varphi(C)$ we obtain S^3. This attaching of a solid torus amounts to a Dehn surgery on K. Since the resulting manifold is S^3 and K is supposed to have Property P, this must be a trivial Dehn surgery.† This is equivalent to saying that φ extends to a diffeomorphism of S^3.

Conversely, suppose that K satisfies (i) and (ii). We need to show that if $\pi_1(K_{1/q}) = 1$, then $q = 0$. By (i) we know that the condition $\pi_1(K_{1/q}) = 1$ implies $K_{1/q} = S^3$. The identity map on $C \subset S^3$ can be read as an embedding into $K_{1/q} = S^3$. This embedding extends to a diffeomorphism of S^3 only if $q = 0$, so property (ii) does indeed ensure $q = 0$. □

Remark 1.7.11 The result above has been phrased in the smooth category, which means that we are implicitly appealing to deep results of Moise and Munkres about the equivalence of the topological, piecewise linear (PL) and smooth classification of 3–manifolds (with or without boundary). The most relevant references are [189] (see also [190]), [194], and [196]. This is the price I have to pay for not introducing PL embeddings or tame topological embeddings. The reader should be warned, however, that the results of [215] invoked in the proof contain intermediate steps that rely on theorems such as the loop theorem, whose usual proofs are not well-suited to differentiable techniques.

We can now deduce an important consequence of Property P. The result is phrased in the topological category favoured by knot theorists, although I use smooth arguments in the proof.

Corollary 1.7.12 (Gordon–Luecke [114]) *If two knots K_1, K_2 in S^3 have homeomorphic complements, then the knots are equivalent, i.e. there is a homeomorphism of S^3 mapping K_1 to K_2.*

Proof According to a result of Edwards [61], two compact 3–manifolds with boundary are homeomorphic if and only if their interiors are homeomorphic. Thus, if $S^3 \setminus K_1$ is homeomorphic to $S^3 \setminus K_2$, then C_1 is homeomorphic

† This argument only requires K to have the weaker property that non-trivial Dehn surgery along it does not produce S^3, see Remark 1.7.13 below.

to C_2, where $C_i := \overline{S^3 \setminus \nu K_i}$. By the result of Munkres [194], the C_i are even diffeomorphic. If the C_i are solid tori, then the K_i are unknots and therefore equivalent. If the C_i are not solid tori, then the K_i are non-trivial knots and hence have Property P. Interpret the diffeomorphism $C_1 \rightarrow C_2$ as an embedding of C_1 into $S^3 = \nu K_2 \cup C_2$. By statement (ii) of the preceding proposition, this embedding extends to a diffeomorphism of S^3. This diffeomorphism has to send the preferred longitude λ_1 on $\partial(\nu K_1)$, up to sign, to the preferred longitude λ_2 on $\partial(\nu K_2)$. Since λ_i is isotopic to K_i in νK_i, this shows the equivalence of K_1 and K_2. □

Remark 1.7.13 Observe that in order to derive statement (ii) of Proposition 1.7.10 we only used the property that non-trivial surgery along K does not produce S^3, not the full Property P. The fact that $K_{1/q}$ can only be the 3–sphere if K is the unknot or $q = 0$ had been proved earlier by Gordon and Luecke [114], so they were the first to arrive at that last corollary.

Contact geometry enters the proof of Theorem 1.7.7 via the notion of symplectic fillings. Recall from Remark 1.1.6 that a contact 3–manifold (M, ξ) is naturally oriented; similarly, a symplectic 4–manifold can be oriented by the volume form ω^2.

Definition 1.7.14 (a) A compact symplectic 4–manifold (W, ω) is called a **weak (symplectic) filling** of the contact manifold (M, ξ) if $\partial W = M$ as oriented manifolds and $\omega|_\xi \neq 0$.

(b) A compact symplectic 4–manifold (W, ω) is called a **strong (symplectic) filling** of the contact manifold (M, ξ) if $\partial W = M$ and there is a Liouville vector field Y defined near ∂W, pointing outwards along ∂W, and satisfying $\xi = \ker(i_Y \omega|_{TM})$.

Observe that in case (b), the condition that $\partial W = M$ as *oriented* manifolds is automatic. Also, it is clear that any strong filling of M is *a fortiori* a weak filling. The converse is false, in general. Eliashberg [71] has given examples of weakly but not strongly fillable contact structures on the 3–torus; in [50] this was extended by Ding and myself to larger classes of 2–torus bundles over S^1.

Theorem 1.7.15 (Eliashberg [73], Etnyre [81]) *Any weak symplectic filling of a contact 3–manifold embeds symplectically into a closed symplectic 4–manifold.*

A proof of this theorem via a surgery description of contact 3–manifolds will be given in Section 6.5, see also [98]. Given this theorem, the

Kronheimer–Mrowka proof of Property P proceeds, very roughly, as follows. Let K be a purported counterexample to Theorem 1.7.7. Then K_1 would be a simply connected 3–manifold. Using the Floer exact triangle, one can deduce the triviality of certain Floer homology groups of K_0 (*sic*). Deep results of Gabai on foliations and Eliashberg–Thurston on so-called confoliations allow one to construct a symplectic manifold $W = K_0 \times [-1, 1]$ that weakly fills a contact structure on its boundary. By the preceding theorem, one obtains a closed symplectic 4–manifold containing K_0 as a separating hypersurface. The triviality of the Floer homology groups of K_0 implies the vanishing of certain gauge theoretic invariants of this closed symplectic 4–manifold. On the other hand, the topological properties of that symplectic manifold can be arranged in such a way that those same gauge theoretic invariants would have to be non-trivial. This is the desired contradiction.

2
Contact manifolds

'And therefore in geometry (which is the only science that it hath pleased God hitherto to bestow on mankind), men begin at settling the significations of their words; which settling of significations, they call *definitions*, and place them in the beginning of their reckoning.'

Thomas Hobbes,
Leviathan

In this chapter we begin with a more systematic introduction to contact topology. In Section 2.1 some further examples of contact manifolds are presented. In Sections 2.2 and 2.3 we introduce two of the most fundamental techniques of Contact Topology, *viz.*, the so-called Moser trick and the concept of contact Hamiltonians. These techniques are then used in Sections 2.5 and 2.6 to prove neighbourhood and isotopy extension theorems. The 'moral' of these theorems is that there are no local invariants in contact geometry: all contact structures look the same in the neighbourhood of a point (Darboux's theorem) or they are determined locally by bundle-theoretic information, e.g. in the neighbourhood of isotropic submanifolds.

2.1 Examples of contact manifolds

Recall the definition of the standard contact structure on $\mathbb{R}^{2n+1}$ given in Example 1.1.5. Here is a further example of a contact form on $\mathbb{R}^{2n+1}$.

Example 2.1.1 On $\mathbb{R}^{2n+1}$, with (r_j, φ_j) denoting polar coordinates in the (x_j, y_j)–plane, $j = 1, \ldots, n$, the following 1–form is a contact form:

$$\alpha_2 := dz + \sum_{j=1}^{n} r_j^2 \, d\varphi_j = dz + \sum_{j=1}^{n} (x_j \, dy_j - y_j \, dx_j).$$

In fact, this contact form is not really 'different' from the standard contact form α_1. The following definition gives a precise notion for the equivalence of contact structures or forms, generalising the concept of a contact transformation (Defn. 1.2.5).

Definition 2.1.2 Two contact manifolds (M_1, ξ_1) and (M_2, ξ_2) are said to be **contactomorphic** if there is a diffeomorphism $f\colon M_1 \to M_2$ with $Tf(\xi_1) = \xi_2$, where $Tf\colon TM_1 \to TM_2$ denotes the differential of f. If $\xi_i = \ker \alpha_i$, $i = 1, 2$, this is equivalent to saying that α_1 and $f^*\alpha_2$ determine the same hyperplane field, and hence equivalent to the existence of a nowhere zero function $\lambda\colon M_1 \to \mathbb{R} \setminus \{0\}$ such that $f^*\alpha_2 = \lambda\alpha_1$. Occasionally one speaks of a **strict contactomorphism** between the **strict contact manifolds** (M_1, α_1) and (M_2, α_2) if $f^*\alpha_2 = \alpha_1$.

Example 2.1.3 The contact manifolds $(\mathbb{R}^{2n+1}, \xi_i = \ker \alpha_i)$, $i = 1, 2$, from Example 1.1.5 and the preceding example are contactomorphic. An explicit contactomorphism f with $f^*\alpha_2 = \alpha_1$ is given by

$$f(\mathbf{x}, \mathbf{y}, z) = \big((\mathbf{x} + \mathbf{y})/2, (\mathbf{y} - \mathbf{x})/2, z + \mathbf{xy}/2\big),$$

where $\mathbf{x}$ and $\mathbf{y}$ stand for $(x_1, \ldots, x_n)$ and $(y_1, \ldots, y_n)$, respectively, and $\mathbf{xy}$ stands for $\sum_j x_j y_j$. Similarly, both these contact structures are diffeomorphic to $\ker(dz - \sum_j y_j\, dx_j)$.

Notation 2.1.4 Any of these contact structures will be referred to as the standard contact structure on $\mathbb{R}^{2n+1}$ and denoted by ξ_{st}. If a specific choice between the different incarnations of ξ_{st} matters for the argument, this will be made explicit.

The following lemma gives a simple example of a general class of contactomorphisms that will be used in Chapter 6.

Lemma 2.1.5 *Let Y be a Liouville vector field on a symplectic manifold (W, ω). Suppose that M_1 and M_2 are hypersurfaces in W transverse to Y, so that $\alpha := i_Y\omega$ defines a contact structure on these hypersurfaces by Lemma/Definition 1.4.5. Assume further that there is a smooth function $\mu\colon W \to \mathbb{R}$ such that the time-1-map, say, of the flow of the vector field μY is a diffeomorphism from M_1 to M_2. Then this diffeomorphism is in fact a contactomorphism from $(M_1, \ker \alpha|_{TM_1})$ to $(M_2, \ker \alpha|_{TM_2})$.*

Proof This follows from

$$\mathcal{L}_{\mu Y}\alpha = d(\alpha(\mu Y)) + i_{\mu Y}\, d\alpha = \mu\alpha$$

and a consideration as in the proof of Lemma 1.5.8. □

Example 2.1.6 On $\mathbb{R}^3$ with cylindrical coordinates $(r, \varphi; z)$, define the 1–form

$$\alpha_{\mathrm{ot}} = \cos r\, dz + r \sin r\, d\varphi.$$

Since the 1–form $r^2\, d\varphi$ and the function

$$r \longmapsto \begin{cases} \dfrac{\sin r}{r} & \text{for } r \neq 0, \\ 1 & \text{for } r = 0, \end{cases}$$

are smooth, so is the 1–form α_{ot}. By computing

$$\alpha_{\mathrm{ot}} \wedge d\alpha_{\mathrm{ot}} = \left(1 + \frac{\sin r}{r} \cos r\right) r\, dr \wedge d\varphi \wedge dz,$$

where the coefficient in parentheses is read as 2 for $r = 0$, we see that α_{ot} is a contact form. The contact structure $\xi_{\mathrm{ot}} = \ker \alpha_{\mathrm{ot}}$ is called the **standard 'overtwisted' contact structure** on $\mathbb{R}^3$. (The terminology 'overtwisted' will be given a formal meaning in Section 4.5.)

Observe that both the standard contact structure ξ_2 (on $\mathbb{R}^3$) and the standard overtwisted contact structure ξ_{ot} are horizontal along the z–axis $\{r = 0\}$, and all the rays perpendicular to the z–axis (with z and φ constant) are tangent to both ξ_2 and ξ_{ot}. For $r \neq 0$, the contact structure ξ_2 is spanned by the vector fields ∂_r and $\partial_\varphi - r^2\, \partial_z$; the contact structure ξ_{ot} is spanned by ∂_r and $\cos r\, \partial_\varphi - r \sin r\, \partial_z$. So both ξ_2 and ξ_{ot} turn counterclockwise as one moves outwards from the z–axis along any such ray. However, while the rotation angle of ξ_2 approaches (but never reaches) $\pi/2$ (i.e. the contact planes never become vertical), the contact planes of ξ_{ot} make infinitely many complete turns as one moves out in radial direction, see Figure 2.1.

Bennequin [26] proved that $(\mathbb{R}^3, \xi_{\mathrm{st}})$ and $(\mathbb{R}^3, \xi_{\mathrm{ot}})$ are in fact *not* contactomorphic. This result can be regarded as the birth of contact topology, for reasons that will become clear when we discuss the classification of contact structures on 3–manifolds. I shall present a proof of Bennequin's theorem in Corollary 6.5.10; see also Remark 4.6.37.

Example 2.1.7 Let $(x_1, y_1, \ldots, x_{n+1}, y_{n+1})$ be Cartesian coordinates on $\mathbb{R}^{2n+2}$. Then the **standard contact structure on the unit sphere** S^{2n+1} in $\mathbb{R}^{2n+2}$ is given by the contact form

$$\alpha_0 := \sum_{j=1}^{n+1} (x_j\, dy_j - y_j\, dx_j).$$

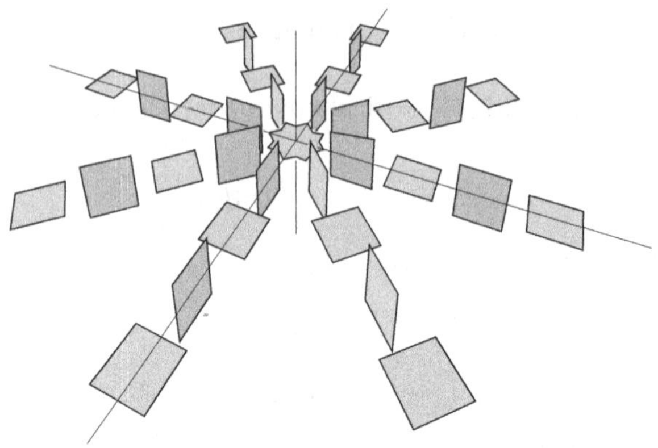

Fig. 2.1. The standard overtwisted contact structure.

Write r for the radial coordinate on $\mathbb{R}^{2n+2}$, that is, $r^2 = \sum_j (x_j^2 + y_j^2)$. One checks easily that $r\,dr \wedge \alpha_0 \wedge (d\alpha_0)^n \neq 0$ for $r \neq 0$. Since S^{2n+1} is a level surface of r (or r^2), this verifies the contact condition.

Alternatively, see also Example 1.4.8, one can define this 1–form as $\alpha_0 := i_Y \omega_{\mathrm{st}}$ with $Y := \sum_j (x_j\, \partial_{x_j} + y_j\, \partial_{y_j}) = r\, \partial_r$ and $\omega_{\mathrm{st}} := \sum_j dx_j \wedge dy_j$ the standard symplectic form on $\mathbb{R}^{2n+2}$. Since Y is, up to a factor 2, a Liouville vector field for ω_{st} transverse to S^{2n+1}, this again shows that α_0 is a contact form on that sphere.

A third description of $\xi_0 := \ker \alpha_0$ is given by regarding S^{2n+1} as the unit sphere in $\mathbb{C}^{n+1}$ with complex structure J (corresponding to complex coordinates $z_j = x_j + \mathrm{i} y_j$, $j = 1, \ldots, n+1$). Then ξ_0 defines at each point $p \in S^{2n+1}$ the complex (i.e. J–invariant) subspace of $T_p S^{2n+1}$, that is,

$$\xi_0 = TS^{2n+1} \cap J(TS^{2n+1}).$$

This follows from the observation that $\alpha_0 = -r\,dr \circ J$. The Hermitian form $d\alpha_0(-, J-)$ on ξ_0 is called the **Levi form** of the hypersurface $S^{2n+1} \subset \mathbb{C}^{n+1}$. The contact condition for ξ corresponds to the positive definiteness of that Levi form, or what in complex analysis is called the **strict pseudoconvexity** of the hypersurface. For more on the question of pseudoconvexity from the contact geometric viewpoint see Section 5.3.

Here is a further useful example of contactomorphic manifolds.

Proposition 2.1.8 *For any point $p \in S^{2n+1}$, the two contact manifolds $(S^{2n+1} \setminus \{p\}, \xi_0)$ and $(\mathbb{R}^{2n+1}, \xi_2)$ are contactomorphic.*

Proof The contact manifold (S^{2n+1}, ξ_0) is a homogeneous space under the natural $\mathrm{U}(n+1)$–action, so we are free to choose $p = (0, \ldots, 0, 1)$. The following argument is inspired by an unpublished note by Erlandsson [78]; for an alternative argument see [97].

Write $\psi\colon S^{2n+1} \setminus \{p\} \to \mathbb{R}^{2n+1}$ for the stereographic projection from p. In terms of Cartesian coordinates

$$(x_1, y_1, \ldots, x_{n+1}, y_{n+1}) \text{ on } \mathbb{R}^{2n+2}$$

and

$$(u_1, v_1, \ldots, u_n, v_n, w) \text{ on } \mathbb{R}^{2n+1},$$

this projection is given by

$$u_j = \frac{x_j}{1-y_{n+1}},\ v_j = \frac{y_j}{1-y_{n+1}},\ j = 1, \ldots, n;\ \ w = \frac{x_{n+1}}{1-y_{n+1}}.$$

The inverse map is given by

$$x_j = \lambda u_j,\ y_j = \lambda v_j,\ j = 1, \ldots, n;\ \ x_{n+1} = \lambda w,\ y_{n+1} = 1 - \lambda$$

with $\lambda = 2/\big(1 + w^2 + \sum_j (u_j^2 + v_j^2)\big)$.

We continue to write (r_j, φ_j) for the polar coordinates in the (u_j, v_j)–plane, so that the contact forms α_0, α_2 defining ξ_0, ξ_2, respectively, are given by

$$\alpha_0 = \sum_{j=1}^{n+1} (x_j\, dy_j - y_j\, dx_j)$$

and

$$\alpha_2 = dw + \sum_{j=1}^{n} r_j^2\, d\varphi_j = dw + \sum_{j=1}^{n} (u_j\, dv_j - v_j\, du_j).$$

The stereographic image of α_0 is

$$
\begin{aligned}
(\psi^{-1})^*\alpha_0 &= \sum_{j=1}^n \lambda u_j(v_j\,d\lambda + \lambda\,dv_j) - \lambda v_j(u_j\,d\lambda + \lambda\,du_j)\\
&\quad + \lambda w(-d\lambda) - (1-\lambda)(w\,d\lambda + \lambda\,dw)\\
&= \lambda^2\Bigg(\sum_{j=1}^n (u_j\,dv_j - v_j\,du_j)\\
&\quad + w\Big(w\,dw + \sum_{j=1}^n (u_j\,du_j + v_j\,dv_j)\Big) + (1-1/\lambda)\,dw\Bigg)\\
&= \lambda^2\Bigg(\sum_{j=1}^n r_j^2\,d\varphi_j + w\sum_{j=1}^n r_j\,dr_j + \frac{1}{2}\Big(1+w^2-\sum_{j=1}^n r_j^2\Big)\,dw\Bigg)\\
&=: \lambda^2\widetilde{\alpha}_2.
\end{aligned}
$$

In order to complete the proof, it (more than) suffices to describe a diffeomorphism f of $\mathbb{R}^{2n+1}$ with $f^*\alpha_2 = \widetilde{\alpha}_2$. Such a diffeomorphism also satisfies $f^*d\alpha_2 = d\widetilde{\alpha}_2$, so it will send the Reeb vector field $\widetilde{R}_2$ of $\widetilde{\alpha}_2$ to the Reeb vector field R_2 of α_2, see also Lemma 2.2.4 below. This observation will guide us in finding a suitable f.

We have $R_2 = \partial_w$, and a simple computation shows that

$$
\widetilde{R}_2 = \frac{2}{1+w^2+\sum_j r_j^2}\Big(\partial_w + \sum_j \partial_{\varphi_j}\Big).
$$

Hence the flow lines of R_2 are the lines parallel to the w–axis, with w as curve parameter; the flow lines of $\widetilde{R}_2$ are the 'helices' determined by the conditions $r_j \equiv \text{const.}$, $\varphi_j = w + \text{const.}$, and by the differential equation

$$
\frac{dw}{dt} = \frac{2}{1+w^2+\sum_j r_j^2} =: g(w). \tag{2.1}
$$

(Writing the right-hand side as a function $g(w)$ is justified by the fact that the r_j are constant along the flow lines of $\widetilde{R}_2$.)

Moreover, we observe that the hyperplane $\{w=0\}$ is a transversal for either flow, and that $d\alpha_2$ and $d\widetilde{\alpha}_2$ restrict to the same form on (the tangent spaces of) this hyperplane. This suggests that the map f should be defined by the condition that it restricts to the identity on $\{w=0\}$ and maps the flow lines of $\widetilde{R}_2$ to those of R_2. It follows that the w–component of $f(\mathbf{r},\boldsymbol{\varphi},w)$ is the time parameter t corresponding to the value $w(t)$ of the solution to the differential equation (2.1) with $w(0)=0$. This parameter can be found by separation of variables, and we obtain

$$f(\mathbf{r}, \boldsymbol{\varphi}, w) = \left(\mathbf{r}, \boldsymbol{\varphi} - w, \int_0^w \frac{d\zeta}{g(\zeta)}\right) = \left(\mathbf{r}, \boldsymbol{\varphi} - w, \frac{1}{2}w\left(1 + \frac{1}{3}w^2 + \sum_j r_j^2\right)\right);$$

here $\boldsymbol{\varphi} - w$ is to be read as $(\varphi_1 - w, \ldots, \varphi_n - w)$. I leave it to the reader to check that this map f is a diffeomorphism satisfying $f^*\alpha_2 = \widetilde{\alpha}_2$. □

Notation 2.1.9 Thanks to this proposition, it is reasonable to write ξ_{st} both for the standard contact structure on $\mathbb{R}^{2n+1}$ and that on S^{2n+1}.

Example 2.1.10 There is a rather less contrived contactomorphism between $(\mathbb{R}^{2n+1}, \xi_2)$ and the lower hemisphere in (S^{2n+1}, ξ_0) given by

$$(u_1, v_1, \ldots, u_n, v_n, w) \longmapsto \frac{(u_1, v_1, \ldots, u_n, v_n, w, -1)}{\sqrt{1 + \mathbf{u}^2 + \mathbf{v}^2 + w^2}}.$$

As mentioned earlier, we shall usually assume that our contact structures are coorientable. For completeness, here is an example of contact structures that are not coorientable, and hence not defined by a global contact form.

Example 2.1.11 Let $M = \mathbb{R}^{n+1} \times \mathbb{R}P^n$. Denote Cartesian coordinates on the $\mathbb{R}^{n+1}$–factor by $(x_0, \ldots, x_n)$ and homogeneous coordinates on the $\mathbb{R}P^n$–factor by $(y_0 : \ldots : y_n)$. Then

$$\xi := \ker\Big(\sum_{j=0}^{n} y_j \, dx_j\Big)$$

is a well-defined hyperplane field on M, because the 1–form on the right-hand side is well defined up to scaling by a non-zero real constant. On the open submanifold $U_k = \{y_k \neq 0\} \cong \mathbb{R}^{n+1} \times \mathbb{R}^n$ of M we have $\xi = \ker \alpha_k$ with

$$\alpha_k := dx_k + \sum_{j \neq k} \left(\frac{y_j}{y_k}\right) dx_j$$

an honest 1–form on U_k. This is the standard contact form of Example 1.1.5, which proves that ξ is a contact structure on M. (In fact, ξ is the natural contact structure on M, regarded as the space of contact elements of $\mathbb{R}^{n+1}$, see Section 1.2.)

If n is even, then M is not orientable, so there can be no global contact form defining ξ (see Remark 1.1.6), i.e. ξ is *not coorientable*. Notice, however, that a contact structure on a manifold of dimension $2n+1$ with n even is always *orientable*: the sign of $(d\alpha)^n|_\xi$ does not depend on the choice of local 1–form α defining ξ.

If n is odd, then M is orientable, so ξ could well be the kernel of a globally defined 1–form. However, since the sign of $\alpha \wedge (d\alpha)^n$, for n odd, is independent of the choice of local 1–form α defining ξ, it is also conceivable that no global contact form exists. This is indeed what happens, as we shall prove presently.

Remark 2.1.12 The previous considerations show that any manifold M of dimension $2n+1$, with n odd, admitting a contact structure (coorientable or not) needs to be orientable. Moreover, given an orientation for M, one can sensibly speak of positive and negative contact structures, even if they are not coorientable.

Proposition 2.1.13 *Let (M, ξ) be the contact manifold of the preceding example. Then TM/ξ can be identified with the canonical line bundle on $\mathbb{R}P^n$ (pulled back to M). In particular, TM/ξ is a non-trivial line bundle, so ξ is not coorientable.*

Proof For given $\mathbf{y} = (y_0 : \ldots : y_n) \in \mathbb{R}P^n$, the vector $y_0\,\partial_{x_0} + \cdots + y_n\,\partial_{x_n} \in T_{\mathbf{x}}\mathbb{R}^{n+1}$ is well defined up to a non-zero real factor (and independent of $\mathbf{x} \in \mathbb{R}^{n+1}$), and hence defines a line $\ell_{\mathbf{y}}$ in $T_{\mathbf{x}}\mathbb{R}^{n+1} \cong \mathbb{R}^{n+1}$. The set

$$\begin{aligned} E &= \{(t, \mathbf{x}, \mathbf{y})\colon\ \mathbf{x} \in \mathbb{R}^{n+1},\ \mathbf{y} \in \mathbb{R}P^n,\ t \in \ell_{\mathbf{y}}\} \\ &\subset T\mathbb{R}^{n+1} \times \mathbb{R}P^n \subset T(\mathbb{R}^{n+1} \times \mathbb{R}P^n) = TM \end{aligned}$$

with projection $(t, \mathbf{x}, \mathbf{y}) \mapsto (\mathbf{x}, \mathbf{y})$ defines a line sub-bundle of TM that restricts to the canonical line bundle over $\{\mathbf{x}\} \times \mathbb{R}P^n \equiv \mathbb{R}P^n$ for each $\mathbf{x} \in \mathbb{R}^{n+1}$. The canonical line bundle over $\mathbb{R}P^n$ is well known to be non-trivial [188, p. 16], so the same holds for E.

Moreover, E is clearly complementary to ξ, i.e. $TM/\xi \cong E$, since

$$\sum_{j=0}^{n} y_j\, dx_j\Big(\sum_{k=0}^{n} y_k\, \partial_{x_k}\Big) = \sum_{j=0}^{n} y_j^2 \neq 0.$$

This proves that ξ is not coorientable. □

To sum up, in the example above we have one of the following two situations:

- If n is odd, then M is orientable; ξ is neither orientable nor coorientable.
- If n is even, then M is not orientable; ξ is not coorientable, but it is orientable.

We have already met isotropic and Legendrian submanifolds of contact manifolds as special classes of submanifolds in contact topology. Another natural condition one can impose on a submanifold is that the ambient contact manifold induce a contact structure on the submanifold.

Definition 2.1.14 Let (M, ξ) be a contact manifold. A submanifold M' of M with contact structure ξ' is called a **contact submanifold** of (M, ξ) if $TM' \cap \xi|_{M'} = \xi'$.

Observe that if $\xi = \ker \alpha$ and $i\colon M' \to M$ denotes the inclusion map, then the condition for (M', ξ') to be a contact submanifold of (M, ξ) is that $\xi' = \ker(i^*\alpha)$. In particular, $\xi' \subset \xi|_{M'}$ is a symplectic sub-bundle with respect to the symplectic bundle structure on ξ given by $d\alpha$.

Remark 2.1.15 A 1–dimensional manifold M' with its unique contact structure (see Remark 1.1.4) is a contact submanifold of (M, ξ) if and only if M' is transverse to ξ. A case of particular interest to us will be transverse knots in contact 3–manifolds.

2.2 Gray stability and the Moser trick

The Gray stability theorem that we are going to prove in this section says that there are no non-trivial deformations of contact structures on closed manifolds. In fancy language, this means that contact structures on closed manifolds have discrete moduli. Here is a preparatory lemma.

Lemma 2.2.1 *Let ω_t, $t \in [0,1]$, be a smooth family of differential k–forms on a manifold M and $(\psi_t)_{t\in[0,1]}$ an isotopy of M. Define a time-dependent vector field X_t on M by $X_t \circ \psi_t = \dot{\psi}_t$, where the dot denotes derivative with respect to t (so that ψ_t is the flow of X_t). Then*

$$\frac{d}{dt}\left(\psi_t^*\omega_t\right)|_{t=t_0} = \psi_{t_0}^*\left(\dot{\omega}_t|_{t=t_0} + \mathcal{L}_{X_{t_0}}\omega_{t_0}\right).$$

Proof For a time-*independent* k–form ω we have

$$\frac{d}{dt}\left(\psi_t^*\omega\right)|_{t=t_0} = \psi_{t_0}^*\left(\mathcal{L}_{X_{t_0}}\omega\right).$$

This follows directly from the definitions, see Appendix B.

We then compute

$$\begin{aligned}\frac{d}{dt}(\psi_t^*\omega_t) &= \lim_{h\to 0}\frac{\psi_{t+h}^*\omega_{t+h}-\psi_t^*\omega_t}{h}\\ &= \lim_{h\to 0}\frac{\psi_{t+h}^*\omega_{t+h}-\psi_{t+h}^*\omega_t+\psi_{t+h}^*\omega_t-\psi_t^*\omega_t}{h}\\ &= \lim_{h\to 0}\psi_{t+h}^*\left(\frac{\omega_{t+h}-\omega_t}{h}\right)+\lim_{h\to 0}\frac{\psi_{t+h}^*\omega_t-\psi_t^*\omega_t}{h}\\ &= \psi_t^*\big(\dot{\omega}_t+\mathcal{L}_{X_t}\omega_t\big).\end{aligned}$$

This is the claimed identity. □

Theorem 2.2.2 (Gray stability theorem) *Let ξ_t, $t\in[0,1]$, be a smooth family of contact structures on a closed manifold M. Then there is an isotopy $(\psi_t)_{t\in[0,1]}$ of M such that*

$$T\psi_t(\xi_0)=\xi_t \quad \text{for each } t\in[0,1].$$

Proof The simplest proof of this result rests on what is known as the **Moser trick**, introduced by Moser [193] in the context of stability results for (equicohomologous) volume and symplectic forms. Gray's original proof [117] was based on deformation theory *à la* Kodaira–Spencer. The idea of the Moser trick is to assume that ψ_t is the flow of a time-dependent vector field X_t. The desired equation for ψ_t then translates into an equation for X_t. If that equation can be solved, the isotopy ψ_t is found by integrating X_t; on a closed manifold the flow of X_t will be globally defined.

Let α_t be a smooth family of 1–forms with $\ker\alpha_t=\xi_t$. The equation in the theorem then translates into

$$\psi_t^*\alpha_t=\lambda_t\alpha_0,$$

where $\lambda_t\colon M\to\mathbb{R}^+$ is a suitable smooth family of smooth functions. Differentiation of this equation with respect to t yields, together with the preceding lemma,

$$\psi_t^*\big(\dot{\alpha}_t+\mathcal{L}_{X_t}\alpha_t\big)=\dot{\lambda}_t\alpha_0=\frac{\dot{\lambda}_t}{\lambda_t}\psi_t^*\alpha_t,$$

or, with the help of Cartan's formula† $\mathcal{L}_X=d\circ i_X+i_X\circ d$ and with $\mu_t:=\frac{d}{dt}(\log\lambda_t)\circ\psi_t^{-1}$,

$$\psi_t^*\big(\dot{\alpha}_t+d(\alpha_t(X_t))+i_{X_t}d\alpha_t\big)=\psi_t^*(\mu_t\alpha_t).$$

† The Cartan formula also holds for time-dependent vector fields, see Appendix B.

If we choose $X_t \in \xi_t$, this equation will be satisfied if

$$\dot{\alpha}_t + i_{X_t} d\alpha_t = \mu_t \alpha_t. \tag{2.2}$$

Plugging in the Reeb vector field R_t of α_t gives

$$\dot{\alpha}_t(R_t) = \mu_t. \tag{2.3}$$

So we can use (2.3) to define μ_t, and then the non-degeneracy of $d\alpha_t|_{\xi_t}$ and the fact that $R_t \in \ker(\mu_t \alpha_t - \dot{\alpha}_t)$ allow us to find a unique solution $X_t \in \xi_t$ of (2.2). □

Remark 2.2.3 For further applications of this theorem it is useful to observe that at points $p \in M$ with $\dot{\alpha}_{t,p}$ identically zero in t we have $X_t(p) \equiv \mathbf{0}$, so such points remain stationary under the isotopy ψ_t.

It may also be worth noticing that the vector field X_t defined by equations (2.2) and (2.3) is, for given ξ_t, independent of the choice of contact forms α_t. Indeed, an equivalent definition of X_t is given by the conditions $X_t \in \xi_t$ and

$$(\dot{\alpha}_t + i_{X_t} d\alpha_t)|_{\xi_t} = 0.$$

If α_t is replaced by $h_t \alpha_t$, with $h_t \colon M \to \mathbb{R} \setminus \{0\}$ a smooth family of smooth functions, the same X_t satisfies the corresponding condition

$$(\dot{h}_t \alpha_t + h_t \dot{\alpha}_t + i_{X_t}(dh_t \wedge \alpha_t + h_t d\alpha_t))|_{\xi_t} = 0.$$

Contact *forms* do *not* satisfy stability, that is, with α_t a smooth family of contact forms one cannot, in general, find an isotopy ψ_t such that $\psi_t^* \alpha_t = \alpha_0$. In order to see this, we formulate the following simple lemma.

Lemma 2.2.4 *If $f \colon (M_1, \alpha_1) \to (M_2, \alpha_2)$ is a strict contactomorphism between two strict contact manifolds, then the Reeb vector fields are related by $Tf(R_1) = R_2$.*

Proof We compute

$$\alpha_2(Tf(R_1)) = f^*\alpha_2(R_1) = \alpha_1(R_1) \equiv 1$$

and

$$d\alpha_2(Tf(R_1), Tf(-)) = f^*(d\alpha_2)(R_1, -) = d\alpha_1(R_1, -) \equiv 0.$$

This proves that $Tf(R_1) = R_2$. □

It follows that the Reeb vector fields of strictly contactomorphic contact forms α_1 and α_2 have the same dynamics; for instance, there has to be a one-to-one correspondence between the closed orbits of the two vector fields.

Example 2.2.5 On $S^3 \subset \mathbb{R}^4$ consider the following smooth family of contact forms:

$$\alpha_t = (x_1\,dy_1 - y_1\,dx_1) + (1+t)(x_2\,dy_2 - y_2\,dx_2),$$

where $t \geq 0$ is a real parameter. The Reeb vector field of α_t is

$$R_t = (x_1\,\partial_{y_1} - y_1\,\partial_{x_1}) + \frac{1}{1+t}(x_2\,\partial_{y_2} - y_2\,\partial_{x_2}),$$

as is proved by a computation similar to that in the proof of Lemma 1.4.9. As we saw there, the flow of R_0 defines the Hopf fibration; in particular, all orbits of R_0 are closed. For $t \in \mathbb{R}^+ \setminus \mathbb{Q}$, on the other hand, R_t has only two periodic orbits: in the same complex notation we used in the proof of Lemma 1.4.9, the flow lines of R_t can be parametrised as

$$\gamma(s) = (\mathrm{e}^{\mathrm{i}s} z_1, \mathrm{e}^{\mathrm{i}s/(1+t)} z_2),$$

so if the periods of the circular motions in the z_1– and z_2–coordinate are not rationally related, the only closed orbits are those through the points $(z_1, z_2) = (1, 0)$ and $(z_1, z_2) = (0, 1)$, respectively. So by the preceding lemma there can be no isotopy with $\psi_t^* \alpha_t = \alpha_0$.

Remark 2.2.6 The condition in Theorem 2.2.2 that M be a closed manifold is essential. Eliashberg [68] has shown that there are non-trivial deformations of contact structures on $S^1 \times \mathbb{R}^2$.

2.3 Contact Hamiltonians

Let X be a vector field on a contact manifold $(M, \xi = \ker\alpha)$. In Definition 1.5.7 and Lemma 1.5.8 we encountered special such vector fields called infinitesimal automorphisms of ξ or α, and we characterised such vector fields in terms of the Lie derivative. The next theorem relates infinitesimal automorphisms of ξ to functions on M; the contact form plays an auxiliary role.

Theorem 2.3.1 *With a fixed choice of contact form α there is a one-to-one correspondence between infinitesimal automorphisms X of $\xi = \ker\alpha$ and smooth functions $H\colon M \to \mathbb{R}$. The correspondence is given by*

- $X \longmapsto H_X = \alpha(X)$;
- $H \longmapsto X_H$, *defined uniquely by* $\alpha(X_H) = H$ *and* $i_{X_H} d\alpha = dH(R_\alpha)\alpha - dH$.

The fact that X_H is uniquely defined by the equations in the theorem follows as in the preceding section from the fact that $d\alpha$ is non-degenerate on ξ and $R_\alpha \in \ker(dH(R_\alpha)\alpha - dH)$.

Proof Let X be an infinitesimal automorphism of ξ. Set $H_X = \alpha(X)$ and write $dH_X + i_X d\alpha = \mathcal{L}_X\alpha = \mu\alpha$ with $\mu\colon M \to \mathbb{R}$. Applying this last equation to R_α yields $dH_X(R_\alpha) = \mu$. So X satisfies the equations $\alpha(X) = H_X$ and $i_X d\alpha = dH_X(R_\alpha)\alpha - dH_X$. This means that we have $X_{H_X} = X$.

Conversely, given $H\colon M \to \mathbb{R}$ and with X_H as defined in the theorem, we have

$$\mathcal{L}_{X_H}\alpha = d(\alpha(X_H)) + i_{X_H} d\alpha = dH(R_\alpha)\alpha,$$

so X_H is an infinitesimal automorphism of ξ. Moreover, it is immediate from the definitions that $H_{X_H} = \alpha(X_H) = H$. □

The following corollary shows that the key statement of Lemma 1.5.8 carries over to time-dependent infinitesimal automorphisms.

Corollary 2.3.2 *Let $(M, \xi = \ker\alpha)$ be a closed contact manifold, and let $H_t\colon M \to \mathbb{R}$, $t \in [0,1]$, be a smooth family of functions. Write $X_t := X_{H_t}$ for the corresponding family of infinitesimal automorphisms of ξ. Then the globally defined flow ψ_t of the time-dependent vector field X_t is a contact isotopy of (M, ξ), that is, $\psi_t^*\alpha = \lambda_t\alpha$ for some smooth family of functions $\lambda_t\colon M \to \mathbb{R}^+$.*

Proof With Lemma 2.2.1 and the preceding proof we have

$$\frac{d}{dt}(\psi_t^*\alpha) = \psi_t^*(\mathcal{L}_{X_t}\alpha) = \psi_t^*(dH_t(R_\alpha)\alpha) = \mu_t\psi_t^*\alpha$$

with $\mu_t = dH_t(R_\alpha) \circ \psi_t$. Since $\psi_0 = \mathrm{id}_M$ (whence $\psi_0^*\alpha = \alpha$) this implies that, with

$$\lambda_t = \exp\Bigl(\int_0^t \mu_s\, ds\Bigr),$$

we have $\psi_t^*\alpha = \lambda_t\alpha$. □

This corollary will be used in Section 2.6 to prove various isotopy extension theorems from isotopies of special submanifolds to isotopies of the ambient contact manifold. In a similar vein, contact Hamiltonians can be used to show that standard general position arguments from differential topology continue to hold in the contact geometric setting. Another application of contact Hamiltonians is a proof of the fact that the contactomorphism group of a connected contact manifold acts transitively on that manifold [124], [29]; a proof of this result will be given below (Cor. 2.6.3). (See [21] for more on the general structure of contactomorphism groups.) The general philosophy behind the use of contact Hamiltonians is that it is much easier to manipulate a function (e.g. multiply it with a bump function) than it would be directly to manipulate the corresponding contactomorphism.

2.4 Interlude: symplectic vector bundles

In Section 1.3 we saw how to construct an ω–compatible complex structure on a symplectic vector space (V, ω), and we proved that the space $\mathcal{J}(\omega)$ of such complex structures is contractible. In the present interlude we extend this result from vector spaces to vector bundles. For further results on symplectic vector bundles see [177]. We also introduce the notion of an almost contact structure, which is the underlying bundle structure of a contact structure.

Definition 2.4.1 A **symplectic vector bundle** (E, ω) over a manifold B is a (smooth) vector bundle $\pi\colon E \to B$ together with a symplectic linear form ω_b on each fibre $E_b = \pi^{-1}(b)$, $b \in B$, with ω_b varying smoothly in b. Formally, this smoothness condition means that the map defined by $b \mapsto \omega_b$ is a smooth section of the bundle $\bigwedge^2 E^* \to B$, the second exterior power of the dual bundle of E.

Example 2.4.2 Given any vector bundle $E \to B$, there is a canonical symplectic bundle structure on the Whitney sum $E \oplus E^*$, defined by

$$\omega_b(X + \eta, X' + \eta') = \eta(X') - \eta'(X) \text{ for } X, X' \in E_b;\ \eta, \eta' \in E_b^*.$$

Definition 2.4.3 A **complex (bundle) structure** on a vector bundle $E \to B$ is a family J_b of complex structures on the fibres E_b, with J_b varying smoothly in b. Again, this smoothness condition can be phrased more formally by saying that the map $b \mapsto J_b$ is a smooth section of the endomorphism bundle $\mathrm{End}\,(E)$.

A complex structure on a symplectic vector bundle (E, ω) is called ω–**compatible** if J_b is ω_b–compatible on E_b for each $b \in B$.

Remarks 2.4.4 (1) A complex bundle structure on the tangent bundle TW of a manifold W is called an **almost complex structure** on W.

(2) Given a vector bundle $E \to B$ of rank (i.e. fibre dimension) $2n$ with a complex structure, one can choose an arbitrary bundle metric on E and then

construct a Hermitian bundle metric by the steps described in Remarks 1.3.9. The bundle can then locally be trivialised by n sections that are orthonormal with respect to this Hermitian bundle metric and linearly independent over $\mathbb{C}$. The transition functions with respect to such local trivialisations take values in the unitary group $\mathrm{U}(n)$. One says that *the structure group of the bundle can be reduced to* $\mathrm{U}(n)$.

Conversely, if the structure group of the bundle E has been reduced to $\mathrm{U}(n)$, that is, if there are local trivialisations $E|_U \cong U \times \mathbb{C}^n$, $U \subset B$, such that the transition functions take values in $\mathrm{U}(n)$, then these transition functions preserve the complex vector space structure. This means that multiplication by i in the $\mathbb{C}^n$–factor of these local trivialisations induces a complex bundle structure J on E.

Proposition 2.4.5 *Let (E, ω) be a symplectic vector bundle over the manifold B. Then the space $\mathcal{J}(\omega)$ of ω–compatible complex bundle structures on E is non-empty and contractible.*

Proof This follows immediately from the proof of the corresponding statement for vector spaces. Let j_b be the map as constructed in the proof of Proposition 1.3.10, associating to any inner product on E_b an ω_b–compatible complex structure. Now choose a smooth bundle metric g on E and define $J_b = j_b(g_b)$ for each $b \in B$. Working in a local trivialisation of the bundle, we see that the smoothness of the map j in the proof of Proposition 1.3.10 implies that J_b depends smoothly on b. The proof of the contractibility of $\mathcal{J}(\omega)$ carries over from that proposition without any changes. □

Remark 2.4.6 This proposition enables us to speak of the **Chern classes** of a symplectic vector bundle (E, ω), defined as the Chern classes of the complex vector bundle (E, J), where J is any ω–compatible complex bundle structure on E.

Returning to contact geometry, we now consider a manifold M with a cooriented contact structure $\xi = \ker \alpha$. As was observed in Remark 1.3.7, the sub-bundle $\xi \subset TM$ together with the 2–form $d\alpha|_\xi$ is then a symplectic vector bundle over M. All the previous considerations carry over to this particular setting. In fact, if α is replaced by $\lambda\alpha$, where $\lambda \colon M \to \mathbb{R}^+$ is some smooth function, then $d(\lambda\alpha)|_\xi = \lambda\, d\alpha|_\xi$, that is, the *conformal class* of the symplectic bundle structure depends only on the (cooriented!) contact structure ξ. This allows us to make the following definition.

Definition 2.4.7 (1) An **almost contact structure** on a $(2n+1)$–dimensional manifold M is a triple (η, J, ϵ) consisting of a cooriented hyperplane field $\eta \subset TM$ with a complex bundle structure J and a choice of oriented (and hence trivial) line sub-bundle $\epsilon \subset TM$ complementary to η and defining the coorientation.

(2) Let (M, ξ) be a contact manifold with cooriented contact structure ξ. A complex bundle structure J on ξ is called **ξ–compatible** if $J_p\colon\, \xi_p \to \xi_p$ is a $d\alpha$–compatible complex structure on ξ_p for each point $p \in M$, where α is any contact form defining $\xi = \ker \alpha$ with the given coorientation.

(3) With (M, ξ) as in (2), an almost contact structure (η, J, ϵ) is called **ξ–compatible** if $\eta = \xi$ as cooriented hyperplane fields and J is ξ–compatible.

In the same way that a complex bundle structure on a real vector bundle induces a reduction of the structure group to the unitary group, an almost contact structure on a $(2n+1)$–dimensional manifold M is equivalent to the reduction of the structure group of the tangent bundle TM to the group $\mathrm{U}(n) \times 1 \subset \mathrm{GL}(2n+1, \mathbb{R})$.

Proposition 2.4.8 *The space of almost contact structures compatible with a cooriented contact structure ξ is non-empty and contractible.*

Proof This follows from Proposition 2.4.5 applied to $(\xi, d\alpha)$ and the fact that the space of trivial line bundles complementary to ξ is non-empty and contractible. □

Again, this implies that we can speak sensibly of the Chern classes of a (cooriented) contact structure.

Remark 2.4.9 The terminology of almost contact structures is a little unfortunate: contact structures that are not coorientable, like those in Example 2.1.11, do not have an underlying almost contact structure.

2.5 Darboux's theorem and neighbourhood theorems

The flexibility of contact structures, which finds its expression in the Gray stability theorem and the possibility to construct contact isotopies via contact Hamiltonians, results in a variety of theorems that can be summed up as saying that there are no local invariants in contact geometry. Such theorems form the theme of the present section.

In contrast with Riemannian geometry, for instance, where the local structure coming from the curvature gives rise to a rich theory, the interesting questions in contact geometry thus appear only at the global level. However, it is actually the local flexibility that allows us to prove strong global theorems, such as the existence of contact structures on certain closed manifolds.

2.5.1 Darboux's theorem

Theorem 2.5.1 (Darboux's theorem) *Let α be a contact form on the $(2n+1)$–dimensional manifold M and p a point on M. Then there are coordinates $x_1, \ldots, x_n, y_1, \ldots, y_n, z$ on a neighbourhood $U \subset M$ of p such that $p = (0, \ldots, 0)$ and*

$$\alpha|_U = dz + \sum_{j=1}^{n} x_j\, dy_j.$$

Remark 2.5.2 Observe that the map $(\mathbf{x}, \mathbf{y}, z) \mapsto (\varepsilon\mathbf{x}, \varepsilon\mathbf{y}, \varepsilon^2 z)$ is a contactomorphism of the standard contact structure ξ_{st} on $\mathbb{R}^{2n+1}$ for any $\varepsilon \in \mathbb{R}^+$. Therefore it is an immediate consequence of the Darboux theorem that there is a contact embedding of the closed unit ball B_{st} in $(\mathbb{R}^{2n+1}, \xi_{\mathrm{st}})$ into $(M, \xi = \ker\alpha)$ which sends the origin to p. Here 'contact embedding of B_{st}' simply means a contactomorphism of a small open neighbourhood of B_{st} in $(\mathbb{R}^{2n+1}, \xi_{\mathrm{st}})$ onto its image in (M, ξ); later we shall encounter a more general concept of contact embeddings.

In fact, by Proposition 2.1.8 and Example 2.1.10 there is a contactomorphism of $(\mathbb{R}^{2n+1}, \xi_{\mathrm{st}})$ with a relatively compact subset of itself, and hence by scaling with a subset of B_{st}. So we can also construct a contactomorphism between $(\mathbb{R}^{2n+1}, \xi_{\mathrm{st}})$ and a neighbourhood of p in (M, ξ).

Proof of Theorem 2.5.1 We may assume without loss of generality that $M = \mathbb{R}^{2n+1}$ and $p = \mathbf{0}$ is the origin of $\mathbb{R}^{2n+1}$. Choose linear coordinates

$$x_1, \ldots, x_n, y_1, \ldots y_n, z$$

on $\mathbb{R}^{2n+1}$ such that

$$\text{on } T_0\mathbb{R}^{2n+1}: \begin{cases} \alpha(\partial_z) = 1, \;\; i_{\partial_z} d\alpha = 0, \\ \partial_{x_j}, \partial_{y_j} \in \ker\alpha \; (j = 1, \ldots, n), \;\; d\alpha = \sum_{j=1}^{n} dx_j \wedge dy_j. \end{cases}$$

This is simply a matter of linear algebra (the normal form theorem for skew-symmetric forms on a vector space).

Now set $\alpha_0 = dz + \sum_j x_j\, dy_j$ and consider the family of 1–forms

$$\alpha_t = (1-t)\alpha_0 + t\alpha, \; t \in [0,1],$$

on $\mathbb{R}^{2n+1}$. Our choice of coordinates ensures that

$$\alpha_t = \alpha, \;\; d\alpha_t = d\alpha \;\text{ at the origin.}$$

Hence, on a sufficiently small neighbourhood of the origin, α_t is a contact form for all $t \in [0,1]$.

We now want to use the Moser trick to find an isotopy ψ_t of a neighbourhood of the origin such that $\psi_t^* \alpha_t = \alpha_0$. This aim seems to be in conflict with our earlier remark that contact forms are not stable, but as we shall see presently, locally this equation can always be solved.

Indeed, differentiating $\psi_t^* \alpha_t = \alpha_0$ (and assuming that ψ_t is the flow of some time-dependent vector field X_t) we find

$$\psi_t^* \left(\dot{\alpha}_t + \mathcal{L}_{X_t} \alpha_t \right) = 0,$$

so X_t needs to satisfy

$$\dot{\alpha}_t + d(\alpha_t(X_t)) + i_{X_t} d\alpha_t = 0. \tag{2.4}$$

Write $X_t = H_t R_t + Y_t$ with R_t the Reeb vector field of α_t, and $Y_t \in \ker \alpha_t$. Inserting R_t in (2.4) gives

$$\dot{\alpha}_t(R_t) + dH_t(R_t) = 0. \tag{2.5}$$

On a neighbourhood of the origin, a smooth family of functions H_t satisfying (2.5) can always be found by integration, provided only that this neighbourhood has been chosen so small that none of the R_t has any closed orbits there. Since $\dot{\alpha}_t$ is zero at the origin, we may require that $H_t(\mathbf{0}) = 0$ and $dH_t|_{\mathbf{0}} = 0$ for all $t \in [0,1]$. Once H_t has been chosen, Y_t is defined uniquely by (2.4), i.e. by

$$\dot{\alpha}_t + dH_t + i_{Y_t} d\alpha_t = 0.$$

Notice that with our assumptions on H_t we have $X_t(\mathbf{0}) = \mathbf{0}$ for all t.

Now define ψ_t to be the local flow of X_t. This local flow fixes the origin, so there it is defined for all $t \in [0,1]$. Since the domain of definition in $\mathbb{R} \times M$ of a local flow on a manifold M is always open (see [38, 8.11]), we can infer† that ψ_t is actually defined for all $t \in [0,1]$ on a sufficiently small neighbourhood of the origin in $\mathbb{R}^{2n+1}$. This concludes the proof of the theorem (strictly speaking, the local coordinates in the statement of the theorem are the coordinates $x_j \circ \psi_1^{-1}$ etc.). □

2.5.2 Isotropic submanifolds

Let $L \subset (M, \xi = \ker \alpha)$ be an isotropic submanifold in a contact manifold with cooriented contact structure. Write $(TL)^\perp \subset \xi|_L$ for the sub-bundle of

† To be absolutely precise, one ought to work with a family α_t, $t \in \mathbb{R}$, where $\alpha_t \equiv \alpha_0$ for $t \leq \varepsilon$ and $\alpha_t \equiv \alpha_1$ for $t \geq 1 - \varepsilon$, i.e. a *technical homotopy* in the sense of [38]. Then X_t will be defined for all $t \in \mathbb{R}$, and the reasoning of [38] can be applied.

$\xi|_L$ that is symplectically orthogonal to TL with respect to the symplectic bundle structure $d\alpha|_\xi$. As we have seen in the preceding symplectic interlude, the conformal class of this symplectic bundle structure only depends on the contact structure ξ, not on the choice of contact form α defining ξ. So the bundle $(TL)^\perp$ is determined by ξ.

The fact that L is isotropic implies $TL \subset (TL)^\perp$. Lemma 1.3.3 allows us to make the following definition, see [241].

Definition 2.5.3 The quotient bundle

$$\mathrm{CSN}_M(L) := (TL)^\perp / TL$$

with the conformal symplectic structure induced by $d\alpha$ is called the **conformal symplectic normal bundle** of L in M.

So the normal bundle $NL := (TM|_L)/TL$ of L in M can be split as

$$NL \cong (TM|_L)/(\xi|_L) \oplus (\xi|_L)/(TL)^\perp \oplus \mathrm{CSN}_M(L). \tag{2.6}$$

Observe that if $\dim M = 2n+1$ and $\dim L = k \leq n$, then the rank of $(TL)^\perp$ is $2n - k$ by Lemma 1.3.1, so the ranks of the three summands in the above splitting are 1, k and $2(n-k)$, respectively. Our aim in this section is to show that a neighbourhood of L in M is determined, up to contactomorphism, by the isomorphism type (as a conformal symplectic bundle) of $\mathrm{CSN}_M(L)$. As a first step, we show that the the topological bundle type of the normal bundle NL is completely determined by the conformal symplectic normal bundle.

The bundle $(TM|_L)/(\xi|_L)$ is a trivial line bundle because ξ is cooriented. The Reeb vector field R_α induces a nowhere vanishing section of this quotient bundle, so it is convenient to identify $(TM|_L)/(\xi|_L)$ with the line sub-bundle $\langle R_\alpha \rangle \subset TM$ spanned by R_α.

The next lemma shows that the second summand in the decomposition (2.6) is determined by the topological type of L alone.

Lemma 2.5.4 *The bundle $(\xi|_L)/(TL)^\perp$ is isomorphic to the cotangent bundle T^*L via the well-defined bundle isomorphism*

$$\begin{array}{rccc} \Psi: & (\xi|_L)/(TL)^\perp & \longrightarrow & T^*L \\ & [Y] & \longmapsto & i_Y d\alpha|_{TL}. \end{array}$$

Proof By the definition of $(TL)^\perp$, the bundle homomorphism Ψ is well defined and injective. Since both bundles have rank k, the result follows. □

In the sequel it will often be convenient to identify all three summands on the right-hand side of (2.6) with sub-bundles of $TM|_L$. The first summand has already been identified with $\langle R_\alpha \rangle$.

Proposition 2.5.5 *Let $J\colon \xi \to \xi$ be a complex bundle structure on ξ compatible with the symplectic bundle structure given by $d\alpha$, as constructed in Proposition 2.4.5. Then the bundle $(\xi|_L)/(TL)^\perp$ is isomorphic to $J(TL)$, and the conformal symplectic normal bundle $\mathrm{CSN}_M(L)$ is isomorphic as a conformal symplectic bundle to $(TL \oplus J(TL))^\perp$.*

Hence, we have a description

$$NL \cong \langle R_\alpha \rangle \oplus J(TL) \oplus (TL \oplus J(TL))^\perp$$

of the normal bundle NL in the desired form. For clarity we shall continue to write $\mathrm{CSN}_M(L)$ for the conformal symplectic normal bundle, but we allow ourselves to interpret it as a sub-bundle of $TM|_L$ whenever convenient.

Example 2.5.6 Take M to be the unit sphere in $\mathbb{R}^{2n+2}$ with its standard contact structure $\xi = \ker \alpha_0$ from Example 2.1.7. Let $L \subset M$ be the $(k-1)$–dimensional sphere defined by the equation $x_1^2 + \cdots + x_k^2 = 1$ (and all other coordinates equal to zero). Recall from that example that the standard complex structure J on $\mathbb{R}^{2n+2}$ determined by $J(\partial_{x_j}) = \partial_{y_j}$, $j = 1, \ldots, n+1$, preserves ξ and is compatible with $d\alpha_0$.

The Reeb vector field R_0 of α_0 is given by

$$R_0 = \sum_{j=1}^{n+1} (x_j\, \partial_{y_j} - y_j\, \partial_{x_j}),$$

which along L reduces to

$$R_0|_L = \sum_{j=1}^{k} x_j\, \partial_{y_j}.$$

It follows that the bundle $\langle R_0|_L \rangle \oplus J(TL)$ is the linear span of $\partial_{y_1}, \ldots, \partial_{y_k}$, and $\mathrm{CSN}_M(L)$ the linear span of $\partial_{x_{k+1}}, \partial_{y_{k+1}}, \ldots, \partial_{x_{n+1}}, \partial_{y_{n+1}}$.

Proof of Proposition 2.5.5 This is fibrewise a matter of linear algebra. Thus, let $p \in L$ be an arbitrary point, and suppose that $X \in (T_pL)^\perp \cap J(T_pL) \subset \xi_p$. Then $JX \in T_pL$, and hence $d\alpha(X, JX) = 0$, which implies $X = \mathbf{0}$. This shows that $(T_pL)^\perp \cap J(T_pL) = \{\mathbf{0}\}$. Moreover, these two subspaces of ξ_p have complementary dimension ($2n - k$ and k, respectively). Hence $J(T_pL)$ is a complement of $(T_pL)^\perp$ in ξ_p, so that $\xi_p/(T_pL)^\perp \cong J(T_pL)$.

In the case of $\mathrm{CSN}_M(L)$ we argue similarly. First of all, we observe that $T_pL \oplus J(T_pL)$ is indeed a direct sum: if $X \in T_pL \cap J(T_pL)$, then $T_pL \ni X =$

JY with $Y \in T_pL \subset (T_pL)^\perp$, whence $d\alpha(Y, JY) = 0$, which implies $Y = \mathbf{0}$ and therefore $X = \mathbf{0}$.

Both T_pL and $(T_pL \oplus J(T_pL))^\perp$ are contained in $(T_pL)^\perp$, and their dimensions k and $2n - 2k$ add up to the dimension of $(T_pL)^\perp$. Hence, in order to show that $\mathrm{CSN}_M(L)_p = (T_pL)^\perp / T_pL$ is isomorphic to $(T_pL \oplus J(T_pL))^\perp$, it suffices to prove that

$$T_pL \cap (T_pL \oplus J(T_pL))^\perp = \{\mathbf{0}\}.$$

Thus, let X be an element of the intersection on the left-hand side. Then in particular $X \in (J(T_pL))^\perp$ and $JX \in J(T_pL)$, so that $d\alpha(X, JX) = 0$, which implies $X = \mathbf{0}$. □

Recall from Example 2.4.2 that on the Whitney sum $TL \oplus T^*L$ (for any manifold L) there is a canonical symplectic bundle structure Ω_L defined by

$$\Omega_{L,p}(X + \eta, X' + \eta') = \eta(X') - \eta'(X) \ \text{ for } X, X' \in T_pL;\ \eta, \eta' \in T_p^*L.$$

Lemma 2.5.7 *The bundle map*

$$\mathrm{id}_{TL} \oplus \Psi\colon\ (TL \oplus J(TL), d\alpha) \longrightarrow (TL \oplus T^*L, \Omega_L)$$

is an isomorphism of symplectic vector bundles.

Proof We need only check that $\mathrm{id}_{TL} \oplus \Psi$ is a symplectic bundle map. Let $X, X' \in T_pL$ and $Y, Y' \in J_p(T_pL)$. Write $Y = J_pZ, Y' = J_pZ'$ with $Z, Z' \in T_pL$. It follows that

$$d\alpha(Y, Y') = d\alpha(JZ, JZ') = d\alpha(Z, Z') = 0,$$

since L is an isotropic submanifold. For the same reason $d\alpha(X, X') = 0$. Hence

$$\begin{aligned} d\alpha(X + Y, X' + Y') &= d\alpha(Y, X') - d\alpha(Y', X) \\ &= \Psi(Y)(X') - \Psi(Y')(X) \\ &= \Omega_L(X + \Psi(Y), X' + \Psi(Y')), \end{aligned}$$

which proves the assertion about the map $\mathrm{id}_{TL} \oplus \Psi$. □

Theorem 2.5.8 *Let (M_i, ξ_i), $i = 0, 1$, be contact manifolds with closed isotropic submanifolds L_i. Suppose there is an isomorphism of conformal symplectic normal bundles $\Phi\colon \mathrm{CSN}_{M_0}(L_0) \to \mathrm{CSN}_{M_1}(L_1)$ that covers a diffeomorphism $\phi\colon L_0 \to L_1$. Then this diffeomorphism ϕ extends to a contactomorphism $\psi\colon \mathcal{N}(L_0) \to \mathcal{N}(L_1)$ of suitable neighbourhoods $\mathcal{N}(L_i)$ of L_i such that the bundle maps $T\psi|_{\mathrm{CSN}_{M_0}(L_0)}$ and Φ are bundle homotopic (as conformal symplectic bundle isomorphisms).*

Corollary 2.5.9 *If $L_i \subset (M_i, \xi_i)$, $i = 0, 1$, are diffeomorphic (closed) Legendrian submanifolds, then they have contactomorphic neighbourhoods.*

Proof The conformal symplectic normal bundle of a Legendrian submanifold has rank 0, so the conditions in the theorem, apart from the existence of a diffeomorphism $\phi\colon L_0 \to L_1$, are void. □

Example 2.5.10 Let $L \subset (M^3, \xi)$ be a Legendrian knot in a contact 3–manifold. Identify a neighbourhood of L with $S^1 \times \mathbb{R}^2$, where $L \equiv S^1 \times \{\mathbf{0}\}$. Then, with S^1–coordinate θ and Cartesian coordinates (x, y) on $\mathbb{R}^2$, the contact structure

$$\cos\theta\, dx - \sin\theta\, dy = 0$$

provides a model for a neighbourhood of L.

Example 2.5.11 Given any manifold L, one can define the space $J^1(L)$ of 1–jets of (germs of) differentiable functions $f\colon L \to \mathbb{R}$. Let $\mathcal{E}(x)$ be the space of germs of differentiable functions at a point $x \in L$. An equivalence relation on $\mathcal{E}(x)$ is given by

$$f_1 \sim f_2 :\Longleftrightarrow f_1(x) = f_2(x) \text{ and } f_1'(x) = f_2'(x).$$

The equivalence class $j_x^1 f$ of a germ $f \in \mathcal{E}(x)$ is called a 1–**jet** at x; see [95] for some basic information on jets. One can identify $J^1(L)$ with the space $\mathbb{R} \times T^*L$ by mapping $j_x^1 f$ to $(f(x), df_x)$.

Let λ be the Liouville form on T^*L introduced in Section 1.4, given in local coordinates $\mathbf{q}$ on L and dual coordinates $\mathbf{p}$ by $\lambda = \mathbf{p}\, d\mathbf{q}$. Writing z for the $\mathbb{R}$–coordinate in $\mathbb{R} \times T^*L$, we have a natural contact structure on the 1–jet space given by $\xi_{\text{jet}} = \ker(dz - \lambda)$.

Any globally defined function $f\colon L \to \mathbb{R}$ gives rise to a Legendrian embedding of L in $(J^1(S^1), \xi_{\text{jet}})$:

$$\begin{array}{ccc} L & \longrightarrow & \mathbb{R} \times T^*L \\ x & \longmapsto & (f(x), df_x); \end{array}$$

this being Legendrian is immediate from the local coordinate description of ξ_{jet} (alternatively, use Lemma 1.4.1). In particular, the identically zero function corresponds to the zero section $L \subset T^*L \subset \mathbb{R} \times T^*L$. By the above theorem, a neighbourhood of $L \subset J^1(L)$ provides a universal model for the neighbourhood of any Legendrian submanifold $L \subset (M, \xi)$.

Proof of Theorem 2.5.8 Choose contact forms α_i for ξ_i, $i = 0, 1$, scaled in such a way that Φ is actually an isomorphism of symplectic vector bundles with respect to the symplectic bundle structures on $\mathrm{CSN}_{M_i}(L_i)$ given by

$d\alpha_i$. Here we think of $\mathrm{CSN}_{M_i}(L_i)$ as a sub-bundle of $TM_i|_{L_i}$ (rather than as a quotient bundle).

We identify $(TM_i|_{L_i})/(\xi_i|_{L_i})$ with the trivial line bundle spanned by the Reeb vector field R_{α_i}. In total, this identifies

$$NL_i = \langle R_{\alpha_i}\rangle \oplus J_i(TL_i) \oplus \mathrm{CSN}_{M_i}(L_i)$$

as a sub-bundle of $TM_i|_{L_i}$.

Let $\Phi_R\colon \langle R_{\alpha_0}\rangle \to \langle R_{\alpha_1}\rangle$ be the obvious bundle isomorphism defined by requiring that $R_{\alpha_0}(p)$ map to $R_{\alpha_1}(\phi(p))$.

Let $\Psi_i\colon J_i(TL_i) \to T^*L_i$ be the isomorphism defined by taking the interior product with $d\alpha_i$. Notice that

$$T\phi \oplus (\phi^*)^{-1}\colon (TL_0 \oplus T^*L_0, \Omega_{L_0}) \to (TL_1 \oplus T^*L_1, \Omega_{L_1})$$

is an isomorphism of symplectic vector bundles. With Lemma 2.5.7 it follows that

$$T\phi \oplus \Psi_1^{-1} \circ (\phi^*)^{-1} \circ \Psi_0\colon (TL_0 \oplus J_0(TL_0), d\alpha_0) \to (TL_1 \oplus J_1(TL_1), d\alpha_1)$$

is an isomorphism of symplectic vector bundles.

Now let

$$\widetilde{\Phi}\colon NL_0 \longrightarrow NL_1$$

be the bundle isomorphism (covering ϕ) defined by

$$\widetilde{\Phi} = \Phi_R \oplus \Psi_1^{-1} \circ (\phi^*)^{-1} \circ \Psi_0 \oplus \Phi.$$

Let $\tau_i\colon NL_i \to M_i$ be tubular maps. Recall that this means the following (I suppress the index i for better readability). The τ are embeddings such that

- under the identification of L with the zero section of NL, the map $\tau|_L$ is the inclusion $L \subset M$, and
- with respect to the splittings $T(NL)|_L = TL \oplus NL = TM|_L$, the differential $T\tau$ induces the identity on NL along L.

Then $\tau_1 \circ \widetilde{\Phi} \circ \tau_0^{-1}\colon \mathcal{N}(L_0) \to \mathcal{N}(L_1)$ is a diffeomorphism of suitable neighbourhoods $\mathcal{N}(L_i)$ of L_i that induces the bundle map

$$T\phi \oplus \widetilde{\Phi}\colon TM_0|_{L_0} \longrightarrow TM_1|_{L_1}.$$

By construction, this bundle map pulls α_1 back to α_0 and $d\alpha_1$ to $d\alpha_0$. Hence, α_0 and $(\tau_1 \circ \widetilde{\Phi} \circ \tau_0^{-1})^*\alpha_1$ are contact forms on $\mathcal{N}(L_0)$ that coincide on $TM_0|_{L_0}$, and so do their differentials.

Now consider the family of 1–forms

$$\beta_t = (1-t)\alpha_0 + t(\tau_1 \circ \widetilde{\Phi} \circ \tau_0^{-1})^*\alpha_1, \;\; t \in [0,1].$$

On $TM_0|_{L_0}$ we have $\beta_t \equiv \alpha_0$ and $d\beta_t \equiv d\alpha_0$. Since the contact condition $\alpha \wedge (d\alpha)^n \neq 0$ is an open condition, we may assume – shrinking $\mathcal{N}(L_0)$ if necessary – that β_t is a contact form on $\mathcal{N}(L_0)$ for all $t \in [0,1]$. By the Gray stability theorem (Thm. 2.2.2) and Remark 2.2.3 following its proof, we find an isotopy ψ_t of $\mathcal{N}(L_0)$, fixing L_0, such that $\psi_t^*\beta_t = \lambda_t\alpha_0$ for some smooth family of smooth functions $\lambda_t\colon \mathcal{N}(L_0) \to \mathbb{R}^+$. In fact, arguing as in the proof of Darboux's theorem (Thm. 2.5.1), we may find an isotopy ψ_t such that $\psi_t^*\beta_t = \alpha_0$.† We conclude that the composition $\psi := \tau_1 \circ \widetilde{\Phi} \circ \tau_0^{-1} \circ \psi_1$ is the desired contactomorphism. □

Remark 2.5.12 With a little more care one can actually achieve $T\psi_1 = \mathrm{id}$ on $TM_0|_{L_0}$, which implies in particular that $T\psi|_{\mathrm{CSN}_{M_0}(L_0)} = \Phi$, see [241]. The key point is the generalised Poincaré lemma, see Appendix A, which allows us to write a closed differential form η, given in a neighbourhood of the zero section of a bundle and vanishing along that zero section, as an exact form $\eta = d\zeta$ with ζ and its partial derivatives with respect to all coordinates (in any chart) vanishing along the zero section. This lemma is applied first to $\eta = d(\beta_1 - \beta_0)$, in order to find (with the symplectic Moser trick on a symplectic hypersurface) a diffeomorphism σ of a neighbourhood of $L_0 \subset M_0$ with $T\sigma = \mathrm{id}$ on $TM_0|_{L_0}$ and such that $d\beta_0 = d(\sigma^*\beta_1)$. It is then applied once again to $\eta' = \beta_0 - \sigma^*\beta_1$. Further details will be provided in Chapter 6, where we need this stronger version of the neighbourhood theorem for isotropic submanifolds (Thm. 6.2.2) in the context of contact surgery.

Example 2.5.13 Let $M_0 = M_1 = \mathbb{R}^3$ with contact forms $\alpha_0 = dz + x\,dy$ and $\alpha_1 = dz + (x+y)\,dy$ and $L_0 = L_1 = \mathbf{0}$ the origin in $\mathbb{R}^3$. Then

$$\mathrm{CSN}_{M_0}(L_0) = \mathrm{CSN}_{M_1}(L_1) = \mathrm{span}\{\partial_x, \partial_y\} \subset T_{\mathbf{0}}\mathbb{R}^3.$$

We take $\Phi = \mathrm{id}_{\mathrm{CSN}}$.

Set $\alpha_t = dz + (x + ty)\,dy$. The Moser trick with $X_t \in \ker \alpha_t$ yields $X_t = -y\,\partial_x$, and hence $\psi_t(x,y,z) = (x - ty, y, z)$. This isotopy satisfies $\psi_t^*\alpha_t = \alpha_0$. Then

$$T\psi_1 = \begin{pmatrix} 1 & -1 & 0 \\ 0 & 1 & 0 \\ 0 & 0 & 1 \end{pmatrix},$$

† Since $\mathcal{N}(L_0)$ is not a closed manifold, ψ_t is *a priori* only a local flow. But on L_0 it is stationary and hence defined for all t. As in the proof of the Darboux theorem we conclude that ψ_t is defined for all $t \in [0,1]$ in a sufficiently small neighbourhood of L_0, so shrinking $\mathcal{N}(L_0)$ once again, if necessary, will ensure that ψ_t is a global flow on $\mathcal{N}(L_0)$.

which does not restrict to Φ on CSN.

However, a different solution (again for the equation $\psi_t^*\alpha_t = \alpha_0$) is given by $X_t = H_t R_t$ with H_t a solution of $\dot{\alpha}_t + dH_t = 0$ (see the proof of Darboux's theorem). This gives $X_t = -y^2\,\partial_z/2$ and $\psi_t(x,y,z) = (x,y,z-ty^2/2)$. Here we get

$$T\psi_1 = \begin{pmatrix} 1 & 0 & 0 \\ 0 & 1 & 0 \\ 0 & -y & 1 \end{pmatrix},$$

hence $T\psi_1|_{T_0\mathbb{R}^3} = \mathrm{id}$, so in particular $T\psi_1|_{\mathrm{CSN}} = \Phi$.

Beware that the preceding example is relatively simple, because $d\alpha_0 = d\alpha_1$; in other words, we need only the second step of the foregoing remark. We shall see in Chapter 6 that for this second step a suitable multiple of the Reeb vector field will always produce the desired flow.

2.5.3 Contact submanifolds

Let $(M', \xi' = \ker\alpha') \subset (M, \xi = \ker\alpha)$ be a contact submanifold, that is, $TM' \cap \xi|_{M'} = \xi'$. As before we write $(\xi')^\perp \subset \xi|_{M'}$ for the symplectically orthogonal complement of ξ' in $\xi|_{M'}$. Since M' is a contact submanifold (so ξ' is a symplectic sub-bundle of $(\xi|_{M'}, d\alpha)$), we have

$$TM' \oplus (\xi')^\perp = TM|_{M'},$$

i.e. we can identify $(\xi')^\perp$ with the normal bundle NM'. Moreover, $d\alpha$ induces a conformal symplectic structure on $(\xi')^\perp$, see Lemma 1.3.4.

Definition 2.5.14 The bundle

$$\mathrm{CSN}_M(M') := (\xi')^\perp$$

with the conformal symplectic structure induced by $d\alpha$ is called the **conformal symplectic normal bundle** of M' in M.

Theorem 2.5.15 *Let (M_i, ξ_i), $i = 0, 1$, be contact manifolds with compact contact submanifolds (M_i', ξ_i'). Suppose there is an isomorphism of conformal symplectic normal bundles $\Phi\colon \mathrm{CSN}_{M_0}(M_0') \to \mathrm{CSN}_{M_1}(M_1')$ that covers a contactomorphism $\phi\colon (M_0', \xi_0') \to (M_1', \xi_1')$. Then ϕ extends to a contactomorphism ψ of suitable neighbourhoods $\mathcal{N}(M_i')$ of M_i' such that $T\psi|_{\mathrm{CSN}(M_0, M_0')}$ and Φ are bundle homotopic (as conformal symplectic bundle isomorphisms).*

Example 2.5.16 A particular instance of this theorem is the case of a transverse knot in a contact manifold (M, ξ), i.e. an embedding $S^1 \hookrightarrow (M, \xi)$ transverse to ξ. Since the symplectic group $\mathrm{Sp}(2n)$ of linear transformations of $\mathbb{R}^{2n}$ preserving the standard symplectic structure $\omega_{\mathrm{st}} = \sum_{j=1}^n dx_j \wedge dy_j$ is connected, see [177, Prop. 2.22], there is only one conformal symplectic $\mathbb{R}^{2n}$–bundle over S^1 up to conformal equivalence. A model for the neighbourhood of a transverse knot is given by

$$\left(S^1 \times \mathbb{R}^{2n}, \xi = \ker\left(d\theta + \sum_{j=1}^n (x_j\, dy_j - y_j\, dx_j)\right)\right),$$

where θ denotes the S^1–coordinate; the theorem says that in suitable local coordinates the neighbourhood of any transverse knot looks like this model.

Proof of Theorem 2.5.15 As in the proof of Theorem 2.5.8 it suffices to find contact forms α_i on M_i and a bundle map $TM_0|_{M_0'} \to TM_1|_{M_1'}$, covering ϕ and inducing Φ, that pulls back α_1 to α_0 and $d\alpha_1$ to $d\alpha_0$; the proof then concludes as there with a stability argument.

For this we need to choose the α_i judiciously. The essential choice is made separately on each M_i, so I suppress the subscript i for the time being. Choose a contact form α' for ξ' on M'. Write R' for the Reeb vector field of α'. Given any contact form α for ξ on M we may first scale it such that $\alpha(R') \equiv 1$ along M'. Then $\alpha|_{TM'} = \alpha'$, and hence $d\alpha|_{TM'} = d\alpha'$. We now want to scale α further such that its Reeb vector field R coincides with R' along M'. To this end we want to find a smooth function $f\colon M \to \mathbb{R}^+$ with $f|_{M'} \equiv 1$ and $i_{R'}d(f\alpha) \equiv 0$ on $TM|_{M'}$. This last equation becomes

$$0 = i_{R'}d(f\alpha) = i_{R'}(df \wedge \alpha + f\, d\alpha) = -df + i_{R'}d\alpha \;\text{ on } TM|_{M'}.$$

Since $i_{R'}d\alpha|_{TM'} = i_{R'}d\alpha' \equiv 0$, such an f exists.

The choices of α_0' and α_1' cannot be made independently of each other; we may first choose α_1', say, and then define $\alpha_0' = \phi^*\alpha_1'$. Then define α_0, α_1 as described and scale Φ such that it is a symplectic bundle isomorphism of

$$((\xi_0')^\perp, d\alpha_0) \longrightarrow ((\xi_1')^\perp, d\alpha_1).$$

Then

$$T\phi \oplus \Phi\colon TM_0|_{M_0'} \longrightarrow TM_1|_{M_1'}$$

is the desired bundle map that pulls back α_1 to α_0 and $d\alpha_1$ to $d\alpha_0$. □

Remark 2.5.17 The condition that $R_i \equiv R_i'$ along M' is necessary for ensuring that $(T\phi \oplus \Phi)(R_0) = R_1$, which guarantees (with the other stated conditions) that $(T\phi \oplus \Phi)^*(d\alpha_1) = d\alpha_0$. The condition $d\alpha_i|_{TM_i'} = d\alpha_i'$ and the described choice of Φ alone would only give $(T\phi \oplus \Phi)^*(d\alpha_1|_{\xi_1}) = d\alpha_0|_{\xi_0}$.

2.5.4 Hypersurfaces

Let S be an oriented hypersurface in a contact manifold $(M, \xi = \ker \alpha)$ of dimension $2n+1$. In a neighbourhood of S in M, which we can identify with $S \times \mathbb{R}$ (and S with $S \times \{0\}$), the contact form α can be written as

$$\alpha = \beta_r + u_r\, dr,$$

where β_r, $r \in \mathbb{R}$, is a smooth family of 1–forms on S and $u_r \colon S \to \mathbb{R}$ a smooth family of functions. The contact condition $\alpha \wedge (d\alpha)^n \neq 0$ then becomes, with the derivative with respect to r denoted by a dot,

$$\begin{aligned} 0 &\neq \alpha \wedge (d\alpha)^n \\ &= (\beta_r + u_r\, dr) \wedge (d\beta_r - \dot{\beta}_r \wedge dr + du_r \wedge dr)^n \\ &= (-n\beta_r \wedge \dot{\beta}_r + n\beta_r \wedge du_r + u_r\, d\beta_r) \wedge (d\beta_r)^{n-1} \wedge dr. \qquad (2.7) \end{aligned}$$

The intersection $TS \cap (\xi|_S)$ determines a distribution (of non-constant rank) of subspaces of TS. If α is written as above, this distribution is given by the kernel of β_0, and hence, at a given $p \in S$, defines either the full tangent space T_pS (if $\beta_{0,p} = 0$) or a 1–codimensional subspace both of T_pS and ξ_p (if $\beta_{0,p} \neq 0$). In the former case, the symplectically orthogonal complement $(T_pS \cap \xi_p)^\perp$ (with respect to the conformal symplectic structure $d\alpha$ on ξ_p) is $\{\mathbf{0}\}$; in the latter case, $(T_pS \cap \xi_p)^\perp$ is a 1–dimensional subspace of ξ_p contained in $T_pS \cap \xi_p$.

From that it is intuitively clear what one should mean by a 'singular 1–dimensional foliation', and we make the following somewhat provisional definition, where the hypersurface S need not be orientable.

Definition 2.5.18 The **characteristic foliation** S_ξ of a hypersurface S in (M, ξ) is the singular 1–dimensional foliation of S defined by the distribution $(TS \cap \xi|_S)^\perp$.

Example 2.5.19 If $\dim M = 3$ and $\dim S = 2$, then $(T_pS \cap \xi_p)^\perp = T_pS \cap \xi_p$ at the points $p \in S$ where $T_pS \cap \xi_p$ is 1–dimensional. Figure 2.2 shows the characteristic foliation of the unit 2–sphere in $(\mathbb{R}^3, \xi_2)$, where ξ_2 denotes the standard contact structure of Example 2.1.1: the only singular points are $(0, 0, \pm 1)$; away from these points the characteristic foliation is spanned by

$$(xz - y)\, \partial_x + (yz + x)\, \partial_y - (x^2 + y^2)\, \partial_z.$$

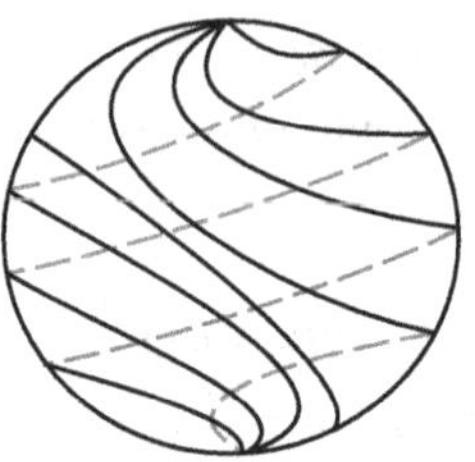

Fig. 2.2. The characteristic foliation on $S^2 \subset (\mathbb{R}^3, \xi_2)$.

The following lemma, where we revert to the assumption that S be orientable, helps to clarify the notion of singular 1–dimensional foliation.

Lemma 2.5.20 *Let β_0 be the 1–form induced on S by a contact form α defining ξ, and let Ω be a volume form on S. Then S_ξ is defined by the vector field X satisfying*

$$i_X\Omega = \beta_0 \wedge (d\beta_0)^{n-1}.$$

Proof First of all, we want to see that $\beta_0 \wedge (d\beta_0)^{n-1} \neq 0$ outside the zeros of β_0. Arguing by contradiction, assume $\beta_{0,p} \neq 0$ and $\beta_0 \wedge (d\beta_0)^{n-1}|_p = 0$ at some $p \in S$. Then $(d\beta_0)^n|_p \neq 0$ by (2.7). On the codimension 1 subspace $\ker \beta_{0,p}$ of T_pS the symplectic form $d\beta_{0,p}$ has maximal rank $2n-2$. It follows that $\beta_0 \wedge (d\beta_0)^{n-1}|_p \neq 0$ after all, a contradiction.

Next we want to show that $X \in \ker \beta_0$. We observe

$$0 = i_X(i_X\Omega) = \beta_0(X)(d\beta_0)^{n-1} - (n-1)\beta_0 \wedge i_X d\beta_0 \wedge (d\beta_0)^{n-2}. \tag{2.8}$$

Taking the exterior product of this equation with β_0 we get

$$\beta_0(X)\beta_0 \wedge (d\beta_0)^{n-1} = 0.$$

By our previous consideration this implies $\beta_0(X) = 0$.

It remains to show that for $\beta_{0,p} \neq 0$ we have

$$d\beta_0(X_p, \mathbf{v}) = 0 \ \text{ for all } \mathbf{v} \in T_pS \cap \xi_p.$$

For $n = 1$ this is trivially satisfied, because in that case $\mathbf{v}$ is a multiple of X_p. I suppress the point p in the following calculation, where we assume $n \geq 2$. From (2.8) and with $\beta_0(X) = 0$ we have

$$\beta_0 \wedge i_X d\beta_0 \wedge (d\beta_0)^{n-2} = 0. \tag{2.9}$$

Taking the interior product with $\mathbf{v} \in TS \cap \xi$ yields

$$-d\beta_0(X, \mathbf{v})\beta_0 \wedge (d\beta_0)^{n-2} + (n-2)\beta_0 \wedge i_X d\beta_0 \wedge i_{\mathbf{v}} d\beta_0 \wedge (d\beta_0)^{n-3} = 0.$$

(Thanks to the coefficient $n-2$ the term $(d\beta_0)^{n-3}$ is not a problem for $n=2$.) Taking the exterior product of that last equation with $d\beta_0$, and using (2.9), we find

$$d\beta_0(X, \mathbf{v})\beta_0 \wedge (d\beta_0)^{n-1} = 0,$$

and thus $d\beta_0(X, \mathbf{v}) = 0$. □

Remark 2.5.21 (1) We can now give a more formal definition of 'oriented singular 1–dimensional foliation' as an equivalence class of vector fields $[X]$, where X is allowed to have zeros, and $[X] = [X']$ if there is a nowhere zero function on all of S such that $X' = fX$. Notice that the non-integrability of contact structures and the reasoning at the beginning of the proof of the lemma imply that the zero set of X does not contain any open subsets of S.

(2) If the contact structure ξ is cooriented rather than just coorientable, so that α is well defined up to multiplication with a *positive* function, then this lemma allows one to give an orientation to the characteristic foliation: changing α to $\lambda\alpha$ with $\lambda\colon M \to \mathbb{R}^+$ will change $\beta_0 \wedge (d\beta_0)^{n-1}$ by a factor λ^n.

(3) A not necessarily orientable singular 1–dimensional foliation can be defined by requiring only that there be a covering of the manifold M by open subsets U_i, and vector fields X_i on U_i that are related by $X_i = f_{ij}X_j$ on $U_i \cap U_j$, with f_{ij} a nowhere zero function. This then covers the case of characteristic foliations S_ξ on non-orientable hypersurfaces S, or induced by a non-coorientable contact structure ξ.

We now restrict attention to surfaces in contact 3–manifolds, where the notion of characteristic foliation has proved to be particularly useful.

The following theorem is due to Giroux [104].

Theorem 2.5.22 (Giroux) *Let S_i be closed surfaces in contact 3–manifolds (M_i, ξ_i), $i = 0, 1$ (with ξ_i cooriented), and $\phi\colon S_0 \to S_1$ a diffeomorphism with $\phi(S_{0,\xi_0}) = S_{1,\xi_1}$ as oriented characteristic foliations. Then there is a contactomorphism $\psi\colon \mathcal{N}(S_0) \to \mathcal{N}(S_1)$ of suitable neighbourhoods $\mathcal{N}(S_i)$ of S_i with $\psi(S_0) = S_1$ and such that $\psi|_{S_0}$ is isotopic to ϕ via an isotopy preserving the characteristic foliation.*

Proof By passing to a double cover, if necessary, we may assume that the S_i are orientable hypersurfaces. Let α_i be contact forms defining ξ_i. Extend ϕ to a diffeomorphism (still denoted ϕ) of neighbourhoods of S_i and consider the contact forms α_0 and $\phi^*\alpha_1$ on a neighbourhood of S_0, which we may identify with $S_0 \times \mathbb{R}$. We may assume that this extended ϕ has been chosen in such a way that α_0 and $\phi^*\alpha_1$ define the same orientation on $S_0 \times \mathbb{R}$.

By rescaling α_1 we may assume that α_0 and $\phi^*\alpha_1$ induce the same form β_0 on $S_0 \equiv S_0 \times \{0\}$, and hence also the same form $d\beta_0$.

Observe that the expression on the right-hand side of Equation (2.7) is linear in $\dot{\beta}_r$ and u_r. This implies that convex linear combinations of solutions of (2.7) (for $n = 1$) with the same β_0 (and $d\beta_0$) will again be solutions of (2.7) for sufficiently small $|r|$. This reasoning applies to

$$\alpha_t := (1-t)\alpha_0 + t\phi^*\alpha_1, \; t \in [0,1].$$

(I hope the reader will forgive the slight abuse of notation, with α_1 denoting both a form on M_1 and its pull-back $\phi^*\alpha_1$ to M_0.) As is to be expected, we now use the Moser trick to find an isotopy ψ_t with $\psi_t^*\alpha_t = \lambda_t\alpha_0$, just as in the proof of Gray stability (Thm. 2.2.2). In particular, we require as there that the vector field X_t we want to integrate to the flow ψ_t lie in the kernel of α_t.

On TS_0 we have $\dot{\alpha}_t \equiv 0$ (thanks to the assumption that α_0 and $\phi^*\alpha_1$ induce the same form β_0 on S_0). In particular, if $\mathbf{v}$ is a vector tangent to S_{0,ξ_0} at a non-singular point of this characteristic foliation, then by Equation (2.2) we have $d\alpha_t(X_t, \mathbf{v}) = 0$, which implies that X_t is a multiple of $\mathbf{v}$, hence tangent to S_{0,ξ_0}. At a singular point of the characteristic foliation, we have $d\alpha_t(X_t, \mathbf{v}) = 0$ for any tangent vector $\mathbf{v}$ to S_0, and hence $X_t = \mathbf{0}$ at these points. This shows that the flow of X_t preserves S_0 and its characteristic foliation. More formally, we have

$$\mathcal{L}_{X_t}\alpha_t = d(\alpha_t(X_t)) + i_{X_t}d\alpha_t = i_{X_t}d\alpha_t,$$

so with $\mathbf{v}$ as above we have $\mathcal{L}_{X_t}\alpha_t(\mathbf{v}) = 0$, which shows that $\mathcal{L}_{X_t}\alpha_t|_{TS_0}$ is a multiple of $\alpha_0|_{TS_0} = \beta_0$. This implies that the (local) flow of X_t changes β_0 by a conformal factor.

Since S_0 is closed, the local flow of X_t restricted to S_0 integrates up to $t = 1$, and so the same is true† in a neighbourhood of S_0. Then $\psi = \phi \circ \psi_1$ will be the desired diffeomorphism $\mathcal{N}(S_0) \to \mathcal{N}(S_1)$. □

As observed previously in the proof of Darboux's theorem for contact *forms*, the Moser trick allows more flexibility if we drop the condition that $\alpha_t(X_t) = 0$. We are now going to exploit this extra freedom to strengthen Giroux's theorem slightly. This will be important when we want to extend isotopies of hypersurfaces.

Theorem 2.5.23 *Under the assumptions of the preceding theorem we can find* $\psi\colon \mathcal{N}(S_0) \to \mathcal{N}(S_1)$ *satisfying the stronger condition* $\psi|_{S_0} = \phi$.

† See the proof (and the footnote therein) of Darboux's theorem (Thm. 2.5.1).

Proof We want to show that the isotopy ψ_t of the preceding proof may be assumed to fix S_0 pointwise. As there, we may assume $\dot{\alpha}_t|_{TS_0} \equiv 0$.

If the condition that X_t be tangent to $\ker \alpha_t$ is dropped, the condition X_t needs to satisfy so that its flow will pull back α_t to $\lambda_t \alpha_0$ is

$$\dot{\alpha}_t + d(\alpha_t(X_t)) + i_{X_t} d\alpha_t = \mu_t \alpha_t, \tag{2.10}$$

where μ_t and λ_t are related by $\mu_t = \frac{d}{dt}(\log \lambda_t) \circ \psi_t^{-1}$, see the proof of the Gray stability theorem (Thm. 2.2.2).

Write $X_t = H_t R_t + Y_t$ with R_t the Reeb vector field of α_t and $Y_t \in \xi_t = \ker \alpha_t$. Then condition (2.10) translates into

$$\dot{\alpha}_t + dH_t + i_{Y_t} d\alpha_t = \mu_t \alpha_t. \tag{2.11}$$

For given H_t one determines μ_t from this equation by inserting the Reeb vector field R_t; the equation then admits a unique solution $Y_t \in \ker \alpha_t$ because of the non-degeneracy of $d\alpha_t|_{\xi_t}$.

Our aim now is to ensure that $H_t \equiv 0$ on S_0 and $Y_t \equiv \mathbf{0}$ along S_0. The latter we achieve by imposing the condition

$$\dot{\alpha}_t + dH_t = 0 \text{ along } S_0 \tag{2.12}$$

(which entails with (2.11) that $\mu_t|_{S_0} \equiv 0$). The conditions $H_t \equiv 0$ on S_0 and (2.12) can be simultaneously satisfied thanks to $\dot{\alpha}_t|_{TS_0} \equiv 0$.

We can therefore find a smooth family of smooth functions H_t satisfying these conditions, and then define Y_t by (2.11). The flow of the vector field $X_t = H_t R_t + Y_t$ then defines an isotopy ψ_t that fixes S_0 pointwise (and thus is defined for all $t \in [0,1]$ in a neighbourhood of S_0). Then $\psi = \phi \circ \psi_1$ will be the diffeomorphism we wanted to construct. □

2.6 Isotopy extension theorems

In this section we show that the isotopy extension theorem of differential topology – an isotopy of a closed submanifold extends to an isotopy of the ambient manifold – remains valid both for isotropic and contact submanifolds of contact manifolds. The neighbourhood theorems proved above provide the key to the corresponding isotopy extension theorems. For simplicity, I assume throughout that the ambient contact manifold M is closed; all isotopy extension theorems remain valid if M has non-empty boundary ∂M, provided the isotopy stays away from the boundary. In that case, the isotopy of M found by extension keeps a neighbourhood of ∂M fixed. A further convention is that our ambient isotopies ψ_t are understood to start at $\psi_0 = \mathrm{id}_M$. Of course, by a **contact isotopy** of a contact manifold (M, ξ) we understand an isotopy consisting of contactomorphisms.

2.6.1 Isotropic submanifolds

In order to formulate the isotopy extension theorem for isotropic submanifolds, we first need to give the obvious definition of the corresponding type of embedding.

Definition 2.6.1 An embedding $j\colon L \to (M, \xi = \ker\alpha)$ is called an **isotropic embedding** if $j(L)$ is an isotropic submanifold of (M, ξ), i.e. everywhere tangent to the contact structure ξ. Equivalently, one needs to require $j^*\alpha \equiv 0$.†

Theorem 2.6.2 *Let $j_t\colon L \to (M, \xi = \ker\alpha)$, $t \in [0,1]$, be an isotopy of isotropic embeddings of a closed manifold L in a contact manifold (M, ξ). Then there is a compactly supported contact isotopy ψ_t of (M, ξ) satisfying $\psi_t \circ j_0 = j_t$.*

Proof Define a time-dependent vector field X_t along $j_t(L)$ by

$$X_t \circ j_t = \frac{d}{dt} j_t.$$

To simplify notation later on, we assume that L is a submanifold of M and j_0 the inclusion $L \subset M$. Our aim is to find a (smooth) family of compactly supported, smooth functions $\widetilde{H}_t\colon M \to \mathbb{R}$ whose Hamiltonian vector field $\widetilde{X}_t$ equals X_t along $j_t(L)$. Recall that $\widetilde{X}_t$ is defined in terms of $\widetilde{H}_t$ by

$$\alpha(\widetilde{X}_t) = \widetilde{H}_t, \quad i_{\widetilde{X}_t} d\alpha = d\widetilde{H}_t(R_\alpha)\alpha - d\widetilde{H}_t,$$

where, as usual, R_α denotes the Reeb vector field of α.

Hence, we need

$$\alpha(X_t) = \widetilde{H}_t, \quad i_{X_t} d\alpha = d\widetilde{H}_t(R_\alpha)\alpha - d\widetilde{H}_t \text{ along } j_t(L). \tag{2.13}$$

Define $H_t\colon j_t(L) \to \mathbb{R}$ by $H_t = \alpha(X_t)$. To satisfy (2.13) we need

$$\widetilde{H}_t = H_t \text{ along } j_t(L). \tag{2.14}$$

This implies

$$d\widetilde{H}_t(\mathbf{v}) = dH_t(\mathbf{v}) \text{ for } \mathbf{v} \in T(j_t(L)).$$

Since j_t is an isotopy of isotropic embeddings, we have $T(j_t(L)) \subset \ker\alpha$. So a prerequisite for (2.13) is that

$$d\alpha(X_t, \mathbf{v}) = -dH_t(\mathbf{v}) \text{ for } \mathbf{v} \in T(j_t(L)). \tag{2.15}$$

† The (equivalent) conditions $Tj(TL) \subset \xi$ and $j^*\alpha \equiv 0$ also make sense for an immersion j, which is then called an **isotropic immersion**.

This condition can be rewritten as

$$d(j_t^* i_{X_t}\alpha) + j_t^*(i_{X_t}d\alpha) \equiv 0,$$

and by Lemma B.1 (page 404) this in turn is equivalent to $\frac{d}{dt}(j_t^*\alpha) \equiv 0$. But this last condition is indeed tautologically satisfied, for the fact that j_t is an isotopy of isotropic embeddings can be written as $j_t^*\alpha \equiv 0$.

This means that we can define $\widetilde{H}_t$ by prescribing the value of $\widetilde{H}_t$ along $j_t(L)$ (with (2.14)) and the differential of $\widetilde{H}_t$ along $j_t(L)$ (with (2.13)), where we are free to impose $d\widetilde{H}_t(R_\alpha) = 0$, for instance. The calculation we just performed shows that these two requirements are consistent with each other. Any function satisfying these requirements along $j_t(L)$ can be smoothed out to zero outside a tubular neighbourhood of $j_t(L)$, and the Hamiltonian flow of this $\widetilde{H}_t$ will be the desired contact isotopy extending j_t.

One small technical point is to ensure that the resulting family of functions $\widetilde{H}_t$ will be smooth in t. To achieve this, we proceed as follows. Set $\hat{M} = M \times [0,1]$ and

$$\hat{L} = \bigcup_{q\in L, t\in[0,1]} (j_t(q), t),$$

so that $\hat{L}$ is a submanifold of $\hat{M}$. Let g be an auxiliary Riemannian metric on M with respect to which R_α is orthogonal to $\ker\alpha$. Identify the normal bundle $N\hat{L}$ of $\hat{L}$ in $\hat{M}$ with a sub-bundle of $T\hat{M}$ by requiring its fibre at a point $(p,t) \in \hat{L}$ to be the g–orthogonal subspace of $T_p(j_t(L))$ in T_pM. Let $\tau\colon N\hat{L} \to \hat{M}$ be a tubular map.

Now define a smooth function $\hat{H}\colon N\hat{L} \to \mathbb{R}$ as follows, where (p,t) always denotes a point of $\hat{L} \subset N\hat{L}$.

- $\hat{H}(p,t) = \alpha(X_t)$,
- $d\hat{H}_{(p,t)}(R_\alpha) = 0$,
- $d\hat{H}_{(p,t)}(\mathbf{v}) = -d\alpha(X_t, \mathbf{v})$ for $\mathbf{v} \in \ker\alpha_p \subset T_pM \subset T_{(p,t)}\hat{M}$,
- $\hat{H}$ is linear on the fibres of $N\hat{L} \to \hat{L}$.

Let $\chi\colon \hat{M} \to [0,1]$ be a smooth function with $\chi \equiv 0$ outside a small neighbourhood $\mathcal{N}_0 \subset \tau(N\hat{L})$ of $\hat{L}$ and $\chi \equiv 1$ in a smaller neighbourhood $\mathcal{N}_1 \subset \mathcal{N}_0$ of $\hat{L}$. For $(p,t) \in \hat{M}$, set

$$\widetilde{H}_t(p) = \begin{cases} \chi(p,t)\hat{H}(\tau^{-1}(p,t)) & \text{for } (p,t) \in \tau(N\hat{L}), \\ 0 & \text{for } (p,t) \notin \tau(N\hat{L}). \end{cases}$$

This is smooth in p and t, and the Hamiltonian flow ψ_t of $\widetilde{H}_t$ (defined globally since $\widetilde{H}_t$ is compactly supported) is the desired contact isotopy. □

As a special case of this theorem, we may take L to be a point. This constitutes a homogeneity result for contact manifolds.

Corollary 2.6.3 *Let (M, ξ) be a connected contact manifold and $p, q \in M$ two (not necessarily distinct) points of M. Let $\gamma\colon [0,1] \to M$ be a smooth path connecting $p = \gamma(0)$ with $q = \gamma(1)$. Then there is a compactly supported contact isotopy $(\psi_t)_{t\in[0,1]}$ of M with $\psi_t(p) = \gamma(t)$. In particular, there is a contactomorphism ψ_1 with $\psi_1(p) = q$. In other words, the group of contactomorphisms of (M, ξ) acts transitively on M.* □

Proofs of this result (without the specification of a curve γ) were first given by Hatakeyama [124] and Boothby [29].

There is also a version of this corollary for a parametric family of contact structures, at least for closed manifolds.

Proposition 2.6.4 *Let ξ_t, $t \in [0,1]$, be a smooth family of contact structures on a closed manifold M, and $\gamma\colon [0,1] \to M$ a smooth path in M. Then there is an isotopy ψ_t of M with $T\psi_t(\xi_0) = \xi_t$ and $\psi_t(\gamma(0)) = \gamma(t)$ for all $t \in [0,1]$.*

Proof By Gray stability (Thm. 2.2.2) there is an isotopy ϕ_t of M with $\xi_t = T\phi_t(\xi_0)$. By the preceding corollary, there is a contact isotopy ψ_t^0 of (M, ξ_0) with $\psi_t^0(\gamma(0)) = (\phi_t)^{-1}(\gamma(t))$. Then $\psi_t := \phi_t \circ \psi_t^0$ is the desired isotopy. □

If M is not closed, we cannot appeal to Gray stability, but we can still prove a local version of the preceding proposition. Write B_{st} for the closed unit ball in $\mathbb{R}^{2n+1}$ with its standard contact structure; the origin of $\mathbb{R}^{2n+1}$ will be denoted by $\mathbf{0}$. Here $2n+1$ is the dimension of M. Recall that by a contact embedding of B_{st} we actually mean such an embedding defined on a small open neighbourhood of B_{st} in $(\mathbb{R}^{2n+1}, \xi_{\mathrm{st}})$.

Lemma 2.6.5 *Let ξ_t, $t \in [0,1]$, be a smooth family of contact structures on a manifold M, and $\gamma\colon [0,1] \to M$ a smooth path in M. Then there is a smooth family of contact embeddings $j_t\colon B_{\mathrm{st}} \to (M, \xi_t)$ with $j_t(\mathbf{0}) = \gamma(t)$ for all $t \in [0,1]$.*

Proof Let ϕ_t be an isotopy of M with $\phi_t(\gamma(0)) = \gamma(t)$. (Simply extend the velocity vector field $\dot\gamma$ along γ to a compactly supported time-dependent vector field on M, and define ϕ_t to be the globally defined flow of this vector field.) Choose a smooth family of contact forms α_t with $\xi_t = \ker\alpha_t$. By precomposing ϕ_t with an isotopy fixing $p := \gamma(0)$ we may assume that $\phi_t^*\alpha_t|_{T_pM} = \alpha_0|_{T_pM}$ and $\phi_t^*(d\alpha_t)|_{T_pM} = d\alpha_0|_{T_pM}$, see the bundle-theoretic arguments in the proof of Theorem 2.5.8. By the argument in the proof of Darboux's theorem (Thm. 2.5.1), we find an isotopy ψ_t of a neighbourhood U of $\gamma(0) \in M$ that fixes this point and satisfies $\psi_t^* \phi_t^* \alpha_t|_U = \alpha_0|_U$. Let $j_0 \colon B_{\mathrm{st}} \to (U, \xi_0)$ be a contact embedding sending $\mathbf{0}$ to $\gamma(0)$, given by a Darboux chart around $\gamma(0)$. Then $j_t := \phi_t \circ \psi_t \circ j_0$ is the desired family of contact embeddings. □

In the above lemma and its proof we assumed implicitly that M is without boundary. However, with the obvious minor modification in the statement, the lemma also holds for manifolds with boundary, since a contact structure can always be extended over a thin collar attached to the boundary (by the openness of the contact condition).

As a further generalisation, we now want to adapt this lemma to a parametric family of paths. This will be relevant for the classification of so-called overtwisted contact structures on 3–manifolds (Section 4.7).

Lemma 2.6.6 *Let ξ_t, $t \in [0,1]$, be a smooth family of contact structures on a manifold M and $\gamma_t \colon [a,b] \to M$, $t \in [0,1]$, a smooth family of paths. Then there exists a partition $a = s_0 < s_1 < \ldots < s_k = b$ of the interval $[a,b]$ and smooth families $j_t^i \colon B_{\mathrm{st}} \to (M, \xi_t)$, $t \in [0,1]$, $i = 1, \ldots, k$, of contact embeddings such that $\gamma_t([s_{i-1}, s_i]) \subset j_t^i(B_{\mathrm{st}})$.*

Proof For each fixed $s \in [a,b]$ there is – by the preceding lemma, applied to the path $t \mapsto \gamma_t(s)$ – a family of contact embeddings

$$j_t^{(s)} \colon B_{\mathrm{st}} \longrightarrow (M, \xi_t), \quad t \in [0,1],$$

with $j_t^{(s)}(\mathbf{0}) = \gamma_t(s)$. The lemma then follows by a simple compactness argument (first in the parameter t, then in the parameter s). □

2.6.2 The contact disc theorem

In this section we record a further observation about contact embeddings of the ball B_{st}. It generalises the disc theorem [132, Thm. 8.3.1] from differential topology and has corresponding consequences, e.g. the uniqueness up to contactomorphism of the connected sum operation for contact manifolds, as described in Chapter 6 below.

Theorem 2.6.7 (Contact disc theorem) *Any two contact embeddings j_i, $i = 0, 1$, of B_{st} into a connected contact manifold (M, ξ) (with $\dim B_{\mathrm{st}} = \dim M = 2n+1$) are contact isotopic. Here it is understood that ξ_{st} and ξ are cooriented, and the j_i are meant to preserve this coorientation, i.e. $Tj_i(\xi_{\mathrm{st}}) = \xi$ as cooriented hyperplane fields.*

Remark 2.6.8 An isotopy j_t, $t \in [0,1]$, of contact embeddings of the closed unit ball B_{st} is, *per definitionem*, an isotopy of contact embeddings into (M,ξ) of some open neighbourhood U of B_{st} in $(\mathbb{R}^{2n+1}, \xi_{\mathrm{st}})$. This defines a time-dependent Hamiltonian vector field X_t over $j_t(U)$, and hence a Hamiltonian function $H_t = \alpha(X_t)$, given a contact form α defining $\xi = \ker \alpha$. One can then find a smooth family $\widetilde{H}_t$ of smooth functions defined on all of M, with $\widetilde{H}_t = H_t$ on $j_t(U')$, with $U' \subset U$ a slightly smaller open neighbourhood of B_{st}. This means that any isotopy of contact embeddings of B_{st} extends to an ambient contact isotopy, defined by the flow of the Hamiltonian vector field $X_{\widetilde{H}_t}$.

The proof of this theorem will take up the rest of this section. The only point where this proof differs significantly from the usual proof of the (topological) disc theorem is the argument for showing that any contact embedding of B_{st} into $(\mathbb{R}^{2n+1}, \xi_{\mathrm{st}})$ is isotopic to its 'linearisation' (suitably understood). Here we have to work with a version of the Alexander trick adapted to the contact setting. Throughout this section we write the standard contact structure ξ_{st} in the form

$$\xi_{\mathrm{st}} = \ker(dz + \mathbf{x}\, d\mathbf{y}).$$

Here $\mathbf{x}, \mathbf{y}$ stand for $(x_1, \ldots, x_n)$ and $(y_1, \ldots, y_n)$ respectively, and $\mathbf{x}\, d\mathbf{y}$ has to be read as $\sum_{k=1}^n x_k\, dy_k$. The same applies to analogous expressions below. In the discussion that follows I shall use $\mathbb{R}^{2n+1}_{\mathrm{st}}$ as shorthand notation for $(\mathbb{R}^{2n+1}, \xi_{\mathrm{st}})$.

Instrumental at various stages in the argument will be the contact dilation

$$\delta_t(\mathbf{x}, \mathbf{y}, z) := (t\mathbf{x}, t\mathbf{y}, t^2 z), \quad t \in \mathbb{R}^+,$$

which is obviously a contactomorphism for ξ_{st}.

(1) Let $j_i \colon B_{\mathrm{st}} \to (M, \xi)$, $i = 0, 1$, be two contact embeddings. Since M is connected, we may, thanks to Corollary 2.6.3, assume after a first contact isotopy that $j_0(\mathbf{0}) = j_1(\mathbf{0})$.

(2) According to Remark 2.5.2 we can find a neighbourhood of $j_0(\mathbf{0}) = j_1(\mathbf{0})$ contactomorphic to $\mathbb{R}^{2n+1}_{\mathrm{st}}$. After isotopies of the form

$$(\mathbf{x}, \mathbf{y}, z; t) \longmapsto j_i \circ \delta_{(1-t+t\varepsilon)}(\mathbf{x}, \mathbf{y}, z),\ i = 0, 1,$$

with $\varepsilon > 0$ chosen sufficiently small, we may assume that both j_0 and j_1 map into that neighbourhood, so that we may regard them as contact embeddings

$$(B_{\mathrm{st}}, \mathbf{0}) \longrightarrow (\mathbb{R}^{2n+1}_{\mathrm{st}}, \mathbf{0}).$$

(3) Given a contact embedding j as just described, we now want to find a contact isotopy to its 'linearisation'. The following considerations will serve

to find the candidate for this 'linearisation'. Observe that the differential $T_0 j$ has no reason, in general, to be a contactomorphism of $\mathbb{R}^{2n+1}_{\mathrm{st}}$. Write j in the form

$$(\mathbf{x}, \mathbf{y}, z) \longmapsto \big(\mathbf{X}(\mathbf{x}, \mathbf{y}, z), \mathbf{Y}(\mathbf{x}, \mathbf{y}, z), Z(\mathbf{x}, \mathbf{y}, z)\big).$$

Then the condition for j to be a contact embedding becomes

$$dZ + \mathbf{X}\, d\mathbf{Y} = \lambda(dz + \mathbf{x}\, d\mathbf{y}),$$

with $\lambda\colon B_{\mathrm{st}} \to \mathbb{R}^+$. This can be rewritten as the following system of equations:

$$\begin{cases} \dfrac{\partial Z}{\partial x_k} + \mathbf{X}\dfrac{\partial \mathbf{Y}}{\partial x_k} & = & 0, & k = 1, \ldots, n, \\ \dfrac{\partial Z}{\partial y_k} + \mathbf{X}\dfrac{\partial \mathbf{Y}}{\partial y_k} & = & \lambda x_k, & k = 1, \ldots, n, \\ \dfrac{\partial Z}{\partial z} + \mathbf{X}\dfrac{\partial \mathbf{Y}}{\partial z} & = & \lambda. \end{cases}$$

By taking various partial derivatives of these equations and evaluating at $\mathbf{0}$ we find (for all $k, l = 1, \ldots, n$ and with δ_{kl} denoting the Kronecker symbol)

$$\begin{cases} \dfrac{\partial^2 Z}{\partial x_k\, \partial x_l}(\mathbf{0}) + \dfrac{\partial \mathbf{X}}{\partial x_k}(\mathbf{0})\dfrac{\partial \mathbf{Y}}{\partial x_l}(\mathbf{0}) & = & 0, \\ \dfrac{\partial^2 Z}{\partial y_k\, \partial y_l}(\mathbf{0}) + \dfrac{\partial \mathbf{X}}{\partial y_k}(\mathbf{0})\dfrac{\partial \mathbf{Y}}{\partial y_l}(\mathbf{0}) & = & 0, \\ \dfrac{\partial^2 Z}{\partial x_k\, \partial y_l}(\mathbf{0}) + \dfrac{\partial \mathbf{X}}{\partial y_l}(\mathbf{0})\dfrac{\partial \mathbf{Y}}{\partial x_k}(\mathbf{0}) & = & 0, \\ \dfrac{\partial^2 Z}{\partial x_k\, \partial y_l}(\mathbf{0}) + \dfrac{\partial \mathbf{X}}{\partial x_k}(\mathbf{0})\dfrac{\partial \mathbf{Y}}{\partial y_l}(\mathbf{0}) & = & \lambda(\mathbf{0})\delta_{kl}. \end{cases}$$

Notice that from the third equation of the first system we have

$$\frac{\partial Z}{\partial z}(\mathbf{0}) = \lambda(\mathbf{0}).$$

Interpret $\mathbf{X}, \mathbf{Y}$ as column vectors, and form the $n \times n$ matrices

$$\begin{aligned} A &= \left(\frac{\partial \mathbf{X}}{\partial x_1}(\mathbf{0}), \ldots, \frac{\partial \mathbf{X}}{\partial x_n}(\mathbf{0})\right) = \left(\frac{\partial \mathbf{X}}{\partial \mathbf{x}}(\mathbf{0})\right), \\ B &= \left(\frac{\partial \mathbf{X}}{\partial y_1}(\mathbf{0}), \ldots, \frac{\partial \mathbf{X}}{\partial y_n}(\mathbf{0})\right) = \left(\frac{\partial \mathbf{X}}{\partial \mathbf{y}}(\mathbf{0})\right), \\ C &= \left(\frac{\partial \mathbf{Y}}{\partial x_1}(\mathbf{0}), \ldots, \frac{\partial \mathbf{Y}}{\partial x_n}(\mathbf{0})\right) = \left(\frac{\partial \mathbf{Y}}{\partial \mathbf{x}}(\mathbf{0})\right), \\ D &= \left(\frac{\partial \mathbf{Y}}{\partial y_1}(\mathbf{0}), \ldots, \frac{\partial \mathbf{Y}}{\partial y_n}(\mathbf{0})\right) = \left(\frac{\partial \mathbf{Y}}{\partial \mathbf{y}}(\mathbf{0})\right). \end{aligned}$$

Since the second partial derivatives of Z commute, the above system gives

$$\left\{\begin{array}{rcl} A^t C & = & C^t A, \\ B^t D & = & D^t B, \\ A^t D - C^t B & = & \lambda(\mathbf{0}) I_n, \end{array}\right.$$

with I_n denoting the $n \times n$ identity matrix. It is a straightforward check that this is equivalent to saying that the linear map

$$\begin{pmatrix} \mathbf{x} \\ \mathbf{y} \end{pmatrix} \longmapsto \begin{pmatrix} A & B \\ C & D \end{pmatrix} \begin{pmatrix} \mathbf{x} \\ \mathbf{y} \end{pmatrix}$$

is (positively) conformally symplectic for the standard symplectic structure on $\mathbb{R}^{2n}$ given by

$$\Omega\left(\begin{pmatrix} \mathbf{x} \\ \mathbf{y} \end{pmatrix}, \begin{pmatrix} \mathbf{x}' \\ \mathbf{y}' \end{pmatrix}\right) = \mathbf{x}\mathbf{y}' - \mathbf{y}\mathbf{x}'.$$

We write $\mathrm{CSp}^+(2n)$ for the group of such $2n \times 2n$ matrices. Since $\mu I_{2n} \in \mathrm{CSp}^+(2n)$ for $\mu \in \mathbb{R} \setminus \{0\}$, and the symplectic group $\mathrm{Sp}(2n)$ of linear maps preserving Ω is connected, see [177, Prop. 2.22], the group $\mathrm{CSp}^+(2n)$ is likewise connected.

Observe that the first partial derivative of Z with respect to z and the second partial derivatives of Z with respect to the x– and y–coordinates (all derivatives taken at the point $\mathbf{0}$) are completely determined by this linear conformally symplectic map.

Lemma 2.6.9 *Let* $\left(\begin{smallmatrix} A & B \\ C & D \end{smallmatrix}\right)$ *be an element of* $\mathrm{CSp}^+(2n)$*, with conformality factor* $\lambda_0 \in \mathbb{R}^+$*. Then the map given by*

$$\begin{array}{rcl} \mathbf{X} & = & A\mathbf{x} + B\mathbf{y} \\ \mathbf{Y} & = & C\mathbf{x} + D\mathbf{y} \\ Z & = & \lambda_0 z - \frac{1}{2}\mathbf{x}^t A^t C\mathbf{x} - \mathbf{x}^t C^t B\mathbf{y} - \frac{1}{2}\mathbf{y}^t B^t D\mathbf{y} \end{array}$$

is a contactomorphism of $\mathbb{R}^{2n+1}_{\mathrm{st}}$.

Proof The map is bijective and its Jacobian is easily seen to have full rank at every point, so the map is a diffeomorphism of $\mathbb{R}^{2n+1}$. Moreover, a straightforward computation gives $dZ + \mathbf{X}\, d\mathbf{Y} = \lambda_0(dz + \mathbf{x}\, d\mathbf{y})$. □

Now return to the given contact embedding $j\colon (B_{\mathrm{st}}, \mathbf{0}) \to (\mathbb{R}^{2n+1}_{\mathrm{st}}, \mathbf{0})$, $(\mathbf{x}, \mathbf{y}, z) \mapsto (\mathbf{X}, \mathbf{Y}, Z)$. Let $S_0 j$ be the contact embedding of B_{st} given by the map from the foregoing lemma, with the conformally symplectic matrix determined by the partial derivatives of j at $\mathbf{0}$ as in the discussion preceding the lemma.

Proposition 2.6.10 (Contact Alexander trick) *There is an isotopy of contact embeddings $(B_{\mathrm{st}}, \mathbf{0}) \to (\mathbb{R}^{2n+1}_{\mathrm{st}}, \mathbf{0})$ between $S_0 j$ and j, given by*

$$(\mathbf{x}, \mathbf{y}, z; t) \longmapsto \begin{cases} \delta_t^{-1} \circ j \circ \delta_t(\mathbf{x}, \mathbf{y}, z) & \text{for } t \in (0,1], \\ S_0 j(\mathbf{x}, \mathbf{y}, z) & \text{for } t = 0. \end{cases}$$

Proof Recall the lemma of Morse, see [153, p. 226], which allows us to write $\mathbf{X}(\mathbf{x}, \mathbf{y}, z)$ in the form

$$\mathbf{X}(\mathbf{x}, \mathbf{y}, z) = A(\mathbf{x}, \mathbf{y}, z)\mathbf{x} + B(\mathbf{x}, \mathbf{y}, z)\mathbf{y} + \mathbf{c}(\mathbf{x}, \mathbf{y}, z)z$$

with $A(\mathbf{0}) = \left(\frac{\partial \mathbf{X}}{\partial \mathbf{x}}(\mathbf{0})\right)$, $B(\mathbf{0}) = \left(\frac{\partial \mathbf{X}}{\partial \mathbf{y}}(\mathbf{0})\right)$, and $\mathbf{c}(\mathbf{0}) = \frac{\partial \mathbf{X}}{\partial z}(\mathbf{0})$. Indeed, simply work out the differentiation in the integrand of

$$\mathbf{X}(\mathbf{x}, \mathbf{y}, z) = \int_0^1 \frac{d}{ds} \mathbf{X}(s\mathbf{x}, s\mathbf{y}, sz)\, ds.$$

For $\mathbf{Y}$ we have a similar expression.

The contact condition implies the vanishing of the first partial derivatives of Z with respect to the x– and y–coordinates at $\mathbf{0}$. So we can iterate the lemma of Morse on these derivatives and obtain an expression for Z containing a linear term in z and quadratic terms in all variables, again with variable coefficients which give the corresponding partial derivatives of Z when evaluated at $\mathbf{0}$.

When we conjugate these expressions with δ_t, we see that in the expressions for $\mathbf{X}$ and $\mathbf{Y}$ only the linear terms in $\mathbf{x}$ and $\mathbf{y}$ survive as t goes to 0; likewise, in the expression for Z only the linear term in z and the quadratic terms in $\mathbf{x}$ and $\mathbf{y}$ survive. From the contact condition one finds that these linear/quadratic expressions are indeed the ones that define $S_0 j$. □

(4) Finally, we want to construct a contact isotopy from $S_0 j$ to the inclusion map $B_{\mathrm{st}} \subset \mathbb{R}^{2n+1}_{\mathrm{st}}$. I mentioned before that the group $\mathrm{CSp}^+(2n)$ is connected. Therefore, any two contact embeddings $B_{\mathrm{st}} \to \mathbb{R}^{2n+1}_{\mathrm{st}}$ as in Lemma 2.6.9 can indeed be connected by an isotopy via contact embeddings of that type. This concludes the proof of Theorem 2.6.7.

2.6.3 Contact submanifolds

In the preceding section we spoke of contact embeddings of B_{st} into an equidimensional contact manifold. We now extend this concept by dropping the condition on dimensions.

Definition 2.6.11 An embedding $j\colon (M',\xi') \to (M,\xi)$ is called a **contact embedding** if $(j(M'), Tj(\xi'))$ is a contact submanifold of (M,ξ), i.e.

$$T(j(M')) \cap \xi|_{j(M')} = Tj(\xi').$$

If $\xi = \ker\alpha$, this can be reformulated as $\ker j^*\alpha = \xi'$.

Theorem 2.6.12 *Let $j_t\colon (M',\xi') \to (M,\xi)$, $t \in [0,1]$, be an isotopy of contact embeddings of the closed contact manifold (M',ξ') in the contact manifold (M,ξ). Then there is a compactly supported contact isotopy ψ_t of (M,ξ) with $\psi_t \circ j_0 = j_t$.*

Proof We follow a slightly different strategy from the one in the isotropic case. Instead of directly finding an extension of the Hamiltonian function $H_t\colon j_t(M') \to \mathbb{R}$, we first use the neighbourhood theorem for contact submanifolds to extend j_t to an isotopy of contact embeddings of tubular neighbourhoods. Again we assume that M' is a submanifold of M and j_0 the inclusion $M' \subset M$. As earlier, NM' denotes the normal bundle of M' in M. We also identify M' with the zero section of NM', and we use the canonical identification

$$T(NM')|_{M'} = TM' \oplus NM'.$$

By the usual isotopy extension theorem from differential topology we find an isotopy

$$\phi_t\colon NM' \to M$$

with $\phi_t|_{M'} = j_t$.

Choose contact forms α, α' defining ξ and ξ', respectively. Define $\alpha_t = \phi_t^*\alpha$. Then $TM' \cap \ker\alpha_t = \xi'$. Let R' denote the Reeb vector field of α'. Analogous to the proof of Theorem 2.5.15, we first find a smooth family of smooth functions $g_t\colon M' \to \mathbb{R}^+$ such that $g_t\alpha_t|_{TM'} = \alpha'$, and then a family $f_t\colon NM' \to \mathbb{R}^+$ with $f_t|_{M'} \equiv 1$ and

$$df_t = i_{R'}d(g_t\alpha_t) \;\text{ on }\; T(NM')|_{M'}.$$

Then $\beta_t = f_t g_t \alpha_t$ is a family of contact forms on NM' representing the contact structures $\ker(\phi_t^*\alpha)$, and with the properties

$$\begin{aligned} \beta_t|_{TM'} &= \alpha', \\ d\beta_t|_{TM'} &= d\alpha', \\ \ker(d\beta_t) &= \langle R'\rangle \text{ along } M'. \end{aligned}$$

The family $(NM', d\beta_t)$ of symplectic vector bundles may be thought of as a symplectic vector bundle over $M' \times [0,1]$, which is necessarily isomorphic to a bundle pulled back from $M' \times \{0\}$ (see [140, Cor. 3.4.4]). In other words, there is a smooth family of symplectic bundle isomorphisms

$$\Phi_t\colon (NM', d\beta_0) \longrightarrow (NM', d\beta_t).$$

Then

$$\mathrm{id}_{TM'} \oplus \Phi_t\colon T(NM')|_{M'} \longrightarrow T(NM')|_{M'}$$

is a bundle map that pulls back β_t to β_0 and $d\beta_t$ to $d\beta_0$.

By the now familiar stability argument we find a smooth family of embeddings

$$\varphi_t\colon \mathcal{N}(M') \longrightarrow NM'$$

for some neighbourhood $\mathcal{N}(M')$ of the zero section M' in NM' with $\varphi_0 =$ inclusion, $\varphi_t|_{M'} = \mathrm{id}_{M'}$ and $\varphi_t^*\beta_t = \lambda_t\beta_0$, where $\lambda_t\colon \mathcal{N}(M') \to \mathbb{R}^+$. This means that

$$\phi_t \circ \varphi_t\colon \mathcal{N}(M') \longrightarrow M$$

is a smooth family of contact embeddings of $(\mathcal{N}(M'), \ker\beta_0)$ in (M,ξ).

Define a time-dependent vector field X_t along $\phi_t \circ \varphi_t(\mathcal{N}(M'))$ by

$$X_t \circ \phi_t \circ \varphi_t = \frac{d}{dt}(\phi_t \circ \varphi_t).$$

This X_t is clearly an infinitesimal automorphism of ξ: by differentiating the equation $\varphi_t^*\phi_t^*\alpha = \mu_t\phi_0^*\alpha$ (where $\mu_t\colon \mathcal{N}(M') \to \mathbb{R}^+$) with respect to t we get

$$\varphi_t^*\phi_t^*(\mathcal{L}_{X_t}\alpha) = \dot{\mu}_t\phi_0^*\alpha = \frac{\dot{\mu}_t}{\mu_t}\varphi_t^*\phi_t^*\alpha,$$

so $\mathcal{L}_{X_t}\alpha$ is a multiple of α (since $\phi_t \circ \varphi_t$ is a diffeomorphism onto its image).

By the theory of contact Hamiltonians, X_t is the Hamiltonian vector field of a Hamiltonian function $\hat{H}_t$ defined on $\phi_t\circ\varphi_t(\mathcal{N}(M'))$. Cut off this function with a bump function so as to obtain $H_t\colon M \to \mathbb{R}$ with $H_t \equiv \hat{H}_t$ near $\phi_t\circ\varphi_t(M')$ and $H_t \equiv 0$ outside a slightly larger neighbourhood of $\phi_t\circ\varphi_t(M')$. Then the Hamiltonian flow ψ_t of H_t satisfies our requirements. □

2.6.4 Surfaces in 3–manifolds

Theorem 2.6.13 *Let* $j_t\colon S \to (M, \xi = \ker\alpha)$, $t \in [0,1]$, *be an isotopy of embeddings of a closed surface* S *in a 3–dimensional contact manifold* (M,ξ). *If all* j_t *induce the same characteristic foliation on* S, *then there is a compactly supported isotopy* $\psi_t\colon M \to M$ *with* $\psi_t \circ j_0 = j_t$.

Proof Extend j_t to a smooth family of embeddings $\phi_t\colon S\times\mathbb{R}\rightarrow M$, and identify S with $S\times\{0\}$. The assumptions say that all $\phi_t^*\alpha$ induce the same characteristic foliation on S. By the proof of Theorem 2.5.23 and in analogy with the proof of Theorem 2.6.12 we find a smooth family of embeddings

$$\varphi_t\colon S\times(-\varepsilon,\varepsilon)\longrightarrow S\times\mathbb{R}$$

for some $\varepsilon>0$ with $\varphi_0=$ inclusion, $\varphi_t|_{S\times\{0\}}=\mathrm{id}_S$ and $\varphi_t^*\phi_t^*\alpha=\lambda_t\phi_0^*\alpha$, where $\lambda_t\colon S\times(-\varepsilon,\varepsilon)\rightarrow\mathbb{R}^+$. In other words, $\phi_t\circ\varphi_t$ is a smooth family of contact embeddings of $(S\times(-\varepsilon,\varepsilon),\ker\phi_0^*\alpha)$ in (M,ξ).

The proof now concludes exactly as the proof of Theorem 2.6.12. □

3
Knots in contact 3–manifolds

'Quem faz nós, quem os desata'
José Saramago,
Fado Adivinha

The study of knots in 3–manifolds is a venerable subject in its own right, the main question being the classification of knots in 3–space up to isotopy. But knots also play an important supporting role in the geometric topology of 3–manifolds, for instance via the surgery construction discussed in Chapter 1.

When we turn to contact 3–manifolds, the situation is analogous. Here there are two distinguished classes of knots: the Legendrian and the transverse ones. Again, these are interesting objects in their own right, and much can be said about the classification of these knots (up to isotopy through knots of the given type). A comprehensive survey about these matters is given by Etnyre [82].

In the present chapter I shall concentrate on those aspects of the theory of Legendrian and transverse knots that are relevant for the construction and classification of contact 3–manifolds via surgical constructions along such knots.

Section 3.1 contains the basic definitions. In Section 3.2 we discuss how Legendrian and transverse knots can be visualised by two kinds of 2–dimensional projections adapted to the contact structure, *viz.*, the front and the Lagrangian projection. These will be used in Section 3.3 for proving that every knot can be approximated both by a Legendrian or by a transverse knot in the same topological isotopy class. Then follows a topological interlude in Section 3.4 on linking numbers and the representation of homology cycles by submanifolds. Finally, Section 3.5 introduces the classical invariants, *viz.*, the Thurston–Bennequin invariant and rotation number of Legendrian knots, and the self-linking number of transverse knots. We discuss relations between these invariants and compute them from the knot projections.

3.1 Legendrian and transverse knots

Recall that a knot in a 3–manifold M is simply an embedding of S^1. (I shall try to be as consistent as possible in distinguishing the actual embedding from its image; usually I write something like γ for the former and K for the latter.) A large part of our discussion, for the time being, will centre on curves parametrised on intervals. I usually write s for a parameter ranging over an interval, and θ for a parameter ranging over $S^1 \equiv \mathbb{R}/2\pi\mathbb{Z}$. We write $\gamma'(s) = T_s\gamma(\partial_s)$ for the tangent or 'velocity' vector of the parametrised curve $s \mapsto \gamma(s)$ at the point $\gamma(s)$.

We have already introduced Legendrian submanifolds of contact manifolds in full generality, so it is clear what is meant by a Legendrian knot. For better reference, I summarise the definitions in the present setting.

Definition 3.1.1 A **Legendrian knot** in a contact 3–manifold (M,ξ) is a Legendrian embedding $\gamma\colon\ S^1 \to M$, that is, an embedding satisfying $\gamma'(\theta) \in \xi_{\gamma(\theta)}$ for all $\theta \in S^1$.

A **transverse knot** in (M,ξ) is an embedding $\gamma\colon\ S^1 \to M$ that is everywhere transverse to ξ, i.e. we require $\gamma'(\theta) \notin \xi_{\gamma(\theta)}$ for all $\theta \in S^1$. If $\xi = \ker\alpha$ is cooriented, one speaks of a **positively** or **negatively transverse knot** depending on whether $\alpha(\gamma'(\theta)) > 0$ or $\alpha(\gamma'(\theta)) < 0$ for all $\theta \in S^1$.

Remark 3.1.2 As always, I adhere to J. H. C. Whitehead's dictum, quoted after [132]: '*Transversal* is a noun; the adjective is *transverse*.'

We shall frequently have to associate with some given Legendrian knot a transverse or a second Legendrian knot (where this associated knot ought to be canonically defined up to transverse or Legendrian isotopy, respectively). This happens, for instance, when we want to prove the analogue for transverse knots of a result already proved for Legendrian knots (see the proof of the transverse case of Theorem 3.3.1 on page 103 below), or when we consider surgeries along such canonically associated knots (see Proposition 6.4.5). For better reference, I collect the relevant definitions here.

Let $\gamma\colon\ S^1 \to K \subset M$ be a Legendrian knot. By Example 2.5.10, a neighbourhood of K in (M,ξ) is contactomorphic to a neighbourhood of the Legendrian knot

$$\theta \longmapsto (\theta, x=0, y=0) \in S^1 \times \mathbb{R}^2,\ \ \theta \in S^1,$$

with contact structure on $S^1 \times \mathbb{R}^2$ given by the contact form

$$\alpha = \cos\theta\, dx - \sin\theta\, dy.$$

We identify γ with this model Legendrian S^1. Observe that the radial vector field $X := x\,\partial_x + y\,\partial_y$ is a contact vector field by Lemma 1.5.8, since $\mathcal{L}_X \alpha = \alpha$, as one computes easily with the help of the Cartan formula.

On the boundary torus $S := \{x^2 + y^2 = \delta\}$, for $\delta \in \mathbb{R}^+$ small, of a thin cylinder around K, there are two distinguished curves† where X is tangent to ξ, namely,

$$\gamma_\pm(\theta) := (\theta, x = \pm\delta \sin\theta, y = \pm\delta\cos\theta), \quad \theta \in S^1.$$

Observe that $\gamma_\pm$ is obtained from γ by pushing it in the direction of the vector field $\pm(\sin\theta\,\partial_x + \cos\theta\,\partial_y)$ tangent to ξ (but transverse to γ'). Moreover, we have $\alpha(\gamma'_\pm) = \pm\delta$. Different choices of $\delta \in \mathbb{R}^+$ yield knots that are isotopic as transverse knots. We therefore call $\gamma_\pm$ *the* positive or negative, respectively, **transverse push-off** of γ.

The torus S is foliated by the Legendrian curves

$$\theta \longmapsto (\theta, x_0, y_0), \quad \theta \in S^1,$$

with $x_0^2 + y_0^2 = \delta$. But there are two further distinguished Legendrian curves on S, along which the contact planes are tangent to S. These are given by

$$\gamma_{\mathrm{L}}(\theta) := (\theta, x = \pm\delta\cos\theta, \mp\delta\sin\theta), \quad \theta \in S^1.$$

Notice that these Legendrian curves lie in the same homotopy class (as curves on S) as $\gamma_\pm$. We call either of them a **Legendrian push-off** of K. By varying the parameter δ we see that these Legendrian push-offs are Legendrian isotopic to K.

Finally, given a transverse knot in a contact 3–manifold, we may think of it (by Example 2.5.16) as the knot

$$\gamma_{\mathrm{T}}(\theta) := (\theta, x = 0, y = 0), \quad \theta \in S^1$$

in $S^1 \times \mathbb{R}^2$ with contact structure given by $d\theta + x\,dy - y\,dx = 0$. This has Legendrian push-offs, but in contrast with the previous cases there is no canonical choice. For instance, for any $k \in \mathbb{Z}^*$ we can form the Legendrian push-off

$$\theta \longmapsto \left(\theta, x = \frac{1}{k}\cos(k^2\theta), y = -\frac{1}{k}\sin(k^2\theta)\right), \quad \theta \in S^1.$$

3.2 Front and Lagrangian projection

For questions such as surgery descriptions of contact 3–manifolds, the (Legendrian or transverse) knots of primary interest are those in S^3 with its standard contact structure. Given any link in S^3, we may assume that it

† These curves will later play a prominent role in the theory of so-called convex surfaces, see Sections 4.6.2 and 4.8.

misses a given point and regard it as a link in $\mathbb{R}^3$. By Proposition 2.1.8 the same is true in contact geometry: any link in S^3 with its standard contact structure may be regarded as a link in $\mathbb{R}^3$ with its standard contact structure $\xi_{\mathrm{st}} = \ker \alpha_{\mathrm{st}}$, where

$$\alpha_{\mathrm{st}} = dz + x\, dy.$$

Remark 3.2.1 The reader should be warned that almost every possible convention for writing this standard contact structure – with a minus instead of a plus sign, or with the roles of x and y reversed – has been used in the literature. Therefore, the knot diagrams we are going to draw below will have a different appearance in other books and papers, depending on whichever school of thought the author follows. I stick to what I believe is one of the more common conventions.

Consider embeddings γ of S^1 or an open interval into $(\mathbb{R}^3, \xi_{\mathrm{st}})$. Write $\gamma(s) = (x(s), y(s), z(s))$. Then

$$\alpha_{\mathrm{st}}(\gamma') = z' + xy',$$

so the condition for a Legendrian curve reads $z' + xy' \equiv 0$; for a positively or negatively transverse curve it becomes $z' + xy' > 0$ or < 0, respectively.

In order to visualise curves or knots in 3–space, one draws their projection onto some plane in $\mathbb{R}^3$. In the contact geometric setting, there are two distinguished types of such projections.

Definition 3.2.2 The **front projection** of a parametrised curve $\gamma(s) = (x(s), y(s), z(s))$ in $(\mathbb{R}^3, \xi_{\mathrm{st}})$ is the curve

$$\gamma_{\mathrm{F}}(s) = (y(s), z(s));$$

its **Lagrangian projection** is the curve

$$\gamma_{\mathrm{L}}(s) = (x(s), y(s)).$$

3.2.1 Legendrian curves

If $\gamma(s) = (x(s), y(s), z(s))$ is a Legendrian curve in $(\mathbb{R}^3, \xi_{\mathrm{st}})$, then $y' = 0$ implies $z' = 0$, so there the front projection has a singular point. Indeed, the curve $s \mapsto (s, 0, 0)$ is an example of a Legendrian curve whose front projection is a single point. We call a Legendrian curve *generic* if $y' = 0$ only holds at isolated points (which we call **cusp points**), and there $y'' \neq 0$.

Lemma 3.2.3 *Let* $\gamma\colon (a,b) \to (\mathbb{R}^3, \xi_{\mathrm{st}})$ *be a Legendrian immersion. Then its front projection* $\gamma_{\mathrm{F}}(s) = (y(s), z(s))$ *does not have any vertical tangencies. Away from the cusp points,* γ *is recovered from its front projection via*

$$x(s) = -\frac{z'(s)}{y'(s)} = -\frac{dz}{dy},$$

i.e. $x(s)$ *is the negative slope of the front projection. The curve* γ *is embedded if and only if* γ_{F} *has only transverse self-intersections.*

By a C^∞*-small perturbation of* γ *we can obtain a generic Legendrian curve* $\widetilde{\gamma}$ *isotopic to* γ*; by a* C^2*-small perturbation we may achieve that the front projection has only semi-cubical cusp singularities, i.e. around a cusp point at* $s = 0$ *the curve* $\widetilde{\gamma}$ *looks like*

$$\widetilde{\gamma}(s) = (s + a, \lambda s^2 + b, -\lambda(2s^3/3 + as^2) + c)$$

with $\lambda \neq 0$*, see Figure 3.1.*

Any regular curve in the (y,z)*–plane with semi-cubical cusps and no vertical tangencies can be lifted to a unique Legendrian curve in* $\mathbb{R}^3$*.*

Fig. 3.1. The cusp of a front projection.

Proof The Legendrian condition is $z' + xy' = 0$. Hence $y' = 0$ forces $z' = 0$, so γ_{F} cannot have any vertical tangencies.

Away from the cusp points, the Legendrian condition tells us how to recover x as the negative slope of the front projection. In particular, a self-intersecting front projection lifts to a non-intersecting curve if and only if the slopes at the intersection point are different, i.e. if and only if the intersection is transverse.

That γ can be approximated in the C^∞–topology by a generic immersion $\widetilde{\gamma}$ follows from the usual transversality theorem (in its simplest form, *viz.*, applied to the function $y(s)$; the function $x(s)$ may be left unchanged, and the new $z(s)$ is then found by integrating the new $-xy'$).

At a cusp point of $\widetilde{\gamma}$ we have $y' = z' = 0$. Since $\widetilde{\gamma}$ is an immersion, this forces $x' \neq 0$, so $\widetilde{\gamma}$ can be parametrised around a cusp point by the x–coordinate, i.e. we may choose the curve parameter s such that the cusp lies at $s = 0$ and $x(s) = s+a$. Since $y''(0) \neq 0$ by the genericity condition, we can write $y(s) = s^2 g(s) + y(0)$ with a smooth function $g(s)$ satisfying $g(0) \neq 0$; this is proved like the lemma of Morse in [153, p. 226]. A C^0–approximation of $g(s)$ by a function $h(s)$ with $h(s) \equiv g(0)$ for s near zero and $h(s) \equiv g(s)$ for $|s|$ greater than some small $\varepsilon > 0$ yields a C^2–approximation of $y(s)$ with the desired form around the cusp point. □

Remark 3.2.4 By continuity, the equation $x = -dz/dy$ also makes sense at generic cusp points. This describes the slope of the tangent line (understood in the obvious sense) at the cusp. So these tangencies cannot be vertical either.

Example 3.2.5 Figure 3.2 shows the left-handed and right-handed trefoil knot. Figure 3.3 gives the front projections of Legendrian realisations of these knots.

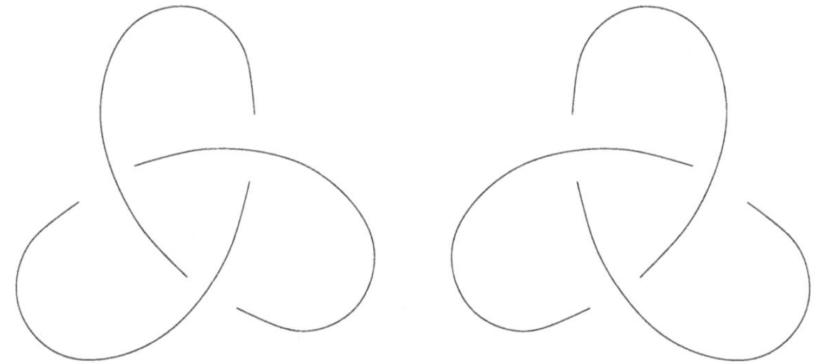

Fig. 3.2. The left and right trefoil knot.

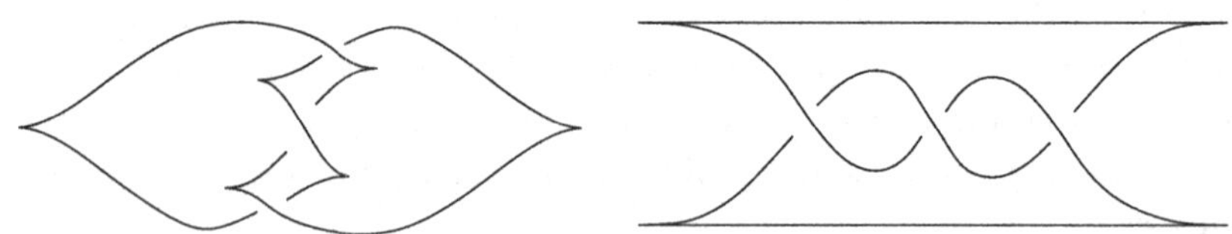

Fig. 3.3. Front projections of Legendrian left and right trefoil.

Here is the corresponding discussion of the Lagrangian projection of Legendrian curves.

Lemma 3.2.6 *Let $\gamma\colon (a,b) \to (\mathbb{R}^3, \xi_{\mathrm{st}})$ be a Legendrian immersion. Then its Lagrangian projection $\gamma_{\mathrm{L}}(s) = (x(s), y(s))$ is also an immersed curve. The curve γ is recovered from γ_{L} via*

$$z(s_1) = z(s_0) - \int_{s_0}^{s_1} x(s)y'(s)\, ds.$$

A Legendrian immersion $\gamma\colon S^1 \to (\mathbb{R}^3, \xi_{\mathrm{st}})$ has a Lagrangian projection that encloses zero area. Moreover, γ is embedded if and only if every loop in γ_{L} (except, in the closed case, the full loop γ_{L}) encloses a non-zero oriented area.

Any curve (defined on an interval) immersed in the (x,y)–plane is the Lagrangian projection of a Legendrian curve in $(\mathbb{R}^3, \xi_{\mathrm{st}})$, unique up to translation in the z–direction. A closed immersed curve γ_{L} in the (x,y)–plane, i.e. an immersion of S^1 in $\mathbb{R}^2$, lifts to a Legendrian immersion of S^1 in $(\mathbb{R}^3, \xi_{\mathrm{st}})$ precisely if $\oint_{\gamma_{\mathrm{L}}} x\, dy = 0$.

Proof The Legendrian condition $z' + xy'$ implies that if $y' = 0$ then $z' = 0$, and hence, since γ is an immersion, $x' \neq 0$. So γ_{L} is an immersion.

The formula for z follows by integrating the Legendrian condition. For a closed curve γ_{L} in the (x,y)–plane, the integral $\oint_{\gamma_{\mathrm{L}}} x\, dy$ computes the oriented area enclosed by γ_{L}. From this observation all the other statements follow. □

Example 3.2.7 Figure 3.4 shows the Lagrangian projection of a Legendrian unknot.

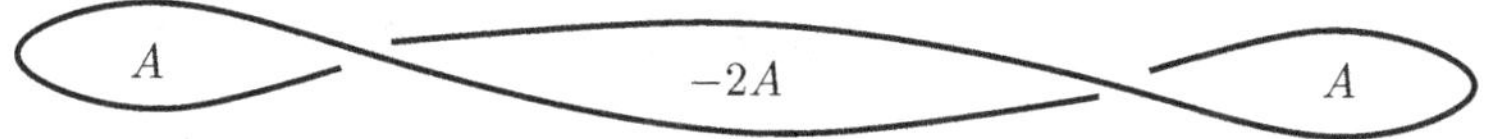

Fig. 3.4. Lagrangian projection of a Legendrian unknot.

The Lagrangian projection has the apparent advantage that it gives rise to immersed curves in $\mathbb{R}^2$ rather than curves with cusp singularities, as is the case with the front projection. However, in most situations the front projection wins hands down, on account of the fact that it allows one to read off directly, even from a qualitative picture, whether the corresponding Legendrian curve is embedded, or whether a front corresponds to a closed curve. In the Lagrangian projection, either of these questions gives rise to an awkward quantitative condition. In Section 6.3.1 we combine both viewpoints in order to derive the classification of Legendre immersions of S^1

in $(\mathbb{R}^3, \xi_{st})$ from the classification of immersions of S^1 in $\mathbb{R}^2$ (known as the Whitney–Graustein theorem) – and vice versa.

3.2.2 Transverse curves

The condition for a parametrised curve $\gamma(s) = (x(s), y(s), z(s))$ in $\mathbb{R}^3$ to be positively transverse to the standard contact structure $\xi = \ker(dz + x\,dy)$ is that $z' + xy' > 0$. Hence,

$$\begin{cases} \text{if } y' = 0, \text{ then } z' > 0, \\ \text{if } y' > 0, \text{ then } x > -z'/y', \\ \text{if } y' < 0, \text{ then } x < -z'/y'. \end{cases}$$

The first statement says that there are no vertical tangencies oriented downwards in the front projection. The second statement says in particular that for $y' > 0$ and $z' < 0$ we have $x > 0$; the third, that for $y' < 0$ and $z' < 0$ we have $x < 0$. This implies that the situations shown in Figure 3.5 are not possible in the front projection of a positively transverse curve. I leave it to the reader to check that all other oriented crossings are possible in the front projection of a positively transverse curve, and that any curve in the (y, z)–plane without the forbidden crossing or downward vertical tangencies admits a lift to a positively transverse curve.

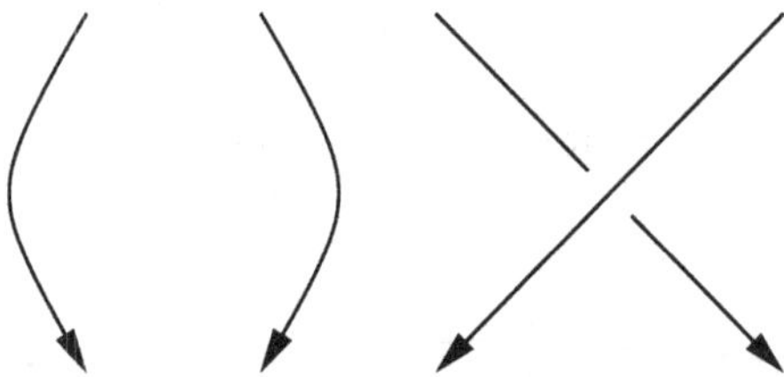

Fig. 3.5. Impossible front projections of positively transverse curve.

Example 3.2.8 Figure 3.6 shows the front projection of a positively transverse trefoil knot.

3.3 Approximation theorems

In this section I present elementary *ad hoc* proofs that every knot in a contact 3–manifold can be C^0–approximated both by a Legendrian and by a (positively or negatively) transverse knot in the same topological isotopy class. As mentioned earlier, there is a general h–principle that allows one to prove

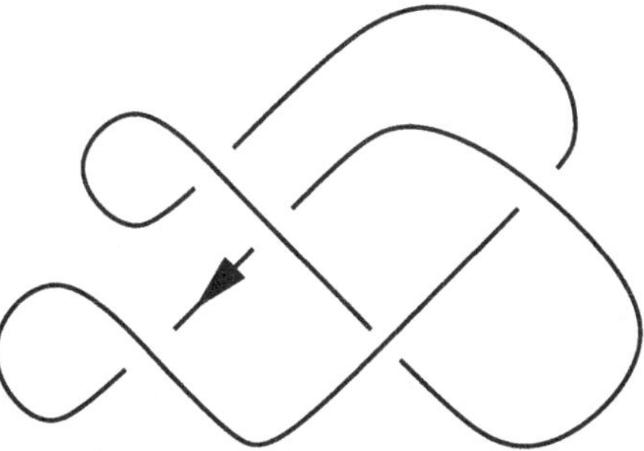

Fig. 3.6. Front projection of a positively transverse trefoil knot.

corresponding statements for isotropic submanifolds in contact manifolds in arbitrary dimension, see Section 6.3.1.

Theorem 3.3.1 *Let $\gamma\colon S^1 \to (M, \xi)$ be a knot in a contact 3–manifold. Then γ can be C^0–approximated by a Legendrian knot isotopic to γ. If ξ is cooriented, γ can be C^0–approximated by a positively as well as a negatively transverse knot isotopic to γ.*

Observe that such a statement has no chance of being true for an integrable tangent 2–plane field.

3.3.1 Legendrian knots

Proof of Theorem 3.3.1 – Legendrian case First of all, we consider a curve γ in standard $\mathbb{R}^3$. In order to find a C^0–close Legendrian approximation of γ, we simply need to choose a C^0–close approximation of its front projection γ_{F} by a regular curve without vertical tangencies and with isolated cusps (we call such a curve a *front*) in such a way, that the slope of the front at the parameter value s is close to $-x(s)$ (see Figure 3.7). Then the Legendrian lift of this front is the desired C^0–approximation of γ.

If γ is defined on a closed interval, then the Legendrian approximation of γ may be assumed to have the same endpoints, simply by choosing the slope of the front (in the above construction) appropriately at the endpoints. If γ is already Legendrian near its endpoints, then the approximation of γ_{F} may be assumed to coincide with γ_{F} near the endpoints, so that the Legendrian lift coincides with γ near the endpoints.

Hence, given a knot in an arbitrary contact 3–manifold, we can cut it (by the Lebesgue lemma) into finitely many little pieces that lie in Darboux charts. There we can use the preceding recipe to find a Legendrian

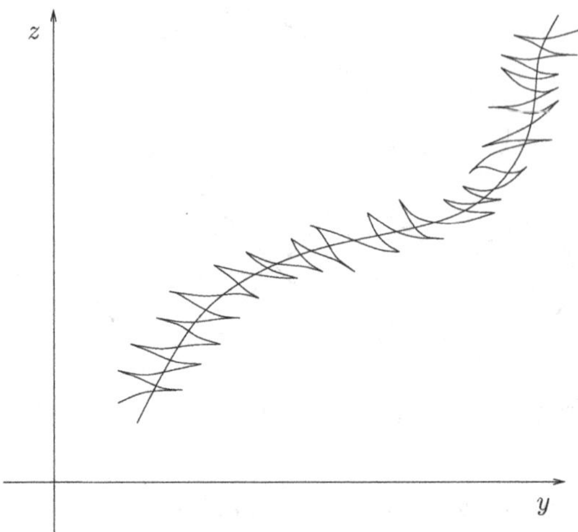

Fig. 3.7. Legendrian C^0–approximation via front projection.

approximation. Since, as just observed, one can find such approximations on intervals with given boundary condition, this procedure yields a Legendrian approximation of the full knot.

Locally (i.e. in $\mathbb{R}^3$) the described procedure does not introduce any self-intersections in the approximating curve, provided we approximate γ_F by a front with only transverse self-intersections. Since the original knot was embedded, the same will then be true for its Legendrian C^0–approximation. Likewise, in the local procedure we can avoid introducing any non-trivial knotting; then the Legendrian C^0–approximation will be (topologically) isotopic to the knot we started with. □

The same result may be derived using the Lagrangian projection.

Alternative proof of Theorem 3.3.1 – Legendrian case Again we consider a curve γ in standard $\mathbb{R}^3$ defined on an interval. The generalisation to arbitrary contact manifolds and closed curves is achieved as in the proof using front projections.

In order to find a C^0–approximation of γ by a Legendrian curve, one only has to approximate its Lagrangian projection γ_L by an immersed curve whose 'area integral'

$$z(s_0) - \int_{s_0}^{s} x\, dy$$

lies as close to the original $z(s)$ as one wishes. This can be achieved by using small loops oriented positively or negatively (see Figure 3.8). If γ_L has self-intersections, this approximating curve can be chosen in such a way that along loops properly contained in that curve the area integral is non-zero, so that again we do not introduce any self-intersections in the Legendrian approximation of γ. □

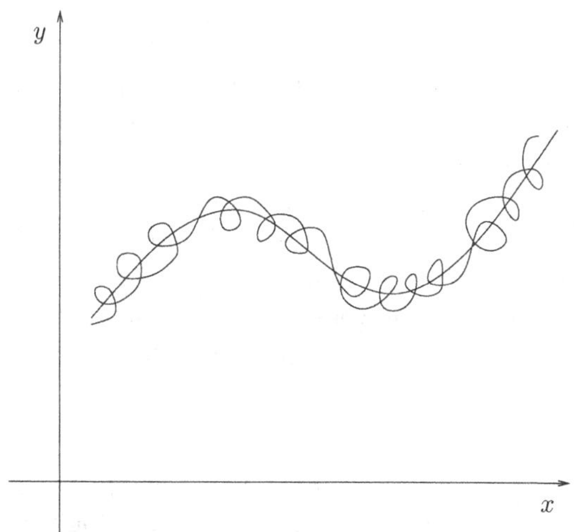

Fig. 3.8. Legendrian C^0–approximation via Lagrangian projection.

3.3.2 Transverse knots

The quickest proof of the transverse case of Theorem 3.3.1 is via the Legendrian case.

Proof of Theorem 3.3.1 – transverse case By the Legendrian case of this theorem, the given knot γ can be C^0–approximated by a Legendrian knot. This, in turn, can be C^0–approximated by a positive or negative transverse push-off, as described on page 95. □

Remark 3.3.2 If γ is defined on a closed interval, the above construction can easily be modified so that the transverse approximation coincides with γ at the endpoints.

The following parametric version of this approximation result will be used in the classification of overtwisted contact structures on 3–manifolds (Section 4.7).

Lemma 3.3.3 *Let ξ_t, $t \in [0,1]$, be a continuous family of cooriented contact structures on a 3–manifold M, and $\gamma_t\colon [a,b] \to M$, $t \in [0,1]$, a continuous family of embeddings. Then, arbitrarily C^0–close to this family of embeddings, one can find a continuous family of embeddings $\widetilde{\gamma}_t\colon [a,b] \to M$ with $\widetilde{\gamma}_t$ positively transverse to ξ_t and $\widetilde{\gamma}_t(a) = \gamma(a)$, $\widetilde{\gamma}_t(b) = \gamma(b)$ for all $t \in [0,1]$.*

Proof As in the non-parametric case, we are going to find the transverse approximation by first constructing a Legendrian approximation. In order for this to work in families, we have to specify canonical constructions in the proof of Theorem 3.3.1.

We begin with the situation that (M, ξ_t) equals $\mathbb{R}^3$ with its standard contact structure $\ker(dz + x\,dy)$ for all $t \in [0,1]$. Given an embedded curve $\gamma\colon [a,b] \to \mathbb{R}^3$, for any given $n \in \mathbb{N}$† a canonical approximation of its front projection γ_{F} by a front can be constructed as follows. We partition the interval $[a,b]$ into n intervals of length $(b-a)/n$. Consider γ on the subinterval $[a, a+(b-a)/n]$; on the other subintervals the construction will be analogous. We prescribe a front by the following data.

- At $s = a$ and $s = a + (b-a)/n$, the front passes through $\gamma_{\mathrm{F}}(s)$ with slope $-x(s)$ (in other words, the corresponding Legendrian curve passes through $\gamma(a)$ and $\gamma(a+(b-a)/n)$).
- At $s = a + (b-a)/3n$ the front has a cusp on the right of slope $-x(a + (b-a)/3n)$ at the point with coordinates
$$y = \max\{y(a), y(a+(b-a)/n)\} + 1/n \quad\text{and}\quad z = z(a + (b-a)/3n).$$
- At $s = a + 2(b-a)/3n$ the front has a cusp on the left of slope $-x(a + 2(b-a)/3n)$ at the point with coordinates
$$y = \min\{y(a), y(a+(b-a)/n)\} - 1/n \quad\text{and}\quad z = z(a + 2(b-a)/3n).$$
- Between these points, the front is given by a canonical choice of curves, determined by their endpoints and the slopes there. This choice can be made in such a way that the resulting front lifts to a smooth curve.

If n is chosen sufficiently large, the resulting Legendre curve will be as C^0–close to γ as we wish. Because the choice in the third point above can be made canonical, this procedure yields a C^0–approximation of a whole family γ_t of curves. (For sufficiently large n the construction will also guarantee that the resulting family consists again of embeddings.)

† My convention is that the natural numbers $\mathbb{N}$ are the positive integers $1, 2, \ldots$

The construction extends to arbitrary M and a family ξ_t of contact structures with the help of Lemma 2.6.6 by performing the above procedure on the curves (in the notation of that lemma) $(j_t^i)^{-1} \circ \gamma_t|_{[s_{i-1}, s_i]}$.

In a similar *ad hoc* fashion, the construction of the transverse approximations can be 'canonised' to allow the approximation in families. □

3.4 Interlude: topology of submanifolds

As a first step towards a classification of Legendrian or transverse knots up to Legendrian or transverse isotopy, respectively, one tries to find numerical invariants under such isotopies. For the definition of such invariants, we need to discuss the linking of knots in 3–manifolds. Textbooks on knot theory traditionally only deal with knots in $\mathbb{R}^3$ or S^3; in order to understand linking in general 3–manifolds we need some topological preparations, in particular concerning the realisation of homology cycles by submanifolds. Similar questions will be addressed in Section 4.2 in the process of classifying 2–plane fields on 3–manifolds. In the present section we also collect some results that will only become relevant there. First we make a short digression on some topological background material which is fairly standard but not explicitly contained in basic textbook references such as Bredon [35].

3.4.1 Hopf's Umkehrhomomorphismus

If $f\colon M^m \to N^n$ is a continuous map between smooth, oriented manifolds, one can define a homomorphism $\varphi\colon H_{n-p}(N) \to H_{m-p}(M)$ on homology classes represented by oriented submanifolds as follows. Given a homology class $[L]_N \in H_{n-p}(N)$ represented by an oriented codimension p submanifold L, replace f by a smooth approximation transverse to L and define $\varphi([L]_N) = [f^{-1}(L)]_M$. Here the submanifold $f^{-1}(L)$ of M is oriented as follows. Since f is transverse to L, the differential Tf induces a fibrewise isomorphism between the normal bundles of $f^{-1}(L)$ and L. The orientations of N and L induce an orientation of the normal bundle of L in N (orientation of L followed by normal orientation is the orientation of N). Under Tf this pulls back to an orientation of the normal bundle of $f^{-1}(L)$ in M, and together with the orientation of M this defines an orientation of $f^{-1}(L)$.

This homomorphism φ is essentially the *Umkehrhomomorphismus* introduced by Hopf [139], except that he worked with combinatorial manifolds of equal dimension and made no assumptions on the homology class. The following theorem, which in spirit is contained in [90], shows that φ is independent of choices (of submanifold L representing a class and smooth transverse approximation to f) and actually a homomorphism of intersection rings. This statement is not as well known as it should be, and I know of a proof in the literature only for the special case where L is a point [115]. In [35] this map is called *transfer map*,† but is only defined indirectly via Poincaré duality (though implicitly the statement of the following theorem is contained in [35], see for instance page 377 there).

Theorem 3.4.1 *Let $f\colon M^m \to N^n$ be a smooth map between closed, oriented manifolds and $L^{n-p} \subset N^n$ a closed, oriented submanifold of codimension p such that f is transverse to L. Write $u \in H^p(N)$ for the Poincaré dual of $[L]_N$, that is, $u \cap [N] = [L]_N$. Then $[f^{-1}(L)]_M = f^*u \cap [M]$. In other words: if u is Poincaré dual to $[L]_N$, then $f^*u \in H^p(M)$ is Poincaré dual to $[f^{-1}(L)]_M$.*

Proof We find (closed) tubular neighbourhoods $W \to L$ and $V = f^{-1}(W) \to f^{-1}(L)$ (considered as disc bundles) such that $f\colon V \to W$ is a fibrewise isomorphism. Write $[V]_0$ and $[W]_0$ for the orientation classes in the homology groups $H_m(V, V \setminus f^{-1}(L))$ and $H_n(W, W \setminus L)$, respectively. We can identify these homology groups with $H_m(V, \partial V)$ and $H_n(W, \partial W)$, respectively. Let $\tau_W \in H^p(W, \partial W)$ and $\tau_V \in H^p(V, \partial V)$ be the Thom classes of these disc bundles, defined by

$$\begin{aligned} \tau_W \cap [W]_0 &= [L]_N, \\ \tau_V \cap [V]_0 &= [f^{-1}(L)]_M. \end{aligned}$$

Notice that $f^*\tau_W = \tau_V$ since $f\colon W \to V$ is fibrewise isomorphic and the Thom class of an oriented disc bundle is the unique class whose restriction to each fibre is a positive generator of $H^p(D^p, \partial D^p)$. Writing $i\colon M \to (M, M \setminus f^{-1}(L))$ and $j\colon N \to (N, N \setminus L)$ for the inclusion maps we have

$$[f^{-1}(L)]_M = \tau_V \cap [V]_0 = f^*\tau_W \cap [V]_0 = f^*\tau_W \cap i_*[M],$$

where we identify $H_m(M, M \setminus f^{-1}(L))$ with $H_m(V, V \setminus f^{-1}(L))$ under the excision isomorphism. Then we have further

$$[f^{-1}(L)]_M = i^*f^*\tau_W \cap [M] = f^*j^*\tau_W \cap [M].$$

So it remains to identify $j^*\tau_W$ as the Poincaré dual u of $[L]_N$. Indeed,

$$j^*\tau_W \cap [N] = \tau_W \cap j_*[N] = \tau_W \cap [W]_0 = [L]_N,$$

where we have used the excision isomorphism between the homology groups $H_n(W, W \setminus L)$ and $H_n(N, N \setminus L)$. □

† More general transfer maps are discussed in [115].

3.4.2 Representing homology classes by submanifolds

We now want to prove various statements about the realisation of homology cycles by submanifolds. Apart from the Hopf *Umkehrhomomorphismus*, the other essential ingredient for proving such results is the homotopy-theoretic description of the homology functor.

Recall that for any abelian group π and natural number n there is a so-called Eilenberg–MacLane space $K(\pi, n)$, a CW–complex with the property that all its homotopy groups vanish except $\pi_n(K(\pi, n))$, which is isomorphic to π. (For $n = 1$, the group π is not required to be abelian for $K(\pi, 1)$ to exist, but for the following statement it will be.) Moreover, there is a natural isomorphism of functors

$$H^n(-;\pi) \cong [-, K(\pi, n)],$$

see [35, VII.12], where brackets denote homotopy classes† of maps. The element in $H^n(M;\pi)$ (with M any topological space) corresponding to a homotopy class $[f]$ of maps $M \to K(\pi, n)$ is given by $f^* u_0$, where u_0 is a so-called **characteristic element** of $H^n(K(\pi, n);\pi)$, that is, any element corresponding to an isomorphism $H_n(K(\pi, n);\mathbb{Z}) \to \pi$.

Proposition 3.4.2 *Let M be a closed, oriented* 3*–manifold. Every homology class $c \in H_1(M;\mathbb{Z})$ is represented by a knot K_c in M.*

Proof Given $c \in H_1(M;\mathbb{Z})$, set $u = PD(c) \in H^2(M;\mathbb{Z})$, where PD denotes the Poincaré duality homomorphism from integral homology to cohomology. We now use the isomorphism

$$H^2(M;\mathbb{Z}) \cong [M, K(\mathbb{Z}, 2)] = [M, \mathbb{C}P^\infty].$$

So u corresponds to a homotopy class of maps $f\colon M \to \mathbb{C}P^\infty$ such that $f^* u_0 = u$, where u_0 is the positive generator of $H^2(\mathbb{C}P^\infty)$ (that is, the one that pulls back to the Poincaré dual of $[\mathbb{C}P^{k-1}]_{\mathbb{C}P^k}$ under the natural inclusion $\mathbb{C}P^k \subset \mathbb{C}P^\infty$). Since $\dim M = 3$, any map $f\colon M \to \mathbb{C}P^\infty$ is homotopic to a smooth map $f_0\colon M \to \mathbb{C}P^1$. Let p be a regular value of f_0. Then

$$PD(c) = u = f_0^* u_0 = f_0^* PD[p]_{\mathbb{C}P^1} = PD[f_0^{-1}(p)]_M$$

by Theorem 3.4.1, and hence $c = [f_0^{-1}(p)]$. So $L_c = f_0^{-1}(p)$ is a link of embedded 1–spheres representing the class c. By forming the connected

† It does not make any difference here whether we work in the pointed category (spaces with base point preserved by maps and homotopies) or not, as long as the functor is only applied to *connected* spaces; see [125, 4.3].

sum of the components of L_c one can obtain a knot K_c representing that same class c. □

Remark 3.4.3 It is important to note that in spite of what we have just said it is not true that $[M, \mathbb{C}P^\infty] = [M, \mathbb{C}P^1]$, since a map $F\colon M \times [0,1] \to \mathbb{C}P^\infty$ with $F(M \times \{0,1\}) \subset \mathbb{C}P^1$ is not, in general, homotopic rel $(M \times \{0,1\})$† to a map into $\mathbb{C}P^1$. However, we do have $[M, \mathbb{C}P^\infty] = [M, \mathbb{C}P^2]$.

Proposition 3.4.4 *Let M be a closed, oriented 3–manifold. Every homology class $c \in H_2(M;\mathbb{Z})$ is represented by a (smoothly embedded) closed, oriented surface $\Sigma_c \subset M$.*

Remark 3.4.5 In general, Σ_c will have several components.

Proof This proof is completely analogous to the preceding one, except that $K(\mathbb{Z}, 2)$ has to be replaced by $K(\mathbb{Z}, 1) = S^1$. That is, for $c \in H_2(M;\mathbb{Z})$ we set $u = PD(c) \in H^1(M;\mathbb{Z})$. Under the isomorphism

$$H^1(M;\mathbb{Z}) \cong [M, K(\mathbb{Z}, 1)] \cong [M, S^1],$$

the class u corresponds to a homotopy class of maps $f\colon M \to S^1$ with $f^*u_0 = u$, where u_0 is a positive generator of $H^1(S^1;\mathbb{Z})$ for a chosen orientation of S^1.

Let p be a regular value of f. Then

$$PD(c) = u = f^*u_0 = f^*PD[p]_{S^1} = PD[f^{-1}(p)]_M,$$

so we can set $\Sigma_c = f^{-1}(p)$. □

Remark 3.4.6 The proofs of the two preceding propositions work for arbitrary dimension of M and homology classes of codimension 2 and 1, respectively.

Proposition 3.4.7 *Let M be a closed 3–manifold (which is not required to be orientable). Every homology class $c \in H_2(M;\mathbb{Z}_2)$ is represented by a smoothly embedded, possibly non-orientable closed surface Σ. Moreover, if $[K] \in H_1(M;\mathbb{Z}_2)$ is a class represented by a smoothly embedded circle $K \subset M$, and the intersection product $[K] \bullet c$ is equal to 1 (over $\mathbb{Z}_2$), then Σ can be chosen to have exactly one transverse intersection point with K.*

Proof For the first part, we argue as before. The Eilenberg–MacLane space

† Here 'rel A' is used as shorthand for 'relative to A', meaning that the subset A stays pointwise fixed during the homotopy.

$K(\mathbb{Z}_2, 1)$ can be realised as the infinite projective space $\mathbb{R}P^\infty$. The characteristic element u_0 is the generator of $H^1(\mathbb{R}P^\infty; \mathbb{Z}_2) = \mathbb{Z}_2$. Notice that this u_0 pulls back to the Poincaré dual of $[\mathbb{R}P^{k-1}]_{\mathbb{R}P^k}$ (over $\mathbb{Z}_2$) under the natural inclusion $\mathbb{R}P^k \subset \mathbb{R}P^\infty$.

Set $u = PD(c) \in H^1(M; \mathbb{Z}_2)$, where PD now denotes the Poincaré duality map from homology to cohomology with $\mathbb{Z}_2$–coefficients. The isomorphism

$$H^1(M; \mathbb{Z}_2) \cong [M, K(\mathbb{Z}_2, 1)] \cong [M, \mathbb{R}P^\infty]$$

tells us that u corresponds to a homotopy class of maps $f\colon M \to \mathbb{R}P^\infty$ such that $f^* u_0 = u$. Because of $\dim M = 3$, any such map f is homotopic to a smooth map $f_0\colon M \to \mathbb{R}P^3$. Moreover, f_0 can be chosen transverse to $\mathbb{R}P^2$. Then

$$PD(c) = u = f_0^* u_0 = f_0^* PD[\mathbb{R}P^2]_{\mathbb{R}P^3} = PD[f_0^{-1}(\mathbb{R}P^2)]_M.$$

Then $\Sigma = f_0^{-1}(\mathbb{R}P^2)$ is an embedded surface in M representing the class c.

If, furthermore, a class $[K]$ is given as described, we can make Σ transverse to K. The fact that the intersection product of $[K]$ and $[\Sigma]$ equals 1 over $\mathbb{Z}_2$ translates into saying that geometrically K and Σ have an odd number of intersection points. By adding a cylinder to Σ around a segment on C between two adjacent intersection points, and removing two discs in Σ around these intersection points, that pair of intersections can be removed† at the cost of increasing the genus of Σ (but without changing the homology class it represents). Iterating this procedure, we can reduce to a single point of intersection. □

The next proposition is an example of a relative version of the preceding results. This will be relevant for the definition of linking numbers of knots in 3–manifolds.

Proposition 3.4.8 *Let K be a nullhomologous oriented knot in a closed, oriented 3–manifold M, that is, $[K]_M = 0 \in H_1(M)$, with integral coefficients for homology understood. Let $c \in H_2(M, K)$ be a relative class that maps to $\partial_* c = [K]$, the generator of $H_1(K)$, under the boundary homomorphism $\partial_*\colon H_2(M, K) \to H_1(K)$; such a class exists by the homology long exact sequence of the pair (M, K). Then there is an embedded compact, oriented surface Σ_c in M with $\partial\Sigma_c = K$ (as oriented manifolds) and $[\Sigma_c]_{(M,K)} = c$.*

Proof As earlier, we write νK for a tubular neighbourhood of K. The pair (M, K) is homotopy equivalent to the pair $(M, \nu K)$, hence

$$H_2(M, K) \cong H_2(M, \nu K) \cong H_2(\overline{M \setminus \nu K}, \partial(\nu K)),$$

† This argument will be described more carefully in the proof of Lemma 8.2.11.

where the second isomorphism is induced by excision and retraction. We identify $H_2(M,K)$ with $H_2(\overline{M\setminus \nu K}, \partial(\nu K))$ under this isomorphism. Moreover, we have a Poincaré duality isomorphism

$$PD\colon\ H_2(\overline{M\setminus \nu K}, \partial(\nu K)) \longrightarrow H^1(\overline{M\setminus \nu K}).$$

As in the proof of Proposition 3.4.4, this implies that $PD(c)$ corresponds to a homotopy class of maps $f\colon\ \overline{M\setminus \nu K} \to S^1$, and we find an embedded surface $\Sigma_c' \subset \overline{M\setminus \nu K}$ as the inverse image of a regular value of f.

Theorem 3.4.1 also holds in the relative case; the same proof goes through without change. (Observe that the Thom class is also defined for neat submanifolds in the sense of [132, Section 1.4], see [35, Section VI.11].) This implies that Σ_c' represents the class c, interpreted as a class in the homology group $H_2(\overline{M\setminus \nu K}, \partial(\nu K))$. Since $\partial_* c = [K]$, it follows that $\partial\Sigma_c \subset \partial(\nu K)$ is a longitude, that is, a simple closed curve isotopic (as an oriented curve in νK) to the spine K of νK, or possibly a collection of parallel longitudes, with some oriented positively and one less oriented negatively. This implies that we can add a collar to Σ_c' so as to obtain a surface $\Sigma_c \subset M$ with $\partial\Sigma_c = K$ and $[\Sigma_c] = c$; if $\partial\Sigma_c'$ has several components, we can join positive and negative longitudes in pairs by an annulus (beginning with a pair of adjacent longitudes), and then add a collar to the one remaining positive longitude. □

Definition 3.4.9 Let K be a nullhomologous knot in a closed, oriented 3–manifold. An embedded connected, compact, orientable surface Σ with boundary $\partial\Sigma = K$ is called a **Seifert surface** for K.

By the preceding proposition, such a surface always exists. While the surface Σ_c constructed in that proposition may have several components, all but one of them will be closed, and the one component with boundary K will be a Seifert surface. Notice that in that proposition we assumed our 3–manifold to be closed. The only non-closed 3–manifold where we want to appeal to the existence of Seifert surfaces is $\mathbb{R}^3$. Here one can apply the proposition by passing to the one-point compactification S^3; the Seifert surface can always be assumed to miss that one extra point.

3.4.3 Linking numbers

Let M be an oriented 3–manifold and $K \subset M$ a nullhomologous oriented knot. We continue to write νK for a tubular neighbourhood of K and μ, λ for (oriented) meridian and a chosen longitude for K.

By excision and homotopy equivalence, we have

$$H_2(M, M \setminus K) \cong H_2(\nu K, \partial(\nu K)),$$

generated by a meridional disc (as follows from considering the homology long exact sequence of the pair $(S^1 \times D^2, T^2)$). The homology exact sequence of the pair $(M, M \setminus K)$ then shows that the kernel of the homomorphism

$$H_1(M \setminus K) \longrightarrow H_1(M)$$

induced by inclusion is isomorphic to $\mathbb{Z}$, generated by the homology class $[\mu]$.

Definition 3.4.10 Let K' be a second nullhomologous oriented knot disjoint from K. Then the homology class $[K']_{M\setminus K}$ lies in the kernel of the above homomorphism, hence

$$[K']_{M\setminus K} = n[\mu]$$

for a uniquely determined $n \in \mathbb{Z}$. This number, which is clearly an isotopy invariant, is called the **linking number** $\mathtt{lk}(K, K')$ of K and K'.

Observe that the linking number is additive in the following sense. Given two disjoint nullhomologous oriented knots K_1', K_2' in the complement of K, the homology class $[K_1' + K_2']_{M\setminus K}$ represented by their connected sum $K_1' + K_2'$ is the sum of the homology classes represented by the two knots, hence

$$\mathtt{lk}(K, K_1' + K_2') = \mathtt{lk}(K, K_1') + \mathtt{lk}(K, K_2').$$

Moreover, it is obvious that $\mathtt{lk}(K, \mu) = 1$. In particular, adding a right-handed twist to K' with respect to K, i.e. passing to the connected sum $K' + \mu$, increases the linking number by 1. Finally, notice that $\mathtt{lk}(K, K')$ changes its sign if the orientation of either K or K' is reversed.

Here is a different interpretation of this linking number. Let Σ' be a Seifert surface for K', which we may assume to be transverse to K. Then K and Σ' intersect in finitely many points; counting these points with sign gives the intersection number $K \bullet \Sigma'$. (If the orientation of K at the intersection point followed by the orientation of Σ' gives the orientation of M, the intersection point counts positively; negatively otherwise.)

Proposition 3.4.11 $\mathtt{lk}(K, K') = K \bullet \Sigma'$.

Proof Set $n = K \bullet \Sigma'$. Figure 3.9 shows that the connected sum $K' - n\mu$ of K' with n copies of $-\mu$ (for $n > 0$) or $-n$ copies of μ (for $n < 0$) bounds a surface disjoint from K. This implies that $[K' - n\mu]_{M\setminus K} = 0$, hence $[K']_{M\setminus K} = n[\mu]$, which by definition means that $\mathtt{lk}(K, K') = n$. □

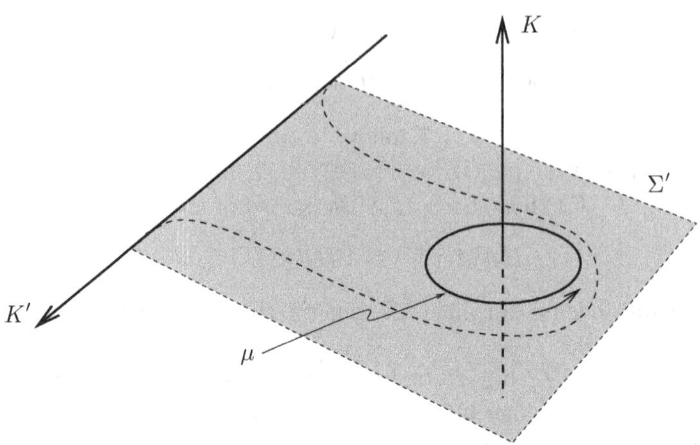

Fig. 3.9. Removing a (positive) intersection point.

Corollary 3.4.12 *The linking number is symmetric, that is, for any two disjoint nullhomologous oriented knots K, K' we have $\mathtt{lk}(K, K') = \mathtt{lk}(K', K)$.*

Proof Let Σ, Σ' be Seifert surfaces for K, K', respectively. We may assume that these surfaces intersect transversely, so the intersection $\Sigma \cap \Sigma'$ consists of a finite collection of circles and arcs with endpoints on K or K'. These endpoints are exactly the intersection points of K with Σ' or K' with Σ. If an arc joins two points of the first kind or two of the second kind, then these points can be seen to have opposite orientation, so they do not contribute to $K \bullet \Sigma'$ or $K' \bullet \Sigma$, respectively. If an arc in $\Sigma \cap \Sigma'$ joins a point of $K \cap \Sigma'$ with a point of $K' \cap \Sigma$, then both points have the same sign, see Figure 3.10. Hence $K \bullet \Sigma' = K' \bullet \Sigma$, and therefore $\mathtt{lk}(K, K') = \mathtt{lk}(K', K)$ by the preceding proposition. □

Remark 3.4.13 The fact that adding a right-handed twist to K' with respect to K increases the linking number by 1 can also be seen from the computation of $\mathtt{lk}(K, K')$ as $K' \bullet \Sigma$, for adding a right-handed twist to K' will produce one additional positive intersection with the Seifert surface Σ.

From this observation it is easy to derive a recipe for computing the linking number of two knots in 3–space. Given disjoint knots K, K' in $\mathbb{R}^3$, represent them by their images under a generic projection onto a 2–plane in $\mathbb{R}^3$, i.e. choose the 2–plane in such a way that the projected knots are immersed except for transverse double points. We call the image of a knot or link under such a projection a *knot* or *link diagram*, respectively.

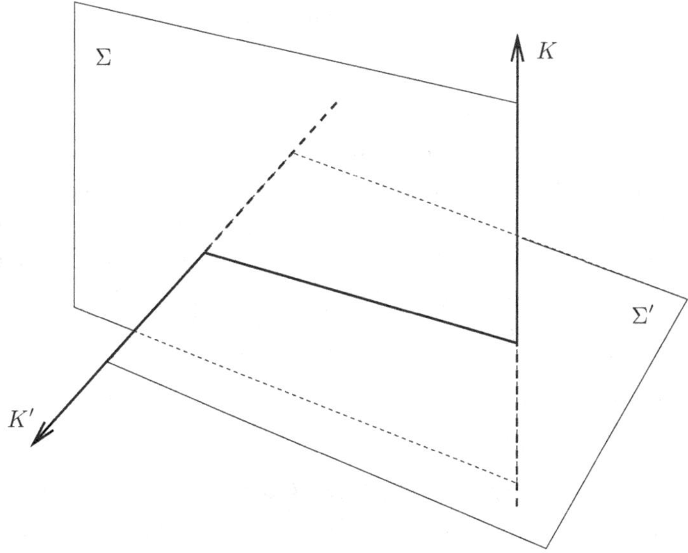

Fig. 3.10. Arc in $\Sigma \cap \Sigma'$ joining two positive intersections.

With each crossing in an oriented knot or link diagram we associate a sign as in Figure 3.11.

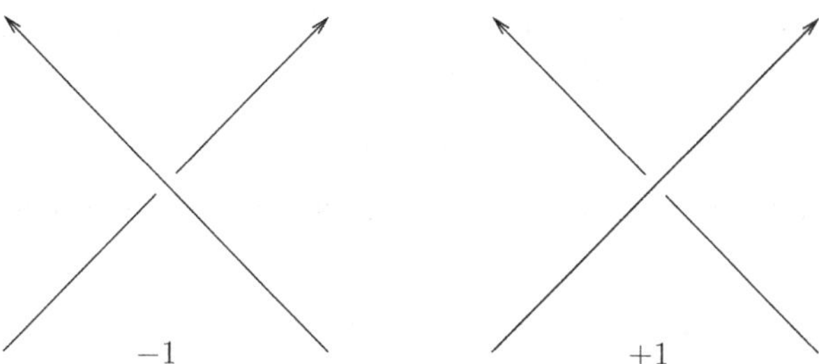

Fig. 3.11. Signs of crossings in a knot or link diagram.

Proposition 3.4.14 *For two disjoint oriented knots $K, K' \subset \mathbb{R}^3$, given by a link diagram, the linking number $\mathtt{lk}(K, K')$ equals the number of crossings where K' crosses under K, counted with sign.*

Proof A positive crossing of K' under K can be turned into an overcrossing by replacing K' with the connected sum $K' - \mu$; analogously, a negative undercrossing can be turned into an overcrossing by passing to $K' + \mu$. Hence, if $n \in \mathbb{Z}$ denotes the total number of times the knot K' crosses under K, then

$K' - n\mu$ crosses over K at all crossings. This implies $\mathtt{lk}(K, K' - n\mu) = 0$, and hence $\mathtt{lk}(K, K') = n$. □

3.5 The classical invariants

After these topological preparations, we now discuss the three invariants – a couple for Legendrian knots and one for transverse knots – that are commonly referred to as the 'classical' ones. The definition of these invariants is given only for homologically trivial knots, although it is possible to extend the definition to some other cases, for instance linear Legendrian curves on the 3–torus with its standard contact structure (Example 1.2.4), see [144] and [101].

3.5.1 Legendrian knots

In the following definitions we make a little more hullabaloo than is strictly necessary for introducing the classical invariants of Legendrian knots. These definitions, however, will come into prominence in our discussion of surgery on contact 3–manifolds.

Definition 3.5.1 Let K be a Legendrian knot in a 3–manifold M with cooriented contact structure ξ. Since $T_pK \subset \xi_p$ for all $p \in K$, the normal bundle $NK = (TM|_K)/TK$ of K in M splits into a Whitney sum of trivial line bundles

$$NK \cong (TM|_K)/(\xi|_K) \oplus (\xi|_K)/TK.$$

The **contact framing** of K is the trivialisation of NK defined by this splitting.

This is a rather roundabout way of expressing something geometrically very simple. In order to define a framing of K, we need to specify a parallel curve to K, which we take as a longitude on the boundary of a tubular neighbourhood of K. The contact framing is the one corresponding to the parallel curve we get by pushing K in a direction transverse to ξ. In the formal definition above, this would correspond to choosing a section of the factor $(TM|_K)/(\xi|_K)$. Alternatively, we may take as longitude a parallel curve obtained by pushing K in a direction tangent to ξ but transverse to K. This would correspond to choosing a section of $(\xi|_K)/TK$, and yields the same overall trivialisation of NK.

Definition 3.5.2 Let K be a homologically trivial knot in an oriented 3–manifold M, and let Σ be a Seifert surface for K. The **surface framing**

of K is the trivialisation of the normal bundle NK corresponding to the parallel curve obtained by pushing K along Σ.

Remark 3.5.3 An equivalent characterisation of the surface framing is that it is determined by a parallel curve that has linking number 0 with K. (On page 43 we referred to this parallel curve as the preferred longitude.) In particular, the definition of the surface framing is independent of the choice of Seifert surface.

Definition 3.5.4 Let K be a homologically trivial Legendrian knot in a contact 3–manifold (M, ξ). The **Thurston–Bennequin invariant** of K, denoted by $\mathtt{tb}(K)$, is the twisting of the contact framing relative to the surface framing of K, with right-handed twists being counted positively. In other words (see Remark 3.4.13), if we choose a vector field along K transverse to ξ and define a parallel knot K' by pushing K along this vector field, then $\mathtt{tb}(K)$ equals the linking number $\mathtt{lk}(K, K')$ of K with K'. (Here K is given any orientation, and K' is given the corresponding orientation that makes K and K' isotopic as oriented knots.)

Remark 3.5.5 It follows from the Legendrian isotopy extension theorem (Thm. 2.6.2) that $\mathtt{tb}(K)$ is indeed invariant under Legendrian isotopies of K. Moreover, observe that $\mathtt{tb}(K)$ does not depend on the choice of orientation for K.

Remark 3.5.6 In order to speak of linking numbers, an orientation for M needs to be chosen. Our convention throughout is that the orientation of a 3–dimensional contact manifold (M, ξ) is the one that turns ξ into a *positive* contact structure, i.e. such that, for any local 1–form α defining $\xi = \ker \alpha$, the 3–form $\alpha \wedge d\alpha$ is a *positive* volume form for M, see Remark 1.1.6.

Example 3.5.7 In $S^3 \subset \mathbb{R}^4$ with its standard contact structure $\xi_{\mathrm{st}} = \ker \alpha$, where

$$\alpha = x_1\, dy_1 - y_1\, dx_1 + x_2\, dy_2 - y_2\, dx_2,$$

consider the Legendrian unknot

$$K = \{x_1^2 + x_2^2 = 1,\ y_1 = y_2 = 0\}.$$

This circle bounds the 2–disc

$$\Sigma = \{x_1^2 + x_2^2 + y_1^2 = 1,\ y_1 \geq 0,\ y_2 = 0\},$$

so the surface framing of K is given by the vector field ∂_{y_1} (or, equivalently, ∂_{y_2}).

In order to compute $\mathtt{tb}(K)$, we choose an orientation of K, say the one defined by the unit tangent vector field $T := x_1\,\partial_{x_2} - x_2\,\partial_{x_1}$. Along K, the volume form $\alpha\wedge d\alpha$ on S^3 takes the form

$$(\alpha\wedge d\alpha)|_K = 2x_1\,dy_1\wedge dx_2\wedge dy_2 + 2x_2\,dy_2\wedge dx_1\wedge dy_1.$$

The inner product of this volume form with T is $-2\,dy_1\wedge dy_2$. So with respect to the chosen orientation of K, the oriented surface framing is given by the ordered basis $\{\partial_{y_2}, \partial_{y_1}\}$.

The vector field $N := x_1\,\partial_{y_2} - x_2\,\partial_{y_1}$ along K lies in ξ_{st}, and together with T it spans the contact planes. Therefore, $\mathtt{tb}(K)$ is given by the number of rotations of N relative to the ordered basis $\{\partial_{y_2}, \partial_{y_1}\}$ as we traverse K once in the chosen positive direction. It is easily seen that this yields $\mathtt{tb}(K) = -1$.

Example 3.5.8 In the front projection picture for Legendrian knots in $(\mathbb{R}^3, \xi_{\mathrm{st}})$, consider the 'shark' K, see Figure 3.12. Since the vector field ∂_z is everywhere transverse to the standard contact structure $\xi_{\mathrm{st}} = \ker(dz + x\,dy)$, a parallel Legendrian knot K' corresponding to the contact framing is simply given by pushing the front projection of the knot in the z–direction. In this particular example we find that $\mathtt{tb}(K) = \mathtt{lk}(K, K') = -2$.

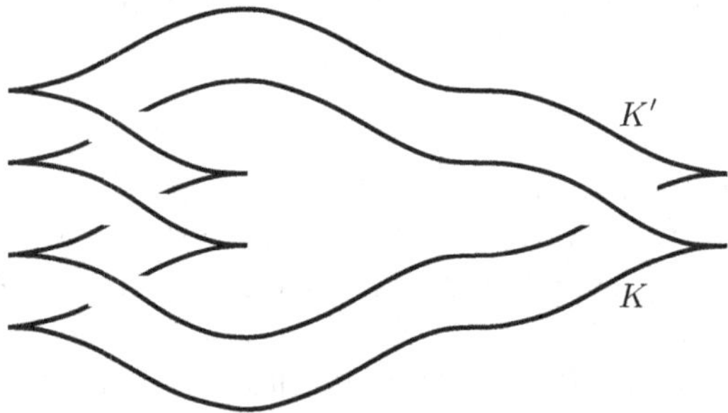

Fig. 3.12. Topological unknot K with $\mathtt{tb}(K) = -2$.

We now want to generalise this example and give a recipe for computing the Thurston–Bennequin invariant of a Legendrian knot in standard $\mathbb{R}^3$ from its front projection. Apart from the cusps, we shall also have to take the self-crossings of the front projection into account. Recall that the **writhe** of an oriented knot diagram is the signed number of self-crossings of the diagram, where the sign of the crossing is given in Figure 3.11. Notice that the writhe of a knot diagram is independent of the chosen orientation of the knot, for if the orientation is reversed, both strands at each crossing reverse their direction, which leaves the sign of the crossing unchanged.

Proposition 3.5.9 *Let K be a Legendrian knot in $(\mathbb{R}^3, \xi_{st})$. Write K_F for the knot diagram of K obtained by the front projection. Then the Thurston–Bennequin invariant of K is given by*

$$\mathtt{tb}(K) = \text{writhe}(K_F) - \frac{1}{2}\#(\text{cusps}(K_F)).$$

Proof By slight abuse of notation we ignore the distinction between K and K_F. Fix an orientation for K. We compute $\mathtt{tb}(K)$ as in the example, i.e. we define a parallel copy K' of K, with the induced orientation, by pushing (the front projection of) K in the z–direction, and then compute $\mathtt{tb}(K)$ as linking number $\mathtt{lk}(K, K')$. That linking number, in turn, we compute by counting the crossings of K' under K with sign.

It is easy to see that a self-crossing of K will contribute a crossing of K' under K of the same sign, a cusp on the right† will give a negative crossing of K' under K, and a cusp on the left will give a crossing of K' over K. Since there are as many cusps on the left as cusps on the right, the claimed formula follows. □

Remark 3.5.10 An alternative way to see that $\mathtt{tb}(K)$ is a Legendrian isotopy invariant, at least for Legendrian knots in $(\mathbb{R}^3, \xi_{st})$, would be to establish the Legendrian analogues of Reidemeister moves for front projection pictures, and then to verify that $\mathtt{tb}(K)$, computed by the above formula, is invariant under those moves, see [82].

Here is the corresponding formula in the Lagrangian projection.

Proposition 3.5.11 *The Thurston–Bennequin invariant $\mathtt{tb}(K)$ of a Legendrian knot K in $(\mathbb{R}^3, \xi_{st})$ is equal to the writhe of its Lagrangian projection.*

Proof Write K_L for the knot diagram of K obtained by the Lagrangian projection. The parallel copy K' of K may be obtained by pushing K in a direction tangent to ξ_{st} but transverse to K. Under the Lagrangian projection, ξ_{st} projects isomorphically onto the (x, y)–plane. This implies that K' can be represented by a diagram K'_L parallel to K_L in the (x, y)–plane. The result now follows from the definitions and Proposition 3.4.14. □

We now come to the definition of the second classical invariant for Legendrian knots, the so-called rotation number. Let K be an oriented Legendrian knot in an oriented contact 3–manifold (M, ξ) with *oriented* contact structure. Let Σ be a Seifert surface for K, oriented in such a way that

† By Remark 3.2.4 we can sensibly speak of left and right cusps.

the given orientation of K equals its orientation as boundary of Σ. Write $c \in H_2(M, K)$ for the relative homology class represented by Σ. (Notice that not all relative homology classes can be represented in this way, since by definition Σ is assumed to be connected, and its orientation is fixed by that of K.) As a surface with boundary, Σ retracts onto its 1–skeleton (which, by the classification of surfaces, is a bouquet of $2g$ copies of S^1). Since the only orientable 2–plane bundle over S^1 is the trivial one, we can find a trivialisation of $\xi|_\Sigma$ (compatible with the given orientation of ξ).

Let $\gamma\colon S^1 \to K \subset M$ be a (regular) parametrisation of K compatible with its orientation. Given a trivialisation $\xi|_\Sigma = \Sigma \times \mathbb{R}^2$ (as oriented bundles†), this induces a map $\gamma'\colon S^1 \to \mathbb{R}^2 \setminus \{\mathbf{0}\}$.

Definition 3.5.12 The **rotation number** $\mathtt{rot}(K, c)$ of the nullhomologous oriented Legendrian knot K relative to the class $c \in H_2(M, K)$ is the degree of that map γ'. In other words, $\mathtt{rot}(K, c)$ counts the number of rotations of the (positive) tangent vector to K relative to the trivialisation of $\xi|_\Sigma$ as we go once around K.

Remark 3.5.13 Here, too, the Legendrian isotopy extension theorem implies that $\mathtt{rot}(K, c)$ is invariant under such isotopies. However, in contrast with the Thurston–Bennequin invariant, the rotation number does depend on the choice of orientation of K. If $\overline{K}$ denotes K with reversed orientation, we have $\mathtt{rot}(K, c) = -\mathtt{rot}(\overline{K}, -c)$. Also, the sign of the rotation number (but not that of $\mathtt{tb}$) depends on the choice of orientation for ξ.

In order to justify our terminology, we need to show that $\mathtt{rot}(K, c)$ is independent of the choices in this construction. For greater clarity, we write $\mathtt{rot}(K, \Sigma)$ for the time being.

Lemma 3.5.14 *The rotation number* $\mathtt{rot}(K, \Sigma)$ *does not depend on the choice of trivialisation of* $\xi|_\Sigma$.

Proof With an appropriate choice of labels $a_1, b_1, \ldots, a_g, b_g$ for the circles making up the 1–skeleton of Σ, the boundary $K = \partial\Sigma$ is homotopic (in Σ) to the word $\prod_{i=1}^{g}[a_i, b_i]$, a product of commutators. The trivialisation of $\xi|_\Sigma$ is determined, up to homotopy, by choosing a trivialisation along the 1–skeleton. Since K is homotopic to a word involving each loop in the 1–skeleton together with its inverse, any two trivialisations of $\xi|_\Sigma$ will have zero relative rotation when going once around K.

Slightly more formally, one may argue as follows. A trivialisation of the

† This will always be understood here when we speak of a 'trivialisation'.

oriented plane field $\xi|_\Sigma$ is determined by the choice of a non-vanishing section. Suppose we have two different trivialisations, corresponding to non-vanishing sections X_1, X_2 of $\xi|_\Sigma$. The difference between the rotation number computed relative to X_1 and that computed relative to X_2 is the degree of the map $S^1 \cong K \to S^1$ associating to $p \in K$ the oriented angle between X_1 and X_2 at p (where we have chosen an auxiliary metric on ξ). This degree can be computed via the induced homomorphism $\mathbb{Z} \cong H_1(K) \to H_1(S^1) \cong Z$ in homology. But the trivialisations are defined over Σ, so the map $K \to S^1$ factors through Σ, and the corresponding factorisation of the induced homomorphism on the first homology group forces that homomorphism to be trivial. □

Proposition 3.5.15 *Let Σ, Σ' be Seifert surfaces for K representing the classes $c, c' \in H_2(M, K)$. Then*

$$\mathtt{rot}(K, \Sigma) - \mathtt{rot}(K, \Sigma') = \langle e(\xi), c - c' \rangle,$$

where $c - c'$ is regarded as a class in $H_2(M)$, and $e(\xi)$ denotes the Euler class of the (oriented) 2–plane bundle ξ.

In particular, by taking $c = c'$ in this proposition we see that the rotation number $\mathtt{rot}(K, \Sigma)$ does not depend on the choice of Seifert surface Σ representing a given class c, which justifies our notation $\mathtt{rot}(K, c)$.

Proof Consider the closed orientable surface $\Sigma - \Sigma'$, obtained by gluing Σ and Σ' (with reversed orientation) along K. By a homotopy we may assume that the trivialisations of $\xi|_\Sigma$ and $\xi|_{\Sigma'}$ coincide along an interval in K; they then define a trivialisation of ξ over $\Sigma - \Sigma'$ with a disc D^2 removed. See Figure 3.13; here $K \cap D^2$ makes up the complementary interval in K. With the indicated orientation of K (which determines that of Σ), the orientation of D^2 is the standard one.

Over D^2 there is a unique trivialisation of ξ up to homotopy, which may be assumed to coincide along $K \cap D^2$ with the trivialisation defined by the positive tangent vector to K. (Homotopically, we can think of $\xi|_{D^2}$ as coinciding with the tangent bundle TD^2 and of the trivialisation as the one given by a constant vector field.) Then the Euler number $\langle e(\xi), c - c' \rangle$ measures the number of rotations of the trivialisation of ξ over $(\Sigma - \Sigma') \setminus D^2$ along ∂D^2 (traversed in counterclockwise direction) relative to this constant vector field.

On the other hand, when computing the difference $\mathtt{rot}(K, \Sigma) - \mathtt{rot}(K, \Sigma')$, our assumptions imply that we need only consider the relative rotations of the appropriate vector fields along $K \cap D^2$. The relevant contribution of

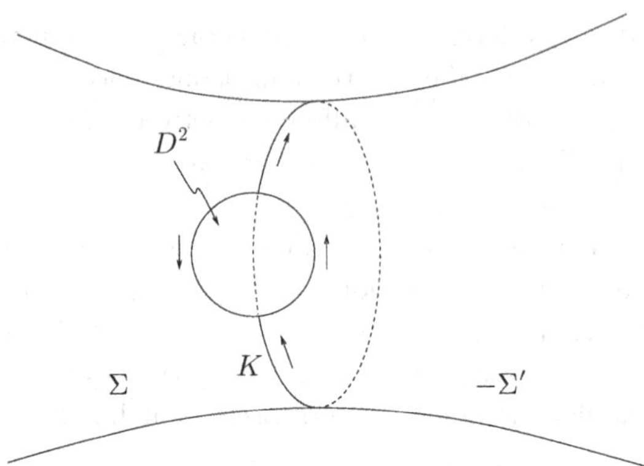

Fig. 3.13. Computing the rotation number.

the term $\mathtt{rot}(K, \Sigma)$ is the number of rotations of the constant vector field relative to the trivialisation of $\xi|_\Sigma$ along the path $K \cap D^2$, which is oriented homotopic rel endpoints to $\partial D^2 \cap \Sigma$ traversed in clockwise direction. This equals the contribution to the Euler number (on the right-hand side of our equation) coming from $\partial D^2 \cap \Sigma$, since both the relative roles of the vector fields and the direction of traversal are reversed.

When comparing the contribution of the term $\mathtt{rot}(K, \Sigma')$ with the contribution to the Euler number from $\partial D \cap \Sigma'$, we notice that only the roles of the vector fields are interchanged, but not the direction of traversal of $\partial D^2 \cap \Sigma'$. This accounts for the minus sign in the equation and proves the proposition. □

Corollary 3.5.16 *If* $e(\xi) = 0$, *then the rotation number* $\mathtt{rot}(K, c)$ *is independent of* c. *Conversely, if* $\mathtt{rot}(K, c)$ *depends on* K *only, for each Legendrian knot* K *in* (M, ξ), *then* $e(\xi) = 0$.

Proof The equality in the preceding proposition may be read as

$$\mathtt{rot}(K, c) - \mathtt{rot}(K, c') = \langle e(\xi), c - c' \rangle.$$

Hence, if $e(\xi) = 0$, then $\mathtt{rot}(K, c) = \mathtt{rot}(K, c')$ for any $c, c' \in H_2(M, K)$ that can be represented by a Seifert surface for K.

Conversely, assume that the rotation number is independent of the choice of relative homology class. Given any absolute class $\tilde{c} \in H_2(M)$, represent it (using Proposition 3.4.4) by a closed, orientable surface $\widetilde{\Sigma}$ embedded in M. Assume that $\widetilde{\Sigma}$ is connected; the general case follows by considering the components of $\widetilde{\Sigma}$ individually. Choose any oriented simple closed curve on $\widetilde{\Sigma}$ separating $\widetilde{\Sigma}$ into two components and isotope it to an oriented Legendrian knot K (Thm. 3.3.1). By the topological isotopy extension theorem, there is an ambient isotopy that will move $\widetilde{\Sigma}$ to a surface, still representing the homology class $\tilde{c}$, that contains K as a separating curve. The splitting of this new surface along K into two components translates into a splitting $\tilde{c} = c - c'$ with $c, c' \in H_2(M, K)$. Hence, the fact that the rotation numbers do not depend on the choice of relative homology class, together with the preceding proposition, yields $\langle e(\xi), \tilde{c}\rangle = 0$. Since the choice of $\tilde{c}$ was arbitrary, this gives $e(\xi) = 0$. □

Remark 3.5.17 The first part of this corollary admits an easy proof without reference to the preceding proposition. The condition $e(\xi) = 0$ is equivalent to saying that ξ is a trivial 2–plane bundle (see the comments after Theorem 4.3.1 below or the more formal argument in the proof of Proposition 8.1.1). So the rotation number of an oriented Legendrian knot K may be computed by counting the number of rotations of the (positive) tangent vector to K relative to a global trivialisation of ξ. Even in that case, however, the condition that K be homologically trivial is necessary for the rotation number not to depend on the choice of global trivialisation.

Whenever $\mathtt{rot}(K, c)$ is independent of c, we shall simply write $\mathtt{rot}(K)$.

Example 3.5.18 Consider the standard Legendrian unknot K in the standard 3–sphere (S^3, ξ_{st}) as in Example 3.5.7. The vector field

$$x_1\,\partial_{x_2} - x_2\,\partial_{x_1} + y_2\,\partial_{y_1} - y_1\,\partial_{y_2}$$

is a nowhere zero section of ξ_{st} that coincides along K with the tangent vector field u to K. This shows that $\mathtt{rot}(K) = 0$.

The contact manifold $(\mathbb{R}^3, \xi_{\mathrm{st}})$ is a further example to which the above corollary applies. The next proposition gives a formula for computing the rotation number of a knot from its front projection. Notice that the orientation of K allows us to speak of cusps being oriented upwards or downwards.

Proposition 3.5.19 *Let K be an oriented Legendrian knot in $(\mathbb{R}^3, \xi_{\mathrm{st}})$, with ξ_{st} cooriented by the vector field ∂_z. Write λ_+ or λ_- for the number (in the front projection of K) of left cusps oriented upwards or downwards, respectively; similarly we write $\rho_\pm$ for the number of right cusps with one or the other orientation. Finally, we write $c_\pm$ for the total number of cusps oriented upwards or downwards, respectively. Then the rotation number of K is given by*

$$\mathtt{rot}(K) = \lambda_- - \rho_+ = \rho_- - \lambda_+ = \frac{1}{2}(c_- - c_+).$$

Proof Observe that the vector fields $e_1 = \partial_x$ and $e_2 = \partial_y - x\,\partial_z$ define a positively oriented trivialisation of ξ_{st}. Therefore, $\mathtt{rot}(K)$ can be computed by counting (with sign) how often the positive tangent vector to K crosses ∂_x as we travel once along K.

Since x equals the negative slope of the front projection, points of K where the (positive) tangent vector equals ∂_x are exactly the left cusps oriented downwards (see Figure 3.14) and the right cusps oriented upwards.

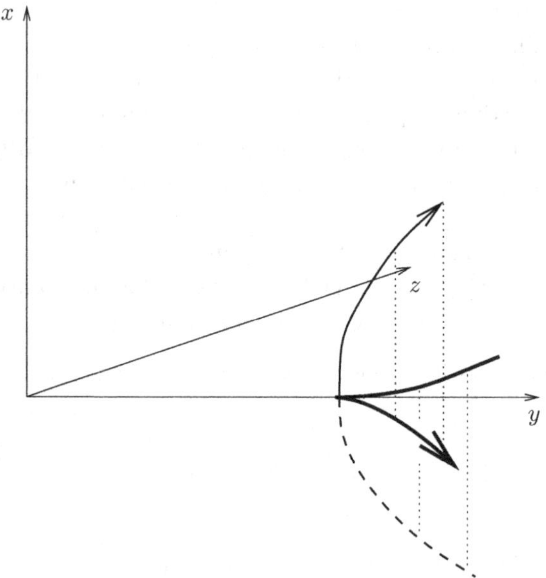

Fig. 3.14. Contribution of a cusp to $\mathtt{rot}(K)$.

At a left cusp oriented downwards, the tangent vector to K, expressed in terms of e_1, e_2, changes from having a negative component in the e_2–direction to a positive one, i.e. such a cusp yields a positive contribution to $\mathtt{rot}(K)$. Analogously, one sees that a right cusp oriented upwards gives a negative contribution to the rotation number. This proves the formula $\mathtt{rot}(K) = \lambda_- - \rho_+$.

The second expression for the rotation number is obtained by counting crossings through $-e_1$ instead; the third expression is found by averaging the first two. □

As mentioned earlier, the front projection is usually more convenient to work with than the Lagrangian projection, so the preceding proposition is important in that respect. However, the rotation number of Legendrian knots in $(\mathbb{R}^3, \xi_{\rm st})$ actually admits a much simpler interpretation in the Lagrangian projection. First recall the following definition:

Definition 3.5.20 The **rotation number** $\texttt{rot}(\overline{\gamma})$ of an immersion $\overline{\gamma}\colon S^1 \to \mathbb{R}^2$ is the degree of its differential $\overline{\gamma}'\colon S^1 \to \mathbb{R}^2 \setminus \{\mathbf{0}\}$.

In other words, $\texttt{rot}(\overline{\gamma})$ can be computed by counting (with sign) how often $\overline{\gamma}'$ crosses ∂_x. Since a global trivialisation of $\xi_{\rm st}$ is given by the vector fields ∂_x and $\partial_y - x\,\partial_z$, the following proposition is immediate.

Proposition 3.5.21 *Let $\gamma\colon S^1 \to K \subset \mathbb{R}^3$ be a regular parametrisation of an oriented Legendrian knot K in $(\mathbb{R}^3, \xi_{\rm st})$ and $\gamma_{\rm L}$ its Lagrangian projection. Then $\texttt{rot}(K) = \texttt{rot}(\gamma_{\rm L})$.* □

Remark 3.5.22 The two preceding propositions allow us to define the rotation number also for Legendrian *immersions* of S^1 in $(\mathbb{R}^3, \xi_{\rm st})$. This will be discussed further in Section 6.3.1.

There are various relations between the Thurston–Bennequin invariant and the rotation number of Legendrian realisations of a given knot type. Arguably the most important one is the *Bennequin inequality*

$$\texttt{tb}(K) + |\texttt{rot}(K)| \leq -\chi(\Sigma),$$

where $\chi(\Sigma)$ denotes the Euler characteristic of a Seifert surface Σ for the Legendrian knot K in $(\mathbb{R}^3, \xi_{\rm st})$. This will be proved (for Legendrian knots in more general contact manifolds) in Section 4.6.5. For the moment, we only prove a very simple relation.

Proposition 3.5.23 *Let K be a Legendrian knot in $(\mathbb{R}^3, \xi_{\rm st})$. Then*

$$\texttt{tb}(K) + \texttt{rot}(K) \equiv 1 \mod 2.$$

Proof From Propositions 3.5.9 and 3.5.19 we have

$$\begin{aligned} \texttt{tb}(K) + \texttt{rot}(K) &= \text{writhe}(K) - \frac{1}{2}(c_- + c_+) + \frac{1}{2}(c_- - c_+) \\ &= \text{writhe}(K) - c_+. \end{aligned}$$

For computing the expression $\text{writhe}(K) - c_+$, we may replace the front projection of K by a smooth knot diagram, with c_+ counting the (isolated) vertical tangencies oriented upwards. If we change a crossing in this diagram, the writhe changes by an even number. We may therefore assume that the diagram represents the unknot, since any knot can be transformed to the unknot by crossing changes (see [209, Thm. 3.8] for a nice proof of that simple fact). By the Reidemeister theorem [209, Thm. 1.7], such a knot diagram can be transformed to the trivial diagram (i.e. a simple closed curve in the plane) by Reidemeister moves and planar isotopies. It is easy to see, from Remark 3.5.24 (2) below, that the number writhe$(K) - c_+$ mod 2 is an invariant of such moves, and for the circle it equals 1. □

Remarks 3.5.24 (1) This congruence relation holds for homologically trivial Legendrian knots in arbitrary contact 3–manifolds, see Remark 4.6.35.

(2) The invariance of writhe$(K) - c_+$ mod 2 under the three Reidemeister moves is a simple check. Under planar isotopies, writhe(K) is clearly invariant. Observe that c_+ mod 2 may be regarded as the mod 2 degree of the Gauß mapping $K \to S^1$. A planar isotopy corresponds to a smooth homotopy of that Gauß map, and the invariance of the mod 2 degree under such a smooth homotopy, while fairly obvious in this particular case, is a general fact; see Chapter 4 of Milnor's little gem [186].

Example 3.5.25 When K is the unknot, we may take the Seifert surface Σ to be a disc. So the Bennequin inequality reads

$$\mathtt{tb}(K) + |\mathtt{rot}(K)| \leq -1.$$

This yields the restriction $\mathtt{tb}(K) \leq -1$. For $\mathtt{tb}(K) = -m$, $m \in \mathbb{N}$, the Bennequin inequality and the parity condition from Proposition 3.5.23 imply that

$$\mathtt{rot}(K) \in \{-m+1, -m+3, \ldots, m-3, m-1\}.$$

Figure 3.15 shows that any combination of tb and rot permitted by these conditions can indeed be realised by a Legendrian unknot. The Legendrian unknot with a front projection diagram as shown, with $c_- = 2n_- + 1$ down cusps and $c_+ = 2n_+ + 1$ up cusps ($n_-, n_+ \in \mathbb{N}_0$) and no crossings, has

$$\mathtt{tb} = -1 - (n_- + n_+), \quad \mathtt{rot} = \frac{1}{2}(n_- - n_+).$$

The example in the picture has $n_- = 2$ and $n_+ = 1$.

It is natural to ask to what extent these classical invariants classify Legendrian knots. In other words, given two Legendrian knots of the same topological knot type and with equal Thurston–Bennequin invariants and rotation numbers, are these knots Legendrian isotopic? This question has generated some deep mathematics. The first result in this direction is due to Eliashberg and Fraser [74].

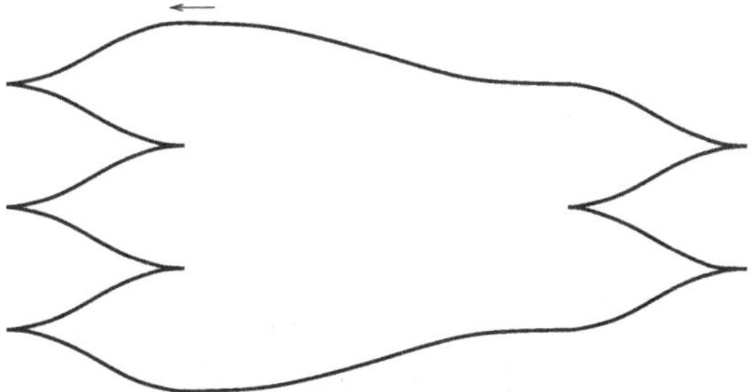

Fig. 3.15. A Legendrian unknot.

Theorem 3.5.26 (Eliashberg–Fraser) *Let K, K' be two topologically trivial Legendrian knots in $(\mathbb{R}^3, \xi_{\mathrm{st}})$. Then they are Legendrian isotopic if and only if* $\mathtt{tb}(K) = \mathtt{tb}(K')$ *and* $\mathtt{rot}(K) = \mathtt{rot}(K')$. □

Remark 3.5.27 In fact, Eliashberg and Fraser prove this theorem for topologically trivial Legendrian knots in arbitrary tight contact 3–manifolds.

This result has been extended to some topologically non-trivial knot types by Etnyre and Honda [85], see also [82] and [53]. In general, however, the classical invariants do not suffice for classifying Legendrian knots in a given knot type.

Chekanov ([45], [46]) found an example of two topologically isotopic Legendrian knots in $(\mathbb{R}^3, \xi_{\mathrm{st}})$ with the same classical invariants which can be shown not to be Legendrian isotopic with the help of a more subtle invariant. This invariant derives from a combinatorial version of what is known as 'contact homology', see [75].

3.5.2 Transverse knots

Now let K be a homologically trivial transverse knot in a contact 3–manifold (M, ξ), with M oriented such that ξ is a positive contact structure. Let Σ be a Seifert surface for K. As in the definition of the rotation number, we make use of the fact that the plane field $\xi|_\Sigma$ is trivial. Choose a non-vanishing section X of $\xi|_\Sigma$ and push K in the direction of X to obtain a parallel copy K' of K. Choose any orientation for K, and give K' the corresponding orientation. The following definition will not depend on this choice.

Definition 3.5.28 The **self-linking number** $\mathtt{sl}(K, \Sigma)$ of the transverse knot K relative to the Seifert surface Σ is the linking number of K and K'.

Remarks 3.5.29 (1) A transverse knot can be regarded as a 1–dimensional contact submanifold in the sense of Definition 2.1.14, see Remark 2.1.15. So the invariance of $\mathtt{sl}(K,\Sigma)$ under isotopies within the class of transverse knots follows from the isotopy extension theorem for contact submanifolds (Thm. 2.6.12).

(2) The fact that $\mathtt{sl}(K,\Sigma)$ is independent of the choice of X (the non-vanishing section of $\xi|_\Sigma$) follows by the argument used in the proof of Lemma 3.5.14. The dependence on the choice of Σ will be discussed presently. In contrast with the notation for rotation numbers, we do not write $\mathtt{sl}(K,c)$, with $c\in H_2(M,K)$ the class represented by Σ, because this would force us to fix (compatible) orientations of Σ and K.

Proposition 3.5.30 *Let Σ,Σ' be oriented Seifert surfaces for K, inducing the same (boundary) orientation of K. Let $c,c'\in H_2(M,K)$ be the relative homology classes represented by Σ,Σ', respectively. Choose a coorientation (and hence orientation) of the contact structure ξ, and write $e(\xi)$ for the Euler class of this oriented 2–plane bundle. Then*

$$\mathtt{sl}(K,\Sigma)-\mathtt{sl}(K,\Sigma')=\mp\langle e(\xi),c-c'\rangle,$$

where, as in Proposition 3.5.15, $c-c'$ is regarded as a class in $H_2(M)$. The sign in this equation depends on whether K is positively or negatively transverse to ξ for the given choice of coorientation.

Proof Given oriented Seifert surfaces Σ,Σ' (and hence an orientation of K), we may without loss of generality choose the coorientation of ξ such that K is positively transverse.

Now the argument is modelled on that for proving Proposition 3.5.15. With the notation as in that proof – refer once more to Figure 3.13 – we may assume again that the trivialisations of $\xi|_\Sigma$ and $\xi|_{\Sigma'}$ (given by the choice of a non-vanishing vector field X in ξ) coincide along the interval $K\setminus D^2$, and hence induce a trivialisation of ξ over $(\Sigma-\Sigma')\setminus D^2$. We may be more specific and require X along $K\setminus D^2$ to be given by the outward normal to Σ.

Homotopically, we may think of $\xi|_{D^2}$ as the trivial bundle of 'horizontal' planes in Figure 3.13, with standard orientation when seen from above, and with trivialisation given by a constant vector field X_0. The Euler number $\langle e(\xi),c-c'\rangle$, as before, measures the number of rotations of X relative to X_0 along ∂D^2 (traversed in a counterclockwise direction).

On the other hand, the self-linking number $\mathtt{sl}(K,\Sigma)$ is given by the linking number $\mathtt{lk}(K,K')$, where K' is obtained by pushing K in the direction of X. By Proposition 3.4.11, this is the same as the intersection number $K'\bullet\Sigma$.

Our choices imply that along the interval $K\setminus D^2$ we get no contribution to this intersection number. Along $K \cap D^2$, which is homotopic rel endpoints to the left half of ∂D^2 in Figure 3.13, but with opposite orientation, this intersection number is given by the rotations of X relative to X_0.

Likewise, $\mathtt{sl}(K, \Sigma') = K' \bullet \Sigma'$ measures the rotations of X relative to X_0 as we travel along $K \cap D^2$, or the right half of ∂D^2 in Figure 3.13, now with the correct orientation, since we are dealing with $-\Sigma'$ in that figure. Thus, $-\mathtt{sl}(K, \Sigma')$ is again the opposite of what we get when we compute the relevant contribution to $\langle e(\xi), c - c' \rangle$.

Putting this together, we have the claimed formula for a positively transverse knot K. □

The following corollary is proved by essentially the same argument as Corollary 3.5.16.

Corollary 3.5.31 *If $e(\xi) = 0$, then the self-linking number $\mathtt{sl}(K, \Sigma)$ does not depend on the choice of Seifert surface Σ; we then simply write $\mathtt{sl}(K)$. Conversely, if $\mathtt{sl}(K, \Sigma)$ depends on K only, for each transverse knot K in (M, ξ), then $e(\xi) = 0$.* □

For transverse knots in $(\mathbb{R}^3, \xi_{\mathrm{st}})$, this invariant, too, can easily be computed from the front projection.

Proposition 3.5.32 *The self-linking number $\mathtt{sl}(K)$ of a transverse knot K in $(\mathbb{R}^3, \xi_{\mathrm{st}})$ is equal to the writhe of its front projection.*

Proof This argument is completely analogous to that used for proving Proposition 3.5.11. We can take $X = \partial_x$ in the definition of $\mathtt{sl}(K)$. This means that K' is obtained from K by pushing it vertically (with respect to the front projection). Hence, by a small isotopy we may assume that the front projection of K' is a parallel curve to the front projection of K. We then observe that each crossing of the front projection of K contributes a crossing of K' underneath K of the corresponding sign. By Proposition 3.4.14 this implies

$$\mathtt{sl}(K) = \mathtt{lk}(K, K') = \mathrm{writhe}\,(K),$$

as claimed. □

Remark 3.5.33 The analogue of Theorem 3.5.26 for transverse knots has been proved by Eliashberg [70]: topologically trivial transverse knots in $(\mathbb{R}^3, \xi_{\mathrm{st}})$ (or in fact an arbitrary tight contact 3–manifold) with the same self-linking number are isotopic as transverse knots.

The next proposition will be relevant for the construction of contact structures on 3–manifolds (see the proof of Theorem 4.3.1 at the end of Section 4.3).

Proposition 3.5.34 *Every integer can be realised as the self-linking number of some transverse link in* $(\mathbb{R}^3, \xi_{\mathrm{st}})$.

Proof Figure 3.16 shows front projections of positively transverse knots with self-linking number ± 3. From that figure the construction principle for realising any odd integer should be clear. With a two-component link any even integer can be realised. □

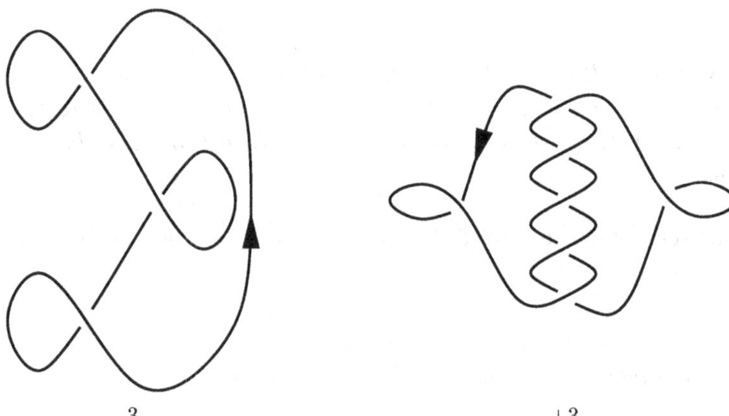

Fig. 3.16. Transverse knots with self-linking number ± 3.

Remark 3.5.35 It is no accident that I do not give an example of a transverse knot with *even* self-linking number, see Remark 4.6.35.

3.5.3 Transverse push-offs

We end this chapter by proving an important relation between the Thurston–Bennequin invariant and rotation number of a homologically trivial Legendrian knot on the one hand, and the self-linking number of its positive or negative transverse push-off (as defined on page 95) on the other.

Proposition 3.5.36 *Let* K *be a homologically trivial oriented Legendrian knot, with compatibly oriented Seifert surface* Σ*, in a contact 3–manifold* (M, ξ)*. Let* $K_{\pm}$ *be the positive or negative, respectively, transverse push-off* *of K. We may regard Σ also as a Seifert surface for $K_\pm$ (since K and $K_\pm$ are topologically isotopic). Then*

$$\mathtt{sl}(K_\pm, \Sigma) = \mathtt{tb}(K) \mp \mathtt{rot}(K, [\Sigma]). \tag{3.1}$$

Proof Let X be a nowhere zero section of $\xi|_\Sigma$. We may (and will) think of X as a section of ξ defined in a neighbourhood of Σ that has Σ as deformation retract. First suppose that $\mathtt{rot}(K, [\Sigma]) = 0$. With $\gamma\colon S^1 \to K$ a positively oriented parametrisation of K, this means that the total number of rotations of γ' relative to X is zero as we traverse K. We may therefore change X by a homotopy so that it is everywhere transverse to γ'. Then $K_\pm$ can be obtained from K by pushing it in the direction of $\pm X$. (We may have to replace X by $-X$ so that the signs correspond nicely.) Let $K'_\pm$ be the knot obtained by pushing $K_\pm$ a little further in the direction of $\pm X$. Then $\mathtt{lk}(K, K_\pm) = \mathtt{lk}(K_\pm, K'_\pm)$. But, by the definitions, the first linking number is precisely $\mathtt{tb}(K)$, the second, $\mathtt{sl}(K_\pm, \Sigma)$.

In the general case, beside the section X of $\xi|_\Sigma$ we also consider a section X_0 of ξ over a neighbourhood of K, transverse to γ' along K. The knots $K_\pm$ can now be defined by pushing K in the direction of $\pm X_0$. The knots $K'_\pm$ we define as before by pushing $K_\pm$ in the direction of $\pm X$, and we let $K^0_\pm$ be the knots obtained by pushing $K_\pm$ further in the direction of $\pm X_0$. Then

$$\mathtt{lk}(K, K_\pm) = \mathtt{lk}(K_\pm, K^0_\pm) = \mathtt{tb}(K) \tag{3.2}$$

and

$$\mathtt{lk}(K_\pm, K'_\pm) = \mathtt{sl}(K_\pm, \Sigma). \tag{3.3}$$

Moreover, $\mathtt{rot}(K, [\Sigma])$ is given by the number of rotations of X_0 relative to X as we traverse K once in positive direction. For the positively transverse knot K_+, rotations of X_0 relative to X contribute positively to $\mathtt{lk}(K_+, K^0_+)$ relative to $\mathtt{lk}(K_+, K'_+)$. For the negatively transverse knot K_-, the sign of this contribution changes. This means that

$$\mathtt{lk}(K_\pm, K^0_\pm) = \mathtt{lk}(K_\pm, K'_\pm) \pm \mathtt{rot}(K, [\Sigma]). \tag{3.4}$$

Equations (3.2), (3.3) and (3.4) imply Equation (3.1). □

Remark 3.5.37 In [26], Bennequin denotes the positively transverse push-off of a Legendrian knot γ by γ^-, the negatively transverse push-off by γ^+. This has led to some sign errors in the literature regarding Equation (3.1).

4
Contact structures on 3–manifolds

'Singularity is almost invariably a clue.'
Arthur Conan Doyle,
The Boscombe Valley Mystery

Throughout the present chapter, M will denote a connected, closed, orientable 3–manifold. Our first major aim here is to prove the one-to-one correspondence between isotopy classes of so-called overtwisted contact structures on M on the one hand, and homotopy classes of tangent 2–plane fields on M on the other.

This programme is carried out in the following steps. In Section 4.1 I present what is essentially Martinet's [175] proof of the existence of a contact structure on every 3–manifold. This construction is based on the surgery description of 3–manifolds due to Lickorish and Wallace (Thm. 1.7.5). The key point is to show how suitable contact forms defined near the boundary of a solid torus can be extended over the whole solid torus. This will enable us to perform Dehn surgeries (along transverse knots) on contact 3–manifolds. For this extension procedure we use an approach due to Thurston and Winkelnkemper [233]; this yields a slight simplification of Martinet's original construction.

In Section 4.2 we show that every orientable 3–manifold is parallelisable and then build on this to classify cooriented tangent 2–plane fields up to homotopy.

In Section 4.3 we study the so-called Lutz twist, a topologically trivial Dehn surgery on a contact manifold (M, ξ) which yields a contact structure ξ' on M that is not, in general, homotopic (as 2–plane field) to ξ. We then use this to show that every tangent 2–plane field on a given 3–manifold M is homotopic to a contact structure. These results are contained in Lutz's thesis [168], albeit in a language that differs markedly from the one used below: for instance, the concept of self-linking number of transverse knots had not been available at the time. Of Lutz's published work, the announcement [167] deals only with the 3–sphere; [169], too, deals only with a restricted problem. I learned the key steps of the construction from an exposition given in Ginzburg's thesis [103], but I have added proofs of many topological details. Sections 4.1 to 4.3 have appeared previously, with minor variations, in [97].

Further proofs of Martinet's theorem are presented in Section 4.4: one by Gonzalo, based on a description of 3–manifolds as branched covers over the 3–sphere, and one by Thurston and Winkelnkemper, based on open book decompositions of 3–manifolds.

In Section 4.5 we discuss the fundamental dichotomy between tight and overtwisted contact structures, introduced by Eliashberg. It turns out that the contact structures found by the Lutz–Martinet construction may always be assumed to be overtwisted. Thus, the proof of the result advertised above is completed (in Section 4.7) by showing that two overtwisted contact structures that are homotopic as plane fields are in fact isotopic as contact structures. This result is due to Eliashberg [64], who in fact proved a stronger statement also involving the higher homotopy groups of the corresponding spaces (a precise formulation will be given below).

This proof relies in an essential way on an understanding of surfaces in contact 3–manifolds and properties of their characteristic foliations. These are investigated systematically in Section 4.6. We include there also the material that will be used later in the classification of tight contact structures: elements of convex surface theory, the elimination lemma for reducing the number of singularities of a characteristic foliation, restrictions on the Euler classes realisable by tight contact structures, and the Bennequin inequality for Legendrian or transverse knots in a tight contact 3–manifold.

Section 4.8 contains a more systematic introduction to convex surface theory, i.e. the theory of surfaces that admit a transverse contact vector field. This concept is central to many developments in modern contact topology, and has proved indispensable for the classification of tight contact structures.

There are at least a couple of approaches towards this classification question. The one I develop in Section 4.9 is tomography, as introduced by Giroux, i.e. the study of contact structures on a product manifold $S\times[-1,1]$, where S is a closed, oriented surface, via the characteristic foliations induced on each $S\times\{z\}$. This is then used in Section 4.10 to classify tight contact structures on S^3, $\mathbb{R}^3$, $S^2\times S^1$, and on the 3–ball.

Section 4.11 contains the proof of Cerf's theorem as advertised in Section 1.7.1 (modulo the technique of filling by holomorphic discs, which is only sketched).

Finally, in Section 4.12 it is shown that the prime decomposition theorem for 3–manifolds has a direct analogue for tight contact 3–manifolds.

4.1 Martinet's construction

The first step towards the classification of overtwisted contact structures on 3–manifolds is the following theorem of Martinet [175].

Theorem 4.1.1 (Martinet) *Every closed, orientable 3–manifold M admits a contact structure.*

In view of the theorem of Lickorish and Wallace and the fact that S^3 admits a contact structure, Martinet's theorem is a direct consequence of the following result.

Theorem 4.1.2 *Let ξ_0 be a contact structure on a 3–manifold M_0. Let M be the manifold obtained from M_0 by a Dehn surgery along a knot K. Then M admits a contact structure ξ which coincides with ξ_0 outside the neighbourhood of K where we perform surgery.*

Proof By Theorem 3.3.1 we may assume that K is positively transverse to ξ_0. Then, by the contact neighbourhood theorem (Example 2.5.16), we can find a tubular neighbourhood νK of K diffeomorphic to $S^1 \times D^2_{\delta_0}$, where K is identified with $S^1 \times \{\mathbf{0}\}$ and $D^2_{\delta_0}$ denotes a disc of radius δ_0, such that the contact structure ξ_0 is given as the kernel of $d\overline{\theta} + \overline{r}^2 d\overline{\varphi}$, with $\overline{\theta}$ denoting the S^1–coordinate and $(\overline{r}, \overline{\varphi})$ polar coordinates on $D^2_{\delta_0}$. Notice that this contact structure is rotationally symmetric about $S^1 \times \{\mathbf{0}\}$, so a transverse knot does not inherit any preferred framing from the contact structure. (We have seen in Definition 3.5.1 that the situation is markedly different for Legendrian knots.)

Now perform a Dehn surgery along K. In order to describe that surgery, we fix an identification of $(\nu K, \xi_0)$ with

$$(S^1 \times D^2_{\delta_0}, \ker(d\overline{\theta} + \overline{r}^2 d\overline{\varphi}))$$

and give K the framing defined by this identification. This means that we cut out $S^1 \times D^2_\delta \subset S^1 \times D^2_{\delta_0}$ for some $\delta < \delta_0$, and we write μ for the meridian $* \times \partial D^2_\delta$ and λ for the longitude $S^1 \times *$, with $*$ denoting a point in S^1 or ∂D^2_δ, respectively.

Now glue back a copy of $S^1 \times D^2_\delta$, whose meridian and longitude we denote by μ_0 and λ_0. If the surgery coefficient of the Dehn surgery with respect to the framing given by λ is $p/q \in \mathbb{Q} \cup \{\infty\}$, this means that the gluing map will be described by

$$\mu_0 \longmapsto p\mu + q\lambda, \quad \lambda_0 \longmapsto m\mu + n\lambda$$

with $\left(\begin{smallmatrix} p & m \\ q & n \end{smallmatrix}\right) \in \mathrm{GL}(2,\mathbb{Z})$, see Section 1.7.2.

Write $(\theta; r, \varphi)$ for the coordinates on the copy of $S^1 \times D^2_\delta$ that we are gluing back. Then the gluing map can be described explicitly by

$$\overline{\theta} = n\theta + q\varphi, \quad \overline{\varphi} = m\theta + p\varphi.$$

So the contact form $d\overline{\theta} + \overline{r}^2 d\overline{\varphi}$ given on $S^1 \times D^2_{\delta_0} \subset M$ pulls back (along $S^1 \times \partial D^2_\delta$) to the contact form

$$d(n\theta + q\varphi) + r^2\, d(m\theta + p\varphi)$$

on the copy of $S^1 \times D^2_\delta$ that we want to glue in. This form is defined on all of $S^1 \times (D^2_\delta \setminus \{\mathbf{0}\})$, and to complete the proof it only remains to find a contact form on $S^1 \times D^2_\delta$ that coincides with this form near $S^1 \times \partial D^2_\delta$.

Lemma 4.1.3 *There is a contact form on $S^1 \times D^2_\delta$ that coincides with $(n + mr^2)\, d\theta + (q + pr^2)\, d\varphi$ near $r = \delta$ and with $\pm d\theta + r^2\, d\varphi$ near $r = 0$.*

Proof We make the ansatz

$$\alpha = h_1(r)\, d\theta + h_2(r)\, d\varphi$$

with smooth functions $h_1(r), h_2(r)$. Then

$$d\alpha = h_1'\, dr \wedge d\theta + h_2'\, dr \wedge d\varphi$$

and

$$\alpha \wedge d\alpha = \begin{vmatrix} h_1 & h_2 \\ h_1' & h_2' \end{vmatrix}\, d\theta \wedge dr \wedge d\varphi.$$

So to satisfy the contact condition $\alpha \wedge d\alpha \neq 0$ all we have to do is to find a parametrised curve

$$r \longmapsto (h_1(r), h_2(r)), \quad 0 \leq r \leq \delta,$$

in the plane satisfying the following conditions:

(1) $h_1(r) = \pm 1$ and $h_2(r) = r^2$ near $r = 0$,
(2) $h_1(r) = n + mr^2$ and $h_2(r) = q + pr^2$ near $r = \delta$,
(3) $(h_1(r), h_2(r))$ is never parallel to $(h_1'(r), h_2'(r))$ for $r \neq 0$.†

Since $np - mq = \pm 1$, the vector (m, p) is not a multiple of (n, q). Figure 4.1 shows possible solution curves for the two cases $np - mq = \pm 1$.

† At $r = 0$, the contact condition is guaranteed by condition 1. The determinant $h_1 h_2' - h_2 h_1'$ vanishes to first order at $r = 0$, as it should, since the volume form really ought to be written as $d\theta \wedge r\, dr \wedge d\varphi$.

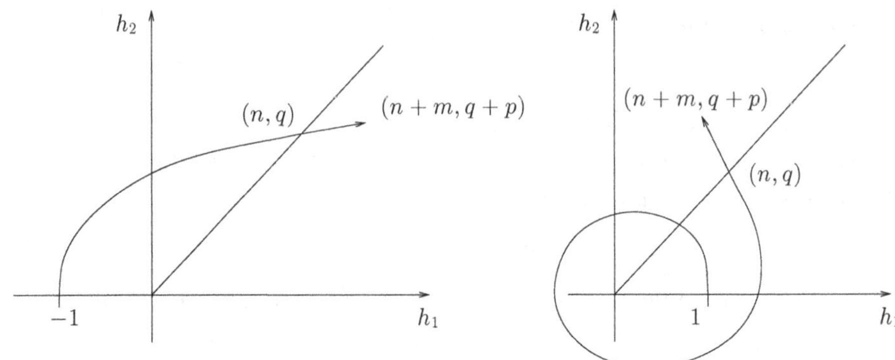

Fig. 4.1. Dehn surgery.

This completes the proof of the lemma and, in consequence, that of Theorem 4.1.2. □

In Section 4.3 we are going to show that in fact every tangent 2–plane field on M is homotopic to a contact structure. Before that, we have to deal with the homotopical classification of such 2–plane fields.

4.2 2–plane fields on 3–manifolds

First we need the following well-known fact. It is unavoidable that at this point we use a little more algebraic or geometric topology than we have done so far. In order to ease the pain, I present three proofs, based on entirely different methods.

Theorem 4.2.1 *Every closed, orientable 3–manifold M is parallelisable, that is, the tangent bundle TM is trivial.*

First Proof – obstruction theoretic The main point will be to show the vanishing of the second Stiefel–Whitney class $w_2(M) = w_2(TM) \in H^2(M;\mathbb{Z}_2)$. Recall the following facts, which can be found in [35]; for the interpretation of Stiefel–Whitney classes as obstruction classes see also [188].

There are Wu classes $v_i \in H^i(M;\mathbb{Z}_2)$ defined by

$$\langle \mathrm{Sq}^i(u), [M]\rangle = \langle v_i \cup u, [M]\rangle$$

for all $u \in H^{3-i}(M;\mathbb{Z}_2)$, where Sq denotes the Steenrod squaring operations. Since $\mathrm{Sq}^i(u) = 0$ for $i > 3-i$, the only (potentially) non-zero Wu classes are $v_0 = 1$ and v_1. The Wu classes and the Stiefel–Whitney classes are related by $w_q = \sum_j \mathrm{Sq}^{q-j}(v_j)$. Hence $v_1 = \mathrm{Sq}^0(v_1) = w_1$, which is the zero class, because M is orientable. We conclude $w_2 = 0$.

Let $V_2(\mathbb{R}^3) = \mathrm{SO}(3)/\mathrm{SO}(1) = \mathrm{SO}(3)$ be the Stiefel manifold of oriented, orthonormal 2–frames in $\mathbb{R}^3$. This is connected, so there exists a section over the 1–skeleton† of M of the 2–frame bundle $V_2(TM)$ associated with TM (with a choice of Riemannian metric on M understood‡). The obstruction to extending this section over the 2–skeleton is equal to w_2, which vanishes as we have just seen. The obstruction to extending the section over all of M lies in $H^3(M; \pi_2(V_2(\mathbb{R}^3)))$, which is the zero group because of $\pi_2(\mathrm{SO}(3)) = 0$. (For that latter fact, recall that SO(3) is diffeomorphic to $\mathbb{R}P^3$, or appeal to the general result that the second homotopy group of any compact Lie group is trivial, see [37, Thm. V.7.1].)

We conclude that TM has a trivial 2–dimensional sub-bundle ϵ^2. The complementary 1–dimensional bundle $\lambda = TM/\epsilon^2$ is orientable and hence trivial since $0 = w_1(TM) = w_1(\epsilon^2) + w_1(\lambda) = w_1(\lambda)$. Thus $TM = \epsilon^2 \oplus \lambda$ is a trivial bundle. □

Second Proof – geometric topological We begin by observing that the 3–sphere is parallelisable. An explicit parallelisation of $S^3 \subset \mathbb{R}^4$ is given by

$$\begin{aligned} X_1(x_0,x_1,x_2,x_3) &= (-x_1,x_0,-x_3,x_2), \\ X_2(x_0,x_1,x_2,x_3) &= (-x_2,x_3,x_0,-x_1), \\ X_3(x_0,x_1,x_2,x_3) &= (-x_3,-x_2,x_1,x_0), \end{aligned}$$

where $\sum_{\nu=0}^3 x_\nu^2 = 1$. This parallelisation is found by identifying $\mathbb{R}^4$ with the quaternions

$$\mathbb{H} = \{x_0 + \mathrm{i}x_1 + \mathrm{j}x_2 + \mathrm{k}x_3 \colon\ x_0,x_1,x_2,x_3 \in \mathbb{R}\},$$

where $\mathrm{i}^2 = \mathrm{j}^2 = \mathrm{k}^2 = \mathrm{ijk} = -1$, and hence $\mathrm{ij} = -\mathrm{ji} = \mathrm{k}$. Then

$$\mathbf{n}(x_1,x_2,x_3,x_4) := (x_1,x_2,x_3,x_4)$$

defines the normal vector field to S^3, and X_1, X_2, X_3 are given as $\mathrm{i}\mathbf{n}$, $\mathrm{j}\mathbf{n}$, $\mathrm{k}\mathbf{n}$.

A theorem of Hilden, Montesinos and Thickstun [130] states that for every closed, orientable 3–manifold M there is a branched covering§ $p\colon M \to S^3$, branched along a simple closed curve in M that bounds an embedded disc. (Moreover, the cover can be chosen 3–fold and simple, i.e. the monodromy representation of $\pi_1(S^3 \setminus K)$, where K – a knot in S^3 – is the downstairs

† Every differential manifold has the homotopy type of a CW–complex, see [132], and in arguments like these it is understood that such a cell decomposition has been chosen.

‡ This is not necessary, of course. One may also work with arbitrary 2–frames without reference to a metric. This does not affect the homotopical data.

§ The notion of a branched covering will be introduced more formally in Section 4.4.1 and, for higher-dimensional manifolds, in Section 7.5.

branch set, represents the meridian of K by a transposition in the symmetric group S_3. This, however, is not relevant for our discussion.)

Now, given M and a branched covering $p\colon M \to S^3$ of the type described, there is a 3–ball $D^3 \subset M$ containing the upstairs branch set in its interior. Outside of† $\mathrm{Int}\,(D^3)$, the covering p is unbranched, so the 3–frame X_1, X_2, X_3 on S^3 can be lifted to a frame field on $M \setminus \mathrm{Int}\,(D^3)$.

The bundle $TM|_{D^3}$ is trivial, so relative to a chosen trivialisation of this bundle, the frame field defined previously determines an element of $\mathrm{SO}(3)$ (see the first footnote in the preceding proof) at each point of ∂D^3. Since $\pi_2(\mathrm{SO}(3)) = 0$, this frame field extends over D^3. □

For the third proof of this theorem we need a preparatory lemma.

Lemma 4.2.2 *Let $\Sigma \subset M$ be an embedded (possibly non-orientable) closed, connected surface and E the total space of the normal disc bundle of Σ in M. Then $w_2(E) = 0$.*

Proof Write η for this normal disc bundle and $\pi\colon E \to \Sigma$ for the bundle projection. The space E retracts to Σ, so

$$w_2(E) = \pi^*\big(w_2(TE|_\Sigma)\big) = \pi^*\big(w_2(T\Sigma \oplus \eta)\big) = \pi^*\big(w_2(\Sigma) + w_1(\Sigma)w_1(\eta)\big).$$

If Σ is orientable, then both $w_1(\Sigma)$ and $w_2(\Sigma)$ vanish (since w_2 is the mod 2 reduction of the Euler class), hence $w_2(E) = 0$ in this case.

If Σ is non-orientable, then Σ is diffeomorphic to a connected sum of h copies of $\mathbb{R}P^2$ for some natural number h, with $H^1(\Sigma;\mathbb{Z}_2) \cong \mathbb{Z}_2^h$, with each summand $\mathbb{Z}_2$ corresponding to a summand $\mathbb{R}P^2$. Then $w_1(\Sigma) = (1, \ldots, 1)$. Since M is orientable, every orientation-reversing loop in Σ for $T\Sigma$ is also orientation-reversing for η, hence $w_1(\eta) = (1, \ldots, 1)$, too. We conclude, identifying $H^2(E;\mathbb{Z}_2) \cong H^2(\Sigma;\mathbb{Z}_2)$ with $\mathbb{Z}_2$, that

$$w_2(E) = \pi^*\big(w_2(\Sigma) + w_1(\Sigma)w_1(\eta)\big) = 2 - h + (1, \ldots, 1)^2 = 2 - h + h = 0,$$

as was to be shown. □

Third Proof of Theorem 4.2.1 – spin structures This proof is due to Kirby [146] and is based on the classification of spin structures. For the present purpose we follow Kirby in defining a **spin structure** on the 3–manifold M as a homotopy class of trivialisations of TM over the 1–skeleton of M that extend over the 2–skeleton. This is in analogy with the definition of an orientation as a homotopy class of trivialisations of TM over the 0–skeleton that extend over the 1–skeleton.

† We write $\mathrm{Int}\,(-)$ for the interior of a manifold with boundary.

The key facts about spin structures (for which a proof can be found in the cited reference) are that a spin structure exists if and only if $w_2(M)$ is the zero class, and spin structures are then in one-to-one correspondence with elements of the cohomology group $H^1(M;\mathbb{Z}_2)$. As in the first proof, we want to show $w_2(M)=0$; the present proof then concludes like the first one.

Arguing by contradiction, we assume that $w_2(M)\neq 0$. Since M has vanishing Euler characteristic, it admits a nowhere vanishing vector field, see [188, §12] or [35, VI.12], that is, the tangent bundle TM splits into a trivial line bundle and a complementary orientable plane bundle. It follows that $w_2(M)$ is the mod 2 reduction of the Euler class of this plane bundle. So the Poincaré dual of $w_2(M)$ in $H_1(M;\mathbb{Z}_2)$ likewise admits an integral lift and can therefore be represented by an embedded circle $K\subset M$. If $[F]$ is a class in $H_2(M\setminus K;\mathbb{Z}_2)$, then the Kronecker product $\langle w_2(M),[F]\rangle$ equals the intersection product $[K]\bullet[F]=0$. This means that under the inclusion map $M\setminus K\rightarrow M$, the class $w_2(M)$ pulls back to $w_2(M\setminus K)=0$. In other words, $M\setminus K$ has a spin structure σ that does not extend over K. Now let $\Sigma\subset M$ be an embedded surface, perhaps non-orientable, that is dual to K in the sense of Proposition 3.4.7. Let p be the single point of intersection between K and Σ.

By Lemma 4.2.2 (and with the notation used there), the total space E of the normal disc bundle over Σ admits a spin structure σ. Spin structures on E are classified by the group $H^1(E;\mathbb{Z}_2)\cong H^1(\Sigma;\mathbb{Z}_2)$, which is isomorphic to the group $H^1(\Sigma\setminus\{p\};\mathbb{Z}_2)$ classifying spin structures on $E|_{\Sigma\setminus\{p\}}$. So the spin structure on $E|_{\Sigma\setminus\{p\}}$ given by the restriction of σ must agree with one on E, which means that σ extends over K, contradicting our choice of σ.

This contradiction gives $w_2(M)=0$, as was to be shown. □

Fix an arbitrary Riemannian metric on M and a trivialisation of the unit tangent bundle $STM\cong M\times S^2$. By associating with a cooriented 2–plane distribution in TM the positively orthonormal vector field, we have a one-to-one correspondence between the following sets, where all maps, homotopies etc., are understood to be smooth.

- Homotopy classes of unit vector fields X on M.
- Homotopy classes of cooriented 2–plane distributions ξ in TM.
- Homotopy classes of maps $f\colon M\rightarrow S^2$.

(Recall that I use the term 2–**plane distribution** synonymously with '2–dimensional sub-bundle of the tangent bundle'.) The map $f=f_\xi$ is the 'Gauß map' associated with the plane field ξ.

Remark 4.2.3 With a Riemannian metric chosen on an oriented 3–manifold M, a cooriented (and hence oriented) tangent 2–plane field ξ can also be interpreted as an almost contact structure: the complex bundle structure on ξ is defined by rotation by $\pi/2$ in positive direction with respect to the metric and orientation on ξ; the complementary trivial line bundle is the bundle orthogonal to ξ.

Let ξ_1, ξ_2 be two arbitrary 2–plane distributions (always understood to be cooriented). By elementary obstruction theory, there is an obstruction

$$d^2(\xi_1, \xi_2) \in H^2(M; \pi_2(S^2)) \cong H^2(M; \mathbb{Z})$$

for ξ_1 to be homotopic to ξ_2 over the 2–skeleton of M and, if $d^2(\xi_1, \xi_2) = 0$ and after homotoping ξ_1 to ξ_2 over the 2–skeleton, an obstruction (which will depend, in general, on that first homotopy)

$$d^3(\xi_1, \xi_2) \in H^3(M; \pi_3(S^2)) \cong H^3(M; \mathbb{Z}) \cong \mathbb{Z}$$

for ξ_1 to be homotopic to ξ_2 over all of M. (The identification of $H^3(M; \mathbb{Z})$ with $\mathbb{Z}$ is determined by the orientation of M.) However, rather than relying on general obstruction theory, we shall interpret d^2 and d^3 geometrically, which will later allow us to give a geometric proof that every homotopy class of 2–plane fields ξ on M contains a contact structure.

The only fact that I want to quote here is that, by the Pontrjagin–Thom construction, homotopy classes of maps $f\colon M \to S^2$ are in one-to-one correspondence with framed cobordism classes of framed (and oriented) links of 1–spheres in M. The general theory can be found in [35] and [153]; a beautiful and elementary account is given in [186].

For given f, the correspondence is defined by choosing a regular value $p \in S^2$ for f and a positively oriented basis $\mathfrak{b}$ of T_pS^2, and associating with it the oriented framed link $(f^{-1}(p), f^*\mathfrak{b})$, where $f^*\mathfrak{b}$ is the pull-back of $\mathfrak{b}$ under the fibrewise bijective map $Tf\colon T(f^{-1}(p))^\perp \to T_pS^2$. The orientation of $f^{-1}(p)$ is the one which together with the frame $f^*\mathfrak{b}$ gives the orientation of M.

For a given framed link L, the corresponding f is defined by projecting a (trivial) disc bundle neighbourhood $L \times D^2$ of L in M onto the fibre $D^2 \cong S^2 \setminus \{p^*\}$, where $\mathbf{0}$ is identified with p, and p^* denotes the antipode of p, and sending $M \setminus (L \times D^2)$ to p^*. Notice that the orientations of M and the components of L determine that of the fibre D^2, and hence determine the map f.

Before proceeding to define the obstruction classes d^2 and d^3, we make another short topological digression.

4.2.1 Cobordism classes of links

We now want to relate elements in $H_1(M;\mathbb{Z})$ to cobordism classes of links in M.

Theorem 4.2.4 *Let M be a closed, oriented 3–manifold. Any homology class $c \in H_1(M;\mathbb{Z})$ is represented by an embedded, oriented link (of 1–spheres) L_c in M. Two links L_0, L_1 represent the same class $[L_0] = [L_1]$ if and only if they are oriented cobordant in M, that is, there is an embedded, oriented surface S in $[0,1] \times M$ with*

$$\partial S = L_1 \sqcup \overline{L}_0 \subset \{1\} \times M \sqcup \{0\} \times M,$$

where $\sqcup$ denotes disjoint union and $\overline{L}$ the link L with reversed orientation.

Proof The first part was already proved in Proposition 3.4.2. We retain the notation used there.

If two links L_0, L_1 are cobordant in M, then clearly

$$[L_0] = [L_1] \in H_1([0,1] \times M;\mathbb{Z}) \cong H_1(M;\mathbb{Z}).$$

For the converse, suppose we are given two links $L_0, L_1 \subset M$ with $[L_0] = [L_1]$. Choose arbitrary framings for these links and use this, as described above, to define smooth maps $f_0, f_1\colon\ M \to S^2$ with common regular value $p \in S^2$ such that $f_i^{-1}(p) = L_i$, $i = 0, 1$. Now identify S^2 with the standardly embedded $\mathbb{C}P^1 \subset \mathbb{C}P^2$. Let $P \subset \mathbb{C}P^2$ be a second copy of $\mathbb{C}P^1$, embedded in such a way that $[P]_{\mathbb{C}P^2} = [\mathbb{C}P^1]_{\mathbb{C}P^2}$ and P intersects $\mathbb{C}P^1$ transversely in p only. This is possible since $\mathbb{C}P^1 \subset \mathbb{C}P^2$ has self-intersection one. Then the maps f_0, f_1, regarded as maps into $\mathbb{C}P^2$, are transverse to P and we have $f_i^{-1}(P) = L_i$, $i = 0, 1$. Hence

$$f_i^* u_0 = f_i^*(PD[P]_{\mathbb{C}P^2}) = PD[f_i^{-1}(P)]_M = PD[L_i]_M$$

is the same for $i = 0$ or 1, and from the identification

$$\begin{array}{ccc} [M, \mathbb{C}P^2] & \xrightarrow{\cong} & H^2(M,\mathbb{Z}) \\ [f] & \longmapsto & f^* u_0 \end{array}$$

we conclude that f_0 and f_1 are homotopic as maps into $\mathbb{C}P^2$.

Let $F\colon\ [0,1] \times M \to \mathbb{C}P^2$ be a homotopy between f_0 and f_1, which we may assume to be constant near 0 and 1. This F can be smoothly approximated by a map $F'\colon\ [0,1] \times M \to \mathbb{C}P^2$ that is transverse to P and coincides with F near $\{0\} \times M$ and $\{1\} \times M$ (since there the transversality condition was already satisfied). In particular, F' is still a homotopy between f_0 and f_1, and $S = (F')^{-1}(P)$ is a surface with $\partial S = L_1 \sqcup \overline{L}_0$. □

Notice that in the course of this proof we have observed that oriented cobordism classes of oriented links in M – equivalently: classes in $H_1(M;\mathbb{Z})$ – correspond to homotopy classes of maps $M \to \mathbb{C}P^2$, whereas oriented framed cobordism classes of oriented framed links correspond to homotopy classes of maps $M \to \mathbb{C}P^1$.

4.2.2 Framed cobordisms

We have seen that if $L_1, L_2 \subset M$ are links with $[L_1] = [L_2] \in H_1(M;\mathbb{Z})$, then L_1 and L_2 are cobordant in M.† In general, however, a given framing on L_1 and L_2 does not extend over the cobordism. The following observation will be useful later on.

Write (S^1, n) for a contractible loop in M with framing $n \in \mathbb{Z}$ (by which we mean that S^1 and a second copy of S^1 obtained by pushing it away in the direction of one of the vectors in the frame have linking number n). When writing $L = L' \sqcup (S^1, n)$ it is understood that (S^1, n) is not linked with any component of L'.

Suppose we have two framed links $L_0, L_1 \subset M$ with $[L_0] = [L_1]$. Let $S \subset [0,1] \times M$ be an embedded surface with

$$\partial S = L_1 \sqcup \overline{L}_0 \subset \{1\} \times M \sqcup \{0\} \times M.$$

With D^2 a small disc embedded in S, the framing of L_1 and L_2 in M extends to a framing of $S \setminus D^2$ in $[0,1] \times M$ (since $S \setminus D^2$ deformation retracts to a 1–dimensional complex containing L_0 and L_1, and over such a complex any orientable 2–plane bundle is trivial). Now we embed a cylinder $[0,1] \times S^1$ into $[0,1] \times M$ such that

$$[0,1] \times S^1 \cap \{0\} \times M = \emptyset,$$

$$[0,1] \times S^1 \cap \{1\} \times M = \{1\} \times S^1,$$

and

$$[0,1] \times S^1 \cap (S \setminus \mathrm{Int}\,(D^2)) = \{0\} \times S^1 = \partial D^2,$$

see Figure 4.2. This shows that L_0 is framed cobordant in M to $L_1 \sqcup (S^1, n)$ for suitable $n \in \mathbb{Z}$.

4.2.3 Definition of the obstruction classes

We are now in a position to define the obstruction classes d^2 and d^3. With a choice of Riemannian metric on M and a trivialisation of STM understood,

† As before, all links and cobordisms are understood to be oriented.

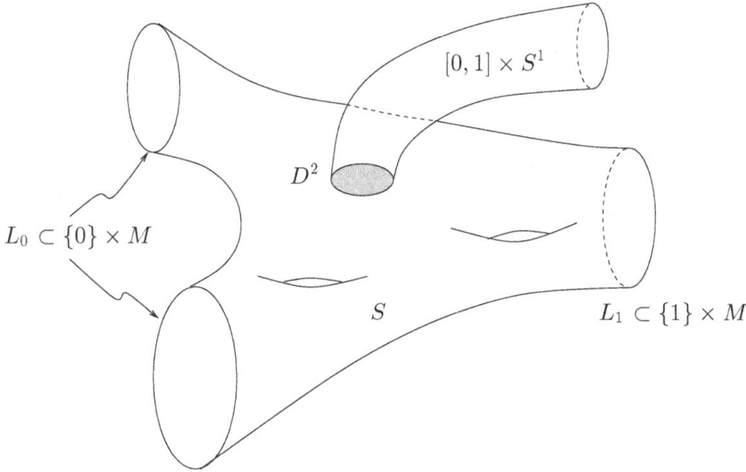

Fig. 4.2. The framed cobordism between L_0 and $L_1 \sqcup (S^1, n)$.

a 2–plane distribution ξ on M defines a map $f_\xi \colon M \to S^2$ and hence an oriented framed link L_ξ as described above. Let $[L_\xi] \in H_1(M;\mathbb{Z})$ be the homology class represented by L_ξ. This depends only on the homotopy class of ξ, since under homotopies of ξ, or choice of different regular values of f_ξ, the cobordism class of L_ξ remains invariant. We define

$$d^2(\xi_1, \xi_2) = PD[L_{\xi_1}] - PD[L_{\xi_2}].$$

With this definition, d^2 is clearly additive, that is,

$$d^2(\xi_1, \xi_2) + d^2(\xi_2, \xi_3) = d^2(\xi_1, \xi_3).$$

The following lemma shows that d^2 is indeed the desired obstruction class.

Lemma 4.2.5 *The 2–plane distributions ξ_1 and ξ_2 are homotopic over the 2–skeleton $M^{(2)}$ of M if and only if $d^2(\xi_1, \xi_2) = 0$.*

Proof Suppose $d^2(\xi_1, \xi_2) = 0$, that is, $[L_{\xi_1}] = [L_{\xi_2}]$. By Theorem 4.2.4 we find a surface S in $[0,1] \times M$ with

$$\partial S = L_{\xi_2} \sqcup \overline{L}_{\xi_1} \subset \{1\} \times M \sqcup \{0\} \times M.$$

From the discussion on framed cobordism above, we know that for suitable $n \in \mathbb{Z}$ we find a *framed* surface S' in $[0,1] \times M$ such that

$$\partial S' = \big(L_{\xi_2} \sqcup (S^1, n)\big) \sqcup \overline{L}_{\xi_1} \subset \{1\} \times M \sqcup \{0\} \times M$$

as framed manifolds.

Hence ξ_1 is homotopic to a 2–plane distribution ξ_1' such that $L_{\xi_1'}$ and L_{ξ_2} differ only by one contractible framed loop (not linked with any other component). Then the corresponding maps f_1', f_2 differ only in a neighbourhood of this loop, which is contained in a 3–ball, so f_1' and f_2 (and hence ξ_1' and ξ_2) agree over the 2–skeleton.

Conversely, if ξ_1 and ξ_2 are homotopic over $M^{(2)}$, we may assume $\xi_1 = \xi_2$ on $M \setminus D^3$ for some embedded 3–disc $D^3 \subset M$ without changing $[L_{\xi_1}]$ and $[L_{\xi_2}]$. Now $[L_{\xi_1}] = [L_{\xi_2}]$ follows from $H_1(D^3, S^2) = 0$. □

Remark 4.2.6 By [224, §37], the obstruction to homotopy between ξ and ξ_0 (corresponding to the constant map $f_{\xi_0} : M \to S^2$) over the 2–skeleton of M is given by $f_\xi^* u_0$, where u_0 is the positive generator of $H^2(S^2; \mathbb{Z})$. So $u_0 = PD[p]$ for any $p \in S^2$, and taking p to be a regular value of f_ξ we have

$$f_\xi^* u_0 = f_\xi^* PD[p] = PD[f_\xi^{-1}(p)] = PD[L_\xi] = d^2(\xi, \xi_0).$$

This gives an alternative way to see that our geometric definition of d^2 does indeed coincide with the one in classical obstruction theory.

Now suppose $d^2(\xi_1, \xi_2) = 0$. We may then assume that ξ_1 and ξ_2 coincide on $M \setminus \mathrm{Int}\,(D^3)$, and we define $d^3(\xi_1, \xi_2)$ to be the Hopf invariant $H(f)$ of the map $f\colon S^3 \to S^2$ defined as $f_1 \circ \pi_+$ on the upper hemisphere and $f_2 \circ \pi_-$ on the lower hemisphere, where π_+, π_- are the orthogonal projections of the upper and lower hemisphere, respectively, onto the equatorial disc, which we identify with $D^3 \subset M$. Here, given an orientation of M, we orient S^3 in such a way that π_+ is orientation-preserving and π_- orientation-reversing; the orientation of S^2 is inessential for the computation of $H(f)$. Recall that $H(f)$ is defined as the linking number of the preimages of two distinct regular values of a smooth map homotopic to f. Since the Hopf invariant classifies homotopy classes of maps $S^3 \to S^2$ (it is in fact an isomorphism $\pi_3(S^2) \to \mathbb{Z}$), this is a suitable definition for the obstruction class d^3. Moreover, the homomorphism property of $H(f)$ and the way addition in $\pi_3(S^2)$ is defined entail the additivity of d^3 analogous to that of d^2.

4.3 The Lutz twist

Consider a 3–manifold M with cooriented contact structure ξ, and an oriented 1–sphere $K \subset M$ embedded transversely to ξ such that the positive orientation of K coincides with the positive coorientation of ξ. Then, as we have seen in Section 4.1, in suitable local coordinates we can identify K with $S^1 \times \{\mathbf{0}\} \subset S^1 \times D^2$ such that $\xi = \ker(d\theta + r^2\, d\varphi)$ and ∂_θ corresponds to the positive orientation of K. Strictly speaking, this formula for ξ holds on

$S^1 \times D^2_\delta$ for some, possibly small, $\delta > 0$. However, to simplify notation we usually work with $S^1 \times D^2$ as local model.

We say that ξ' is obtained from ξ by a **Lutz twist** along K, and write $\xi' = \xi^K$, if on $S^1 \times D^2$ the new contact structure ξ' is defined by

$$\xi' = \ker(h_1(r)\, d\theta + h_2(r)\, d\varphi)$$

with $(h_1(r), h_2(r))$ as in Figure 4.3, and ξ' coincides with ξ outside the solid torus $S^1 \times D^2$.

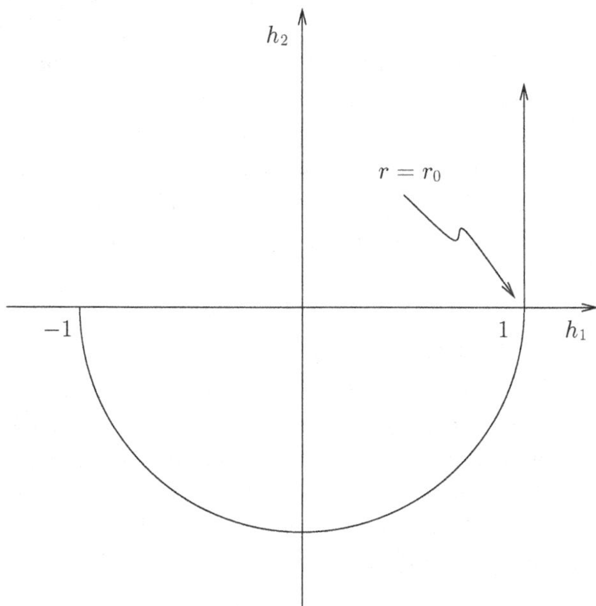

Fig. 4.3. Lutz twist.

More precisely, $(h_1(r), h_2(r))$ is required to satisfy the conditions

(1) $h_1(r) = -1$ and $h_2(r) = -r^2$ near $r = 0$,
(2) $h_1(r) = 1$ and $h_2(r) = r^2$ near $r = 1$,
(3) $(h_1(r), h_2(r))$ is never parallel to $(h_1'(r), h_2'(r))$ for $r \neq 0$.

This is the same as applying the construction of Section 4.1 to the topologically trivial Dehn surgery given by the gluing map $\mu_0 \mapsto -\mu$, $\lambda_0 \mapsto -\lambda$.

It will be useful later on to understand more precisely the behaviour of the map $f_{\xi'}\colon M \to S^2$. For the definition of this map, we assume – this assumption will be justified below – that on $S^1 \times D^2$ the map f_ξ was defined in terms of the standard metric $d\theta^2 + du^2 + dv^2$ (with (u, v) Cartesian coordinates on D^2 corresponding to the polar coordinates (r,φ)) and the trivialisation $\partial_\theta, \partial_u, \partial_v$ of $T(S^1 \times D^2)$. Since ξ' is spanned by ∂_r and $h_2(r)\,\partial_\theta - h_1(r)\,\partial_\varphi$ (or, for $r=0$, by ∂_u, ∂_v), a vector positively orthogonal to ξ' is given by $h_1(r)\,\partial_\theta + h_2(r)\,\partial_\varphi$, which makes sense even for $r=0$. Observe that the ratio $h_1(r)/h_2(r)$ (for $h_2(r)\neq 0$) is a strictly monotone decreasing function, for by the third condition above we have

$$(h_1/h_2)' = (h_1'h_2 - h_1h_2')/h_2^2 < 0.$$

This implies that any value on S^2 other than $(1,0,0)$ (corresponding to ∂_θ) is regular for the map $f_{\xi'}$; the value $(1,0,0)$ is attained along the torus $\{r=r_0\}$, with $r_0>0$ determined by $h_2(r_0)=0$, and hence not regular.

If $S^1\times D^2$ is endowed with the orientation defined by the volume form $d\theta\wedge r\,dr\wedge d\varphi = d\theta\wedge du\wedge dv$ (so that ξ and ξ' are positive contact structures) and $S^2\subset\mathbb{R}^3$ is given its 'usual' orientation defined by the volume form $x\,dy\wedge dz + y\,dz\wedge dx + z\,dx\wedge dy$, then $f_{\xi'}^{-1}(-1,0,0) = S^1\times\{\mathbf{0}\}$ with orientation given by $-\partial_\theta$, since $f_{\xi'}$ maps the slices $\{\theta\}\times D^2_{r_0}$ in an orientation-reversing manner onto S^2.

More generally, for any $p\in S^2\setminus\{(1,0,0)\}$ the preimage $f_{\xi'}^{-1}(p)$ (inside the domain $\{(\theta,r,\varphi)\colon\ h_2(r)<0\} = \{r<r_0\}$) is a circle $S^1\times\{\mathbf{u}\}$, $\mathbf{u}\in D^2$, with orientation given by $-\partial_\theta$.

We are now ready to prove what is commonly known as the Lutz–Martinet theorem (see the references given in the introduction to this chapter).

Theorem 4.3.1 (Lutz–Martinet) *Every cooriented tangent 2–plane field on a closed, orientable 3–manifold is homotopic to a contact structure.*

The idea of the proof is to start with an arbitrary contact structure, and then to perform suitable Lutz twists. First we deal with homotopy over the 2–skeleton. One way to proceed would be to prove directly, with notation as above, that $d^2(\xi^K,\xi) = -PD[K]$. However, it is somewhat easier to compute $d^2(\xi^K,\xi)$ in the case where ξ is a trivial 2–plane bundle and the trivialisation of STM is adapted to ξ. Since I would anyway like to present an alternative argument for computing the effect of a Lutz twist on the Euler class of the contact structure, and thus relate $d^2(\xi_1,\xi_2)$ to the Euler classes of ξ_1 and ξ_2, it seems opportune to do this first and use it to show the existence of a contact structure with Euler class zero. In the next section we shall actually discuss a direct geometric proof of the existence of a contact structure with Euler class zero (Remark 4.4.2).

Recall that the Euler class $e(\xi)\in H^2(B;\mathbb{Z})$ of a 2–plane bundle over a complex B (of arbitrary dimension) is the obstruction to finding a nowhere zero section of ξ over the 2–skeleton of B. Since $\pi_i(S^1)=0$ for $i\geq 2$, all higher obstruction groups $H^{i+1}(B; \pi_i(S^1))$ are trivial, so a 2–dimensional orientable bundle ξ is trivial if and only if $e(\xi) = 0$, no matter what the dimension of B.

Now let ξ be an arbitrary cooriented 2–plane distribution on an oriented 3–manifold M. Then $TM \cong \xi \oplus \epsilon$, where ϵ denotes a trivial line bundle. Hence $w_2(\xi) = w_2(\xi \oplus \epsilon) = w_2(TM) = 0$, and since $w_2(\xi)$ is the mod 2 reduction of $e(\xi)$ we infer that $e(\xi)$ has to be even.

Proposition 4.3.2 *For any even element $e \in H^2(M;\mathbb{Z})$ there is a contact structure ξ on M with $e(\xi) = e$.*

Proof Start with an arbitrary contact structure ξ_0 on M with $e(\xi_0) = e_0$ (which we know to be even). Given any even $e_1 \in H^2(M;\mathbb{Z})$, represent the Poincaré dual of $(e_0 - e_1)/2$ by a collection of embedded oriented circles positively transverse to ξ_0.† Choose a section of ξ_0 transverse to the zero section of ξ_0, that is, a vector field in ξ_0 with generic zeros. We may assume that there are no zeros on the curves representing $PD^{-1}(e_0 - e_1)/2$. Now perform a Lutz twist as described above along these curves, and call ξ_1 the resulting contact structure. It is easy to see that, in the local model for the Lutz twist, a constant vector field tangent to ξ_0 along $\partial(S^1 \times D^2_{r_0})$ extends to a vector field tangent to ξ_1 over $S^1 \times D^2_{r_0}$ with zeros of index +2 along $S^1 \times \{\mathbf{0}\}$ (Figure 4.4). So the vector field in ξ_0 extends to a vector field in ξ_1 with new zeros of index +2 along the curves representing $PD^{-1}(e_1 - e_0)/2$ (*sic*; notice that a Lutz twist along a positively transverse knot K turns K into a negatively transverse knot). Since the self-intersection class of M in the total space of a vector bundle is Poincaré dual to the Euler class of that bundle, see [35, Prop. VI.12.8], this proves $e(\xi_1) = e(\xi_0) + e_1 - e_0 = e_1$. □

We now fix a contact structure ξ_0 on M with $e(\xi_0) = 0$ and give M the orientation induced by ξ_0 (i.e. the one for which ξ_0 is a positive contact structure). Moreover, we fix a Riemannian metric on M and define X_0 as the vector field positively orthonormal to ξ_0. Since ξ_0 is a trivial plane bundle, X_0 extends to an orthonormal frame X_0, X_1, X_2, hence a trivialisation of STM, with X_1, X_2 tangent to ξ_0 and defining the orientation of ξ_0. With these choices, ξ_0 corresponds to the constant map $f_{\xi_0} : M \to (1,0,0) \in S^2$.

Proposition 4.3.3 *Let $K \subset M$ be an embedded, oriented circle positively transverse to ξ_0. Then $d^2(\xi_0^K, \xi_0) = -PD[K]$.*

† Here by $(e_0 - e_1)/2$ I mean some class whose double equals $e_0 - e_1$; in the presence of 2–torsion there is of course a choice of such classes.

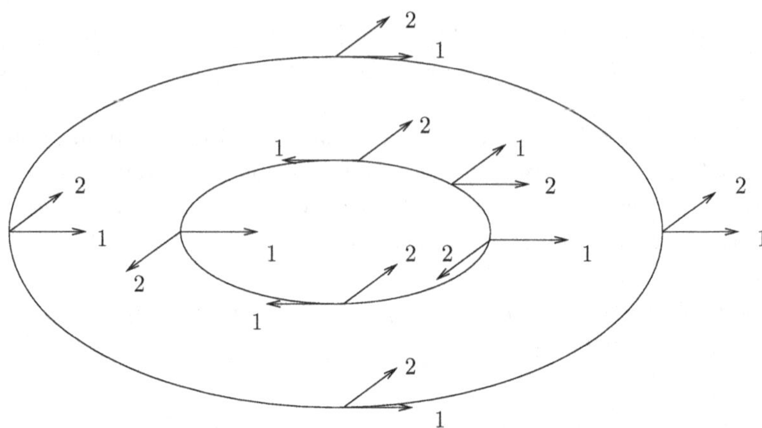

Fig. 4.4. Effect of Lutz twist on Euler class.

Proof Identify a tubular neighbourhood of $K \subset M$ with $S^1 \times D^2$ with framing defined by X_1, and ξ_0 given in this neighbourhood as the kernel of $d\theta + r^2\,d\varphi = d\theta + u\,dv - v\,du$. We may then change the trivialisation X_0, X_1, X_2 by a homotopy, fixed outside $S^1 \times D^2$, such that $X_0 = \partial_\theta$, $X_1 = \partial_u$ and $X_2 = \partial_v$ near K; this does not change the homotopical data of 2–plane distributions computed via the Pontrjagin–Thom construction. Then f_{ξ_0} is no longer constant, but its image still does not contain the point $(-1,0,0)$.

Now perform a Lutz twist along $K \times \{\mathbf{0}\}$. Our discussion at the beginning of this section shows that $(-1,0,0)$ is a regular value of the map $f_\xi\colon M \to S^2$ associated with $\xi = \xi_0^K$ and $f_\xi^{-1}(-1,0,0) = \overline{K}$. Hence, by definition of the obstruction class d^2 we have $d^2(\xi_0^K, \xi_0) = -PD[K]$. □

Remark 4.3.4 From this proposition, the additivity of d^2, and the proof of Proposition 4.3.2, we conclude that

$$2d^2(\xi, \xi') = e(\xi) - e(\xi')$$

for any contact structures (or in fact arbitrary 2–plane distributions) ξ, ξ' on M.

Proof of Theorem 4.3.1 Let η be a 2–plane distribution on M, and let ξ_0 be the contact structure on M with $e(\xi_0) = 0$ that we fixed earlier on. According to our discussion in Section 3.4.2 and Theorem 3.3.1, we can find an oriented knot K positively transverse to ξ_0 with $-PD[K] = d^2(\eta, \xi_0)$. Then $d^2(\eta, \xi_0) = d^2(\xi_0^K, \xi_0)$ by the preceding proposition, and therefore $d^2(\xi_0^K, \eta) = 0$.

We may then assume that $\eta = \xi_0^K$ on $M \setminus D^3$, where we choose D^3 so small that ξ_0^K is in Darboux normal form there (and identical with ξ_0). By Proposition 3.5.34 we can find a link K' in D^3 transverse to ξ_0^K with self-linking number $\mathtt{sl}(K')$ equal to $d^3(\eta, \xi_0^K)$.

Now perform a Lutz twist of ξ_0^K along each component of K' and let ξ be the resulting contact structure. Since this does not change ξ_0^K over the 2–skeleton of M, we still have $d^2(\xi, \eta) = 0$.

Observe that $f_{\xi_0^K}|_{D^3}$ does not contain the point $(-1,0,0) \in S^2$, and – since $f_{\xi_0^K}(D^3)$ is compact – there is a whole neighbourhood $U \subset S^2$ of $(-1,0,0)$ not contained in $f_{\xi_0^K}(D^3)$. Let $f\colon S^3 \to S^2$ be the map used to compute $d^3(\xi, \xi_0^K)$, that is, f coincides on the upper hemisphere with $f_\xi|_{D^3}$ and on the lower hemisphere with $f_{\xi_0^K}|_{D^3}$. By the discussion on page 143 *et seq.*, the preimage $f^{-1}(u)$ of any $u \in U \setminus \{(-1,0,0)\}$ will be a push-off of $\overline{K}'$ determined by the trivialisation of $\xi_0^K|_{D^3} = \xi_0|_{D^3}$. So the linking number of $f^{-1}(u)$ with $f^{-1}(-1,0,0)$, which is by definition the Hopf invariant $H(f) = d^3(\xi, \xi_0^K)$, will be equal to $\mathtt{sl}(K')$. By our choice of K' and the additivity of d^3 this implies $d^3(\xi, \eta) = 0$. So ξ is a contact structure that is homotopic to η as a 2–plane distribution. □

4.4 Other proofs of Martinet's theorem

This section will provide further evidence that contact structures are very flexible topological objects: virtually any known structure theorem for 3–manifolds can be used to prove the existence of a contact structure on every (closed, orientable) 3–manifold.

4.4.1 Branched covers

The notion of branched covers can be defined in various degrees of generality, see [88], [209], [215]. It turns out that 3–manifolds can be presented as branched cover of the 3–sphere in a very special way. We describe this in the following theorem, which is due independently to Hilden [129] and Montesinos [191]. In Section 4.2 I alluded to a stronger branched covering theorem, obtained by these authors in collaboration with Thickstun. For the present purposes, the original theorem will be sufficient. For a more general notion of branched covers (in higher dimensions) see Section 7.5.

Theorem 4.4.1 (Hilden, Montesinos) *Let M be a closed, connected, orientable 3–manifold. There is a 3–fold branched covering $p\colon M \to S^3$,* *branched along a knot $K \subset S^3$. This means that*

$$p|_{M\setminus p^{-1}(K)}\colon\; M \setminus p^{-1}(K) \longrightarrow S^3 \setminus K$$

is a (smooth) 3–fold covering in the usual sense. The set $p^{-1}(K)$ consists of two knots K_1 and K_2 in M. Along K_1, the covering is unbranched; along K_2, the branching index is 2. One can identify neighbourhoods of K_2 and K with $S^1 \times D^2$ (where K or K_2, respectively, becomes identified with $S^1 \times \{\mathbf{0}\}$) with S^1–coordinate θ and polar coordinates r, φ on the D^2–factor in such a way that p is given by $(\theta, r, \varphi) \mapsto (\theta, r, 2\varphi)$ in these coordinates. □

Notice that the local description of the branched covering near K_2 is not smooth. It turns out that this description is more convenient for lifting a contact form on S^3 to one on M than the smooth one (with r replaced by r^2 on the right). For a very readable proof of the Hilden–Montesinos theorem see [209, §23].

Gonzalo [112] observed that the Hilden–Montesinos theorem allows one to give a very simple proof of Martinet's theorem (Thm. 4.1.1). Start with any contact form α on S^3 defining the standard contact structure ξ. By Theorem 3.3.1 we may assume that K is transverse to ξ. The local coordinate description of the branched covering near K_2 and K defines a framing of K (given by $\varphi = 0$, say) and thus a trivialisation of the (conformal symplectic) normal bundle of K in (S^3, ξ). By Theorem 2.5.15 we may assume that $\alpha = d\theta + r^2\, d\varphi$ near K.

Outside $K_2 \subset M$ the branched covering map p is a local diffeomorphism, so there the 1–form $p^*\alpha$ is obviously a contact form. In the coordinates (θ, r, φ) near K_2, this 1–form is given (*a priori* only for $r > 0$) by $d\theta + 2r^2\, d\varphi$. This explicit description shows that $p^*\alpha$ extends smoothly as a contact form over K_2.

Remark 4.4.2 According to the stronger theorem of Hilden, Montesinos and Thickstun [130], one may in addition assume that K_2 bounds an embedded disc in M, so in particular we find a 3–ball $D^3 \subset M$ that contains K_2 in its interior. Over $M \setminus \mathrm{Int}\,(D^3)$, the contact structure ξ found by the above construction is trivial, as a 2–plane bundle, since there it is the lift of the standard contact structure on S^3. Likewise, $\xi|_{D^3}$ admits a trivialisation. Since $\pi_2(\mathrm{SO}(2)) = 0$, the trivialisation given over $M \setminus \mathrm{Int}\,(D^3)$ extends over D^3, see the second proof of Theorem 4.2.1 on page 135.

4.4.2 Open books

According to a theorem of Alexander [10], every closed, connected, orientable 3–manifold M admits a so-called open book decomposition. The following theorem is phrased in terms of a variant of the definition of open books that is now often referred to as an *abstract* open book.

Given a topological space W and a homeomorphism $\phi\colon W \rightarrow W$, the **mapping torus** $W(\phi)$ is the quotient space obtained from $W \times [0, 2\pi]$ by identifying $(x, 2\pi)$ with $(\phi(x), 0)$ for each $x \in W$. If W is a differential manifold and ϕ a diffeomorphism equal to the identity near the boundary ∂W, then $W(\phi)$ is in a natural way a differential manifold with boundary $\partial W \times S^1$.

Theorem 4.4.3 (Alexander) *Let M be a closed, connected, orientable 3–manifold. Then there is a compact, orientable surface Σ with boundary $\partial\Sigma = S^1$ and a diffeomorphism $\phi\colon \Sigma \rightarrow \Sigma$, equal to the identity near the boundary, such that M is diffeomorphic to*

$$M(\phi) := \Sigma(\phi) \cup_{\mathrm{id}} (\partial\Sigma \times D^2),$$

i.e. the manifold obtained by gluing $\Sigma(\phi)$ and a solid torus $\partial\Sigma \times D^2$ along their boundary, using the identity map to identify $\partial(\Sigma(\phi)) = \partial\Sigma \times S^1$ with $\partial(\partial\Sigma \times D^2)$. □

For a proof of this theorem see [215]; see also [83].

Write $K \subset M$ for the knot corresponding to $\partial\Sigma \times \{\mathbf{0}\}$ under this identification. Then we can define a fibration $p\colon M \setminus K \rightarrow S^1 := \mathbb{R}/2\pi\mathbb{Z}$ by

$$p([x, \varphi]) = [\varphi] \quad \text{for} \quad [x, \varphi] \in \Sigma(\phi)$$

and

$$p(\theta, r\mathrm{e}^{\mathrm{i}\varphi}) = [\varphi] \quad \text{for} \quad (\theta, r\mathrm{e}^{\mathrm{i}\varphi}) \in \partial\Sigma \times D^2 \subset \partial\Sigma \times \mathbb{C}.$$

This gives rise to the following definition.

Definition 4.4.4 *An* **open book decomposition** *of an n–dimensional manifold N consists of a codimension 2 submanifold B, called the* **binding**, *and a (smooth, locally trivial) fibration $p\colon N \setminus B \rightarrow S^1$. The fibres $p^{-1}(\varphi)$, $\varphi \in S^1$, are called the* **pages**. *Moreover, it is required that the binding B have a trivial tubular neighbourhood $B \times D^2$ in which p is given by the angular coordinate in the D^2–factor.*

Observe that the closure of each page is a codimension 1 submanifold of N with boundary B. This is illustrated in Figure 4.5. A nice little survey of the many applications of open book decompositions is given by Winkelnkemper in an appendix to Ranicki's book [212].

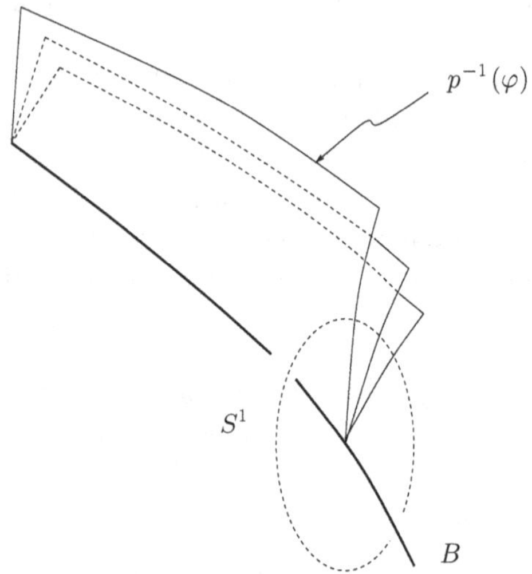

Fig. 4.5. An open book near the binding.

We have seen how an abstract open book (Σ, ϕ), where Σ may in principle be a manifold of arbitrary dimension with non-empty (but not necessarily connected) boundary, gives rise to an open book decomposition with pages diffeomorphic to the interior of Σ and binding $\partial\Sigma$.

Conversely, the abstract open book may be recovered from an open book decomposition as follows. Define Σ as the intersection of the page $p^{-1}(0)$, say, with the complement of an open tubular neighbourhood $B \times \mathrm{Int}\,(D^2_{1/2})$. Choose an auxiliary Riemannian metric on N such that the vector field ∂_φ on $B \times (D^2 \setminus \{\mathbf{0}\})$ is orthogonal to the pages. Extend this to a vector field on $N \setminus B$ orthogonal to the pages that projects under the differential Tp to the vector field ∂_φ on S^1; this extends to a smooth vector field on N vanishing along B. By slight abuse of notation, we write ∂_φ for the vector field on N thus defined. The time-2π-map of the flow of this vector field sends Σ to itself; this 'monodromy' map is the desired ϕ.

Example 4.4.5 Figure 4.6 shows an open book decomposition of S^3 with binding S^1 and the pages (open) 2–discs. Here we regard S^3 as $\mathbb{R}^3 \cup \{\infty\}$, and one has to rotate the picture around the axis (which passes through the point ∞) as shown to get $\mathbb{R}^3 \cup \{\infty\}$. The two thickened points yield the binding S^1 under this rotation; each indicated arc yields a page. The two horizontal half-lines emanating from the binding in this figure really constitute an arc passing through ∞, and this arc, too, becomes a disc under rotation. The monodromy map $\phi\colon D^2 \to D^2$ is the identity.

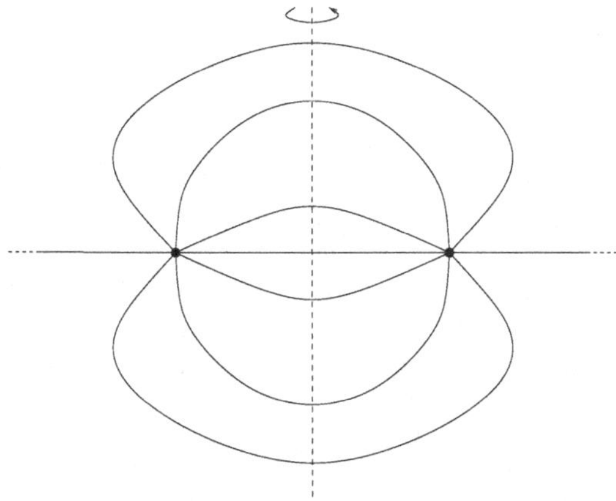

Fig. 4.6. The 3–sphere as open book.

It appears that the original paper by Alexander is vague on the question of the connectedness of the binding. The sources [215] and [83] cited above leave it as an exercise (the latter with copious hints) to show that an open book decomposition of a 3–manifold may always be assumed to have a connected binding. Proofs of this result can be found in [197] and [172].

Here is a slight variation on a simple argument of Thurston and Winkelnkemper [233] for producing a contact structure on any (abstract) open book; in conjunction with Alexander's theorem this constitutes yet another proof of Martinet's theorem.

Choose an orientation for Σ, and give $\partial\Sigma$ the induced boundary orientation. For ease of notation we assume $\partial\Sigma$ to be connected. Let $\theta \in S^1$ be the coordinate along $\partial\Sigma$, and s a collar parameter for $\partial\Sigma$ in Σ, with $\partial\Sigma = \{s = 0\}$, and $s < 0$ in the interior of Σ. We assume that this parameter s has been chosen in such a way that the diffeomorphism $\phi\colon \Sigma \to \Sigma$ is the identity on $[-2, 0] \times \partial\Sigma \subset \Sigma$.

Lemma 4.4.6 *The set of* 1*–forms* β *on* Σ *satisfying*

(i) $\beta = \mathrm{e}^s\, d\theta$ *on* $[-3/2, 0] \times \partial\Sigma \subset \Sigma$,

(ii) $d\beta$ *is an area form on* Σ *of total area* 2π,

is non-empty and convex.

Proof Start with an arbitrary 1–form β_0 on Σ with $\beta_0 = \mathrm{e}^s\, d\theta$ on the collar

$[-2,0] \times \partial\Sigma$. Then, by Stokes's theorem,

$$\int_\Sigma d\beta_0 = \int_{\partial\Sigma} \beta_0 = \int_{\partial\Sigma} d\theta = 2\pi.$$

Let ω be any area form on Σ of total area 2π and with $\omega = \mathrm{e}^s ds \wedge d\theta$ on the collar; notice that

$$\int_{[-2,0]\times\partial\Sigma} \mathrm{e}^s ds \wedge d\theta = 2\pi(1-\mathrm{e}^{-2}) < 2\pi.$$

Then

$$\int_\Sigma \omega - d\beta_0 = 0 \quad \text{and} \quad \omega - d\beta_0 \equiv 0 \text{ on } [-2,0]\times\partial\Sigma.$$

The compactly supported de Rham cohomology of the open surface

$$\Sigma \setminus \left(\left[-\frac{3}{2},0\right]\times\partial\Sigma\right)$$

is isomorphic to $\mathbb{R}$, and the isomorphism is given by integration of 2–forms over the surface, see [31, Cor. 5.8]. It follows that $\omega - d\beta_0 = d\beta_1$ for some 1–form β_1 on Σ that vanishes on $[-3/2,0]\times\partial\Sigma$. Then the 1–form $\beta := \beta_0 + \beta_1$ is an element of the set of 1–forms described in the lemma, so that set is non-empty.

Given any other 1–form β' satisfying conditions (i) and (ii), one readily checks that $(1-t)\beta + t\beta'$ does also satisfy those conditions for any $t \in [0,1]$. □

Now let β be any 1–form as just described. Let $\mu\colon [0,2\pi] \to [0,1]$ be a smooth function that is identically 1 near $\varphi = 0$ and identically 0 near $\varphi = 2\pi$. Then the 1–form

$$\mu(\varphi)\beta + (1-\mu(\varphi))\phi^*\beta$$

on $\Sigma\times[0,2\pi]$ descends to a 1–form on the mapping torus $\Sigma(\phi)$. We continue to denote this 1–form by β. Notice that the restriction of this β to (the tangent bundle of) the fibre of $\Sigma(\phi)$ over any $\varphi \in S^1 = \mathbb{R}/2\pi\mathbb{Z}$ satisfies conditions (i) and (ii) of Lemma 4.4.6, and in fact we have $\beta = \mathrm{e}^s d\theta$ on the collar

$$\left[-\frac{3}{2},0\right]\times\partial\Sigma(\phi) = \left[-\frac{3}{2},0\right]\times\partial\Sigma\times S^1.$$

Now a contact form α on $\Sigma(\phi)$ can be defined by setting $\alpha = \beta + C\,d\varphi$ for some sufficiently large positive constant C, such that in the expression

$$\alpha\wedge d\alpha = (\beta + C\,d\varphi)\wedge d\beta$$

the non-zero term $d\varphi\wedge d\beta$ dominates.

It remains to extend this α as a contact form over $\partial\Sigma \times D^2$. In order to ensure smoothness of the resulting 1–form, we modify the definition of the open book. The $M(\phi)$ in Alexander's theorem can be represented as

$$M(\phi) \cong \big(\Sigma(\phi) + (\partial\Sigma \times D^2_2)\big)/\sim,$$

where, in the quotient space on the right-hand side, we identify

$$(s, \theta, \varphi) \in [-1, 0] \times \partial\Sigma \times S^1$$

with

$$(\theta, r = 1 - s, \varphi) \in \partial\Sigma \times [1, 2] \times S^1 \subset \partial\Sigma \times D^2_2.$$

Mimicking the Lutz twist, we make the ansatz $\alpha = h_1(r)\,d\theta + h_2(r)\,d\varphi$ on $\partial\Sigma \times D^2_2$. The boundary conditions in the present situation may be taken to be

(1) $h_1(r) = 2$ and $h_2(r) = r^2$ near $r = 0$,
(2) $h_1(r) = \mathrm{e}^{1-r}$ and $h_2(r) = C$ for $r \in [1, 2]$.

The condition $\alpha \wedge d\alpha \neq 0$ translates, as before, into

(3) $(h_1(r), h_2(r))$ is never parallel to $(h_1'(r), h_2'(r))$ for $r \neq 0$.

Figure 4.7 shows a curve $r \mapsto (h_1(r), h_2(r))$, $r \in [0, 2]$, that satisfies these conditions. This concludes the proof of Martinet's theorem via open books.

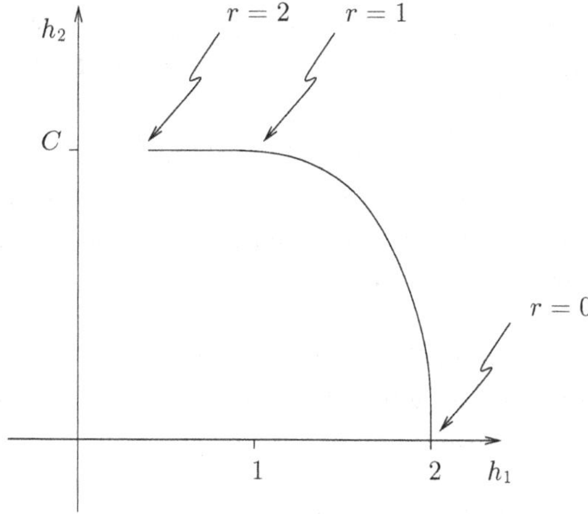

Fig. 4.7. Extension of the contact structure over the binding.

The contact structure obtained via this construction has two special features that we summarise in the following definition. We phrase this definition in such a way that it applies to contact manifolds and open books of arbitrary dimension.

Definition 4.4.7 *Let M be an odd-dimensional oriented manifold with an open book decomposition (B,p), where B is likewise oriented. The pages $p^{-1}(\varphi)$ are oriented by the requirement that the induced orientation on the boundary of (the closure of) each page coincides with the orientation of B. This is the same as saying that ∂_φ together with the orientation of the page gives the orientation of M.*†

A contact structure $\xi = \ker\alpha$ on M is said to be **supported** *by the open book decomposition (B,p) of M if*

(o) the contact form α induces the positive orientation of M,
(i) the 2–form $d\alpha$ induces a symplectic form on each page, defining its positive orientation, and
(ii) the 1–form α induces a positive contact form on B.

In our construction above, we see that α induces the 1–form $2\,d\theta$ on the binding $\partial\Sigma \times \{\mathbf{0}\}$. This is condition (ii). The 2–form $d\alpha$ restricts to the area (hence symplectic) form $d\beta$ on (the tangent bundle of) any fibre of the mapping torus $\Sigma(\phi)$. On the collar $[-2,0] \times \partial\Sigma$ this area form is given by $d\beta = \mathrm{e}^s ds \wedge d\theta$, and ∂_s is the outward normal vector along $\partial\Sigma$. This partially verifies condition (i).

Inside the tubular neighbourhood $\partial\Sigma \times D^2_2$, the pages are given by the condition $\varphi = \text{const.}$ (and $r > 0$). Here the restriction of $d\alpha$ is the 2–form $h_1'(r)\,dr \wedge d\theta$, so we need to ensure that $h_1'(r) < 0$. This is not a problem; for instance, we may choose $h_1(r) = 2 - r^2$ near $r = 0$.

Another observation is worth making: the curve $r \mapsto (h_1(r), h_2(r))$, with $r \in [0,2]$, may be assumed not to pass the h_2–axis, that is, $h_1(r) \neq 0$ for all $r \in [0,2]$. In the 3–dimensional setting this is not essential (and the Thurston–Winkelnkemper ansatz lacked that feature), but it is crucial when one tries to generalise this construction to higher dimensions – as we shall do in Section 7.3 – because then $d\theta$ will have to be replaced by a 1–form on the boundary $\partial\Sigma$ of a higher-dimensional page Σ, and powers of h_1 will enter the contact condition $\alpha \wedge (d\alpha)^n \neq 0$.

Example 4.4.8 Regard S^3 as the subset of $\mathbb{C}^2$ given by

$$S^3 = \{(z_1, z_2) \in \mathbb{C}^2 \colon\ |z_1|^2 + |z_2|^2 = 1\}.$$

† If $\mathfrak{b}$ is a positive basis for (the tangent space to a point of) $B \equiv \partial\Sigma$, then $\mathfrak{b}, \partial_r, \partial_\varphi$ is a positive basis for $B \times D^2 \subset M$. The orientation for Σ is given by $-\partial_r, \mathfrak{b}$, since $-\partial_r$ is the outer normal of Σ. The orientations $-\partial_r, \mathfrak{b}, \partial_\varphi$ and $\mathfrak{b}, \partial_r, \partial_\varphi$ coincide, since B is odd-dimensional.

Write the standard contact form α in polar coordinates as

$$\alpha = r_1^2\,d\varphi_1 + r_2^2\,d\varphi_2.$$

(1) Set

$$B = \{(z_1, z_2) \in S^3 \colon\ z_1 = 0\},$$

and consider the fibration

$$\begin{array}{rccc} p\colon & S^3 \setminus B & \longrightarrow & S^1 \subset \mathbb{C} \\ & (z_1, z_2) & \longmapsto & \dfrac{z_1}{|z_1|}. \end{array}$$

In polar coordinates this map is given by

$$p\colon\ (r_1 \mathrm{e}^{\mathrm{i}\varphi_1}, r_2 \mathrm{e}^{\mathrm{i}\varphi_2}) \longmapsto \varphi_1 \in S^1 = \mathbb{R}/2\pi\mathbb{Z}.$$

As vector field ∂_φ in the construction described on page 150 we may take ∂_{φ_1}. This shows that (B, p) is an open book with pages the open 2–discs

$$p^{-1}(\varphi) = \{|z_2| < 1,\ z_1 = \sqrt{1 - |z_2|^2}\,\mathrm{e}^{\mathrm{i}\varphi}\},$$

and monodromy the identity map.

The contact form α restricts to $d\varphi_2$ along B, and $d\alpha$ to $r_2\,dr_2 \wedge d\varphi_2$ on the tangent spaces to each page. This shows that $\ker\alpha$ is supported by the open book (B, p).

(2) Set

$$B' = \{(z_1, z_2) \in S^3 \colon\ z_1 z_2 = 0\}.$$

This set B' is the union of two Hopf fibres (see the proof of Lemma 1.4.9)

$$K_i = \{(z_1, z_2) \in S^3 \colon\ z_i = 0\},\ \ i = 1, 2.$$

With the orientation of K_1, K_2 given by regarding these circles as the boundary of the discs

$$D_i = \{(z_1, z_2) \in D^4 \colon\ z_i = 0\},\ \ i = 1, 2,$$

we have $\mathtt{lk}(K_1, K_2) = 1$. The link $B' = K_1 \sqcup K_2$ is called the **positive Hopf link**.

Consider the fibration

$$\begin{array}{rccc} p'\colon & S^3 \setminus B' & \longrightarrow & S^1 \subset \mathbb{C} \\ & (z_1, z_2) & \longmapsto & \dfrac{z_1 z_2}{|z_1 z_2|}. \end{array}$$

In polar coordinates this map is given by

$$p'\colon\ (r_1 \mathrm{e}^{\mathrm{i}\varphi_1}, r_2 \mathrm{e}^{\mathrm{i}\varphi_2}) \longmapsto \varphi_1 + \varphi_2 \in S^1 = \mathbb{R}/2\pi\mathbb{Z}.$$

The pages $(p')^{-1}(\varphi)$ of the open book (B', p') can be parametrised as

$$\begin{array}{ccccc} (0,1) & \times & S^1 & \longrightarrow & (p')^{-1}(\varphi) \subset S^3 \setminus B \\ (r_1 & , & \varphi_1) & \longmapsto & (r_1 e^{i\varphi_1}, \sqrt{1-r_1^2}\, e^{i(\varphi-\varphi_1)}). \end{array}$$

Using this, one can easily check that $\ker \alpha$ is supported by the open book (B', p').

The collection of these parametrisations, as φ ranges over S^1, gives a trivialisation of the bundle $p' \colon S^3 \setminus B' \to S^1$. Beware that this does not imply that the monodromy map ϕ of this open book is the identity map – in order to describe ϕ one has to take into account the boundary condition given by how the pages are 'attached' to the binding.

Here is how to determine this monodromy map. Choose a smooth function $[0,1] \to [0,1]$, $r \mapsto \delta(r)$ that is identically 1 near $r = 0$ and identically 0 near $r = 1$. Define a flow $\phi_t \colon S^3 \to S^3$ in polar coordinates by

$$\phi_t \colon \begin{cases} \varphi_1 \longmapsto \varphi_1 + \delta(r_1) \cdot t, \\ \varphi_2 \longmapsto \varphi_2 + (1 - \delta(r_1)) \cdot t. \end{cases}$$

Then $p' \circ \phi_t(z_1, z_2) = p'(z_1, z_2) + t$, so this flow is transverse to the pages of the open book and maps pages to pages. Moreover, near $r_i = 0$ we have $\dot{\phi}_t = \partial_{\varphi_i}$, $i = 1, 2$. This implies that the monodromy of the open book is given by the map $\phi_{2\pi}$. In terms of the parametrisation of the pages by (r_1, φ_1) as above, this map is, up to isotopy rel boundary, a right-handed Dehn twist along the core circle $\{r_1 = 1/2\}$ of the page, see Figure 4.8.

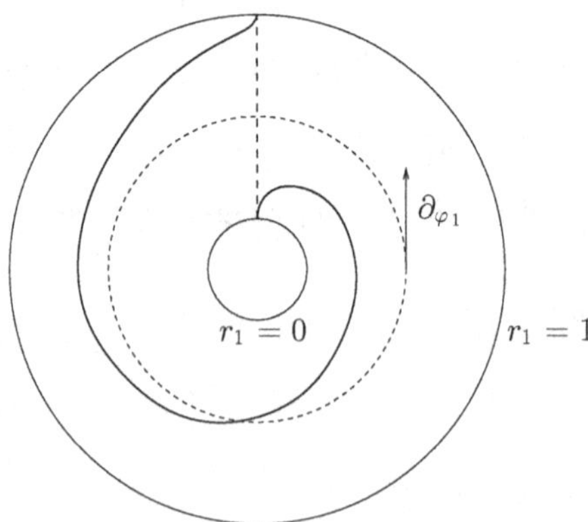

Fig. 4.8. A right-handed Dehn twist.

Open books on 3–manifolds are not merely an alternative means for constructing contact structures. As discovered by Giroux [106], there is in fact a one-to-one correspondence – on any given closed, orientable 3–manifold

– between contact structures up to isotopy and open book decompositions up to 'stabilisation'. This stabilisation of a given open book decomposition consists in adding a 1–handle to the page, and changing the monodromy by composing it with a right-handed Dehn twist along a simple closed curve going exactly once over the 1–handle. The preceding example is an instance of this.

Here is the simplest aspect of this correspondence.

Proposition 4.4.9 *Let $\xi_i = \ker \alpha_i$, $i = 0, 1$, be two contact structures supported by the same open book decomposition (B, p) of the closed, oriented 3–manifold M. Then ξ_0 and ξ_1 are isotopic.*

Proof Let $\partial\Sigma \times D^2_\varepsilon$ be a neighbourhood of the binding where – in terms of the coordinates (θ, r, φ) on any component of this neighbourhood – we have $\alpha_i(\partial_\theta) > 0$. This is possible by condition (ii) in Definition 4.4.7. Choose a function $h = h(r)$ with essentially the same features as h_2 in the foregoing construction, i.e. $h(0) = 0$, $h' \geq 0$, $h(r) = r^2$ near $r = 0$, and $h \equiv 1$ outside $\partial\Sigma \times D^2_{\varepsilon/2}$, say. For any $R \in \mathbb{R}^+_0$, set

$$\alpha_{i,R} = \alpha_i + Rh(r)\, d\varphi.$$

Then

$$\alpha_{i,R} \wedge d\alpha_{i,R} = \alpha_i \wedge d\alpha_i + Rh(r)\, d\varphi \wedge d\alpha_i + Rh'(r)\, \alpha_i \wedge dr \wedge d\varphi.$$

The orientation assumptions in Definition 4.4.7 imply that

$$\alpha_i \wedge d\alpha_i > 0, \quad h(r)\, d\varphi \wedge d\alpha_i \geq 0, \quad \alpha_i \wedge h'(r)\, dr \wedge d\varphi \geq 0,$$

hence $\alpha_{i,R} \wedge d\alpha_{i,R} > 0$. In particular, all $\alpha_{i,R}$ (for fixed $i \in \{0, 1\}$ and arbitrary $R \in \mathbb{R}^+_0$) define isotopic contact structures.

Observe that in fact $h(r)\, d\varphi \wedge d\alpha_i > 0$ away from the binding, and $\alpha_i \wedge h'(r)\, dr \wedge d\varphi > 0$ near the binding. With this observation it is a straightforward check that the linear interpolation $(1-t)\alpha_{0,R} + t\alpha_{1,R}$ is a contact form for any $t \in [0, 1]$, provided R has been chosen sufficiently large. This yields the desired isotopy between ξ_0 and ξ_1. □

4.5 Tight and overtwisted

In this section we are going to discuss a fundamental dichotomy of contact structures on 3–manifolds, introduced by Eliashberg [64], namely, the division of contact structures into *tight* and *overtwisted* ones. At first sight, the definition of theses types of contact structures looks slightly peculiar. We shall see later what motivated this definition, and why it has proved seminal for the development of 3–dimensional contact topology.

Recall the definition of the standard overtwisted contact structure ξ_{ot} on $\mathbb{R}^3$ given in Example 2.1.6. This was described by the equation (in cylindrical coordinates)

$$\cos r\, dz + r \sin r\, d\varphi = 0.$$

Let Δ be the disc $\{z = 0, r \leq \pi\} \subset \mathbb{R}^3$. The boundary $\partial\Delta$ of this disc is a Legendrian curve for ξ_{ot}, in fact, the disc Δ is tangent to ξ_{ot} along its boundary. This means that the characteristic foliation $\Delta_{\xi_{\mathrm{ot}}}$ consists of all radial lines, with singular points at the origin and at all boundary points (Figure 4.9).

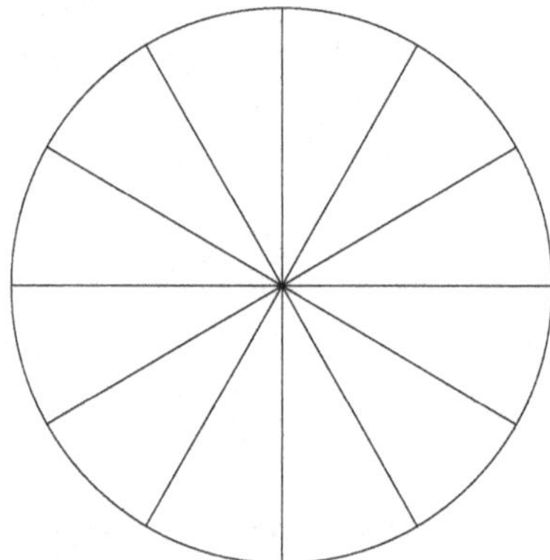

Fig. 4.9. Characteristic foliation on the disc with singular boundary points.

If the interior of Δ is pushed up slightly, the singular points at the boundary can be made to disappear. Only the singular point at the centre remains, and the characteristic foliation now looks as in Figure 4.10, with $\partial\Delta$ a closed leaf of the characteristic foliation Δ_ξ. (We shall prove this presently by an explicit calculation in a related case.) We call Δ (in its perturbed or unperturbed form) the **standard overtwisted disc**.

For our discussion in the following chapter, it is useful to describe the properties of Δ in terms of the contact framing and the surface framing of Legendrian knots (Defns. 3.5.1 and 3.5.2).

Definition 4.5.1 An embedded disc Δ in a contact manifold (M, ξ) is an **overtwisted disc** if its boundary $\partial\Delta$ is a Legendrian curve for ξ whose surface framing coincides with its contact framing, i.e. $\mathtt{tb}(\partial\Delta) = 0$, and the

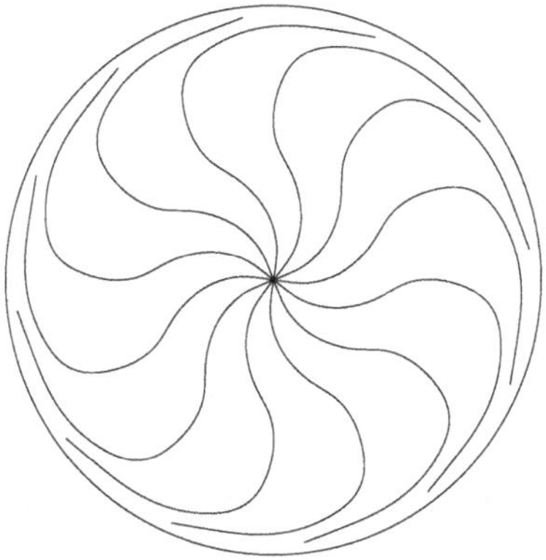

Fig. 4.10. Characteristic foliation on an overtwisted disc after perturbation.

characteristic foliation Δ_ξ contains a unique singular point in the interior of Δ.

Translated into the original definitions, this means that if $\Delta \subset (M,\xi)$ is an overtwisted disc, then the two parallel curves to $\partial\Delta$ we obtain by pushing $\partial\Delta$ either along Δ or transverse to ξ have the same linking number with $\partial\Delta$, in other words, they are isotopic as curves on the boundary of a tubular neighbourhood of $\partial\Delta$. This allows us to isotope Δ rel boundary to a disc (which we continue to call Δ) whose boundary remains tangent to ξ, but such that Δ is now transverse to ξ along $\partial\Delta$. This is the situation when there are no singular points of Δ_ξ along $\partial\Delta$.

Definition 4.5.2 A contact structure ξ on a 3–manifold M is called **overtwisted** if it contains an overtwisted disc. A contact structure is called **tight** if it is not overtwisted.

In fact, the condition that there be but one singular point in the interior of the overtwisted disc turns out to be superfluous. This will be the content of the elimination lemma proved in Section 4.6.3. Whenever there is an embedded disc Δ in a contact 3–manifold such that the contact framing of $\partial\Delta$ equals the surface framing it inherits from Δ, then by a C^0–small isotopy (fixed on the boundary) one can cancel pairs of elliptic and hyperbolic points in the characteristic foliation. Consequently, Δ can be isotoped to an overtwisted disc as described, or it contains such a disc in the interior.

We now want to show how a suitable Lutz twist turns any given contact structure ξ into an overtwisted one, without changing the homotopy class of ξ as a 2–plane field.

Consider a 'full' Lutz twist as follows. Let (M, ξ) be a contact 3–manifold and $K \subset M$ a knot transverse to ξ. As before, identify K with $S^1 \times \{\mathbf{0}\} \subset S^1 \times D^2 \subset M$ such that $\xi = \ker(d\theta + r^2\, d\varphi)$ on $S^1 \times D^2$. Now define a new contact structure ξ' as in Section 4.3, with $(h_1(r), h_2(r))$ now as in Figure 4.11, that is, the boundary conditions are now

$$h_1(r) = 1 \text{ and } h_2(r) = r^2 \text{ for } r \in [0, \varepsilon] \cup [1 - \varepsilon, 1]$$

for some small $\varepsilon > 0$.

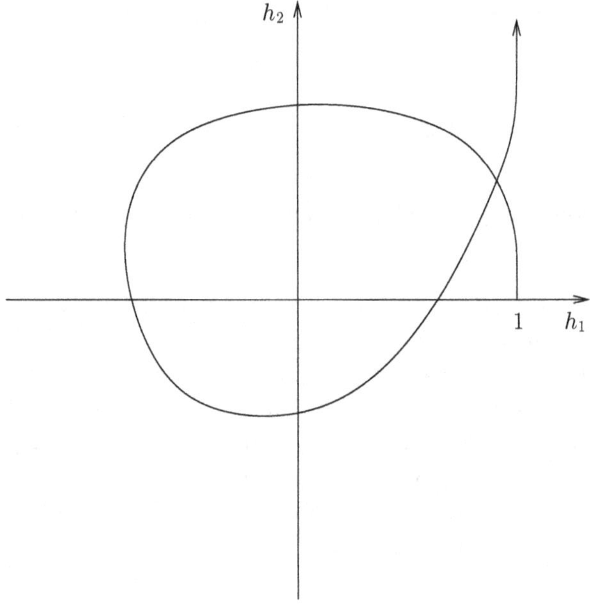

Fig. 4.11. A full Lutz twist.

Lemma 4.5.3 *A full Lutz twist does not change the homotopy class of ξ as a 2–plane field.*

Proof Let $(h_1^t(r), h_2^t(r))$, $r, t \in [0, 1]$, be a homotopy of paths such that

(1) $h_1^0 \equiv 1$, $h_2^0(r) = r^2$,
(2) $h_i^1 \equiv h_i$, $i = 1, 2$,
(3) $h_i^t(r) = h_i(r)$ for $r \in [0, \varepsilon] \cup [1 - \varepsilon, 1]$.

Let $\chi : [0, 1] \to \mathbb{R}$ be a smooth function which is identically 0 near $r = 0$ and $r = 1$, and $\chi(r) > 0$ for $r \in [\varepsilon, 1 - \varepsilon]$. Then

$$\alpha_t = t(1 - t)\chi(r)\, dr + h_1^t(r)\, d\theta + h_2^t(r)\, d\varphi$$

is a homotopy from $\alpha_0 = d\theta + r^2\,d\varphi$ to $\alpha_1 = h_1(r)\,d\theta + h_2(r)\,d\varphi$ through non-zero 1–forms. This homotopy stays fixed near $r = 1$, and so it defines a homotopy between ξ and ξ' as 2–plane fields. □

Let r_0 be the smaller of the two positive radii with $h_2(r_0) = 0$, and consider the embedding

$$\begin{array}{rccc} \phi: & D^2_{r_0} & \longrightarrow & S^1 \times D^2 \\ & (r,\varphi) & \longmapsto & (\theta(r), r, \varphi), \end{array}$$

where $\theta(r)$ is a smooth function with $\theta(r_0) = 0$, $\theta(r) > 0$ for $0 \leq r < r_0$, and $\theta'(r) = 0$ only for $r = 0$. Then $\phi(\partial D^2_{r_0})$ is a Legendrian curve whose contact framing and surface framing are both given by the vector field ∂_θ, see Figure 4.12.

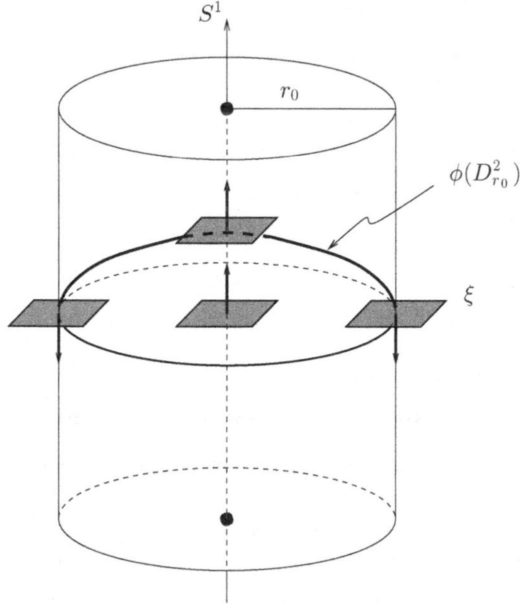

Fig. 4.12. An overtwisted disc.

We may require in addition that $\theta(r) = \theta(0) - r^2$ near $r = 0$. Then $\Delta = \phi(D^2_{r_0})$ is an overtwisted disc. Indeed,

$$\phi^*(h_1(r)\,d\theta + h_2(r)\,d\varphi) = h_1(r)\theta'(r)\,dr + h_2(r)\,d\varphi$$

is a differential 1–form on $D^2_{r_0}$ that vanishes only for $r = 0$, and along $\partial D^2_{r_0}$ the vector field ∂_φ tangent to the boundary lies in the kernel of this 1–form. This implies that the characteristic foliation on the embedded disc Δ has a unique singular point in the interior. Notice that the leaves of this foliation are the integral curves of the vector field $h_1(r)\theta'(r)\,\partial_\varphi - h_2(r)\,\partial_r$, so this is indeed the situation depicted in Figure 4.10.

Recall that in the radially symmetric standard contact structure ξ_{st} of Example 2.1.1, the twisting angle of the contact planes asymptotically approaches $\pi/2$ as r goes to infinity. In Section 4.6 we are going to show that ξ_{st} is in fact tight. This will prove the claim made earlier that $(\mathbb{R}^3, \xi_{st})$ and $(\mathbb{R}^3, \xi_{ot})$ are not contactomorphic.

By contrast, any contact manifold which has been constructed using at least one (simple) Lutz twist contains a cylindrical region where the contact planes twist by more than π in radial direction (at the smallest positive radius r_0 with $h_2(r_0) = 0$ the twisting angle has reached π).

We have shown the following.

Proposition 4.5.4 *Let ξ be a contact structure on M. By a full Lutz twist along any transversely embedded circle, one obtains an overtwisted contact structure ξ' that is homotopic to ξ as a 2–plane distribution.* □

Together with Theorem 4.3.1 this yields the existence of an *overtwisted* contact structure in every homotopy class of 2–plane distributions on any given closed, orientable 3–manifold. In Section 4.7 we are going to prove Eliashberg's much stronger homotopy classification of overtwisted contact structures.

4.6 Surfaces in contact 3–manifolds

We now want to take a more systematic look at surfaces in contact 3–manifolds, with a view towards using them as a tool in the classification of contact structures. Obviously some of the material on hypersurfaces in contact manifolds of arbitrary dimension (Section 2.5.4) will be relevant here. I am going to reiterate some of the arguments from that section in the special 3–dimensional setting to spare the reader from having to leaf back and forth. Throughout I assume that M is a 3–manifold with *oriented and cooriented* contact structure $\xi = \ker \alpha$ (with $d\alpha$ defining the orientation of ξ), and $S \subset M$ an oriented surface embedded in M. Occasionally we allow S to have boundary, but then there will be some control over the boundary, e.g. if it consists of Legendrian curves. All results can typically be proved for non-orientable surfaces by passing to a double cover.

As in Section 2.5.4 we identify a neighbourhood of S in M with $S \times \mathbb{R}$, and S with $S \times \{0\}$, where we write z† for the $\mathbb{R}$–coordinate. We make this

† In the 3–dimensional setting this notation z seems to be preferable to r.

identification compatible with orientations: the orientation of S followed by the natural orientation of $\mathbb{R}$ gives the orientation of M (induced by ξ). We write the contact form α as

$$\alpha = \beta_z + u_z\, dz, \tag{4.1}$$

where β_z, $z \in \mathbb{R}$, is a smooth family of 1–forms on S, and $u_z \colon S \to \mathbb{R}$, a smooth family of functions. Then

$$d\alpha = d\beta_z - \dot{\beta}_z \wedge dz + du_z \wedge dz, \tag{4.2}$$

where the dot denotes the derivative with respect to z. Thus, the contact condition becomes

$$u_z\, d\beta_z + \beta_z \wedge (du_z - \dot{\beta}_z) > 0, \tag{4.3}$$

meaning that the 2–form on the left is a positive area form on S.

4.6.1 The characteristic foliation

Recall that by a singular (and oriented) 1–dimensional foliation on the surface S we mean an equivalence class of vector fields, where two vector fields X, X' are called equivalent if there is a smooth function $f \colon S \to \mathbb{R}^+$ such that $X' = fX$.

We now fix an area form Ω on S, compatible with the given orientation. This allows us to make the following definition, where we set $\beta := \beta_0$.

Definition 4.6.1 The **characteristic foliation** S_ξ on S induced by the contact structure ξ is the singular 1–dimensional foliation represented by the vector field X that is defined by

$$i_X \Omega = \beta. \tag{4.4}$$

Notice that different choices of area form Ω and different choices of contact form α for the cooriented contact structure ξ differ by a positive function, so the equivalence class of X as a singular 1–dimensional foliation does indeed depend on S and ξ only. Furthermore, observe that by (4.4) and (4.1) the points $p \in S$ with $X_p = 0$ are precisely those points of S where the contact plane ξ_p coincides with the tangent plane T_pS; if $X_p \neq \mathbf{0}$, then the line in T_pS spanned by X_p equals $T_pS \cap \xi_p$, since $\alpha_p(X_p) = \beta_p(X_p) = 0$.

Definition 4.6.2 The **divergence** $\mathrm{div}_\Omega(X)$ of a vector field X on S with respect to the area form Ω is defined by $\mathcal{L}_X \Omega = \mathrm{div}_\Omega(X)\Omega$; using the Cartan formula we can rewrite this as

$$\mathrm{div}_\Omega(X)\Omega = d(i_X \Omega). \tag{4.5}$$

We write div instead of div_Ω if there can be no confusion about which area form we are using.

For a smooth function $f\colon S \to \mathbb{R}^+$, both $d(i_{fX}\Omega)$ and $d(i_X(f\Omega))$ equal

$$d(fi_X\Omega) = df \wedge i_X\Omega + f\,\text{div}(X)\Omega.$$

This implies that the condition

$$\text{div}_\Omega(X)(p) \neq 0 \text{ at all points } p \in S \text{ where } X(p) = \mathbf{0} \tag{4.6}$$

is independent both of the choice of Ω (for defining the divergence) and of X (representing a given singular foliation). By the defining Equation (4.4) for the characteristic foliation, this means further that condition (4.6) is independent of the choice of contact form representing a given contact structure.

This condition turns out to characterise those singular foliations on S that can arise as the characteristic foliation of some contact structure ξ.

Lemma 4.6.3 *A vector field X on S represents the characteristic foliation of some contact structure defined near S if and only if it satisfies condition (4.6).*

Proof First assume that X represents the characteristic foliation of the contact structure $\xi = \ker\alpha$ (defined near S), where we write α as in Equation (4.1). If $p \in S$ is a point with $X_p = \mathbf{0}$, then, as remarked above, $\xi_p = T_pS$, and hence, with (4.4),

$$(d(i_X\Omega))_p = (d\beta)_p = d\alpha|_{\xi_p} \neq 0, \tag{4.7}$$

which by (4.5) implies $\text{div}(X)(p) \neq 0$.

Conversely, let X be a vector field on S that satisfies (4.6). Define a 1–form on S by $\beta := i_X\Omega$, and a function $u\colon S \to \mathbb{R}$ by $d\beta = u\Omega$. Observe that the condition (4.6) on X, together with Equation (4.5), implies that $u(p) \neq 0$ at all points $p \in S$ with $\beta(p) = 0$. The fact that S is oriented allows us to choose a 1–form γ on S satisfying $\beta \wedge \gamma \geq 0$, and $(\beta \wedge \gamma)(p) > 0$ if $\beta(p) \neq 0$. Then set

$$\beta_z = \beta + z(du - \gamma)$$

and

$$\alpha = \beta_z + u\,dz.$$

Then

$$u\,d\beta_0 + \beta_0 \wedge (du - \dot{\beta}_z|_{z=0}) = u^2\Omega + \beta \wedge \gamma > 0.$$

Thus, by (4.3), α defines a contact structure near S. □

The argument simplifies if X has nowhere zero divergence, which of course can only happen if the surface is open, i.e. non-compact or with boundary. We record this for future reference.

Lemma 4.6.4 *Let Ω be an area form on an open surface S, and X a vector field on S with non-vanishing divergence with respect to Ω. Then $i_X\Omega + dz$ is an $\mathbb{R}$–invariant contact form on $S\times\mathbb{R}$, with characteristic foliation on the 'horizontal' slices given by X.*

Proof We compute

$$(i_X\Omega + dz)\wedge d(i_X\Omega + dz) = \operatorname{div}(X)\,dz\wedge\Omega \neq 0.$$

The statement about the characteristic foliation is obvious. □

Recall Theorem 2.5.22, which says that, if S is closed, the characteristic foliation S_ξ determines the germ of the contact structure ξ near S, in a sense made precise there. As it stands, that theorem requires the characteristic foliation to be known up to diffeomorphism. The diffeomorphism classification of singular vector fields, unfortunately, is a very subtle issue.† Thus, for Theorem 2.5.22 to be useful one needs methods for manipulating the characteristic foliation.

Our first aim will be to show that by perturbing the surface S one may achieve that the characteristic foliation S_ξ belongs to a certain class of 'generic' singular foliations. We then show that, again after a suitable perturbation of S, the 'generic' singularities may be assumed to be diffeomorphic to certain standard models.

In Section 4.6.2 we formulate the basic notions about convex surfaces in contact 3–manifolds, and $\mathbb{R}$–invariant (or 'vertically' invariant) contact structures on $S\times\mathbb{R}$. This will be one tool used for manipulating the characteristic foliation.

In Section 4.6.3 we then discuss the elimination lemma, which gives a method for reducing the number of singularities in S_ξ by perturbing S. In Section 4.8, finally, we return in more detail to convex surface theory. We show that, under suitable 'genericity' assumptions, one can tell from some simple information whether a given (abstract) singular foliation on S can be realised as a characteristic foliation S_ξ by, you've guessed it, perturbing S. In particular, this allows one to show that homeomorphic characteristic foliations on two copies of S (embedded in two contact manifolds) become

† The flows of two singular vector fields on a given manifold are called *topologically conjugate* if there is a homeomorphism of the manifold that sends the flow lines of one vector field onto those of the other. For the classification of flows on surfaces up to this equivalence see [207].

diffeomorphic after perturbing one of the two embeddings. I should put in a caveat that the homeomorphism in question is supposed to send each so-called hyperbolic point to a hyperbolic point *of the same sign*. The notions of 'hyperbolic point' and 'sign' will be defined presently; the exact statement of this last result is given in Proposition 4.8.14.

In practice, this means that if one wants to use Theorem 2.5.22 for cutting and pasting contact 3–manifolds along surfaces, one may do so, provided only the characteristic foliations 'look the same', with the caveat given above. Convex surface theory goes even further than that; it shows that the essential information (for determining the germ of a contact structure near a surface) is already contained in a simple collection of curves on the surface, the so-called *dividing set.*

Before we come to all that, we need to give meaning to the term 'generic', and we start doing so by considering the types of singularities that can occur in a characteristic foliation. Later we shall also discuss how small (e.g. C^∞–small or C^0–small) the various perturbations of S can be chosen.

Since, for the time being, the considerations are local, we take S to be an open neighbourhood of the origin $\mathbf{0}$ in $\mathbb{R}^2$ with its standard orientation. Let $X(x,y) = a(x,y)\,\partial_x + b(x,y)\,\partial_y$ be a (smooth) vector field on S with an isolated zero in $\mathbf{0}$. The singular foliation determined by X is given by the flow lines of the differential equation $\dot{\mathbf{x}} = X(\mathbf{x})$ with $\mathbf{x} := (x,y)$; this flow has $\mathbf{0}$ as an isolated fixed point. Set $\Omega = dx \wedge dy$. Then

$$\operatorname{div}(X)(\mathbf{0}) = a_x(\mathbf{0}) + b_y(\mathbf{0}),$$

with subscripts denoting partial derivatives, equals the trace of the matrix

$$A := \begin{pmatrix} a_x(\mathbf{0}) & a_y(\mathbf{0}) \\ b_x(\mathbf{0}) & b_y(\mathbf{0}) \end{pmatrix}$$

describing the linearised differential equation $\dot{\mathbf{x}} = A\mathbf{x}$ at the point $\mathbf{0}$.

The isolated zero or 'singular point' $\mathbf{0}$ of X is called **non-degenerate**† if any eigenvalue of A (there may be one or two) has non-vanishing real part. A non-degenerate singular point of X is called **elliptic** if there is only one eigenvalue, or two eigenvalues with real parts of equal sign; it is called **hyperbolic** if there are two real eigenvalues of opposite sign. (Since A is real, this covers all cases.)

Recall that the **index** of the isolated singular point $\mathbf{0}$ of X is defined as

† In the dynamical systems literature – good texts would be [203] or [214] – such singular points are called *hyperbolic*, but I follow the convention in the contact world to reserve that term for one particular geometric type of non-degenerate singularity.

the degree of the map

$$\begin{array}{ccc} S^1_\varepsilon & \longrightarrow & S^1 \\ \mathbf{x} & \longmapsto & X(\mathbf{x})/\|X(\mathbf{x})\|, \end{array}$$

where S^1_ε is a small circle around $\mathbf{0}$ not containing any other zeros of X, see [186].

In the special case that X describes the characteristic foliation S_ξ induced on $S \subset \mathbb{R}^2$ by some cooriented contact structure $\xi = \ker\alpha$ in a neighbourhood of the origin in $\mathbb{R}^2 \times \mathbb{R}$, there is one further item of terminology. At the isolated singular point $\mathbf{0} \in S$ of X, the contact plane ξ_0 coincides with T_0S. The **sign** of the isolated singular point is set to be $+1$ or -1, depending on whether the orientation of ξ_0 (given by $d\alpha|_{\xi_0}$) does or does not coincide with the orientation of T_0S (given by Ω). By Equation (4.7), this sign is determined by whether $(\mathcal{L}_X\Omega)(\mathbf{0})$ is a positive or negative multiple of $\Omega(\mathbf{0})$, i.e. by the sign of the divergence $\mathrm{div}_\Omega(X)(\mathbf{0})$.

Remark 4.6.5 If $\mathbf{0}$ is a non-degenerate singular point of X, then in a neighbourhood of $\mathbf{0}$ the flow φ_t of the differential equation $\dot{\mathbf{x}} = X(\mathbf{x})$ is C^0–conjugate to the linear flow $(\mathbf{x}, t) \mapsto \mathrm{e}^{At}\mathbf{x}$, i.e. there is a neighbourhood U of $\mathbf{0}$ and a homeomorphism h of U such that

$$\varphi_t(h(\mathbf{x})) = h(\mathrm{e}^{At}\mathbf{x}), \quad \text{provided } \mathrm{e}^{At}\mathbf{x} \in U.$$

This is the content of the so-called Hartman–Grobman theorem, see [214, p. 155] or [203, §2.4], which holds in fact in all dimensions. In dimension 2, a C^2–flow is even C^1–conjugate to its linearisation near a non-degenerate fixed point, see [123]. Consequently (see the following examples), elliptic and hyperbolic singularities are determined up to C^1–diffeomorphism by the eigenvalues of their linearisation.

Examples 4.6.6 (1) The linear models for elliptic points are given by the following real matrices:

(i) $\begin{pmatrix} \lambda_1 & 0 \\ 0 & \lambda_2 \end{pmatrix}$, $\lambda_1\lambda_2 > 0$, see Figure 4.13 (a) for the case $\lambda_1 = \lambda_2$,

(ii) $\begin{pmatrix} \lambda & 1 \\ 0 & \lambda \end{pmatrix}$, $\lambda_{1,2} = \lambda \neq 0$,

(iii) $\begin{pmatrix} \mu & -\nu \\ \nu & \mu \end{pmatrix}$, $\lambda_{1,2} = \mu \pm \mathrm{i}\nu$, $\mu, \nu \neq 0$.

The index of an elliptic singularity is equal to $+1$. An elliptic point is called a **source** if the flow moves away from the singularity in forward time; if the flow moves towards the singularity one speaks of a **sink**. The former happens for $\mathrm{Re}(\lambda_1), \mathrm{Re}(\lambda_2) > 0$, the latter for $\mathrm{Re}(\lambda_1), \mathrm{Re}(\lambda_2) < 0$. By the characterisation of the sign in terms of $\mathcal{L}_X \Omega$, we see that a source has sign $+1$; the sign of a sink is -1.

By a straightforward calculation one sees that in case (i) with $\lambda_1 \neq \lambda_2$, both coordinate axes are flow lines and all the other flow lines approach the singularity (in forward or backward time) along the coordinate axis corresponding to the λ_i of greater absolute value. In case (ii), all flow lines approach the singularity along the x–axis. In case (iii), the flow lines do not approach the singularity from a definite direction. See [214, p. 107] for the terminology used to distinguish these cases.

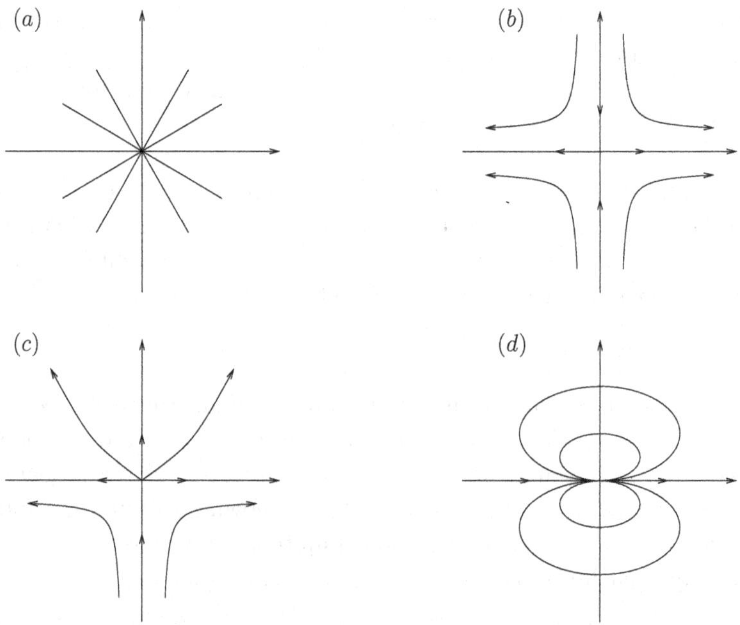

Fig. 4.13. Singular points of planar flows.

(1′) The vector field $X(x,y) := x\,\partial_x + y^3\,\partial_y$ has an isolated singularity at the origin $\mathbf{0}$ that looks topologically like a source (index $+1$, sign $+1$), but now this singularity is degenerate. Because of $\mathrm{div}(X)(\mathbf{0}) = 1$, it can nonetheless be the vector field describing a characteristic foliation. The recipe in the proof of Lemma 4.6.3 (with $\gamma := -x\,dx - y\,dy$) shows that X describes the characteristic foliation on $\mathbb{R}^2 \times \{0\} \subset \mathbb{R}^3$ induced by the contact form

$$\alpha = (xz - y^3)\,dx + (x + 7yz)\,dy + (1 + 3y^2)\,dz.$$

Of course, by Lemma 4.6.4, a simpler example of a contact form with this property is

$$\alpha = -y^3\,dx + x\,dy + dz.$$

If we modify the coefficient y^3 of ∂_y by a C^∞–small perturbation to a smooth function in y with an isolated zero of order 1 at $y = 0$, the singularity of X becomes a non-degenerate source. This perturbation translates into a corresponding C^∞–small perturbation of the contact structure. We shall argue below how for a closed surface S such a perturbation of the contact structure translates, by Gray stability, into a C^∞–small perturbation of S.

(2) The linear model for a hyperbolic singularity is given by a real diagonal matrix with eigenvalues λ_1, λ_2 of opposite sign. Figure 4.13 (b) depicts the situation with $\lambda_1 > 0 > \lambda_2$. The index of a hyperbolic singularity is -1. The sign of the singularity is determined by the dominating eigenvalue. Indeed, in the linear model we have

$$X(x,y) = \lambda_1 x\,\partial_x + \lambda_2 y\,\partial_y,$$

hence $\mathcal{L}_X\Omega = (\lambda_1+\lambda_2)\Omega$. If X describes the characteristic foliation induced by a contact structure, then $\lambda_1+\lambda_2 \neq 0$, since this sum equals the divergence of X in the singular point. Notice that the sign of a hyperbolic point is only a C^1–invariant, not a C^0–invariant; one cannot read off that sign from a topological picture.

The two flow lines approaching the hyperbolic point (in Figure 4.13 (b) these are the positive and negative y–axis) are called the **stable separatrices** of that point; the two flow lines emanating from the hyperbolic point (the positive and negative x–axis in our example) are the **unstable separatrices**.

(3) The vector field $X(x,y) := x\,\partial_x + y^2\,\partial_y$ has an isolated degenerate singularity of index 0 at the origin, see Figure 4.13 (c). The divergence of X at that singular point is equal to 1. By Lemma 4.6.4 we find that X describes the characteristic foliation on $\mathbb{R}^2\times\{0\}\subset\mathbb{R}^3$ induced by the contact form

$$\alpha = -y^2\,dx + x\,dy + dz.$$

By a C^∞–small perturbation we can change the function y^2 to a function without zeros and thus get rid of this singularity.

(4) The vector field $X(x,y) := (x^2-y^2)\,\partial_x + 2xy\,\partial_y$ is a typical model for an index 2 singularity, see Figure 4.13 (d). This vector field has divergence 0 at the singular point $\mathbf{0}$, so it does not describe the characteristic foliation of any contact structure.

(5) The vector field $X(x,y) := x\,\partial_y - y\,\partial_x$ has a degenerate singularity

of index $+1$ at $\mathbf{0}$. Here the degeneracy manifests itself in the fact that every circle centred at $\mathbf{0}$ is a closed orbit of the flow of X, so we have closed orbits 'accumulating' at the singularity. Such a degenerate elliptic singularity, where the linearisation at the singular point has two purely imaginary eigenvalues, has divergence zero, so this cannot occur in the characteristic foliation induced by a contact structure.

We want to exclude the degeneracy phenomenon displayed by that last example not only in the neighbourhood of singular points, but also in the neighbourhood of closed leaves of the singular foliation (i.e. periodic orbits of the vector field X defining the characteristic foliation). Identify the neighbourhood of a periodic orbit γ of X with $S^1 \times (-\varepsilon, \varepsilon)$, where γ is identified with $S^1 \times \{0\}$. (This is possible since S was assumed to be orientable.) Fix a point $(p_0, 0)$ on γ, and consider the transversal $\{p_0\} \times (-\varepsilon, \varepsilon)$ to γ. Given a point (p_0, x) on this transversal, let $(p_0, h(x))$ be the first intersection of the flow line through (p_0, x) (in forward time) with the transversal; this will be defined for x sufficiently small. The map $x \mapsto h(x)$ is the so-called **Poincaré return map** of the periodic orbit γ, see Figure 4.14.

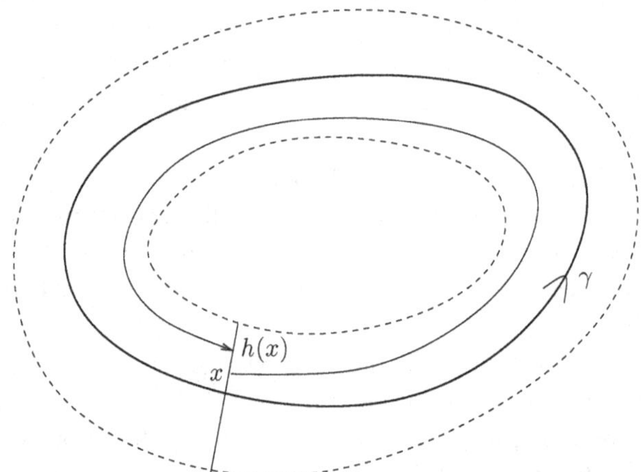

Fig. 4.14. The Poincaré return map h of a periodic orbit γ.

A periodic orbit γ is called **non-degenerate** if its Poincaré return map h satisfies the condition $h'(0) \neq 1$. This precludes in particular the existence of a sequence of closed orbits accumulating at γ. We call such a periodic orbit **attracting** if $h'(0) < 1$; if $h'(0) > 1$ it is said to be **repelling**.

Example 4.6.7 Consider $S^1 \times \mathbb{R}^2$ with (x, z) Cartesian coordinates on the $\mathbb{R}^2$–factor, and y mod 1 the coordinate along $S^1 \equiv \mathbb{R}/\mathbb{Z}$. The circle γ given by $\{x = z = 0\}$ (parametrised by y) is Legendrian for the contact structure $\xi = \ker(dz + x\,dy)$. By Theorem 2.5.8, a neighbourhood of this circle in $(S^1 \times \mathbb{R}^2, \xi)$ provides a universal model for the neighbourhood of a Legendrian circle in any contact 3–manifold.

Consider the cylinder $S := S^1 \times \mathbb{R}$ over this Legendrian circle given by $\{x = z\}$. Observe that the surface framing of γ with respect to S equals its contact framing, since $\partial_x + \partial_z$ is transverse to ξ. The cylinder S is not tangent to ξ along γ, so γ is a closed orbit of the characteristic foliation S_ξ.

The condition for a curve $s \mapsto (x(s), y(s), z(s))$ to be Legendrian for ξ and tangent to S is given by

$$z' + xy' = 0 \text{ and } x' = z'.$$

This implies (for a regular curve) $y' \neq 0$, so we may take y as the curve parameter. We then find that the general solution is given by

$$s \mapsto (x_0 \mathrm{e}^{-s}, s \bmod 1, x_0 \mathrm{e}^{-s}).$$

On the transversal $\{y = 0, x = z\}$ for γ, which we can parametrise by x, the Poincaré return map h is given by $h(x) = x/\mathrm{e}$. Thus $h'(0) = 1/\mathrm{e} \neq 1$, so γ is non-degenerate.

Because of the universality mentioned above, this example shows in particular the following: Whenever we have a surface with a Legendrian boundary such that the contact framing coincides with the surface framing, then by perturbing the surface, keeping the boundary fixed, we can ensure that the boundary is a non-degenerate periodic orbit of the characteristic foliation.

We recall one further notion from dynamical systems. Let φ_t be a globally defined flow. The α– or ω–**limit set** of a flow line $t \mapsto \varphi_t(\mathbf{x}_0)$ (or of the point $\mathbf{x}_0$) is the set of points that can be written as $\lim_{k\to\infty} \varphi_{t_k}(\mathbf{x}_0)$ for some sequence $t_1, t_2, \ldots$ going to $-\infty$ or $+\infty$, respectively.

We can single out a class of vector fields with certain nice properties.

Definition 4.6.8 A vector field X (or its associated singular 1–dimensional foliation) on a closed, orientable surface S is said to be of **Morse–Smale type** if it has the following properties.

(i) There are finitely many singularities and closed orbits, all of which are non-degenerate.

(ii) The α– and ω–limit set of each flow line is either a singular point or a closed orbit.

(iii) There are no trajectories connecting hyperbolic points.†

† Beware that what we call a 'hyperbolic point', a dynamic person would call a 'saddle', as a subclass of (dynamically) hyperbolic points; see the footnote on page 166.

Remarks 4.6.9 (1) There are certain redundancies in this definition, see the discussion in [205]. The non-degeneracy of singular points implies that they lie isolated. Since the singular points form a closed subset of S, and S is compact, this entails that there are only finitely many of them.

The non-degeneracy of the closed orbits likewise implies that they lie isolated. This alone does not guarantee that their number is finite, since they might accumulate at a union of singular points and connecting trajectories. Those singular points, however, would have to be hyperbolic, so this situation is ruled out by (iii).

(2) Particular examples of Morse–Smale vector fields on surfaces are given by the gradient flow of Morse functions, provided there are no connecting orbits between hyperbolic points; see [214, Section 8.12], which also gives the corresponding definitions and statements in higher dimensions.

Peixoto ([205], [206]) has shown the following two important theorems.

(1) Morse–Smale vector fields on closed, orientable surfaces are *structurally stable*, that is, given a Morse–Smale vector field X on a closed surface S and an $\varepsilon > 0$, for any vector field X' sufficiently C^1–close to X there is a homeomorphism of S that is ε–close to the identity (in the C^0–topology) sending the trajectories of X onto those of X'. Conversely, any structurally stable vector field is Morse–Smale.
(2) Morse–Smale vector fields of class C^r, $1 \leq r \leq \infty$, on a given closed, orientable surface are open and dense (with respect to the C^r–topology) in the space of all vector fields of class C^r on that surface.

Remark 4.6.10 Peixoto stated his results without the orientability assumption on the surface. His proof, however, turned out to contain a gap in the non-orientable case. As a consequence, property (2) of Morse–Smale vector fields remains largely conjectural for non-orientable surfaces. For the current state of this conjecture see [121]. For a textbook account see [203].

After this brief dynamical excursion, we can now clarify what we want to understand by a ‘generic’ characteristic foliation. Property (2) is stating formally what one means by saying that vector fields (on closed, orientable surfaces) are generically of Morse–Smale type.

Proposition 4.6.11 *Let S be a closed, orientable surface in a contact 3–manifold $(M, \xi = \ker\alpha)$. Then there is a surface S' isotopic and C^∞–close to S such that the characteristic foliation S'_ξ is of Morse–Smale type.*

Proof By the (C^1–)openness of the contact condition and Peixoto’s theorem, we can choose an arbitrarily C^∞–small 1–form β, compactly supported near S, such that $\xi_t := \ker(\alpha + t\beta)$ is a contact structure for all $t \in [0,1]$, and such that the characteristic foliation S_{ξ_1} is of Morse–Smale type. By Gray stability (Thm. 2.2.2) there is an isotopy $(\psi_t)_{t\in[0,1]}$ of M such that $T\psi_1(\xi) = \xi_1$. Moreover, the proof of that theorem shows that ψ_1 is C^∞–close to the identity. Then $S' := \psi_1^{-1}(S)$ is the desired surface. □

The following proposition will hardly ever be used directly, but I mention it for completeness; notably (b) is an oft-quoted folklore result.

Proposition 4.6.12 *(a) The flows of any two singularities of the same type (elliptic sink, elliptic source, or hyperbolic) are topologically conjugate (as oriented flows). Here the sign of the hyperbolic singularity is irrelevant.*

(b) Given two isolated singularities $p \in S_\xi$, $p' \in S'_{\xi'}$ of the same type and sign in the characteristic foliations of two surfaces $S \subset (M,\xi)$, $S' \subset (M',\xi')$, there is an embedding $\psi\colon S \to M$ isotopic (via an isotopy supported near p) and arbitrarily C^1–close to the inclusion $S \subset M$ such that the characteristic foliation $(\psi(S))_\xi$ has an isolated singularity at $\psi(p)$ that is C^1–conjugate to the singularity at $p' \in S'_{\xi'}$.

Proof (a) We construct a homeomorphism between the flows near two singularities of the same type by mapping the singularity to the singularity, and a transversal to the first flow (outside the singular point) homeomorphically onto such a transversal for the second flow. Finally, we map flow lines through corresponding points of the transversals homeomorphically onto one another with the help of the respective flow. For a hyperbolic singularity, a little more care is necessary. Here are the details, first for elliptic singularities.

The flow φ_t near an elliptic source $\mathbf{0} \in \mathbb{R}^2$ is, up to C^1–conjugation, of the form $\varphi_t(\mathbf{x}) = \mathrm{e}^{At}\mathbf{x}$, where A is one of the matrices in Example 4.6.6 (1) with $\mathrm{Re}(\lambda_1), \mathrm{Re}(\lambda_2) > 0$. Suppose we are given two such flows – corresponding to matrices A, A' (which need not be of the same type (i), (ii) or (iii)) – both defined in the region $\{\|\mathbf{x}\| < 2\varepsilon\}$, say, for some $\varepsilon > 0$. A homeomorphism between these two flows is given by sending $\mathbf{0}$ to $\mathbf{0}$, and $\mathrm{e}^{At}\mathbf{x}$ to $\mathrm{e}^{A't}\mathbf{x}$ for $\|\mathbf{x}\| = \varepsilon$. For a sink the argument is completely analogous.

Near a hyperbolic point $\mathbf{0} \in \mathbb{R}^2$ we may assume, up to C^1–conjugation, that the flow φ_t is of the form $\varphi_t(\mathbf{x}) = \mathrm{e}^{At}\mathbf{x}$, with A a diagonal matrix with diagonal entries $\lambda_1 > 0 > \lambda_2$. Given two such flows, define a homeomorphism as follows. On the diagonals $\{(x,y) \in \mathbb{R}^2 \colon y = \pm x\}$ and the coordinate axes we take the homeomorphism to be the identity. On a segment

$$\{(x,y) \in \mathbb{R}^2 \colon x = x_0 > 0, 0 < y < x_0\},$$

define the homeomorphism sending that segment to itself by flowing a point back under the first flow until it meets the diagonal, and then flowing it forward under the second flow until it meets the segment again. The definition on other segments of this kind cut out by the diagonals and the coordinate axes is analogous.

(b) Again we start with an elliptic source (with sign $+1$). Since the characteristic foliation determines the germ of the contact structure along the surface, it suffices to prove the following: The inclusion $\mathbb{R}^2 \equiv \mathbb{R}^2 \times \{0\} \subset (\mathbb{R}^3, \xi = \ker\alpha)$, with

$$\alpha = dz + x\,dy - y\,dx = dz + r^2\,d\varphi,$$

can be C^1–approximated by an embedding $\psi\colon \mathbb{R}^2 \to \mathbb{R}^3$, identical to the inclusion outside a small neighbourhood of $\mathbf{0} \in \mathbb{R}^2$, such that $\psi^*\alpha$ defines a singular foliation with an elliptic source at $\mathbf{0}$ corresponding to any given linear model.

Suppose we want to realise the linear model (1i) in Example 4.6.6. Since we are free to scale the vector field defining the characteristic foliation, we may assume without loss of generality that $\lambda_1 + \lambda_2 = 2$ and $\lambda_1 \geq \lambda_2 > 0$. Set $c = (\lambda_1 - \lambda_2)/4$, so that $\lambda_1 = 1 + 2c$ and $\lambda_2 = 1 - 2c$ with $c \in [0, 1/2)$. Choose a bump function $h\colon \mathbb{R}^+_0 \to [0,1]$ supported near $r = 0$ and with $h(r) \equiv 1$ in a smaller neighbourhood of $r = 0$. Observe that

$$\begin{aligned} dz + \lambda_1 x\,dy - \lambda_2 y\,dx &= dz + x\,dy - y\,dx + 2c(x\,dy + y\,dx) \\ &= dz + r^2\,d\varphi + c \cdot d\big(r^2 \sin(2\varphi)\big). \end{aligned}$$

Then define $\psi\colon \mathbb{R}^2 \to \mathbb{R}^3$ in terms of polar coordinates on $\mathbb{R}^2$ by $z(r,\varphi) = c \cdot h(r) r^2 \sin(2\varphi)$. This function will be as C^1–close to the function $z \equiv 0$ as we wish, provided we choose the support of h small enough. We compute

$$\psi^*\alpha = c \cdot h(r)\, d\big(r^2 \sin(2\varphi)\big) + c \cdot h'(r) r^2 \sin(2\varphi)\, dr + r^2\, d\varphi,$$

which has the desired form near $r = 0$.

Notice that we did not introduce any new singularities into the characteristic foliation, for the coefficient of $d\varphi$ in $\psi^*\alpha$ is given by

$$r^2\big(2c \cdot h(r) \cos(2\varphi) + 1\big),$$

which does not vanish for $r \neq 0$.

For the model (1ii) we may assume, by scaling the vector field and the x–coordinate relative to the y–coordinate, that the characteristic foliation is given by $(x + y/\lambda)\,\partial_x + y\,\partial_y$ with $0 < 1/\lambda < 2$. So the form we want to induce near $\mathbf{0}$ is

$$dz + x\,dy - y\,dx + \frac{1}{\lambda} y\,dy.$$

Choose h as before and define

$$\psi(x,y) = \Big(x, y, h(x^2+y^2)\cdot\frac{1}{2\lambda}y^2\Big).$$

Then $\psi^*\alpha$ is as desired near $\mathbf{0}$. I leave it to the reader to check that there are no new singularities; this uses the condition $|h/\lambda| < 2$.

In the remaining cases I also leave it to the reader to verify that the perturbation has the desired properties. In order to realise (1iii), it is convenient to start from

$$\alpha = dz + \mu r^2\,d\varphi;$$

this can be achieved by rescaling the r–coordinate. Then the function $z(r,\varphi) := -h(r)\cdot\nu r^2/2$ will define the desired perturbation.

Finally, suppose we start from

$$\alpha = dz + \lambda_1 x\,dy - \lambda_2 y\,dx$$

with $\lambda_1 > 0 > \lambda_2$ and $\lambda_1 + \lambda_2 > 0$, defining a positive hyperbolic singularity at $\mathbf{0} \in \mathbb{R}^2$. A singularity of the same kind corresponding to eigenvalues $\lambda_1' > 0 > \lambda_2'$ can be realised by perturbing the inclusion $\mathbb{R}^2 \to \mathbb{R}^3$ as follows. By scaling we may assume $\lambda_1' + \lambda_2' = \lambda_1 + \lambda_2$. It suffices to deal with the case where $\lambda := \lambda_1 - \lambda_1' = \lambda_2' - \lambda_2 > 0$. Then a solution is given by $z(x,y) := -\lambda h(x^2+y^2)\cdot xy$. □

Remark 4.6.13 Having C^1–conjugate characteristic foliations is enough for cut-and-paste constructions. The gluing will initially produce a contact form of class C^1, which can then be C^∞–approximated by a smooth contact form.

We now show how the Euler characteristic $\chi(S)$ and the Euler class $e(\xi)$ relate to the characteristic foliation.

Proposition 4.6.14 *Let $S \subset (M,\xi)$ be a closed, oriented surface with characteristic foliation S_ξ of Morse–Smale type. Write $e_\pm$ and $h_\pm$ for the number of positive (or negative, respectively) elliptic and hyperbolic singularities of S_ξ, respectively. Then*

$$\chi(S) = (e_+ - h_+) + (e_- - h_-)$$

and

$$\langle e(\xi), [S]\rangle = (e_+ - h_+) - (e_- - h_-).$$

Proof Write $j\colon S \to M$ for the inclusion of S in M. Then the left-hand side in the second equation can be written formally correct as

$$\langle e(\xi), j_*[S]\rangle = \langle e(j^*\xi), [S]\rangle.$$

Likewise, $\chi(S)$ should be regarded as $\langle e(TS), [S]\rangle$. With an area form Ω on S defining the orientation, we define the characteristic foliation by a vector field X as in Definition 4.6.1. This vector field defines a section of either bundle $j^*\xi$ and TS whose Euler number we want to compute.

Recall that for an oriented vector bundle over a closed, oriented manifold, the self-intersection class of that manifold (regarded as the zero section) in the total space of the bundle is the Poincaré dual of the Euler class of the bundle, see [35, Prop. VI.12.8]. For a 2–plane bundle over a surface this translates into saying that the Euler class of the bundle, evaluated on the fundamental class of the surface, equals the sum of indices of a section of the bundle with non-degenerate zeros (such a section is transverse to the zero section). When the bundle in question is the tangent bundle, this is the Poincaré–Hopf index theorem. The proof given in [35, p. 387] shows that it would in fact be enough to have isolated, but possibly degenerate zeros.

The index of an elliptic singularity of X relative to TS is $+1$, that of a hyperbolic singularity, -1, as discussed in Examples 4.6.6. This gives the formula for $\chi(S)$.

Relative to $j^*\xi$, the index at a negative singularity $p \in S$ – where the contact plane ξ_p coincides with T_pS, but with opposite orientation – is the opposite of what it was relative to TS. This yields the second formula. □

Here is a lengthy example of two characteristic foliations.

Example 4.6.15 Consider $\mathbb{R}^2$ with Cartesian coordinates (x, y), and identify the 2–torus T^2 with $\mathbb{R}^2/(2\mathbb{Z})^2$. We take the set

$$\{(x,y) \in \mathbb{R}^2 \colon \ -1 \leq x, y \leq 1\}$$

as fundamental domain for T^2. On $T^2 \times \mathbb{R}$, with z denoting the $\mathbb{R}$–coordinate, we consider the 1–forms

$$\alpha_1 := \sin(\pi y)\, dx + 2\sin(\pi x)\, dy + \big(2\cos(\pi x) - \cos(\pi y)\big)\, dz$$

and

$$\begin{aligned} \alpha_2 \quad &:= \quad \sin(\pi y)\, dx + \Big(1 - \frac{1}{K}\cos(\pi x)\Big)\sin(\pi x)\, dy \\ &\qquad + \Big(\cos(\pi x) - \frac{1}{K}\cos(2\pi x) - \cos(\pi y)\Big)\, dz, \end{aligned}$$

where $K \in \mathbb{R}^+$. We compute

$$d\alpha_1 = \pi\big(2\cos(\pi x) - \cos(\pi y)\big)\, dx\wedge dy + \pi\sin(\pi y)\, dy\wedge dz + 2\pi\sin(\pi x)\, dz\wedge dx,$$

hence

$$\alpha_1 \wedge d\alpha_1 = \pi\big[\sin^2(\pi y) + 4\sin^2(\pi x) + \big(2\cos(\pi x) - \cos(\pi y)\big)^2\big]\, dx\wedge dy\wedge dz.$$

We have $2\cos(\pi x) - \cos(\pi y) \neq 0$ for $x, y \in \mathbb{Z}$, which implies $\alpha_1 \wedge d\alpha_1 > 0$.

For α_2 we compute

$$\begin{aligned} d\alpha_2 &= \pi\Big(\cos(\pi x) - \frac{1}{K}\cos(2\pi x) - \cos(\pi y)\Big)\, dx \wedge dy \\ &\quad + \pi \sin(\pi y)\, dy \wedge dz \\ &\quad + \pi\Big(\sin(\pi x) - \frac{2}{K}\sin(2\pi x)\Big)\, dz \wedge dx, \end{aligned}$$

which yields

$$\begin{aligned} \alpha_2 \wedge d\alpha_2 &= \pi\Big[\sin^2(\pi y) + \sin^2(\pi x)\Big(1 - \frac{1}{K}\cos(\pi x)\Big) \\ &\quad + \Big(\cos(\pi x) - \cos(\pi y) - \frac{1}{K}\cos(2\pi x)\Big)^2 \\ &\quad - \frac{2}{K}\sin(\pi x)\sin(2\pi x)\Big(1 - \frac{1}{K}\cos(\pi x)\Big)\Big]\, dx \wedge dy \wedge dz. \end{aligned}$$

For $K > 1$, the first three summands inside the square brackets are non-negative and do not vanish simultaneously. Since they depend only on the coordinates x, y ranging over the compact set T^2, their sum is bounded away from zero. The last summand becomes small for K large, hence $\alpha_2 \wedge d\alpha_2 > 0$ for $K \gg 1$. In the sequel, we shall always assume that K has been chosen large enough so that α_2 is a contact form. We write $\xi_i = \ker \alpha_i$, $i = 1, 2$ for the contact structures defined by these two contact forms.

We now consider the surface $S := T^2 \times \{0\}$. We use the area form $\Omega := dx \wedge dy$ for determining the vector fields X_1, X_2 defining the characteristic foliations S_{ξ_1}, S_{ξ_2}, respectively. We find

$$\begin{aligned} X_1 &= 2\sin(\pi x)\,\partial_x - \sin(\pi y)\,\partial_y, \\ X_2 &= \Big(1 - \frac{1}{K}\cos(\pi x)\Big)\sin(\pi x)\,\partial_x - \sin(\pi y)\,\partial_y. \end{aligned}$$

Both characteristic foliations look qualitatively as depicted in Figure 4.15. We have one positive elliptic point $E_+ = (0,1)$ and one negative elliptic point $E_- = (1,0)$. In order to determine the sign of the two hyperbolic points $H_0 = (0,0)$ and $H_1 = (1,1)$, we compute the divergence of X_1 and X_2. We have

$$\mathcal{L}_{X_1}\Omega = d\big(2\sin(\pi x)\, dy + \sin(\pi y)\, dx\big) = \pi\big(2\cos(\pi x) - \cos(\pi y)\big)\, dx \wedge dy$$

and

$$\begin{aligned} \mathcal{L}_{X_2}\Omega &= d\Big[\Big(1 - \frac{1}{K}\cos(\pi x)\Big)\sin(\pi x)\, dy + \sin(\pi y)\, dx\Big] \\ &= \pi\Big[\Big(1 - \frac{1}{K}\cos(\pi x)\Big)\cos(\pi x) + \frac{1}{K}\sin^2(\pi x) - \cos(\pi y)\Big]\, dx \wedge dy. \end{aligned}$$

Hence

$$\begin{array}{rclrcl}\operatorname{div}(X_1)(H_0) & = & \pi, & \operatorname{div}(X_1)(H_1) & = & -\pi,\\ \operatorname{div}(X_2)(H_0) & = & -\pi/K, & \operatorname{div}(X_2)(H_1) & = & -\pi/K.\end{array}$$

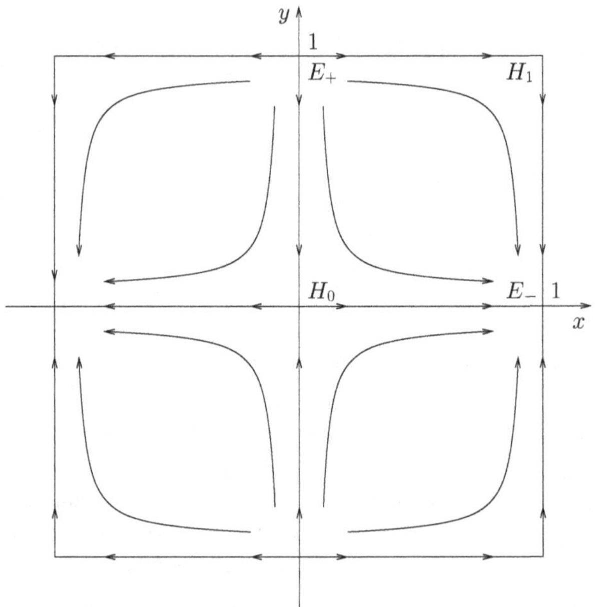

Fig. 4.15. The characteristic foliation on $S = T^2 \times \{0\}$.

Thus, the characteristic foliation S_{ξ_1} has a positive and a negative hyperbolic singularity, while S_{ξ_2} has two negative hyperbolic singularities. In either case, the equation $0 = \chi(S) = (e_+ - h_+) + (e_- - h_-)$ is satisfied.

The two characteristic foliations S_{ξ_1} and S_{ξ_2} are plainly homeomorphic, see the argument in the proof of Proposition 4.6.12 (a). However, they cannot be diffeomorphic, even after perturbing one copy of S in one of the two contact manifolds $(T^2 \times \mathbb{R}, \xi_i)$. Indeed, by Proposition 4.6.14 we have

$$\langle e(\xi_1), [S]\rangle = 0 \neq 2 = \langle e(\xi_2), [S]\rangle.$$

These Euler numbers are invariant under isotopies of S, so by Theorem 2.5.22 (the characteristic foliation determines the germ of the contact structure), S_{ξ_1} cannot be diffeomorphic to S'_{ξ_2} for any perturbation S' of S. This illustrates the caveat made earlier that a reasonable concept of 'homeomorphic characteristic foliations' should include the requirement that the homeomorphism respect the signs of hyperbolic points.

4.6.2 Convex surfaces

In some of the examples of the previous section we saw how useful a statement such as Lemma 4.6.4 may be for writing down explicit contact forms. In the present section we collect a few more basic facts about vertically invariant contact structures on $S \times \mathbb{R}$ and a related concept, *viz.* convex surfaces. Here the emphasis will be on those results that we want to use in the next section for manipulating the characteristic foliation. The real power of convex surface theory will be explained in Section 4.8 below.

Definition 4.6.16 A **convex surface** S in (M, ξ) is an embedded surface with the property that there is a contact vector field (see Defn. 1.5.7) defined near and transverse to S.

Remark 4.6.17 This terminology, introduced by Giroux [104], is – as Giroux himself admits – a little unfortunate, since there is really no distinction between 'convex' and 'concave' in the present context. The notion 'con' has been suggested but seems unlikely to catch on, at least in the French literature.

Example 4.6.18 The vector field $Y := x\,\partial_x + y\,\partial_y + 2z\,\partial_z$ is a contact vector field for the standard contact structure $\xi_2 = \ker \alpha_2$ on $\mathbb{R}^3$, where $\alpha_2 = dz + x\,dy - y\,dx$, since $\mathcal{L}_Y \alpha_2 = 2\alpha_2$. This vector field is transverse to S^2, so the unit sphere is a convex surface in $(\mathbb{R}^3, \xi_2)$.

As before, our surface S is always understood to be oriented, although most statements carry over to non-orientable surfaces.

Lemma 4.6.19 *A closed surface $S \subset (M, \xi = \ker \alpha)$ is convex if and only if there is an embedding $\Psi\colon\ S \times \mathbb{R} \rightarrow M$ (with image inside an arbitrarily small prescribed neighbourhood of S) such that $p \mapsto \Psi(p, 0)$ defines the inclusion of S in M, and $\Psi^*\alpha$ determines a vertically invariant (i.e. $\mathbb{R}$–invariant) contact structure on $S \times \mathbb{R}$.*

Proof Suppose that S is a convex surface. Let Y be a contact vector field defined near and transverse to S. We can consider the corresponding Hamiltonian function $H := \alpha(Y)$, which will be defined in some tubular neighbourhood of S. Multiply H by a bump function that is supported inside this tubular neighbourhood and identically 1 on a smaller tubular neighbourhood. Then take the new Y to be the contact vector field corresponding to this new Hamiltonian function.

The flow ψ_t of Y will then be globally defined. It thus gives rise to an embedding

$$\begin{array}{rcccl} \Psi\colon & S & \times & \mathbb{R} & \longrightarrow & M \\ & (p & , & t) & \longmapsto & \psi_t(p). \end{array}$$

Observe that $T\Psi_{(p,t)}(\partial_t) = \dot{\psi}_t(p) = Y(\psi_t(p))$. Hence $\ker(\Psi^*\alpha)$ is a vertically invariant contact structure on $S \times \mathbb{R}$.

Conversely, given an embedding Ψ as described in the lemma, then $T\Psi(\partial_t)$ defines a contact vector field transverse to S. □

Remark 4.6.20 If S is not closed, one may apply the preceding argument to any open and relatively compact subset of S.

We now want to show that, in the situation of Lemma 4.6.19, we may always work with a vertically invariant contact *form*. In the situation of the lemma we have $\mathcal{L}_{\partial_t}(\Psi^*\alpha) = \mu\Psi^*\alpha$ for some smooth function $\mu\colon S \times \mathbb{R} \rightarrow \mathbb{R}$. Given a smooth function $\lambda\colon S \times \mathbb{R} \rightarrow \mathbb{R}^+$, we have, by Equation (1.1) in the proof of Lemma 1.5.8,

$$\mathcal{L}_{\partial_t}(\lambda\Psi^*\alpha) = \Big(\frac{\dot{\lambda}}{\lambda} + \mu\Big) \cdot (\lambda\Psi^*\alpha).$$

Hence, with

$$\lambda(p,t) := \exp\Big(-\int_0^t \mu(p,s)\, ds\Big),$$

the contact form $\lambda\Psi^*\alpha$ will be ∂_t–invariant.

Write this contact form as $\beta_t + u_t\, dt$ in our usual notation. Notice that $\beta_0 = \alpha|_{TS}$, since $\lambda(p,0) \equiv 1$. With the help of the Cartan formula (or directly from the definition of $\mathcal{L}_{\partial_t}$), one finds

$$0 = \mathcal{L}_{\partial_t}(\beta_t + u_t\, dt) = \dot{\beta}_t + \dot{u}_t\, dt,$$

hence $\dot{\beta}_t \equiv 0$ and $\dot{u}_t \equiv 0$. So with $u := u_0$ we have $\lambda\Psi^*\alpha = \beta + u\, dt$.

The contact condition (4.3) simplifies in this $\mathbb{R}$–invariant situation to

$$u\, d\beta + \beta \wedge du > 0. \tag{4.8}$$

In the sequel, the embedding Ψ will usually be understood. That is, we regard $S \times \mathbb{R}$ as a subset of M (contained inside an arbitrarily small neighbourhood of $S \subset M$) with contact structure $\xi = \ker(\beta + u\, dz)$.

This gives us a simple means for showing that certain characteristic foliations *cannot* arise on convex surfaces. Let X be the vector field given by $i_X\Omega = \beta$, defining the characteristic foliation S_ξ on a convex surface S. We can rewrite the contact condition (4.8) in terms of X as follows. First of all, we have $du \wedge \Omega \equiv 0$ for dimensional reasons. Hence

$$0 = du(X)\Omega - du \wedge i_X\Omega = X(u)\Omega + \beta \wedge du. \tag{4.9}$$

Secondly, $d\beta = \mathrm{div}_\Omega(X)\Omega$. Hence (4.8) translates into

$$u \,\mathrm{div}_\Omega(X) - X(u) > 0. \tag{4.10}$$

Example 4.6.21 Consider S^3 with its standard contact structure, which we write in the form

$$r_1^2\, d\varphi_1 + r_2^2\, d\varphi_2 = 0,$$

where (r_i, φ_i) denote polar coordinates in the (x_i, y_i)–plane, $i = 1, 2$. In these coordinates, the Hopf vector field is given by $r_1^2\,\partial_{\varphi_1} + r_2^2\,\partial_{\varphi_2}$. Consider an invariant torus S of this vector field, given by the equations $r_1^2 = c$ and $r_2^2 = 1 - c$ for some $c \in (0, 1)$. Then

$$\beta := \alpha|_{TS} = c\, d\varphi_1 + (1 - c)\, d\varphi_2.$$

With $\Omega := d\varphi_1 \wedge d\varphi_2$ as area form on S we have

$$X = (1 - c)\,\partial_{\varphi_1} - c\,\partial_{\varphi_2}.$$

This vector field has zero divergence, since β is closed. Thus, if S were a convex surface, Equation (4.10) would imply that the function u is strictly decreasing along the flow lines of X, which is impossible.

When trying to modify a given characteristic foliation S_ξ by perturbing the surface S, we are often not working with a closed surface, but with an open subset $U \subset S$ whose closure is a compact surface with boundary. The following proposition provides us with a particularly useful tool for such modifications. Notice that any oriented hypersurface $U \subset M$ whose closure is a surface with boundary can be thought of as a piece of a closed surface S: simply take a second copy U' of U, obtained by pushing U in normal direction, and then join the corresponding boundary components of U and U' by annuli. We may therefore apply Theorems 2.5.22 and 2.5.23 to such surfaces U, i.e. we may conclude that the germ of the contact structure ξ near U is determined by the characteristic foliation U_ξ. (In order to avoid problems at the boundary points of U in the corresponding argument, one may smooth out the contact Hamiltonian to zero near ∂U in the proofs of the mentioned theorems. So from $U_{\xi_1} = U_{\xi_2}$ we may then only conclude that the germs of two contact structures ξ_1 and ξ_2 coincide near a slightly shrunk U. We shall ignore this subtlety, since it is irrelevant in applications.)

Proposition 4.6.22 *Let $U \subset (M, \xi)$ be a hypersurface as just described, with area form Ω. Assume that there is a vector field X on U defining the characteristic foliation U_ξ, having positive divergence with respect to Ω.*

(a) There is a contact embedding of $(U \times \mathbb{R}, \ker(i_X\Omega + dz))$ into any arbitrarily small neighbourhood of U in (M, ξ).

(b) Let X_t, $t \in [0,1]$, with $X_0 = X$, be a smooth family of singular vector fields on U, constant (in t) near ∂U, and all with positive divergence with respect to Ω. Then by a C^0–small perturbation of U that keeps a neighbourhood of ∂U pointwise fixed, one can realise the singular foliation $\mathfrak{F}_1$ on U given by X_1 as the characteristic foliation of the perturbed surface. More precisely, there is an embedding $\psi \colon U \to M$ into an arbitrarily C^0–small neighbourhood of U, coinciding with the inclusion $U \subset M$ near ∂U, such that the characteristic foliation $(\psi(U))_\xi$ coincides with $\psi(\mathfrak{F}_1)$.

Proof (a) By Lemma 4.6.4, $i_X\Omega + dz$ is an $\mathbb{R}$–invariant contact structure on $U \times \mathbb{R}$ inducing the characteristic foliation U_ξ on $U \equiv U \times \{0\}$. By Theorem 2.5.22 (and the comments preceding this proposition), the germs of ξ and $\ker(i_X\Omega + dz)$ near U coincide. In particular, the vector field ∂_z corresponds to a contact vector field for ξ near U, making this a convex surface. So the result follows from Lemma 4.6.19 and the remark following it.

(b) By (a) we may work on the contact manifold $(U \times \mathbb{R}, \ker(i_X\Omega + dz))$. Any perturbation of $U \equiv U \times \{0\}$ inside this $U \times \mathbb{R}$, fixed near ∂U, then translates into a C^0–small perturbation of U inside (M, ξ).

Set $\beta_t := i_{X_t}\Omega$ and $\alpha_t := \beta_t + dz$. Again by Lemma 4.6.4, this defines a family of (vertically invariant) contact forms. Moreover, α_t does not depend on t near ∂U. Therefore, by Gray stability (Thm. 2.2.2), we find an isotopy ψ_t, identical near ∂U, such that $\ker(\psi_1^*\alpha_0) = \ker\alpha_1$; the vector field we need to integrate to get this isotopy is $\mathbb{R}$–invariant and compactly supported in the U–factor. Then $\psi := \psi_1|_U$ is the desired embedding. □

We now apply this technique to a simple example. The upshot of the following argument is that near an elliptic point of the characteristic foliation one can always specify any two leaves emanating from the elliptic point and require that they meet, after a perturbation of the surface, at an angle of 180°. When we later apply the elimination lemma to such an elliptic point (and a further hyperbolic point), these two leaves will form part of a single leaf after the elimination.

First, though, we reiterate a short calculation to see how the divergence of a vector field behaves under rescaling. This will allow us later, for suitable singular foliations, to choose defining vector fields with positive divergence.

Thus, let Ω be an area form on a surface S. For a smooth function $f\colon S \to \mathbb{R}^+$, we compute

$$\operatorname{div}(fX)\Omega = d(i_{fX}\Omega) = df \wedge i_X\Omega + f\operatorname{div}(X)\Omega.$$

For dimensional reasons we have $df \wedge \Omega \equiv 0$ on S. By taking the inner product of this identity with X we obtain

$$X(f)\Omega - df \wedge i_X\Omega = 0,$$

hence

$$\operatorname{div}(fX) = X(f) + f\operatorname{div}(X). \tag{4.11}$$

When we are in a situation where $\operatorname{div}(X)$ is bounded from below by some constant $-C$, then by solving the differential inequality $X(f) > Cf$ along any flow line of X we obtain a vector field fX with positive divergence. Of course it will depend on the topology of the flow whether such a solution does indeed exist.

Now back to the perturbation near an elliptic point. Consider the $\mathbb{R}$–invariant contact structure $\xi = \ker(dz + r^2\, d\varphi)$ on $\mathbb{R}^2 \times \mathbb{R}$ in obvious notation. The characteristic foliation $\mathfrak{F}_0 := \left(\mathbb{R}^2 \times \{0\}\right)_\xi$ is given by the radial lines, with an elliptic point at the origin.

Lemma 4.6.23 *Let $h_r\colon S^1 \to S^1$, $r \in [1,2]$, be a technical isotopy between the identity map h_1 and any given orientation-preserving diffeomorphism h_2 of S^1. Consider the singular foliation $\mathfrak{F}_1$ of $\mathbb{R}^2$ given by the radial lines for $r \leq 1$ and $r \geq 2$, and by the flow lines of the isotopy on the annulus $1 \leq r \leq 2$, i.e. the lines $r \mapsto (r, h_r(\varphi)) \subset \mathbb{R}^+ \times S^1$, $r \in [1,2]$, for any $\varphi \in S^1$ (Figure 4.16).*

Then there is an isotopy of the inclusion $\mathbb{R}^2 \xrightarrow{\equiv} \mathbb{R}^2 \times \{0\} \subset \mathbb{R}^2 \times \mathbb{R}$ to an embedding ψ, only moving points in the disc $\{r < 2\}$, such that the characteristic foliation $(\psi(\mathbb{R}^2))_\xi$ (with ξ as given before the lemma) coincides with $\psi(\mathfrak{F}_1)$.

Proof We take $\Omega = r\,dr \wedge d\varphi$ as area form on $\mathbb{R}^2$. The foliation $\mathfrak{F}_0$ by radial lines is given by the vector field $X_0 := r\,\partial_r$, which has positive divergence with respect to Ω.

The leaves of $\mathfrak{F}_1$ are transverse to the level sets of r, so this foliation can be defined by a vector field $X_1' := r\,\partial_r + a(r,\varphi)\,\partial_\varphi$, with the function $a(r,\varphi)$ supported in the annulus $1 < r < 2$. The vector field $X_1 := f(r)X_1'$ will have positive divergence with respect to Ω, provided we choose a function $f\colon \mathbb{R}_0^+ \to \mathbb{R}^+$ with $f'(r)$ sufficiently large on the annulus $1 < r < 2$. In

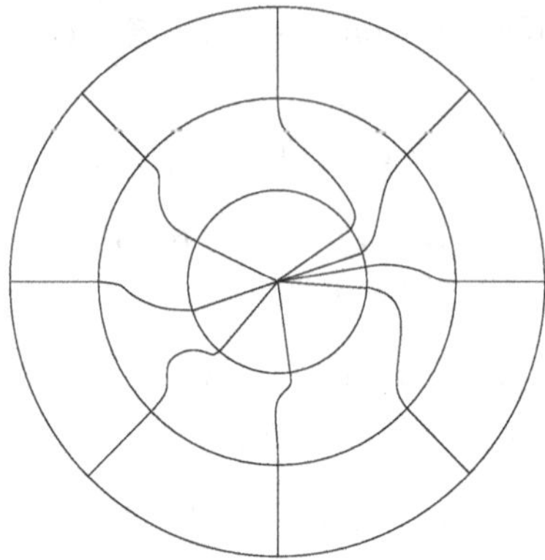

Fig. 4.16. The singular foliation $\mathfrak{F}_1$.

terms of the function $a(r,\varphi)$, the estimate one needs to satisfy is $2+rf'(r)+\frac{\partial a}{\partial\varphi}f(r)>0$; this follows from Equation (4.11). We may therefore impose the additional condition that $f(r)$ be identically equal to some sufficiently large positive constant c for $r\geq 2$.

Then $X_t := (1-t)cX_0 + tX_1$ has positive divergence with respect to Ω for all $t\in[0,1]$. Now apply part (b) of Proposition 4.6.22. □

4.6.3 The elimination lemma

The elimination lemma, first proved by Giroux and later refined by Fuchs, gives a method for reducing the number of singularities of a characteristic foliation S_ξ by perturbing the surface S inside the ambient contact manifold (M,ξ). Although some ideas of convex surface theory are present in the proof of this lemma, I defer the discussion of this general method to Section 4.8.

Throughout it will be assumed that S_ξ is of Morse–Smale type.

We say that an elliptic and a hyperbolic point of S_ξ are **in elimination position** if they have the same sign and are connected by a separatrix of the hyperbolic point. This terminology anticipates the fact that such a pair of singular points can be removed. In the proofs below (or at least in our illustrations) we only deal with a pair of positive points in elimination position; the negative case is completely analogous. First we choose a suitable annulus-shaped neighbourhood of such a pair.

Lemma 4.6.24 *Let x_{e} and x_{h} be an elliptic and a hyperbolic point, respectively, of the characteristic foliation S_ξ. If they are in elimination position, one can find an annulus $A\subset S$ with the following properties.*

- *The only singularities of A_ξ are x_{e} and x_{h}.*
- *The characteristic foliation A_ξ does not have any closed leaves.*
- *The characteristic foliation A_ξ is transverse to the boundary ∂A of A.*

Proof Saying that x_{e} and x_{h} are in elimination position means that there is a stable separatrix of x_{h} coming from x_{e}. We have to consider various cases, depending on the behaviour of the other stable separatrix of x_{h}.

(1) If both separatrices of x_{h} come from x_{e}, we may take A to be a neighbourhood of the union of the two singular points and the two separatrices, see Figure 4.17.

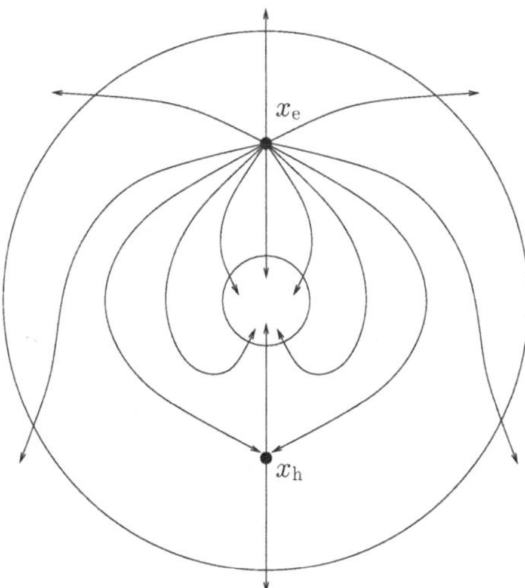

Fig. 4.17. An elimination pair, case (1).

(2) If the second stable separatrix of x_{h} comes from a second source x'_{e}, we may take A to be the annulus formed by a disc neighbourhood of the three singular points and the two separatrices, with a disc neighbourhood of x'_{e} removed, see Figure 4.18.

(3) If the second stable separatrix of x_{h} comes from a closed orbit γ_0, we first consider a thin annulus A' around γ_0. Let x be the unique intersection point between $\partial A'$ and the second stable separatrix of x_{h}. Then the annulus A can be defined as a neighbourhood of the set consisting of the component of $\partial A'$ containing x, and the segment between x_{e} and x along the two separatrices, see Figure 4.19 (where we have taken one component of ∂A to coincide with the component of $\partial A'$ containing x).

In all cases, the characteristic foliation A_ξ has the desired properties. □

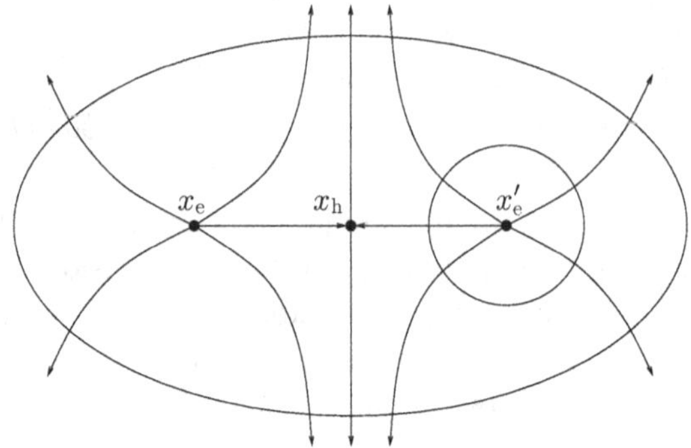

Fig. 4.18. An elimination pair, case (2).

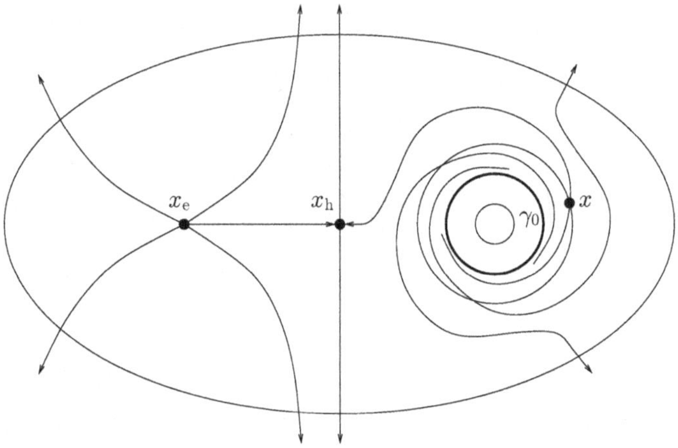

Fig. 4.19. An elimination pair, case (3).

As before, we assume that an area form Ω has been fixed on S, and that the characteristic foliation is defined by a vector field X as in Equation (4.4) of Definition 4.6.1. If we replace the contact form α defining ξ by $f\alpha$, with $f\colon M \to \mathbb{R}^+$ some smooth function, X will be replaced by fX.

In the case of positive singularities x_e, x_h considered here, the divergence of X with respect to Ω is positive at the singular points. We want to achieve positivity on all of A by choosing f suitably. When computing the divergence of fX, we regard f as a function on S, in particular when we consider the differential df. Therefore we write $f\colon S \to \mathbb{R}^+$ in the next lemma; any such function extends to a smooth function on M.

Lemma 4.6.25 *For a suitable smooth function $f\colon S \to \mathbb{R}^+$ the divergence of fX will be positive on all of A.*

Proof Our goal is achieved by choosing a function f that is constant near the singular points x_{e} and x_{h}, and grows fast enough along the flow lines of X so as to satisfy $X(f) + f\mathrm{div}(X) > 0$. This is possible in all three cases considered in the preceding proof; see Equation (4.11) and the comment following it. □

We now replace α by $f\alpha$, where f is a smooth extension to M of the function found in the preceding lemma. We continue to write α for this new contact form, and X for the induced vector field on S with positive divergence on A.

As before we write $\alpha = \beta_z + u_z\, dz$ on a neighbourhood $S \times \mathbb{R}$ of S in M, so that $\beta := \beta_0 = i_X\Omega = \alpha|_{TS}$. By Proposition 4.6.22 we may identify an arbitrarily small (vertical) neighbourhood of A in (M, ξ) with the contact manifold $(A \times \mathbb{R}, \ker(\beta + dz))$.

The following statement is worthy of being called a theorem, but it is commonly referred to as the elimination *lemma*.

Lemma 4.6.26 (Elimination Lemma) *Let $S \subset (M, \xi)$ be a closed, oriented surface, embedded in such a way that the characteristic foliation S_ξ is of Morse–Smale type. Assume that there are an elliptic point x_{e} and a hyperbolic point x_{h} in this characteristic foliation which have the same sign and are connected by a separatrix γ of x_{h}. Then there is an arbitrarily C^0–small isotopy $\psi_t\colon S \to M$, $t \in [0, 1]$, with ψ_0 the inclusion of S in M, fixed on the separatrix γ and outside an arbitrarily small neighbourhood U of it, such that the characteristic foliation $(\psi_1(S))_\xi$ has no singularities on $\psi_1(U)$.*

Proof The idea of the proof is very simple. We have seen above that we may identify a neighbourhood of U in (M, ξ) with $(U \times \mathbb{R}, \ker(\beta + dz))$. We may now replace the inclusion $U \equiv U \times \{0\} \subset U \times \mathbb{R}$ by the graph of a function $g\colon U \to \mathbb{R}$, compactly supported in U. The linear interpolation between the zero function and g defines an isotopy ψ_t of embeddings as desired. Moreover, the characteristic foliation $(\psi_1(U))_\xi$ is given by the vector field $X' = X + X_g$ on U defined by

$$i_{X+X_g}\Omega = \beta + dg,$$

where Ω denotes the area form on S as before. Our aim therefore has to be to find a function g such that $\beta + dg \neq 0$ on all of U. The condition that γ remain fixed under the isotopy translates into g being identically 0 along γ. Notice that the C^0–smallness of the isotopy is guaranteed by our identification of a small neighbourhood of U in M with $U \times \mathbb{R}$; we do not need to choose $|g|$ small. However, if one succeeds in choosing $|g|$ small (as in the second part of this proof below), there is no need to appeal to Proposition 4.6.22, for then Theorem 2.5.22 will suffice.

(1) Now to the details. I first present the beautiful argument due to Giroux [104]. In this argument we use in an essential way the annulus-shaped neighbourhood A of γ constructed above. This argument does not quite prove the elimination lemma in the form stated, since γ may not stay fixed under the isotopy, and the perturbation is non-stationary over A, not just over a small neighbourhood of the separatrix γ.

With this proviso, what we do is the following. Fix a foliation of A by circles parallel to the boundary ∂A. Observe that X will be transverse to this foliation near ∂A. Choose a smooth function $g\colon A \to (-\infty, 0]$, identically 0 near ∂A and constant along the leaves of the foliation by circles. Since $dg(X_g) = \Omega(X_g, X_g) = 0$, the vector field X_g will be tangent to this foliation. Hence, near ∂A this guarantees that $X' = X + X_g$ does not have any zeros. Away from ∂A we can ensure that dg is large enough to dominate β, so that there $X' \neq \mathbf{0}$ as well. (Notice that the minimum of g needs to be chosen near one of the two components of ∂A.)

(2) The following argument of Fuchs, which proves the elimination lemma in the form stated, was presented by Eliashberg in [70].

Choose local Cartesian coordinates (x, y) on U such that the square

$$\{(x, y)\colon\ |x| \leq 2, |y| \leq 2\}$$

is contained in U and the characteristic foliation looks as in Figure 4.20. After the perturbation, we would like the characteristic foliation to look as in Figure 4.21.

Slightly more specifically, we want that $x_{\mathrm{e}} = (-1, 0)$, $x_{\mathrm{h}} = (1, 0)$, and that γ is given by the horizontal line segment between these two points. Write $\Omega = a(x, y)\, dx \wedge dy$ in these coordinates and $X(x, y) = X_1(x, y)\, \partial_x + X_2(x, y)\, \partial_y$.

We now make the ansatz $g(x, y) = g_1(x) g_2(y)$. Then

$$dg = g_1'(x) g_2(y)\, dx + g_1(x) g_2'(y)\, dy,$$

hence

$$X_g(x, y) = \frac{1}{a(x, y)} \big(g_1(x) g_2'(y)\, \partial_x - g_1'(x) g_2(y)\, \partial_y \big).$$

Choose g_1, g_2 as illustrated in Figure 4.22. We now discuss the behaviour of X_g in various parts of the square $\{|x| \leq 2,\ |y| \leq 2\}$.

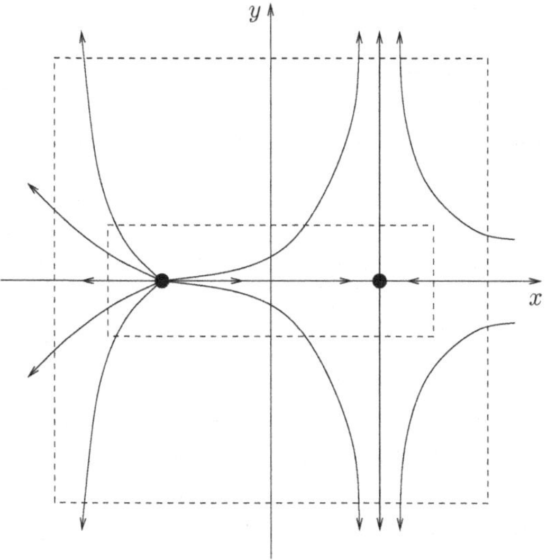

Fig. 4.20. An elimination pair in 'standard' form.

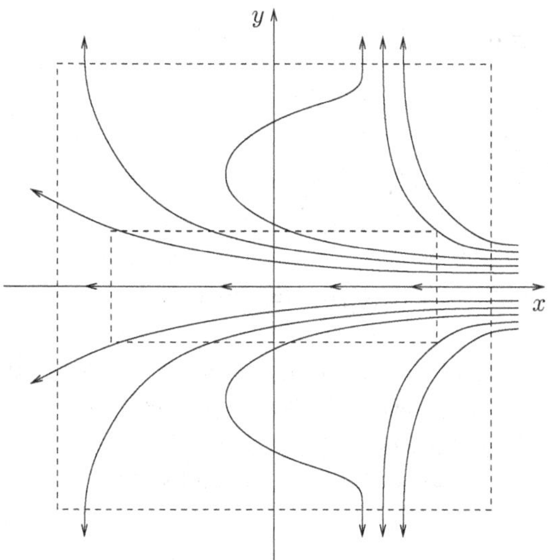

Fig. 4.21. After the elimination.

(i) $|x| \leq 3/2$, $y = 0$. Here

$$X_g(x, 0) = \frac{g_2'(0)}{a(x, 0)}\, \partial_x.$$

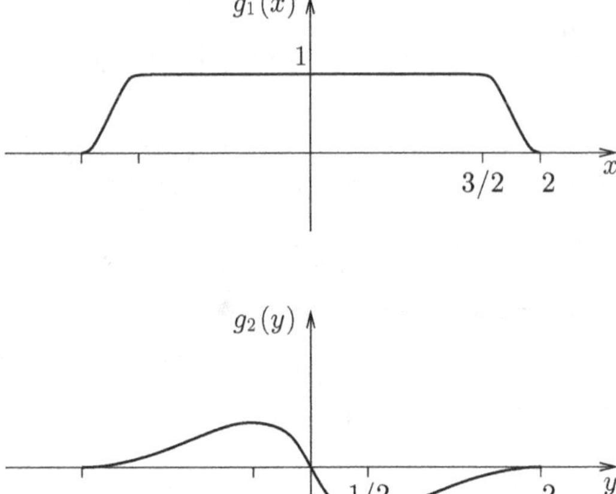

Fig. 4.22. The functions g_1, g_2.

We can ensure $X' \neq \mathbf{0}$ on this line segment if we choose g_2 such that

$$g_2'(0) < -\max_{|x|\leq 1}\bigl(a(x,0)\cdot X_1(x,0)\bigr).$$

(On the two segments $\{1 < |x| \leq 3/2,\ y = 0\}$ both X and X_g have a negative ∂_x–component, so here $X' \neq \mathbf{0}$ anyway.)

(ii) $|x| \leq 3/2$, $0 < |y| \leq 1/2$. Here

$$X_g(x,y) = \frac{g_2'(y)}{a(x,y)}\,\partial_x$$

is a multiple of ∂_x, while X has a non-trivial ∂_y–component. Hence $X' \neq \mathbf{0}$.

(iii) $3/2 \leq |x| \leq 2$, $|y| \leq 1/2$. In these two rectangles, the vector field X has a negative ∂_x–component. As suggested by Figure 4.22, we require $g_2'(y) \leq 0$ for $|y| \leq 1/2$. Then X_g has a non-positive ∂_x–component. This guarantees $X' \neq \mathbf{0}$.

(iv) $|x| \leq 2$, $1/2 \leq |y| \leq 2$. In these two rectangles, the norm $\|X\|$ is bounded from below, and so is the function a in the coordinate description of the area form Ω. With g_1 given, we can choose g_2 with $|g_2|$ arbitrarily small on $|y| \leq 2$ and $|g_2'|$ arbitrarily small on $1/2 \leq |y| \leq 2$. This enables us to ensure that $\|X_g\| < \|X\|$, hence $X' \neq \mathbf{0}$.

The fact that we may choose $|g|$ as small as we like means that we can make the perturbation C^0–small. The condition that the separatrix γ be fixed under the perturbation is satisfied because of $g = 0$ along γ. □

Remarks 4.6.27 (1) An analogous construction allows one to introduce a pair of singular points (one elliptic, one hyperbolic) of the same sign into a given characteristic foliation. For instance, one can break up a repelling or attracting closed orbit by introducing two positive or negative singularities, respectively, without changing the remaining structure of the foliation.

(2) The elimination procedure does not destroy the Morse–Smale property of the characteristic foliation.

(3) Notice that this elimination of an elliptic and a hyperbolic point of the same sign is consistent with the formulæ in Proposition 4.6.14.

Here is a useful addendum to the elimination lemma. Let γ' be the leaf of the characteristic foliation in Figure 4.20 emanating from x_{e} and meeting the separatrix γ at an angle of 180° at that elliptic point. Then, after the elimination, γ' and the two separatrices of x_{h} lie on the same leaf of the new characteristic foliation. By Lemma 4.6.23 we are completely free to specify the leaf γ' in this construction by changing the characteristic foliation in a neighbourhood of the elliptic point.

As advertised earlier, we can now use the elimination lemma to give a more useful characterisation of overtwisted contact structures on 3–manifolds.

Proposition 4.6.28 *Let Δ be an embedded disc in a contact 3–manifold (M, ξ) such that its boundary $\partial\Delta$ is a Legendrian curve for ξ whose surface framing coincides with its contact framing, i.e. $\mathtt{tb}(\partial\Delta) = 0$. Then ξ is an overtwisted contact structure.*

Proof By Example 4.6.7 we may perturb Δ, keeping its boundary fixed, such that $\partial\Delta$ is a non-degenerate closed orbit of the characteristic foliation Δ_ξ. The result of Peixoto [205] also applies in this situation (see Remark 2 at the end of his paper). This means that we may perturb Δ in its interior to make Δ_ξ of Morse–Smale type.

We may assume that the characteristic foliation does not contain any closed leaves in the interior of Δ, otherwise we replace Δ by a smaller disc bounded by such a leaf. (The fact that there are no singular points on a closed leaf means that its surface framing coincides with the contact framing.) By the Poincaré–Hopf index theorem, the sum of the indices of the singular points of Δ_ξ equals 1. Therefore, if there are no hyperbolic points, then Δ_ξ only contains a single elliptic point, in which case Δ is an overtwisted disc.

If, on the other hand, there *are* hyperbolic points, our aim has to be to remove them using the elimination lemma. We may assume that the closed orbit $\partial\Delta$ is an attracting *limit cycle* of the characteristic foliation, i.e. the neighbouring orbits approach the closed orbit in forward time. The case of a repelling orbit is completely analogous.

(i) Given a positive hyperbolic point in Δ_ξ, its stable separatrices have to come from positive elliptic points (which may coincide), since the only limit cycle in Δ_ξ is $\partial\Delta$, which is attracting. The hyperbolic point is in elimination position with either of these elliptic points.

(ii) Given a negative hyperbolic point with an unstable separatrix ending at a negative elliptic point, we can again eliminate them both.

(iii) The only case remaining is that we have a negative hyperbolic point x_{h} with both unstable separatrices γ, γ' attracted by the limit cycle $\partial\Delta$. By Remark 4.6.27 (1) we may introduce an additional pair of negative singular points, one elliptic, the other hyperbolic. By Lemma 4.6.23 we can do this in such a way that γ and γ' meet at an angle of $180°$ at the new elliptic point x_{e}, and such that the new hyperbolic point lies outside the circle made up by γ and γ' (together with the two singular points $x_{\mathrm{h}}, x_{\mathrm{e}}$), see Figure 4.23.

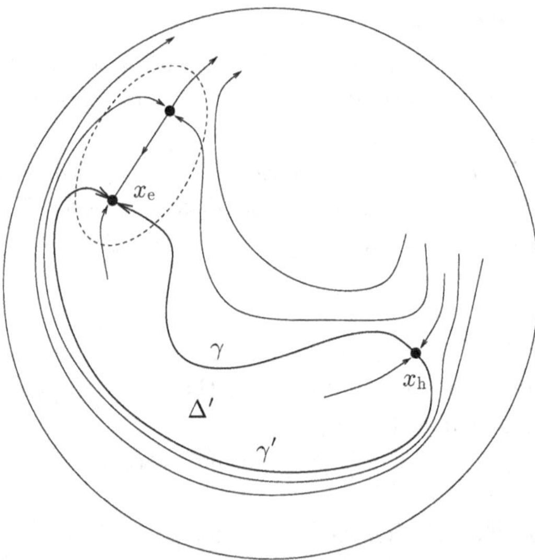

Fig. 4.23. Finding an overtwisted disc.

Now apply the elimination lemma to the singular points x_{e}, x_{h} and the separatrix γ. After the elimination (as in part (2) of the proof of Lemma 4.6.26), γ and γ' will form a closed orbit of the characteristic foliation. The characteristic foliation of the disc $\Delta' \subset \Delta$ bounded by this orbit contains fewer hyperbolic points than Δ_ξ. Since there are no singular points of the characteristic foliation on $\partial\Delta'$, the surface framing of this Legendrian curve coincides with its contact framing.

Thus, in all three cases we can reduce the number of hyperbolic points, possibly by passing to a smaller disc. By iterating this procedure we find an overtwisted disc. □

Remark 4.6.29 The original definition of a tight contact structure, see [69], required not only that there be no *overtwisted* discs, but that there be no embedded discs as in the proposition. Then, by definition, overtwisted contact structures are not tight, but there might well have been contact structures that are neither tight nor overtwisted. The proposition, due to Eliashberg in the paper just quoted, says that contact structures that are not overtwisted are tight (in this older sense). Nowadays it is more common, as I have done in Definition 4.5.2, to define tight contact structures directly as the non-overtwisted ones.

4.6.4 Genus bounds

According to the Lutz–Martinet theorem, any homotopy class of tangent 2–plane fields on a given closed, orientable 3–manifold is represented by a contact structure. Moreover, the same remains true if we restrict ourselves to overtwisted contact structures, see the closing remarks in Section 4.5. In particular, any even element $e \in H^2(M;\mathbb{Z})$ can be realised as the Euler class of an overtwisted contact structure, see Proposition 4.3.2.

The aim of the present section is to show that the world of tight contact structures is notably different. Specifically, we are going to prove that only finitely many elements in $H^2(M;\mathbb{Z})$ can be realised as the Euler class of a tight contact structure. This result is due to Eliashberg [69].

The essential part of this argument is contained in the genus bound given by the following theorem of Eliashberg [69].

Theorem 4.6.30 *Let (M,ξ) be a tight contact 3–manifold and S a closed, connected, oriented surface embedded in M. Then the Euler class $e(\xi) \in H^2(M;\mathbb{Z})$ satisfies the inequality*

$$|\langle e(\xi),[S]\rangle| \leq \begin{cases} 0 & \textit{if}\, S = S^2, \\ -\chi(S) & \textit{otherwise.} \end{cases}$$

The right-hand side of this inequality can also be written more succinctly as $\max(0,-\chi(S))$. Thurston [231] called this the 'negative part of the Euler characteristic'. In that paper (p. 119) he proved the same inequality for transversely orientable codimension 1 foliations without Reeb components.

The proof of Theorem 4.6.30, for which I follow the exposition given in [80], is rather involved.† Therefore I first want to draw your attention to the following corollary, which is the central result of this section.

Corollary 4.6.31 *On any given closed, orientable 3–manifold M, only finitely many elements of $H^2(M;\mathbb{Z})$ can be realised as the Euler class of a tight contact structure.*

Proof By the universal coefficient theorem we have, with $\mathbb{Z}$–coefficients for (co-)homology understood,

$$H^2(M) \cong \mathrm{Hom}(H_2(M), \mathbb{Z}) \oplus \mathrm{Ext}(H_1(M), \mathbb{Z}).$$

Any compact differentiable manifold has the homotopy type of a finite *CW*–complex and hence finitely generated (co-)homology; this follows from Morse theory. This implies the finiteness of the Ext term, and hence that the Euler class $e(\xi) \in H^2(M)$ of a contact structure ξ is determined up to finite ambiguity by its values on a finite set of generators of $H_2(M)$. By Proposition 3.4.4, these generators can be realised as embedded surfaces. Thus, by Theorem 4.6.30, the set of permissible values of $e(\xi)$ on each generator is finite if ξ is tight. □

Remark 4.6.32 The first example of a 3–manifold that does not admit any tight contact structure (for either orientation) was found by Etnyre and Honda [86]. Their example is a connected sum of two copies of the Poincaré homology sphere, with opposite orientations.

Remark 4.3.4 tells us that the Euler class of a tangent 2–plane field η on M determines the homotopy class of the restriction of η to the 2–skeleton of M up to finite ambiguity, resulting from 2–torsion in $H^2(M)$. Moreover, any two tight contact structures on the 3–ball that coincide near the boundary of the 3–ball are in fact isotopic (see Theorem 4.10.1 below). Even so, this does not allow one to conclude immediately that only finitely many homotopy classes of tangent 2–plane fields can be realised by tight contact structures.‡ It turns out that the homotopy class (as tangent 2–plane field) of a tight contact structure over the 2–skeleton does not determine its extension over M; this follows from examples of non-homotopic tight contact structures on homology 3–spheres described by Lisca [162].

The paper [49] by Colin, Giroux and Honda contains an announcement that their methods do indeed yield a proof of the homotopical finiteness of tight contact structures.

† An alternative proof, based on convex surface theory, will be presented on page 243 below.

‡ The corresponding claim in [69] was 'a bit previous', to borrow a phrase from [155].

Proof of Theorem 4.6.30 First of all, by Proposition 4.6.11 we may perturb S so that the characteristic foliation S_ξ is of Morse–Smale type. (We continue to call the surface S after this and further perturbations below.) Recall the notation and statement of Proposition 4.6.14. From that proposition we have

$$\langle e(\xi), [S]\rangle + \chi(S) = 2(e_+ - h_+).$$

We are going to show that the tightness of ξ implies that we can cancel every positive elliptic point by a perturbation of S, unless $S = S^2$ and $e_+ = e_- = 1$ and $h_+ = h_- = 0$. In the latter case, we have the equality $\langle e(\xi), [S]\rangle = 0$; in the former, the inequality $\langle e(\xi), [S]\rangle + \chi(S) \leq 0$. Alternatively, we can cancel every negative elliptic point (unless we are again in the exceptional case). This gives $-\langle e(\xi), [S]\rangle + \chi(S) \leq 0$, and hence the claimed inequality.

Assume that S_ξ contains a positive elliptic point x_{e}^+. By Remark 4.6.27 (1) we may, again by perturbing S, assume that there are no attracting closed leaves in S_ξ – at the price of introducing additional negative elliptic and hyperbolic points. This guarantees that all flow lines whose α–limit is x_{e}^+ will have a hyperbolic or a negative elliptic point as ω–limit. Let U be the set of all points whose α–limit is x_{e}^+. This is an open disc in S; see the arguments used for proving [132, Thm. 6.2.2]. Let B denote the closure of U, and set $\partial B := B \setminus U$. Observe that all the singularities of S_ξ that lie in B, except x_{e}^+, actually lie in ∂B.

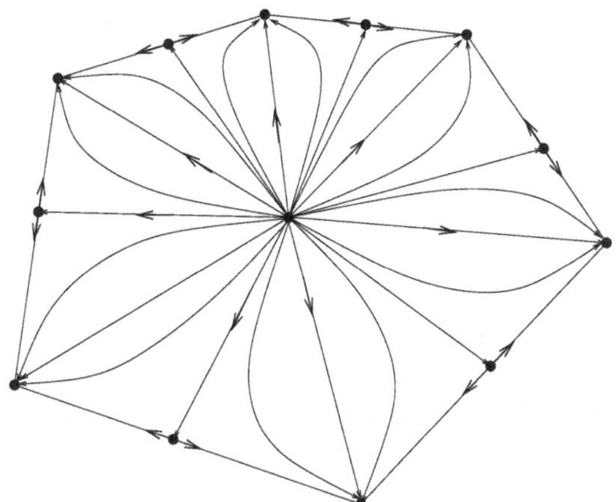

Fig. 4.24. A Legendrian polygon B.

Refer to Figure 4.24 for the following discussion. Between any two flow lines with α–limit x_{e}^+ and ω–limit two different negative elliptic points, there must be a separatrix of a hyperbolic point. Hence, if there are no hyperbolic points in ∂B, then ∂B consists of a single negative elliptic point and, consequently, S is a 2–sphere.†

If ∂B contains a hyperbolic point, then the unstable separatrices of that singularity are also contained in ∂B; those separatrices have a negative elliptic point in ∂B as ω–limit. Thus, B is what one may call a **Legendrian polygon**: the edges of B consist of pairs of separatrices; the vertices of B are negative elliptic points. (The edges are smooth in the hyperbolic points, so these points are not regarded as vertices.) In general, however, B is *not* an *embedded* polygon in S, for several vertices or pairs of edges may happen to be identified. An example of a Legendrian pentagon with four edges identified in pairs is shown in Figure 4.25; here (in contrast with the pictures below) the shaded region is the part of $S = T^2$ *outside* the Legendrian pentagon. (This shaded region has to contain further singularities, e.g. a single positive elliptic point.)

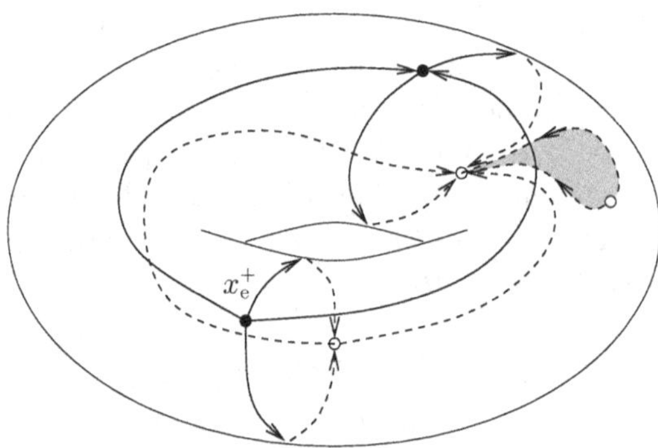

Fig. 4.25. A Legendrian pentagon on the 2–torus.

If one of the hyperbolic points in ∂B is positive, then we can cancel it together with x_e^+ by the elimination lemma (Lemma 4.6.26). This may create a closed leaf, but only one that is repelling, so we can iterate the argument for the other positive elliptic points. That way we can either remove all positive elliptic singularities in S_ξ (in which case we are done), or we find a Legendrian polygon B with a positive elliptic singularity in the interior and ∂B containing only negative singularities.

Let us consider that latter case. We want to show that then ξ would be overtwisted, contradicting the assumption of the theorem. If $B \subset S$ were embedded,‡ we could cancel the singularities along ∂B in pairs by the

† This is proved like Reeb's sphere theorem [132, Thm. 6.2.4].
‡ Here I think of B as an abstract polygon. Saying that 'B is embedded' thus means that $B \subset S$ is actually an embedded polygon.

elimination lemma, turning ∂B into a smooth Legendrian curve.† The result would be an embedded disc with Legendrian boundary, the surface framing of which coincides with the contact framing. So ξ would be overtwisted by Proposition 4.6.28.

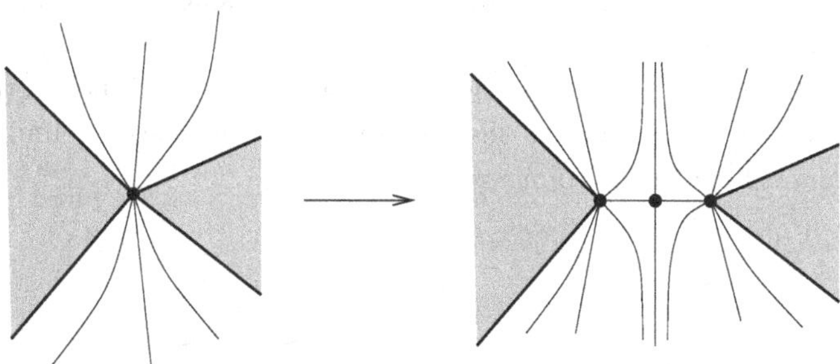

Fig. 4.26. Separating two vertices by creating a 'double-sun'.

If two vertices of B (but not their adjacent edges) are identified in S, one can separate them by the following method. Perturb S so that the characteristic foliation develops a 'double-sun' singularity as in Figure 4.26. This can be achieved by a procedure analogous to the proof of the elimination lemma (or rather its converse).

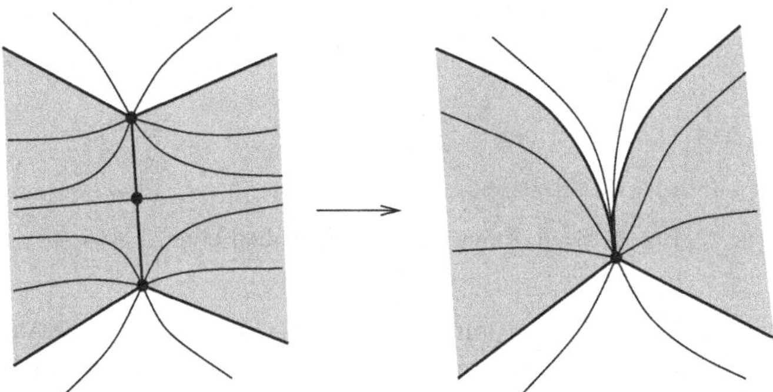

Fig. 4.27. Cancelling an identified pair of edges.

If two edges of B are identified in S to a single edge with *distinct* vertices as in Figure 4.27, we can cancel these edges with the help of the elimination lemma as shown in that figure.‡ Likewise, we can of course cancel any edge with two distinct vertices that is not identified with any other edge.

† Use Lemma 4.6.23 in order to obtain a smooth Legendrian curve after the elimination.
‡ The presence of other identified vertices or edges in that figure would not prevent this cancellation.

Continuing in this fashion, we arrive at a Legendrian polygon in S with all vertices identified to a single point (like in the example in Figure 4.25).

If there is no edge left, then S is the 2–sphere, and the only singular points in S_ξ are a pair of elliptic points of opposite sign. If there is only one edge left, then B is embedded, and by eliminating the two singularities on ∂B we obtain an overtwisted disc.

If there are at least two non-identified edges, we may create a new vertex by the 'double-sun' construction and then kill one of those edges by eliminating an elliptic and a hyperbolic point, see Figure 4.28.†

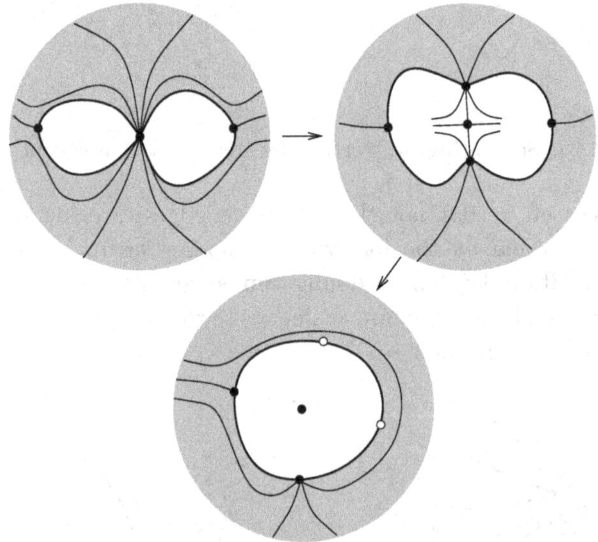

Fig. 4.28. Removing a non-identified edge.

After iterating these constructions, the process will either have terminated, or there is a pair of identified edges. A typical situation is depicted in Figure 4.25. Now proceed as follows. Consider such a pair of identified edges; they form a closed loop in S (since there is only one vertex). Such a loop is non-separating, for points on either side of it are connected by flow lines with x_{e}^+. Let $\mathcal{N}$ be a neighbourhood of the chosen loop. We may assume that the flow lines of S_ξ are all transverse to $\partial\mathcal{N}$, directed inwards. Now double $\mathcal{N}$, as it were, push down one copy of $\mathcal{N}$ (by an isotopy fixing one component of $\partial\mathcal{N}$), and push up the second copy of $\mathcal{N}$ (fixing the other

† The presence of other edges terminating at the single vertex in the centre of the first picture does not affect this construction.

component of $\partial\mathcal{N}$), see Figure 4.29. (Since S is an oriented surface, the notions 'up' and 'down' in M with respect to S are meaningful.)

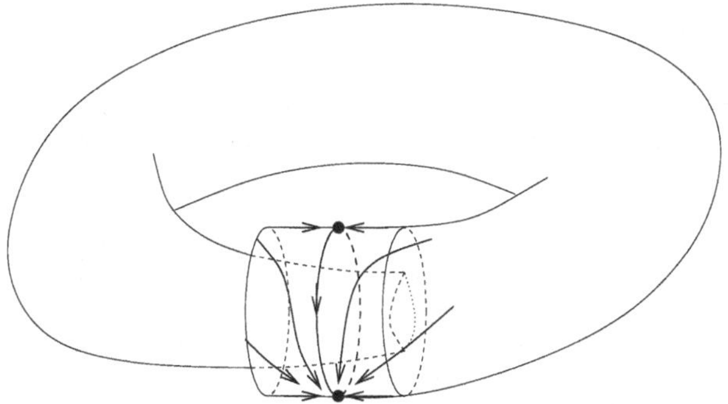

Fig. 4.29. Separating edges by cutting S.

For all practical purposes, this new surface is just as good as the initial closed surface S. For instance, the structural stability result for Morse–Smale vector fields by Peixoto (p. 172) holds true for compact surfaces with boundary, provided the vector field is transverse to the boundary. Thus, in particular, the described modification of S can be performed in such a way that the characteristic foliation in the two copies of $\mathcal{N}$ is homeomorphic to the initial one.

Applied to the example of Figure 4.25, this procedure yields a Legendrian polygon on a cylinder as shown in Figure 4.30.

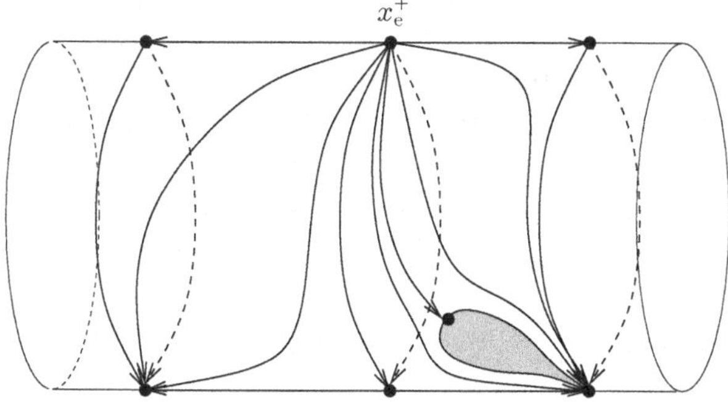

Fig. 4.30. The Legendrian pentagon from Figure 4.25 on the cylinder.

We can now continue the procedure as before, for instance by cancelling the lower edge in Figure 4.30, and then separating vertices again. The process will terminate, since we are reducing the number of edges in the Legendrian polygon at each step. □

The following result, a version of which was stated without proof in [69], describes a related situation. In contrast with the situation in the preceding proof, it deals with an embedded polygon right from the start, but also with hyperbolic vertices.

Lemma 4.6.33 *Let S be an oriented surface in a tight contact 3–manifold (M, ξ). Then S cannot contain any embedded polygon all of whose vertices are non-degenerate singular points of the same sign in the characteristic foliation S_ξ, and whose edges are separatrices of hyperbolic points in S_ξ. (Figure 4.31 shows such a polygon with all vertices hyperbolic.)*

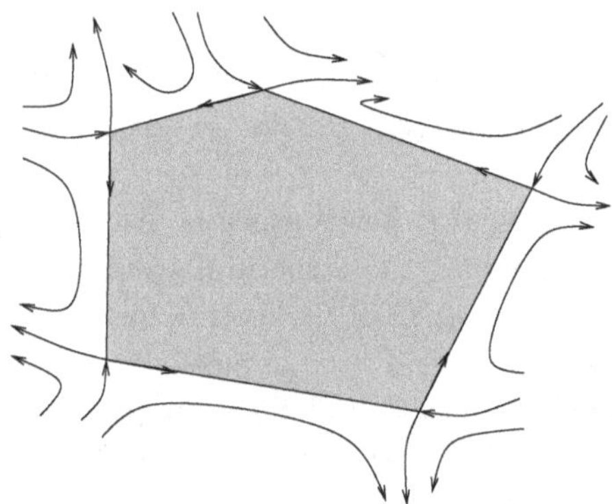

Fig. 4.31. A Legendrian polygon with hyperbolic vertices.

Proof Assume that such a polygon existed. We are going to show that then ξ would be overtwisted. There are two types of hyperbolic vertices in such a purported polygon: 'honest' vertices as in Figure 4.31, where the adjacent edges are given by one stable and one unstable separatrix, and 'pseudo-vertices' as in Figure 4.24, where the adjacent edges are a pair of stable or a pair of unstable separatrices.

First we consider the honest hyperbolic vertices. Recall the local model of a hyperbolic point from Examples 4.6.6 (2). This tells us that we can write the contact structure locally as

$$\xi = \ker(dz + \lambda_1 x\, dy - \lambda_2 y\, dx)$$

on $\mathbb{R}^2 \times \mathbb{R}$, with the origin $\mathbf{0}$ in $\mathbb{R}^2 \times \{0\}$ a hyperbolic point of the characteristic foliation. Here $\lambda_1 > 0 > \lambda_2$ and $\lambda_1 + \lambda_2 > 0$, say.†

† For $\lambda_1 + \lambda_2 < 0$ one should put a minus sign in front of dz in order for ξ to define the usual orientation of $\mathbb{R}^3$.

We think of $\mathbf{0}$ as a vertex in our Legendrian polygon, of the positive x–axis and the positive y–axis as the edges meeting at that vertex, and of the first quadrant as a neighbourhood of that vertex inside the polygon. The surface framing defined by the first quadrant may be thought of as being given by the vector field ∂_z along the two half-axes.

Observe that, analogous to the proof of Lemma 3.2.6, the z–coordinate of a Legendrian curve in $\mathbb{R}^2 \times \mathbb{R}$ with a given projection into the (x, y)–plane can be recovered by integrating the 1–form $-(\lambda_1 x\, dy - \lambda_2 y\, dx)$ along the projected curve. Along a closed curve, this integral computes (with sign) the enclosed area up to a (non-zero!) factor $-(\lambda_1 + \lambda_2)$. Along the coordinate axes, this 1–form integrates to zero.

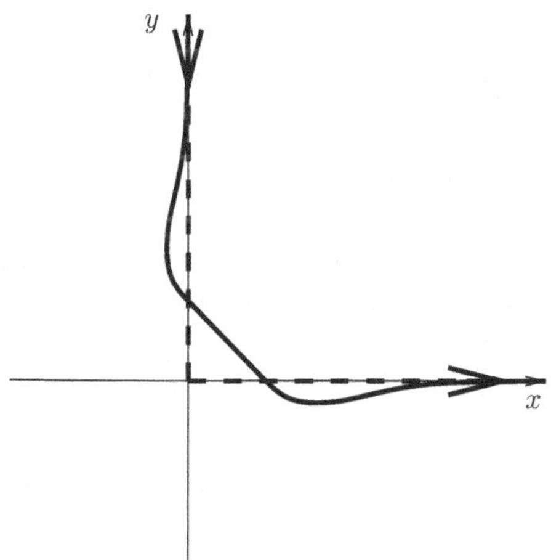

Fig. 4.32. Smoothing a hyperbolic vertex of a Legendrian polygon.

Now replace the two half-axes by a smooth curve as shown in Figure 4.32, with total 'area integral' equal to zero. This defines a perturbation of the polygonal disc near one of its vertices to a disc with smooth boundary.

We are then left with elliptic vertices and hyperbolic pseudo-vertices only. By the sign condition, any elliptic vertex must be adjacent to two hyperbolic pseudo-vertices (which may coincide in the case of a 2–gon). Therefore, as in the proof of Theorem 4.6.30, all elliptic vertices can be eliminated together with one hyperbolic pseudo-vertex each.

After performing this construction, we have a smoothly embedded disc Δ in (M, ξ) with Legendrian boundary $\partial\Delta$, with $\partial\Delta$ possibly containing finitely many hyperbolic pseudo-vertices. The fact that all these hyperbolic points are of the same sign implies that the surface framing of $\partial\Delta$ coincides with its contact framing. According to Proposition 4.6.28, this means that ξ must be overtwisted. □

4.6.5 The Bennequin inequality

Next we want to derive similar bounds for the classical invariants of Legendrian and transverse knots in tight contact 3–manifolds. We shall discuss that such bounds, conversely, can be used as a tightness criterion.

Theorem 4.6.34 (Transverse Bennequin Inequality) *Let K be a homologically trivial transverse knot in a tight contact 3–manifold (M, ξ). Let Σ be a Seifert surface for K. Then the self-linking number of K satisfies the inequality*

$$\mathtt{sl}(K, \Sigma) \leq -\chi(\Sigma).$$

Proof Let α be a contact form defining $\xi = \ker \alpha$ as oriented and cooriented plane field. Choose a (regular) parametrisation $\gamma\colon S^1 \to K \subset M$ (and hence an orientation for K) which makes K positively transverse to ξ, that is, $\alpha(\gamma') > 0$. Give Σ the orientation so that the boundary orientation of $\partial\Sigma = K$ is consistent with this choice. From Definition 4.6.1 we then see that the vector field X on S defining the characteristic foliation Σ_ξ satisfies, along $\partial\Sigma$,

$$\Omega(X, \gamma') = i_X\Omega(\gamma') = \alpha(\gamma') > 0,$$

so X is pointing outwards along $\partial\Sigma$.

Form the abstract closed, oriented surface $S := \Sigma \cup_K D^2$. Observe that the orientation of K is the negative of the boundary orientation of D^2. We may extend ξ to a plane field on S such that X extends to a section with only one zero over the interior of D^2, of sink type. Although this is not essential, it is convenient to think of $\xi|_{D^2}$ as a plane field over $D^2 \subset \mathbb{R}^3$ that is transverse to ∂D^2 and rotates to the left as we move in radial direction (as a contact structure would do), until at the centre of D^2 it coincides with the tangent plane to D^2, but with opposite orientation. Then, as in the case of a characteristic foliation induced by a contact structure, this sink gives a contribution -1 to the Euler class of the plane bundle. By the argument used to prove Proposition 4.6.14 we have

$$e_0 := \langle e(\xi), [S] \rangle = (e_+ - h_+) - (e_- - h_-) - 1.$$

Let X_0 be a nowhere vanishing section of $\xi|_\Sigma$. Since the vector field X defines the surface framing of K, the self-linking number $\mathtt{sl}(K, \Sigma)$ is given by counting the number of rotations of X_0 relative to X as we once traverse K in positive direction.

Over D^2 there is a unique trivialisation of ξ up to homotopy. Relative to this trivialisation, the number of rotations of X_0 as we traverse K in positive direction is $-e_0$, since K is oriented as the negative boundary of D^2. The corresponding number of rotations of X is $+1$ (*sic*). Hence

$$\mathtt{sl}(K,\Sigma) = -e_0 - 1 = -\big[(e_+ - h_+) - (e_- - h_-)\big].$$

Similarly, since X is pointing outwards along $\partial\Sigma$, we have

$$\chi(\Sigma) = (e_+ - h_+) + (e_- - h_-).$$

Hence

$$\mathtt{sl}(K,\Sigma) + \chi(\Sigma) = 2(e_- - h_-). \tag{4.12}$$

The fact that X is pointing outwards along $\partial\Sigma$ ensures that the α–limit set of any negative elliptic point in Σ_ξ stays away from $\partial\Sigma$. Therefore, the argument from the preceding section allows us to eliminate all negative elliptic points – it is only at this stage that we use the tightness of ξ. This gives the desired inequality. □

Remark 4.6.35 Equation (4.12) explains the remark made earlier (Remark 3.5.35) about the parity of the self-linking number. Together with Proposition 3.5.36, this also gives – as promised in Remark 3.5.24 – the relation between `tb` and `rot` from Proposition 3.5.23 in arbitrary contact 3–manifolds.

Theorem 4.6.36 (Legendrian Bennequin Inequality) *Let K be a homologically trivial oriented Legendrian knot in a tight contact 3–manifold (M,ξ). Let Σ be a Seifert surface for K, oriented compatibly. Then the Thurston–Bennequin invariant and the rotation number of K satisfy the inequality*

$$\mathtt{tb}(K) + |\mathtt{rot}(K,[\Sigma])| \leq -\chi(\Sigma).$$

Proof This follows from the transverse Bennequin inequality and Proposition 3.5.36, applied to both the positive and the negative transverse push-off of K. □

Remark 4.6.37 Bennequin [26] proved these inequalities for the standard contact structure ξ_{st} on $\mathbb{R}^3$ (or S^3) by direct geometric methods, prior to the introduction of the concept 'tight'. It is easy to see that the Bennequin inequalities fail in overtwisted contact manifolds. For instance, the Legendrian boundary $K = \partial\Delta$ of an overtwisted disc Δ has $\mathtt{tb}(K) = 0$ and $\mathtt{rot}(K, [\Delta]) = \pm 1$. Thus, Bennequin's proof can actually be read as a tightness proof for ξ_{st}.

In the generalised form stated here, the Bennequin inequalities are due to Eliashberg [70].

4.7 The classification of overtwisted contact structures

At first glance, the classification of overtwisted contact structures on 3–manifolds appears to be a hopeless task. After all, one can make an arbitrary number of topologically trivial Lutz twists as described in Section 4.5. While this does not change the topology of the manifold, nor the homotopy class of the contact structure as a 2–plane field, it leads to an increasing collection of overtwisted discs.

On the other hand, tight contact structures are typically those with a simple global description. We shall see, for instance, that the standard contact structures on $\mathbb{R}^3$ and S^3 are tight (Cor. 6.5.10). So is the natural contact structure on the 3–torus $T^3 = S^1 \times S^1 \times S^1$ (see Example 1.2.4).

As so often, first appearances are deceiving: it is in fact the classification of overtwisted contact structures that turns out to be manageable. The classification of tight structures, on the other hand, is extremely intricate even on manifolds with a simple global description (such as surface bundles over S^1), and beyond the reach of even the most advanced current methods on more complicated 3–manifolds. We shall get a taste of these intricacies when classifying tight contact structures on S^3 and some other simple 3–manifolds in Section 4.10.

4.7.1 Statement of the classification result

In essence, Eliashberg's [64] classification of overtwisted contact structures says that the isotopy classification of overtwisted contact structures on a given closed 3–manifold coincides with the homotopy classification of tangent 2–plane fields.

Here is the precise formulation. As always in this chapter, M denotes a (connected) closed, orientable 3–manifold. We now fix an orientation for M and an embedded (closed) oriented 2–disc $\Delta \subset M$. Furthermore, we fix a point $\mathbf{0}_\Delta$ in the interior of Δ, which we call the *centre* of Δ. (We may always choose an embedding $\phi\colon D^2 \xrightarrow{\cong} \Delta \subset M$ such that $\phi(\mathbf{0}) = \mathbf{0}_\Delta$.)

We say that a contact structure ξ on M is **overtwisted along** Δ if (M, ξ) contains Δ as a standard overtwisted disc centred at $\mathbf{0}_\Delta$ as described at the beginning of Section 4.5. We restrict our attention to *cooriented* and *positive* contact structures, i.e. those contact structures ξ for which $\alpha \wedge d\alpha$ is a volume form compatible with the chosen orientation for M for some (and hence any) contact form α defining $\xi = \ker \alpha$. It will be understood that the orientations of ξ and Δ agree at the singular point $\mathbf{0}_\Delta$ of the characteristic foliation Δ_ξ.

Notation 4.7.1 We write $\Xi^{\mathrm{ot}}(M, \Delta)$ for the space of cooriented, positive contact structures on M that are overtwisted along Δ,† and we write $\mathrm{Distr}\,(M, \Delta)$ for the space of cooriented 2–plane distributions on M that are tangent to Δ at $\mathbf{0}_\Delta$ (again with matching orientations). Both spaces will be equipped with the C^∞–topology (as spaces of sections of the unit cotangent bundle of M, for instance).

Recall that a **weak homotopy equivalence** between two topological spaces X, Y is a (continuous) map from X to Y, say, that induces a bijection $\pi_0(X) \to \pi_0(Y)$ between the sets of path components, and isomorphisms $\pi_k(X) \to \pi_k(Y)$, $k \in \mathbb{N}$, on all homotopy groups (with base points in corresponding path components).

Theorem 4.7.2 (Eliashberg) *The inclusion map*

$$i_\Delta \colon \Xi^{\mathrm{ot}}(M, \Delta) \longrightarrow \mathrm{Distr}\,(M, \Delta)$$

is a weak homotopy equivalence.

Remark 4.7.3 The spaces $\Xi^{\mathrm{ot}}(M, \Delta)$ and $\mathrm{Distr}\,(M, \Delta)$ have the homotopy type of (infinite) *CW*–complexes (this follows from [182]), so by the Whitehead theorem (see [35]), i_Δ is in fact a homotopy equivalence. See also the remark on page 62 of [77].

We are only going to prove Theorem 4.7.2 on the level of π_0, where we can formulate it in a slightly more attractive way. (The full proof of the theorem is then essentially a matter of notational complication, not of any conceptually new ideas.)

Notation 4.7.4 We write $\Xi^{\mathrm{ot}}(M)$ for the space of positive, cooriented overtwisted contact structures on M and $\mathrm{Distr}\,(M)$ for the space of cooriented tangent 2–plane distributions. Again it is understood that both spaces are equipped with the C^∞–topology.

† The characteristic foliation Δ_ξ is meant to be exactly the same for all contact structures $\xi \in \Xi^{\mathrm{ot}}(M, \Delta)$.

Corollary 4.7.5 *The inclusion map*

$$i\colon\ \Xi^{\mathrm{ot}}(M) \longrightarrow \mathrm{Distr}\,(M)$$

induces a bijection on path components.

Translated into the everyday language of a mathematician in the street, surjectivity on π_0 means that every tangent 2–plane field is homotopic to an overtwisted contact structure; injectivity on π_0 is the same as saying that two overtwisted contact structures that are homotopic as tangent 2–plane fields are homotopic via overtwisted contact structures (and hence isotopic by Gray's stability theorem).†

Proof of Corollary 4.7.5 First we are going to prove the surjectivity on π_0. On $S^3 \subset \mathbb{R}^4$ we have the standard contact forms

$$\alpha_\pm = x\,dy - y\,dx \pm (z\,dt - t\,dz),$$

defining opposite orientations. By performing the construction of Section 4.5, starting either from the contact structure $\ker\alpha_+$ or from $\ker\alpha_-$, we can ensure that we obtain a *positive* contact structure on the oriented manifold M in any homotopy class of 2–plane fields. By an isotopy of the plane field and an additional topologically trivial Lutz twist we can guarantee that Δ is an overtwisted disc, so we also get the corresponding surjectivity on π_0 in Theorem 4.7.2.

We reduce the injectivity on π_0 in the corollary to that in the theorem. Consider the following commutative diagram, where the vertical maps are also inclusions.

$$\begin{array}{ccc} \Xi^{\mathrm{ot}}(M,\Delta) & \xrightarrow{\ i_\Delta\ } & \mathrm{Distr}\,(M,\Delta) \\ {\scriptstyle j_{\mathrm{c}}}\big\downarrow & & \big\downarrow{\scriptstyle j_{\mathrm{d}}} \\ \Xi^{\mathrm{ot}}(M) & \xrightarrow{\ \ i\ \ } & \mathrm{Distr}\,(M) \end{array}$$

First of all, observe that the map on π_0 induced by j_{c} is surjective: this follows immediately from the fact that any two embeddings of a 2–disc into M are isotopic, so one can always find an isotopy of M that moves any given overtwisted disc to the disc Δ.

Secondly, we claim that the map j_{d} induces an isomorphism on π_0. I give a

† In order to apply Gray stability, one needs a smooth path $(\xi_t)_{t\in[0,1]}$, of contact structures. Thanks to general approximation results from differential topology, see [132, Chapter 2], a given continuous path can always be approximated rel $\{0,1\}$ by a smooth path.

proof of this claim appealing to general properties of Serre fibrations; below I present an elementary *ad hoc* argument. Consider the evaluation map

$$\mathbf{ev}\colon \operatorname{Distr}(M) \longrightarrow S^2$$

at the point $\mathbf{0}_\Delta$. In other words, by fixing a trivialisation $ST^*M \cong M \times S^2$ of the unit cotangent bundle we may regard elements of $\operatorname{Distr}(M)$ as maps $f\colon M \to S^2$, and the map $\mathbf{ev}$ simply sends f to $\mathbf{ev}(f) := f(\mathbf{0}_\Delta)$.

With respect to the trivialisation $ST^*M \cong M \times S^2$, the (cooriented) tangent plane to Δ at $\mathbf{0}_\Delta$ corresponds to a point $p_0 \in S^2$, and the fibre of the map $\mathbf{ev}$ over p_0 is

$$\mathbf{ev}^{-1}(p_0) = \operatorname{Distr}(M, \Delta).$$

Furthermore, the evaluation map can be shown to be a Serre fibration, see [192]. The claim now follows from the homotopy exact sequence for Serre fibrations.

A little diagram chase in the commutative diagram above then shows that the injectivity of the map induced by i_Δ (on π_0) entails the injectivity of the map induced by i. □

Here is an argument for the relevant part of the above proof that avoids having to appeal to the Serre fibration property of the evaluation map **ev**.

Lemma 4.7.6 *The inclusion map*

$$j_{\mathrm{d}}\colon \operatorname{Distr}(M, \Delta) \to \operatorname{Distr}(M)$$

induces a bijection on connected components.

Proof Given a trivialisation $ST^*M \cong M \times S^2$, surjectivity of the map induced by j_{d} on π_0 translates into saying that any smooth map $f\colon M \to S^2$ is homotopic to a smooth map sending $\mathbf{0}_\Delta$ to p_0 (in the notation of the preceding proof). That latter statement is obviously true: simply compose f with a homotopy of smooth maps $S^2 \to S^2$ beginning at the identity map of S^2 and ending at a map sending $f(\mathbf{0}_\Delta)$ to p_0.

Injectivity on π_0 translates into the following statement. Given a continuous family of smooth maps $f_t\colon M \to S^2$, $t \in [0,1]$, with

$$f_0(\mathbf{0}_\Delta) = f_1(\mathbf{0}_\Delta) = p_0,$$

there is a continuous family of smooth maps $g_s\colon S^2 \to S^2$, $s \in [0,1]$, with

- $g_0 = f_0$, $g_1 = f_1$,
- $g_s(\mathbf{0}_\Delta) = p_0$ for all $s \in [0,1]$.

In order to construct such a family of maps, we first observe that by a homotopy rel $\{0,1\}$ of the family $(f_t)_{t\in[0,1]}$ we may always assume that $f_t(\mathbf{0}_\Delta)$ is never equal to the antipodal point $-p_0$. It is then easy to find a family $\psi_s\colon S^2\to S^2$, $s\in[0,1]$, of smooth maps with the property that

- $\psi_0=\mathrm{id}_{S^2}$,
- $\psi_s(p_0)=p_0$ for all $s\in[0,1]$,
- $\psi_1(f_t(\mathbf{0}_\Delta))=p_0$ for all $t\in[0,1]$.

Then define

$$g_s=\begin{cases}\psi_{3s}\circ f_0, & 0\le s\le 1/3,\\ \psi_1\circ f_{3s-1}, & 1/3\le s\le 2/3,\\ \psi_{3-3s}\circ f_1, & 2/3\le s\le 1.\end{cases}$$

This is the desired continuous family of smooth maps. □

4.7.2 Outline of the argument

Our aim is to show that

$$(i_\Delta)_\#\colon\ \pi_0\big(\Xi^{\mathrm{ot}}(M,\Delta)\big)\longrightarrow\pi_0\big(\mathrm{Distr}\,(M,\Delta)\big)$$

is injective. In other words, given a continuous family of plane fields $\xi_t\in\mathrm{Distr}\,(M,\Delta)$, $t\in[0,1]$, with $\xi_0,\xi_1\in\Xi^{\mathrm{ot}}(M,\Delta)$, we want to find a family of contact structures $\xi_t''\in\Xi^{\mathrm{ot}}(M,\Delta)$ with $\xi_0''=\xi_0$, $\xi_1''=\xi_1$.

In a first step, we homotope the family $(\xi_t)_{t\in[0,1]}$ rel $\{0,1\}$ to a family (ξ_t') of plane fields that satisfy the contact condition outside a finite number of disjoint embedded balls $B^0,B^1,\ldots,B^m$, where B^0 is a ball around Δ, and the B^i, $i=1,\ldots,m$, are balls contained in Darboux charts for $\xi_0=\xi_0'$ and $\xi_1=\xi_1'$. Moreover, it is required that we keep some control over the characteristic foliations $(\partial B^i)_{\xi_t'}$. To this end, we consider the ξ_t in the neighbourhood of the 2–skeleton of a suitable simplicial decomposition of M; the mentioned balls will constitute the complement of that neighbourhood.

In a second step, we connect these balls to a single ball B_t, by tubes (depending on t) transverse to ξ_t'. This is where Lemma 3.3.3 will come into play. We continue to keep control over the characteristic foliations $(\partial B_t)_{\xi_t'}$.

In the third and final step, we use the information about $(\partial B_t)_{\xi_t'}$ for extending the contact structures $\xi_t'|_{M\setminus B_t}$ to a continuous family (ξ_t'') of contact structures on M with $\xi_0''=\xi_0$, $\xi_1''=\xi_1$. It is at this ingenious step where the fact that B^0 was a neighbourhood of the overtwisted disc Δ (for ξ_0 and ξ_1) is used in an essential way.

4.7.3 Characteristic foliations on spheres

In this section we discuss the types of characteristic foliations that will arise on the boundary of the balls B^i and B_t in the argument sketched above.

Definition 4.7.7 An oriented singular 1–dimensional foliation on S^2 is said to be **simple** if it has the following properties, see Figure 4.33.

- There are exactly two singular points, one source and one sink, which we are going to call the **north pole** and **south pole**, respectively.
- There are finitely many closed leaves, not of necessity non-degenerate.

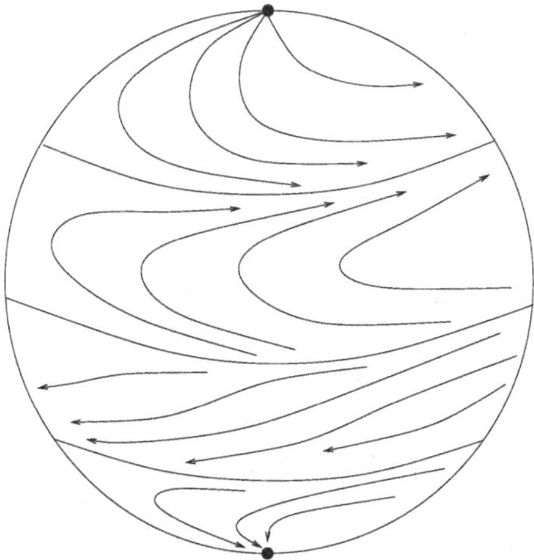

Fig. 4.33. A simple foliation on S^2.

On the 2–sphere, the Jordan curve theorem imposes certain restrictions on the topology of the flow of a singular vector field, as summarised in the so-called Poincaré–Bendixson theorem; for a proof see [203].

Theorem 4.7.8 (Poincaré–Bendixson) *Let X be a smooth vector field on S^2 with a finite number of singularities. Then the α– and ω–limit of each flow line of X is either a singular point, a closed orbit, or a poly-cycle made up of singular points and connecting orbits.* □

In particular, in the interior of a disc bounded by a periodic orbit there must be at least one singularity.† As a consequence, the closed leaves of a simple foliation on S^2 necessarily constitute parallels between the two poles.

† This also follows from the Poincaré–Hopf index theorem.

Definition 4.7.9 A simple foliation on S^2 is called **almost horizontal** if all its closed leaves are oriented from west to east, see Figure 4.34.

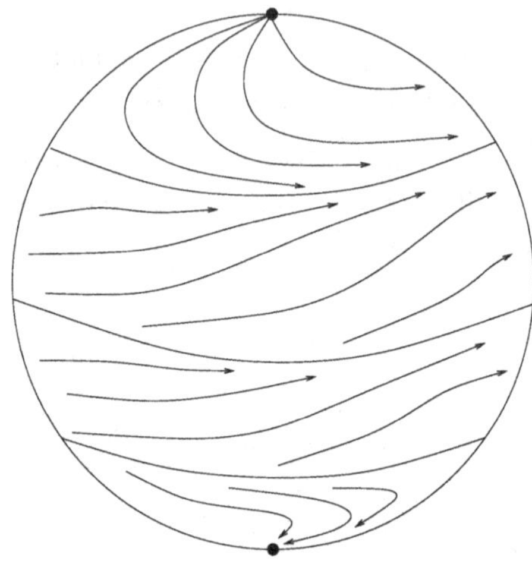

Fig. 4.34. An almost horizontal foliation on S^2.

With the standard orientations of $\mathbb{R}^3$ and $S^2 \subset \mathbb{R}^3$, a simple foliation $\mathfrak{F}$ on S^2 induced by a cooriented 2–plane field ξ on $\mathbb{R}^3$ (with the orientation of $\mathfrak{F}$ given by the convention of Definition 4.6.1) is almost horizontal if and only if ξ is cooriented from south to north along the closed leaves of $\mathfrak{F}$. An almost horizontal foliation is characterised among simple foliations by the existence of a positive transversal connecting the south with the north pole.

Examples 4.7.10 (1) The characteristic foliation induced on the unit sphere $S^2 \subset \mathbb{R}^3$ by the standard contact structure, see Example 2.5.19, is almost horizontal. This characteristic foliation has no closed leaves, and a transversal to the foliation is given by any flow line of the vector field

$$-(y + xz)\,\partial_x + (x - yz)\,\partial_y + (x^2 + y^2)\,\partial_z .$$

(2) In $\mathbb{R}^3$ with the overtwisted contact structure $\xi_{\mathrm{ot}} = \ker \alpha_{\mathrm{ot}}$, with

$$\alpha_{\mathrm{ot}} = \cos r\, dz + r \sin r\, d\varphi,$$

see Example 2.1.6, consider the 2–sphere of radius r_0 centred at the origin, with $\pi < r_0 < 2\pi$. The induced characteristic foliation is depicted (qualitatively) in Figure 4.35. There are two closed leaves oriented from east to west, so the foliation is simple, but not almost horizontal. We shall refer to this characteristic foliation $\left(S^2(r_0)\right)_{\xi_{\mathrm{ot}}}$ as $\mathfrak{F}^0$.

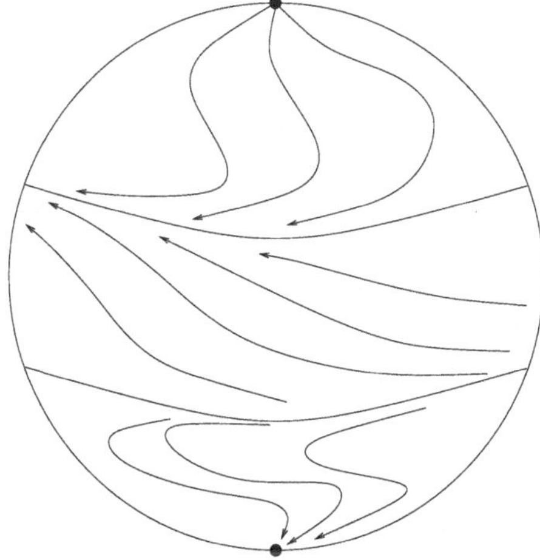

Fig. 4.35. The characteristic foliation $\mathfrak{F}^0$ on $S^2(r_0)$ in $(\mathbb{R}^3, \xi_{\mathrm{ot}})$.

The qualitative behaviour (and in fact its topological type, see Proposition 4.7.11 below) of a simple foliation $\mathfrak{F}$ can be encoded in a diagram as follows. Let $\mathfrak{f}_1, \ldots, \mathfrak{f}_n$ be the closed leaves of $\mathfrak{F}$, ordered from south to north. The **diagram** of $\mathfrak{F}$ consists of distinct points $p_1, \ldots, p_n$ on a line, together with the following information.

- If $\mathfrak{f}_i$ is oriented from west to east, then p_i is labelled with a plus sign; with a minus sign otherwise.
- The line segment between two adjacent points p_i and p_{i+1} is labelled with an arrow from the point representing a repelling closed leaf to the point representing an attracting closed leaf, where the attributes repelling/attracting refer to the restriction of $\mathfrak{F}$ to the band between $\mathfrak{f}_i$ and $\mathfrak{f}_{i+1}$.† The diagrams for the simple foliations depicted in Figures 4.33, 4.34 and 4.35 are shown in Figure 4.36.

Proposition 4.7.11 *Up to homeomorphism, a simple foliation on S^2 is determined by its diagram.*

Proof This plausible statement is implied by the classification of (non-singular) foliations on the annulus that are tangent to the boundary and contain no closed leaves in the interior, as described in [126, Prop. 4.2.12]. The arguments used in the the proof of that cited proposition also show

† Since the closed leaves may be degenerate, they may be repelling on one side and attracting on the other.

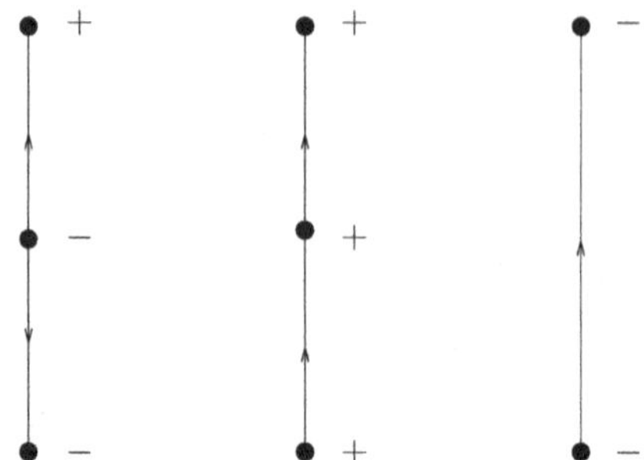

Fig. 4.36. Diagrams of simple foliations on S^2.

that a foliation on the disc – tangent to the boundary, with only one non-degenerate singular point and no closed leaves in the interior – is unique up to homeomorphism.† □

In the special case of an almost horizontal foliation $\mathfrak{F}$, we may fix a transversal connecting the poles, which we identify with the interval $[-1, 1]$, with -1 corresponding to the south pole and 1, to the north pole. Given $x \in [-1, 1]$, consider the leaf $\mathfrak{f}$ of $\mathfrak{F}$ passing through x. Starting at x, follow $\mathfrak{f}$ in its positive direction until it intersects $[-1, 1]$ again for the first time in a point $h(x)$; the existence of such a further intersection point follows from the Poincaré–Bendixson theorem and the fact that the transversal $[-1, 1]$ intersects all closed leaves of $\mathfrak{F}$. The diffeomorphism h of $[-1, 1]$ is the **holonomy along the leaves of** $\mathfrak{F}$.

The fixed points of h are the poles ± 1 and those $p_i \in [-1, 1]$ where a closed leaf $\mathfrak{f}_i$ of $\mathfrak{F}$ intersects the transversal. Moreover, the holonomy diffeomorphism determines the diagram and hence the topological type of $\mathfrak{F}$. Indeed, the segment between two fixed points $-1 < p_i < p_{i+1} < 1$ is oriented from p_i to p_{i+1} precisely if $h(x) > x$ on the interval (p_i, p_{i+1}).

Thus, given a diffeomorphism h of the interval $[-1, 1]$, fixing the boundary and with finitely many fixed points in the interior, we can sensibly speak of the almost horizontal foliation $\mathfrak{F}(h)$, determined up to homeomorphism by the requirement that its holonomy map be h.

In the second step of the proof of the classification result, we connect a finite collection of balls by tubes into one single ball. This amounts to the following operation on the characteristic foliations on the boundaries of these balls.

† The text [126] can be recommended as an immensely readable source of information about the topological classification of foliations on surfaces.

Definition 4.7.12 Let $\mathfrak{F}, \mathfrak{F}'$ be simple foliations on two copies S, S' of S^2, respectively. The **connected sum** $\mathfrak{F}\#\mathfrak{F}'$ is the simple foliation on the 2–sphere $S\#S'$ constructed as follows. Choose an orientation-preserving embedding $\varphi\colon D^2 \to S$ and an orientation-reversing embedding $\varphi'\colon D^2 \to S'$, with the north pole of $\mathfrak{F}$ contained in the interior of $\varphi(D^2)$, and the south pole of $\mathfrak{F}'$ in the interior of $\varphi'(D^2)$, and with $\varphi(\partial D^2)$ and $\varphi'(\partial D^2)$ transverse to the respective foliation. Form the connected sum $S\#S'$ as the identification space

$$\big(S \setminus \varphi(\mathrm{Int}\,(D^2)\big) \cup_{\partial D^2} \big(S' \setminus \varphi'(\mathrm{Int}\,(D^2)\big);$$

$\mathfrak{F}\#\mathfrak{F}'$ is then given (unique up to homeomorphism) by smoothing the resulting foliation on $S\#S'$.

In the final step of the proof of the classification result we want to extend a contact structure defined near the boundary of a ball to the interior of that ball. The next lemma tells us that the extendibility depends only on the characteristic foliation on the boundary of the ball.

Lemma 4.7.13 *Let ξ be a contact structure defined near the boundary $S = \partial B$ of a 3–ball B, inducing a simple characteristic foliation S_ξ. Whether or not ξ extends over B as a contact structure is a question only about the topological type of S_ξ.*

Proof What we need to show is the following. Given two contact structures ξ, ξ' defined near S, with S_ξ homeomorphic to $S_{\xi'}$ and ξ' extendible over B as a contact structure, then ξ is also extendible.

After isotoping ξ' we may assume that S_ξ and $S_{\xi'}$ have the same poles and the same closed leaves. We may now argue with Proposition 4.8.14 below that there is a C^0–small perturbation S' of S (inside the neighbourhood of S where ξ' is defined) with $S'_{\xi'}$ diffeomorphic to S_ξ; see the remark following this proof.

This implies by Theorem 2.5.22 that the germs of ξ and ξ' near S and S', respectively, are diffeomorphic. Since $\xi'|_{S'}$ extends over the ball bounded by S', the contact structure $\xi|_S$ extends over B. □

Remark 4.7.14 Proposition 4.8.14 is formulated for characteristic foliations of Morse–Smale type, but the argument there can easily be adapted to simple foliations on S^2. Foliations of this kind fail to be Morse–Smale only near degenerate closed leaves. In a neighbourhood of such leaves we can use Example 4.6.7 for an *ad hoc* argument to show that there is a C^0–small perturbation S'' of S such that S_ξ and $S''_{\xi'}$ are diffeomorphic in those neighbourhoods. In the complement we may now apply the argument from the proof of Proposition 4.8.14.

4.7.4 Construction near the 2–skeleton

The following result is the central tool of the present section. It tells us that a family of plane fields can be turned into a family of contact structures (relative to a subset where the contact condition was already satisfied), provided there is a field of directions along which one can twist the plane fields. This field of directions will be given by the intersection of the original plane field with some foliation. This will enable us to control the 'norm' (defined below) of the resulting contact structures and hence the characteristic foliations they induce on the ∂B^i as described in the outline.

Proposition 4.7.15 *Consider a bounded domain $U \subset \mathbb{R}^3$ and a closed subset $A \subset U$. Let $\mathfrak{P}$ be an oriented 2–dimensional foliation on the closure $\overline{U}$ of U. Let $(\xi_t)_{t\in[0,1]}$, be a continuous family of (cooriented) 2–plane distributions on $\overline{U}$ with the following properties:*

- ($\xi 1$) *ξ_0 and ξ_1 are contact structures;*
- ($\xi 2$) *ξ_t is a contact structure in a neighbourhood of A for each $t \in [0,1]$;*
- ($\mathfrak{P}1$) *ξ_t is transverse to $\mathfrak{P}$ for each $t \in [0,1]$;*
- ($\mathfrak{P}2$) *for any leaf $\mathfrak{p}$ of $\mathfrak{P}$, the 1–dimensional foliation $\mathfrak{p}_{\xi_t}$, for any $t \in [0,1]$, consists of curves that hit A at most once.*

Then there exists a continuous family $(\xi'_t)_{t\in[0,1]}$ of contact structures on $\overline{U}$ such that

- *$\xi'_0 = \xi_0$ and $\xi'_1 = \xi_1$ on $\overline{U}$;*
- *$\xi'_t|_A = \xi_t|_A$ for all $t \in [0,1]$.*

Proof Let $\mathfrak{L}_t$ be the 1–dimensional foliation of $\overline{U}$ formed by the leaves of $\mathfrak{p}_{\xi_t}$, where $\mathfrak{p}$ runs through the leaves of $\mathfrak{P}$. The usual proof of the so-called flow box theorem or tubular flow theorem [203, Thm. 2.1.1] allows one to construct a covering $U_t^1, \ldots, U_t^n$ of $\overline{U}$, depending continuously on $t \in [0,1]$, with the following properties:

- each U_t^i consists of whole leaves of $\mathfrak{L}_t$;
- each U_t^i can be diffeomorphically identified with a subset of $\mathbb{R}^3$ – the identification varying continuously in t – in such a way that $\mathfrak{L}_t|_{U_t^i}$ is given by lines parallel to the x–axis.

With this identification understood, we can write $\xi_t|_{U_t^i}$ as

$$\xi_t|_{U_t^i} = \ker\big(b_t^i(x,y,z)\, dy + c_t^i(x,y,z)\, dz\big),$$

with $b_t^i(x,y,z), c_t^i(x,y,z)$ smooth in (x,y,z) and continuous in t. The contact condition then becomes

$$c_t^i \frac{\partial b_t^i}{\partial x} - b_t^i \frac{\partial c_t^i}{\partial x} > 0,$$

which is a condition along each leaf of $\mathfrak{L}_t$, analogous to that in the Lutz–Martinet construction (proof of Lemma 4.1.3). Geometrically, this condition means that the contact planes $\xi_t(x,y,z)$ turn to the left as one moves in the x–direction, while the planes $\mathfrak{P}(x,y,z)$ in these local coordinates do not turn at all for (y,z) fixed.

Thanks to the condition ($\mathfrak{P}2$) that each leaf of $\mathfrak{L}_t$ hit A at most once, it is possible to modify the b_t^i, c_t^i on $\overline{U}\setminus A$ (and away from $t=0$ and $t=1$), step by step for $i=1,\ldots,n$, such that this contact condition will be satisfied everywhere (and such that ξ_t remains transverse to $\mathfrak{P}$). Notice that these modification do not change the foliations $\mathfrak{L}_t$, so after the ith step we can still work with U_t^{i+1} at the $(i+1)$st step. □

On $\mathbb{R}^3$ we consider the infinite cubical complex whose individual cubes are the standard cubes of edge length 1 with vertices on the integer lattice $\mathbb{Z}^3$. By a periodic deformation of this standard complex and by a simplicial subdivision of each cube in a prescribed way – for instance, a barycentric subdivision – we obtain a simplicial complex $\mathcal{C}_\infty$ such that

($\mathcal{C}1$) the angle between non-incident† 1– or 2–simplices with a common vertex assumes a minimum $\alpha > 0$;

($\mathcal{C}2$) the distance between disjoint 0–, 1– or 2–simplices assumes a minimum $\delta > 0$;

($\mathcal{C}3$) the diameter of the simplices (of any dimension) assumes a maximum d.

Now consider compact subsets $A \subset B \subset \mathbb{R}^3$ and a continuous family $(\xi_t)_{t\in[0,1]}$ of 2–plane fields on B with the properties ($\xi1$) and ($\xi2$) as in the preceding proposition, with B taking the role of $\overline{U}$. After scaling B (but not $\mathcal{C}_\infty$) linearly with a large constant, and after pulling back ξ_t to this scaled B, we can find finite subcomplexes $\mathcal{C}_A \subset \mathcal{C}_B \subset \mathcal{C}_\infty$ such that

- $A \subset \mathcal{C}_A$, $B \subset \mathcal{C}_B$;
- (the rescaled) ξ_t is defined on $\mathcal{C}_B$ and contact on $\mathcal{C}_A$;
- each simplex of $\mathcal{C}_B$ not completely contained in $\mathcal{C}_A$ has at most one face belonging to $\mathcal{C}_A$.‡

† For two simplices of the same dimension to be non-incident is the same as to be distinct (but not necessarily disjoint); a 1–simplex is non-incident with a 2–simplex if it is not a face of that 2–simplex.

‡ This can be guaranteed, for instance, by taking $\mathcal{C}_A$ to consist of all simplices in a cubical subcomplex of the original cubical complex.

Write $\measuredangle$ (with the relevant arguments) for the angle between two planes or lines in $\mathbb{R}^3$. Assume that we have

$$\measuredangle(\xi_t(p), \xi_{t'}(p)) < \frac{\alpha}{16} \text{ for all } t, t' \in [0,1] \text{ and } p \in \mathcal{C}_B. \tag{4.13}$$

This assumption can be made without loss of generality: by compactness of $\mathcal{C}_B$, we can partition the interval $[0,1]$ into finitely many intervals where (4.13) is satisfied; the extension character (relative to the parameter t) of the result we are going to formulate then implies that if we can prove the result on any of these subintervals, then it also holds on $[0,1]$.

As in Section 4.2, we can describe a cooriented 2–plane distribution ξ on B in terms of a smooth 'Gauß map' $f_\xi\colon B \to S^2$, associating with each point $p \in B$ the vector $f_\xi(p) \in S^2 \subset \mathbb{R}^3$ positively orthonormal to $\xi(p)$.

Definition 4.7.16 The **norm** of ξ is

$$\|\xi\| = \max_{p \in B} \|T_p f_\xi\|.$$

A 1– or 2–dimensional simplex σ of $\mathcal{C}_B$ not completely contained in $\mathcal{C}_A$ will be called *special* if

$$\measuredangle(\sigma, \xi_t(p)) < \frac{\alpha}{4} \text{ for some } t \in [0,1] \text{ and } p \in \sigma. \tag{4.14}$$

After rescaling B and, correspondingly, the ξ_t, we may assume that

$$\|\xi_t\| < \frac{\alpha}{16d} \text{ for all } t \in [0,1].$$

Then, for any *non-special* 1– or 2–dimensional simplex σ, and any 3–simplex σ' with $\sigma \subset \partial\sigma'$,

$$\measuredangle(\sigma, \xi_t(p')) > \frac{\alpha}{8} \text{ for any } t \in [0,1] \text{ and } p' \in \sigma'. \tag{4.15}$$

Indeed, given such σ, p', t, choose any $p \in \sigma$, and compute

$$\begin{aligned} \measuredangle(\sigma, \xi_t(p')) &\geq \measuredangle(\sigma, \xi_t(p)) - \measuredangle(\xi_t(p), \xi_t(p')) \\ &\geq \frac{\alpha}{4} - \|\xi_t\| \cdot \|p - p'\| \\ &> \frac{\alpha}{4} - \frac{\alpha}{16} > \frac{\alpha}{8}. \end{aligned}$$

Next we want to show that non-incident special simplices are isolated. Arguing by contradiction, assume that σ, σ' were non-incident special simplices with a common vertex. Choose t, p and t', p' that satisfy the respective condition (4.14). Then, with (4.13) and (4.14),

$$\begin{aligned}\measuredangle(\sigma,\sigma') &\leq \measuredangle(\sigma,\xi_t(p)) + \measuredangle(\xi_t(p),\xi_{t'}(p)) \\ &\quad + \measuredangle(\xi_{t'}(p),\xi_{t'}(p')) + \measuredangle(\xi_{t'}(p'),\sigma') \\ &< \frac{\alpha}{4} + \frac{\alpha}{16} + \|\xi_{t'}\| \cdot \|p-p'\| + \frac{\alpha}{4} \\ &< \frac{\alpha}{4} + \frac{\alpha}{16} + \frac{\alpha}{8} + \frac{\alpha}{4} < \alpha,\end{aligned}$$

contradicting condition $(\mathcal{C}1)$.

A similar calculation shows that if σ is a *special* simplex, then

$$\measuredangle(\sigma,\xi_t(p)) < \frac{\alpha}{2} \text{ for all } t \in [0,1] \text{ and } p \in \sigma. \tag{4.16}$$

We now want to use Proposition 4.7.15 for perturbing the family (ξ_t) into a new family (ξ_t') of 2–plane fields such that

- $\xi_0' = \xi_0$ and $\xi_1' = \xi_1$ on B;
- $\xi_t' = \xi_t$ near A;
- ξ_t' is a contact structure near the 2–skeleton of $\mathcal{C}_B$.

To this end, we define an auxiliary foliation $\mathfrak{P}_\sigma$ by parallel 2–planes near each 0–, 1– or 2–simplex σ of $\mathcal{C}_B$ with the property that the leaves of $\mathfrak{P}_\sigma$ are

- perpendicular to $\xi_t(p)$ for some $t \in [0,1]$ and some $p \in \sigma$;
- parallel to σ if $\dim \sigma = 1$, orthogonal to σ if $\dim \sigma = 2$.

The distinction between special and non-special simplices is relevant for making sure that condition $(\mathfrak{P}2)$ in Proposition 4.7.15 is satisfied. Roughly speaking, the 1–dimensional foliations $\mathfrak{L}_t$ have leaves that make a large angle with non-special simplices, but a small angle with special simplices. So the non-special simplices are less problematic; here we perturb the plane fields in a direction 'very much' transverse to the simplex. In the case of the special simplices, the perturbation is in a direction close to the simplex. But condition $(\mathfrak{P}2)$ will now be satisfied thanks to the special simplices being isolated. Now to the details.

(1) Change ξ_t in disjoint neighbourhoods of all special simplices and all vertices not belonging to a special simplex. Hypothesis $(\mathfrak{P}1)$ of Proposition 4.7.15 will be satisfied (also for the non-special simplices treated below), provided we choose the neighbourhoods in question small enough. Hypothesis $(\mathfrak{P}2)$ is satisfied by condition $(\mathcal{C}1)$, Equation (4.16), and the fact that each simplex in question has at most one face belonging to $\mathcal{C}_A$.

The foliations $\mathfrak{P}_\sigma$ have norm zero. By scaling B and the ξ_t, we can make the norms $\|\xi_t\|$, $t \in [0,1]$, as small as we like. Therefore, if that first perturbation and the ones below are performed within an ε–neighbourhood of the respective simplex, for some prescribed ε, we can guarantee that the perturbed 2–plane distributions have likewise arbitrarily small norm. This means, in particular, that after each successive perturbation all the above estimates on angles and norms still hold and can be used to control the next stage of the perturbation. With our rescaling of B we can ensure from the start that the ξ_t are contact structures in an ε–neighbourhood of $\mathcal{C}_A$. This means that we only have to modify the ξ_t in neighbourhoods of the simplices in $\mathcal{C}_B$ not completely contained in $\mathcal{C}_A$ in order to, ultimately, obtain plane fields ξ'_t that are contact in an ε–neighbourhood of the 2–skeleton of $\mathcal{C}_B$.

(2) Extend the perturbation to an ε_1–neighbourhood of the non-special 1–simplices that are not the face of a special 2–simplex, where we choose

$$\varepsilon_1 < \frac{\delta}{4} \tan \frac{\alpha}{8}. \tag{4.17}$$

Figure 4.37 shows such a 1–simplex σ, drawn in the plane $\mathfrak{p}$ of $\mathfrak{P}_\sigma$ containing σ. We may assume that the ξ_t are contact structures in an ε_1–neighbourhood of the two endpoints of σ; there we do not want to perturb the ξ_t any further (and that neighbourhood plays the role of A in Proposition 4.7.15). Condition (4.17) then ensures with (4.15) that ($\mathfrak{P}2$) will be satisfied, so that Proposition 4.7.15 can indeed be applied to carry out the desired perturbation.

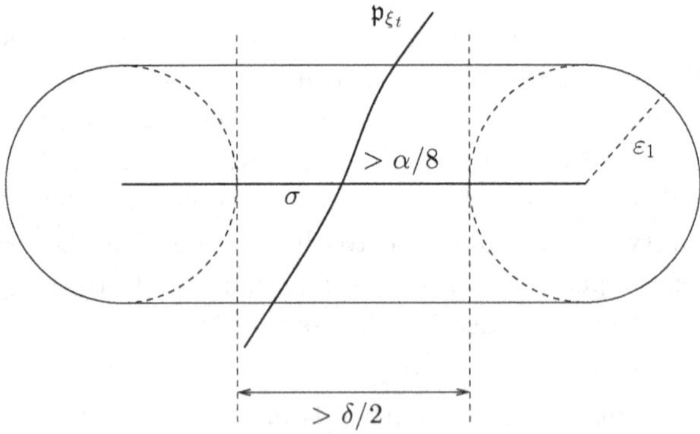

Fig. 4.37. Perturbation near a non-special 1–simplex.

(3) Extend the perturbation to an ε–neighbourhood of the non-special 2–simplices, where we take

$$\varepsilon < \frac{\varepsilon_1}{2} \tan \frac{\alpha}{8}. \tag{4.18}$$

Figure 4.38 shows such a 2–simplex (in the plane $P \subset \mathbb{R}^3$ containing σ) and the lines of intersection of P with two planes $\mathfrak{p}, \mathfrak{p}'$ of $\mathfrak{P}_\sigma$. This time, the role of A in Proposition 4.7.15 is taken by the $(\varepsilon_1/2)$–neighbourhood of $\partial\sigma$. A line segment such as $\sigma \cap \mathfrak{p}'$ stays inside the ε_1–neighbourhood of $\partial\sigma$, where the ξ_t have already been perturbed into contact structures. A line segment such as $\sigma \cap \mathfrak{p}$, on the other hand, has a segment of length greater than ε_1 outside of the $(\varepsilon_1/2)$–neighbourhood of $\partial\sigma$. With (4.15), condition (4.18) suffices to guarantee ($\mathfrak{P}2$).

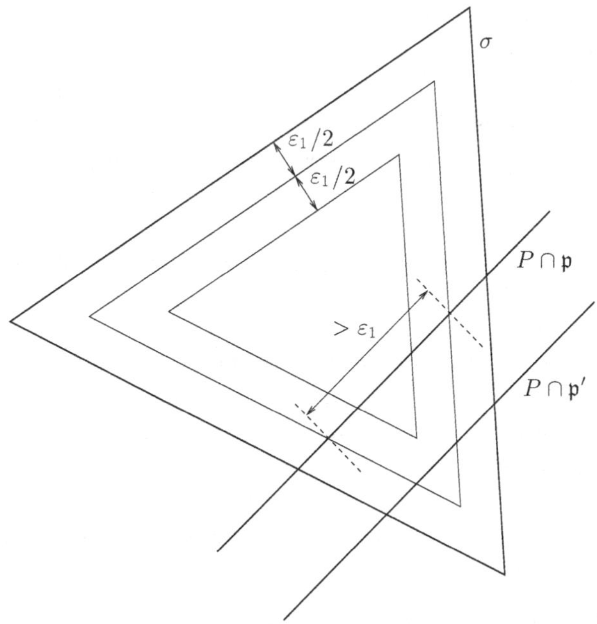

Fig. 4.38. Perturbation near a non-special 2–simplex.

Later, when we want to control the characteristic foliations on the ∂B^i (as described in the outline), we need to estimate the normal curvatures of the ∂B^i against the norm of the ξ_t'. The following two lemmata provide the relevant estimates.

Lemma 4.7.17 *Let a simplicial complex $\mathcal{C}_\infty$ as above and an $\varepsilon > 0$ be given. Then, for any 3–simplex σ of $\mathcal{C}_\infty$, there exists a ball B_σ embedded in the interior of σ, with boundary ∂B_σ contained in an ε–neighbourhood of $\partial\sigma$ and normal curvatures of ∂B_σ greater than or equal to some constant $c = c(\mathcal{C}_\infty, \varepsilon) > 0$.*

Proof Let σ be a 3–simplex of $\mathcal{C}_\infty$. Let $\sigma' \subset \sigma$ be a smaller copy of this simplex with distance ε from the boundary $\partial\sigma$.† Let S be the unique 2–sphere circumscribed around σ'. If S is entirely contained in the interior of σ, take B_σ to be the ball bounded by S. Otherwise, modify S to an ovaloid (i.e. a surface of positive Gaussian curvature) contained in an ε–neighbourhood of $\partial\sigma$ in the interior of σ as follows.

Consider a spherical cap of S cut off by the plane through one of the faces of σ'. Replace this cap by one that stays on one side of the plane through the corresponding face of σ, e.g. by scaling the cap linearly in the direction orthogonal to these planes. The resulting surface fails to be smooth along a circle, but smoothing it there will only increase the normal curvatures. Iterate this process with the other three faces, and take B_σ to be the ball bounded by the resulting ovaloid. By compactness, there is a lower bound for the normal curvatures of this ovaloid. Since we assumed $\mathcal{C}_\infty$ to result from a periodic perturbation of the standard complex, there will be only finitely many shapes of 3–simplices in $\mathcal{C}_\infty$, and hence a lower bound for the normal curvatures of all ovaloids obtained by the construction above. □

Lemma 4.7.18 *Let $S \subset \mathbb{R}^3$ be an embedded 2–sphere, cooriented by the outer normal vector, with all normal curvatures greater than or equal to some $k > 0$. Let ξ be a cooriented 2–plane field defined near S with norm $\|\xi\| < k$. Then, possibly after a C^∞–small perturbation of ξ, the characteristic foliation S_ξ is almost horizontal.*

Proof Let f_S, f_ξ be the Gauß maps $S \to S^2$ of S and ξ, respectively. Then, for any $q \in S$ and $\mathbf{w} \in T_qS$, we have

$$\|T_q f_S(\mathbf{w})\| \geq k\|\mathbf{w}\| \;\text{ and }\; \|T_q f_\xi(\mathbf{w})\| < k\|\mathbf{w}\|.$$

In particular, f_S is a local diffeomorphism. Since S is compact and S^2 connected, the map f_S is a covering map. But S^2 is simply connected, so f_S is in fact a diffeomorphism.‡

Remark 4.7.19 Under the natural identification of T_qS with $T_{f_S(q)}S^2$ – both may be regarded as the 2–dimensional subspace of $\mathbb{R}^3$ orthogonal to $f_S(q)$ – and with respect to the inner product induced from $\mathbb{R}^3$, the differential T_qf_S is a self-adjoint linear map with two positive eigenvalues $k_1, k_2 \geq k$. With the sign conventions of classical differential geometry, see [181], where the normal curvatures are defined as the eigenvalues of the

† We assume implicitly that ε is small relative to the size of σ.
‡ This statement is known as Hadamard's theorem on ovaloids; see [181, Thm. 6.2] for a proof using the Gaussian curvature instead of our argument with coverings.

negative differential of the Gauß map, our choice of f_S would, strictly speaking, correspond to all normal curvatures being bounded above by $-k < 0$.

Define $f\colon S^2 \to S^2$ by $f = f_\xi \circ f_S^{-1}$. Then

$$\|T_p f(\mathbf{v})\| < \|\mathbf{v}\| \text{ for all } p \in S^2 \text{ and } \mathbf{v} \in T_p S^2. \tag{4.19}$$

Let q^+ be a singular point of S_ξ (which has to exist by the Poincaré–Hopf index theorem). Without loss of generality assume that q^+ is a positive point, that is, $f_S(q^+) = f_\xi(q^+)$. We think of q^+ as the north pole of S.

For $0 < \theta < \pi$, consider the set of points

$$C_\theta = \{q \in S\colon \measuredangle(f_S(q), f_S(q^+)) = \theta\}.\dagger$$

This is a simple closed curve, corresponding under f_S to a circle of latitude on S^2. From the estimate (4.19) we have

$$\measuredangle(f_\xi(q), f_\xi(q^+)) < \theta \text{ along } C_\theta. \tag{4.20}$$

It follows that

$$0 < \measuredangle(f_\xi(q), f_S(q)) < \pi$$

for all points q on the closed 'northern hemisphere' $S^+ := \{\theta \leq \pi/2\}$ of S with q^+ removed. In particular, $f_\xi(q) \neq \pm f_S(q)$. This means that q^+ is the only singular point in this hemisphere. The estimate $0 < \measuredangle(f_\xi(q), f_S(q))$ does in fact hold for all $q \in S \setminus \{q^+\}$, so any further singular point of S_ξ has to be negative. Obviously, the same argument that precludes the existence of two positive singular points also rules out the existence of more than one negative singular point.

The characteristic foliation S_ξ can be defined by the vector field X on S given by $X_q := f_\xi(q) \times f_S(q)$. On $S^+ \setminus \{q^+\}$, this vector field always has a positive west-east component, and in particular it never points straight towards the west.

Let $\gamma_\theta\colon S^1 \to S^2$ be a regular parametrisation of a circle of latitude on S^2, oriented in positive direction around the north pole (i.e. from west to east). Then, thanks to the description of Tf_S in the remark above, the parametrised curve $f_S^{-1} \circ \gamma_\theta$ around q^+ has a tangent vector that, likewise, never points straight towards the west. Thus X does not complete a full turn with respect to that tangent vector, which implies that the index of X at q^+ is $+1$.

Once again invoking the Poincaré–Hopf index theorem, we now see that S_ξ has precisely two singular points $q^\pm$, both of index $+1$, one of them positive, the other negative.

† Here $\measuredangle$ denotes an angle between vectors, and thus takes values in $[0, \pi]$.

From (4.20) we deduce that, for $q \in S^+ \setminus \{q^+\}$, the orthogonal projection of $f_\xi(q)$ onto T_qS gives a vector Y_q orthogonal to X_q and with a positive south–north component. The same is true in the 'southern hemisphere' about q^-, with that point regarded as the south pole.† Notice that X_q, Y_q is a positively oriented basis for T_qS.

Choose a point q on S in the intersection of these two closed hemispheres. Then the flow line of Y through q is defined on $\mathbb{R}$, and it has $q^\pm$ as the ω– or α–limit, respectively. This flow line constitutes a positive transversal to the characteristic foliation.

Together with the Poincaré–Bendixson theorem, these considerations establish that S_ξ is an almost horizontal foliation, but possibly of a 'degenerate' type, that is, the singular points $q^\pm$ may be degenerate, and there may be infinitely many closed leaves. A simple example where the described degenerate phenomena occur for a general 2–plane distribution ξ is $S = S^2$ and ξ a constant plane field on $\mathbb{R}^3$.

By the proof of Proposition 4.6.11, however, a C^∞–small perturbation of ξ (which may be assumed to preserve the property that the norm of ξ is smaller than the minimal normal curvature of S) allows us to change S_ξ into Morse–Smale type. In that case, $q^\pm$ are non-degenerate (elliptic), likewise the closed leaves. So closed leaves can accumulate neither at a closed leaf, nor at a singular point, which means that they are finite in number. □

As seen in Example 4.6.6 (1′), a singular point like the one of the vector field $X(x, y) := x\,\partial_x + y^3\,\partial_y$ at the origin $\mathbf{0} \in \mathbb{R}^2$ can arise in the characteristic foliation induced by a contact structure. The C^∞–small perturbation $X_{-\varepsilon}(x, y) := x\,\partial_x + (y^3 - \varepsilon y)\,\partial_y$, with $\varepsilon > 0$ small, has a positive hyperbolic point at $\mathbf{0}$ and positive elliptic points at $(0, \pm\sqrt{\varepsilon})$. In the preceding proof, however, we control the norm of the plane field relative to the minimal normal curvature of S during the perturbation, and this guarantees that the perturbation really yields a single non-degenerate elliptic point. Such a perturbation might give us for instance $X_\varepsilon(x, y) = x\,\partial_x + (y^3 + \varepsilon y)\,\partial_y$, which has a positive elliptic point at $\mathbf{0}$ and no further singularities.

Below we intend to apply this lemma to a parametric family of contact structures. Unfortunately, the Morse–Smale property is not generic for parametric families. For instance, in a 1–parametric family of foliations an attracting and a repelling closed leaf can merge into a degenerate closed leaf and then disappear. So in the parametric situation we need to modify the final part of the above proof.

† Beware that our 'northern' and 'southern' hemispheres are not complementary unless $f_S(q^-)$ happens to be the antipodal point of $f_S(q^+)$.

In fact, it is essential that we need the parametric form of this lemma for contact structures only, not for arbitrary 2–plane fields. Here is a simple example how the parametric form of Lemma 4.7.18 may fail for general plane fields.

Example 4.7.20 Take $S = S^2 \subset \mathbb{R}^3$ and

$$\xi_t = \ker\bigl(dz + t(x\,dy - y\,dx)\bigr),\ t \in [-1, 1],$$

oriented by $(dx \wedge dy)|_{\xi_t}$ (and cooriented by the defining 1–form). The induced characteristic foliations have singular points at the north and south pole $q^{\pm} = (0, 0, \pm 1)$ of S^2. The vector field on S^2 defining the characteristic foliation $S^2_{\xi_t}$ has positive divergence at q^+ for $t > 0$, negative divergence for $t < 0$, and zero divergence for $t = 0$. So there can be no small perturbation to a family of plane fields all inducing a characteristic foliation with a non-degenerate elliptic point.

The problem illustrated by this example lies with the notion of 'sign' of a singular point. For an oriented plane field ξ tangent to an oriented surface S at some point $q \in S$, we can speak of a positive or negative singularity of S_ξ, depending on whether the orientation of ξ_q does or does not coincide with that of T_qS. This sign does not change at q^+ in our example. The contact geometric definition of 'sign' given on page 167, however, requires that we orient ξ_t for $\pm t > 0$ by $\pm(dx \wedge dy)|_{\xi_t}$. Notice that the orientation of $\mathbb{R}^3$ induced by the contact structures ξ_t, $t \neq 0$, depends on the sign of t.

I now want to argue that Lemma 4.7.18 remains valid for parametric families $(\xi_t)_{t\in[0,1]}$ of *contact structures*. The first issue to deal with is that we would like the deformation to be rel $\{0, 1\}$. This we ensure below by taking the B^i to be balls in Darboux charts for ξ_0 and ξ_1. The standard contact structure of a 3–dimensional Darboux chart is tight (see Corollary 6.5.10 below), and hence Proposition 4.6.28 precludes the existence of any closed leaves in $(\partial B_i)_{\xi_0}$ and $(\partial B_i)_{\xi_1}$. So these characteristic foliations are already of the desired type. In order to avoid problems with the deformation near $t = 0$ and $t = 1$, we reparametrise the family (ξ_t) in such a way that it is constant in t near these boundary points of the parameter interval. We may also assume without loss of generality that this family is smooth in t.

Next we consider the two singular points. Write q_t^+ for the positive singular point in S_{ξ_t}, and X_t for the vector field defining that characteristic foliation. Then $\mathrm{div}(X_t)(q_t^+) > 0$ for all t. The only way for q_t^+ to be degenerate is that the linearisation of X_t at q_t^+ be of the form $\left(\begin{smallmatrix}\lambda_t & 0\\ 0 & 0\end{smallmatrix}\right)$ with $\lambda_t > 0$. This can be perturbed to the non-degenerate family $\left(\begin{smallmatrix}\lambda_t & 0\\ 0 & \varepsilon\end{smallmatrix}\right)$ with $\varepsilon > 0$. This translates into a perturbation of (ξ_t) into a family (ξ_t') of contact structures inducing a characteristic foliation $S_{\xi'_t}$ with a non-degenerate positive elliptic singularity at q_t^+. Thanks to the considerations in the proof of Lemma 4.7.18, we do not create any additional singularities, provided the perturbation is sufficiently C^∞–small.

We are now in a situation where the closed leaves of the S_{ξ_t} may only accumulate away from the poles $q_t^{\pm}$. In order to show that we can prevent such an accumulation by a further perturbation of the family (ξ_t), we study the characteristic foliations S_{ξ_t} in terms of their holonomy map $h_t\colon [-1,1] \to [-1,1]$. (Such a holonomy map can be defined because we have a transversal to S_{ξ_t}.) A deformation of this family of holonomy maps h_t translates into a deformation of the ξ_t. Closed leaves of S_{ξ_t} correspond to fixed points of h_t; these in turn correspond to zeros of the function $x \mapsto g_t(x) := h_t(x) - x$.

It is clear that we cannot avoid the occurrence of degenerate zeros of g_t, corresponding to degenerate closed leaves.

Example 4.7.21 Consider the family of functions $g_t\colon [-1,1] \to \mathbb{R}$, $t \in [0,1]$, given by

$$g_t(x) = \kappa(x^2-1)\cdot\Big(1+8t\big(x^2-\frac{1}{4}\big)\Big),$$

with $\kappa > 0$ a constant chosen sufficiently small for the condition $g_t' \geq -1$ to hold, so that the map $x \mapsto h_t(x) = g_t(x) + x$ is a strictly monotone mapping from the interval $[-1,1]$ to itself.

No matter how one interpolates between g_0 and g_1 (with fixed endpoints), either one of these endpoints becomes a degenerate zero at some stage during the interpolation, or one of the intermediate functions has a degenerate zero at an interior local maximum of value 0.

For the example in hand, $(x,t) = (0,1/2)$ is the only point where both $g_t(x)$ and $g_t'(x)$ vanish. In fact, $g''_{1/2}(0) = -8\kappa \neq 0$, and this condition alone is sufficient to guarantee that the degenerate zero of $g_{1/2}$ at $x=0$ is isolated. In particular, each g_t has but finitely many (here: two, three, or four) zeros. Moreover, with the dot denoting the derivative with respect to t, we have $\dot{g}_{1/2}(0) = 2\kappa \neq 0$.

The so-called 1–prolongation of the map $(x,t) \mapsto g_t(x)$, associating with each (t,x) the 1–jet of that map, is given by

$$\begin{array}{rccccc} j^1g\colon & [-1,1] & \times & [0,1] & \longrightarrow & [-1,1]\times[0,1]\times\mathbb{R}^3 \\ & (x & , & t) & \longmapsto & \big(x,t,g_t(x),g_t'(x),\dot{g}_t(x)\big). \end{array}$$

According to the jet transversality theorem† [132, Thm. 3.2.8], the map $(x,t) \mapsto g_t(x)$ may be C^∞–approximated by a map whose 1–prolongation is transverse to the submanifold

$$L := [-1,1] \times [0,1] \times \{0\} \times \{0\} \times \mathbb{R} \subset J^1([-1,1] \times [0,1], \mathbb{R}).$$

We have

$$\begin{aligned} \frac{\partial j^1 g}{\partial x}(x,t) &= \big(1, 0, g_t'(x), g_t''(x), \dot{g}_t'(x)\big), \\ \frac{\partial j^1 g}{\partial t}(x,t) &= \big(0, 1, \dot{g}_t(x), \dot{g}_t'(x), \ddot{g}_t(x)\big). \end{aligned}$$

Thus, the condition for $j^1 g$ to be transverse to L can be written as

$$g_t'(x)\dot{g}_t'(x) - \dot{g}_t(x) g_t''(x) \neq 0$$

at points (x,t) where $g_t(x) = g_t'(x) = 0$. This is equivalent to saying that both $\dot{g}_t(x)$ and $g_t''(x)$ have to be non-zero at such points. This means that degenerate zeros of $g_t(x)$ are not only isolated in terms of the variable x (which is all we need for our application), but also in terms of the variable t. So the simple Example 4.7.21 actually describes the generic situation.

4.7.5 Proof of the classification result

We are now ready to prove Theorem 4.7.2. As indicated previously, we only want to prove injectivity of $(i_\Delta)_\#$ on the level of π_0. Recall the outline of the argument in Section 4.7.2. Thus, let $\xi_t \in \mathrm{Distr}\,(M, \Delta)$, $t \in [0,1]$, be a continuous family of plane fields, with $\xi_0, \xi_1 \in \Xi^{\mathrm{ot}}(M, \Delta)$. We need to find a family $\xi_t'' \in \Xi^{\mathrm{ot}}(M, \Delta)$ with $\xi_0'' = \xi_0$, $\xi_1'' = \xi_1$.

First Step – Where we leave holes The plane fields ξ_t all coincide at the centre of the disc Δ, and ξ_0, ξ_1 coincide near Δ by Theorem 2.5.22. This allows us, by a first homotopy rel $\{0,1\}$ of the family (ξ_t), to assume that all ξ_t actually coincide near Δ. Then we can find an embedded ball $B^0 \subset (M, \xi_t)$ contactomorphic to the ball $\{(z/\delta)^2 + r^2 \leq (\pi + \delta)^2\}$ in $(\mathbb{R}^3, \xi_{\mathrm{ot}})$ for some small $\delta > 0$. In particular, the characteristic foliation $(\partial B^0)_{\xi_t}$ is the foliation $\mathfrak{F}^0$ from Figure 4.35. We are going to write B_{ot} for this ball (for some fixed δ) and call it the **standard overtwisted ball**.

Because of the extension character of the result discussed on page 217, we may proceed as follows:

† In [132] the jet transversality theorem is formulated for manifolds without boundary. In our situation, the boundary of $[-1,1] \times [0,1]$ does not cause any problems, because near it the transversality condition already holds by our previous considerations.

- consider the ξ_t (for all t simultaneously) on small subsets of M contained in a Darboux chart for each t;
- perturb them there to a family of plane fields satisfying the contact condition in the neighbourhood of the 2–skeleton of an auxiliary simplicial complex, relative to the subset of M where this contact property had been achieved by a previous perturbation.

For these perturbations, we always work on a subset B of $\mathbb{R}^3$, relative to some subset $A \subset B$, and our simplicial complex is $\mathcal{C}_\infty$. With the scaling argument from the preceding section, Lemmata 4.7.17 and 4.7.18 (together with the latter's extension to families) allow us to conclude that there exist disjoint embedded 3–balls $B^1, \ldots, B^m$ in $M \setminus B^0$ such that the family $(\xi_t)_{t\in[0,1]}$, is homotopic rel $\{0,1\}$ to a family (ξ'_t) with the following properties for all $t \in [0,1]$:

- ξ'_t coincides with ξ_t on B^0,
- ξ'_t is contact on $M \setminus (B^1 \cup \ldots \cup B^m)$,
- the characteristic foliations $\mathfrak{F}^i_t := (\partial B^i)_{\xi'_t}$ are almost horizontal for $i = 1, \ldots, m$.

Then, by Proposition 4.6.28, the characteristic foliations $(\partial B^i)_{\xi_0}$ and $(\partial B^i)_{\xi_1}$, $i = 1, \ldots, m$, necessarily look as in Figure 2.2. By Theorem 4.10.1 below, this in fact implies that

- all the (B^i, ξ_0) and (B^i, ξ_1), $i = 1, \ldots, m$, are isomorphic to convex balls in $(\mathbb{R}^3, \xi_{\mathrm{st}})$.

Remark 4.7.22 Strictly speaking, it is not necessary to invoke Theorem 4.10.1 at this point, and doing so may be regarded as attacking starlings with rice puddings fired from catapults.† For the steps that follow it suffices that the B^i, $i = 1, \ldots, m$, are contained in Darboux charts for ξ_0 and ξ_1, for we may then identify them with balls in the subset $\{r < \pi/2\}$ of $(\mathbb{R}^3, \xi_{\mathrm{ot}})$; observe that this subset is contactomorphic to $(\mathbb{R}^3, \xi_{\mathrm{st}})$. Why it is helpful that these balls can be represented in such a manner will shortly become clear.

Second Step – Where we connect the holes Recall that our construction so far guarantees that the characteristic foliation $(\partial B^0)_{\xi'_t}$ coincides with $\mathfrak{F}^0$ for all $t \in [0,1]$.

† This expression, which I owe to Charles Thomas, originates from [180]. It is closer in spirit to the German 'Mit Kanonen auf Spatzen schießen' than idiomatic equivalents of that German phrase like 'to crack nuts with a sledgehammer' or 'to break butterflies on a wheel'.

Lemma 4.7.23 *There exists a family of embedded closed 3–balls $B_t \subset M$, $t \in [0,1]$, such that for all t we have*

- $B_t \supset B^0 \cup B^1 \cup \cdots \cup B^m$,
- *the characteristic foliation $(\partial B_t)_{\xi'_t}$ is homeomorphic to the connected sum $\mathfrak{F}^0 \# \mathfrak{F}^1_t \# \ldots \# \mathfrak{F}^m_t$,*
- *the balls (B_0, ξ_0) and (B_1, ξ_1) are contactomorphic to the standard overtwisted ball B_{ot}.*

Proof Lemma 3.3.3 allows us to connect the north pole of B^{i-1} with the south pole of B^i, $i = 1, \ldots, m$, by a path γ^i_t (in the complement of the balls) transverse to ξ'_t, with γ^i_t disjoint from γ^j_t for $i \neq j$. By Example 2.5.16, the characteristic foliation on the boundary of a thin tubular neighbourhood around these transverse paths consists of spirals. Thus, if we form the boundary connected sum $B^0 \#_{\mathrm{b}} B^1 \#_{\mathrm{b}} \ldots \#_{\mathrm{b}} B^m$ using these tubes, we obtain an embedded ball B_t with a characteristic foliation $(\partial B_t)_{\xi'_t}$ as described.

The final statement follows from the fact that the (B^i, ξ_0) and (B^i, ξ_1), $i = 1, \ldots, m$, are standard tight balls (see the next remark). □

Remark 4.7.24 In that last argument, we use the following observation: given some contact structure ξ on a 3–ball B, the boundary connected sum with a standard ball B_{st} (in the notation of Remark 2.5.2), using a tube between two poles as described, yields a ball contactomorphic to (B, ξ). That this is indeed the case can be deduced from Theorem 4.10.1. But again there is a simple way to avoid having to quote that deep theorem. All that is really needed in the third step below is that (B_0, ξ_0) and (B_1, ξ_1) can be realised in $(\mathbb{R}^3, \xi_{\mathrm{ot}})$ by taking the connected sum of B_{ot} with disjoint balls contained in the subset $\{r < \pi/2\}$ of $\mathbb{R}^3$, and it is easily seen that such a realisation is possible.

Notice that the foliation $\mathfrak{F}_t := \mathfrak{F}^1_t \# \ldots \# \mathfrak{F}^m_t$, as a connected sum of almost horizontal foliations, is again almost horizontal.

Third Step – Where we fill the hole As is transparent from Figure 4.39, the foliations $\mathfrak{F}^0 \# \mathfrak{F}_t$, $t \in [0,1]$, can be realised (up to homeomorphism) as the characteristic foliations induced by ξ_{ot} on a continuous family of surfaces of revolution S_t in $\mathbb{R}^3$.

Figure 4.40 shows the characteristic foliation on the particular surface of revolution in Figure 4.39.

The contact structure ξ_{ot} on $\mathbb{R}^3$ defines the extension of the germ of ξ_{ot} near S_t to a contact structure on the ball in $\mathbb{R}^3$ bounded by S_t. By Lemma 4.7.13, this implies that the germ of ξ'_t near ∂B_t extends to a contact struc-

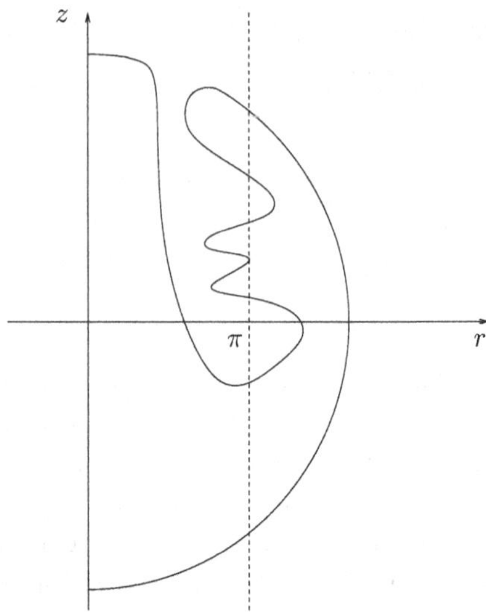

Fig. 4.39. Realising $\mathfrak{F}^0\#\mathfrak{F}_t$ on a surface of revolution S_t in $(\mathbb{R}^3, \xi_{\mathrm{ot}})$.

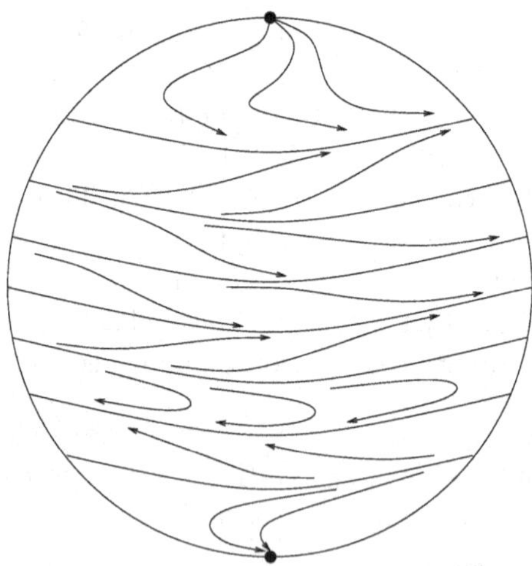

Fig. 4.40. The characteristic foliation on S_t corresponding to Figure 4.39.

ture on B_t. We call $(\xi_t'')_{t\in[0,1]}$, the resulting family of contact structures on M.

By the third point of Lemma 4.7.23 (or by Remark 4.7.24 if you prefer not to invoke Theorem 4.10.1), one can ensure that for $t = 0, 1$ this extension gives us the original contact structures ξ_0, ξ_1.

This concludes the proof of Theorem 4.7.2 on the level of π_0.

Remarks 4.7.25 (1) Tangent 2–plane fields on the 3–ball, fixed along the boundary, are classified by the Hopf invariant, see Section 4.2.3. Since the contact structure ξ_t'' coincides with ξ_t on B_t for $t = 0, 1$, and the extensions $\xi_t''|_{B_t}$ are continuous in t by construction, it follows that the resulting path $(\xi_t'')_{t\in[0,1]}$ of contact structures from ξ_0 to ξ_1 is actually homotopic rel $\{0, 1\}$ to the original path $(\xi_t)_{t\in[0,1]}$.

(2) In order to prove Theorem 4.7.2 in full generality, one needs to replace the parameter space pair $\big([0, 1], \{0, 1\}\big)$ by $(S^k, \emptyset)$ for surjectivity on π_k, and by (D^{k+1}, S^k) for injectivity on π_k. The proof above carries through unchanged. In the discussion at the end of Section 4.7.4, one needs to work with jets whose order equals the dimension of the parameter space.

4.8 Convex surface theory

Recall from Section 4.6.2 that a convex surface S in (M, ξ) is a surface admitting a transverse contact vector field. We take up the notation from Section 4.6. Most of the results of the present section are from Giroux's thesis [104]. The class notes of Etnyre [84], which delve much deeper into convex surface theory, were an invaluable guide in the preparation of the present section.

I shall discuss convex surface theory for closed surfaces only. This is good enough for some simple applications, like a criterion for overtwistedness (Prop. 4.8.13) or the result, alluded to earlier, that homeomorphic characteristic foliations on two surfaces can be made diffeomorphic by a suitable perturbation of one of the surfaces (Prop. 4.8.14). I also present a simple alternative proof of Theorem 4.6.30. For an alternative proof of the Bennequin inequalities, or in order really to be useful for classification questions of tight contact structures, one needs to develop convex surface theory for surfaces with Legendrian boundary. I hope that readers who have persevered thus far will be well prepared to approach the notes by Etnyre [84] and thence the sources where that theory was initiated, like Honda's paper [137]. A good survey on the use of convex surface theory for the classification of tight contact structures is the one by Ghiggini and Schönenberger [102].

It will turn out that all the pertinent information about a given convex surface is contained in the set we define next.

Definition 4.8.1 The **dividing set** Γ_S of a convex surface $S \subset (M, \xi)$ (corresponding to a given contact vector field Y) is the set of points $p \in S$ where Y is tangent to ξ.

Example 4.8.2 The dividing set of the convex surface $S^2 \subset (\mathbb{R}^3, \xi_2)$ from Example 4.6.18 is the equator $\{z = 0\}$, because (in the notation of that example) $\alpha_2(Y) = 2z$.

Thus, in the contactomorphic identification of a neighbourhood of S in (M, ξ) with $(S \times \mathbb{R}, \ker(\beta + u\,dz))$ as in Section 4.6.2, we have

$$\Gamma_S = \{p \in S\colon\ u(p) = 0\}.$$

By the contact condition (4.8) we have $du \neq 0$ along the zero set Γ_S of u, so the dividing set is a 1–dimensional submanifold of S, i.e. a collection of embedded circles (remember that we assume S to be closed in this discussion). Moreover, this dividing set turns out to be transverse to the characteristic foliation, in particular it does not meet the singularities of S_ξ. For otherwise there would be a point $p \in \Gamma_S = u^{-1}(0)$ where the tangent line $T_p\Gamma_S$ is contained in the kernel of β (either because we are at a singular point of S_ξ, where $\beta = 0$, or because $T_p\Gamma_S$ coincides with $T_pS_\xi = \ker\beta$). For any vector $\mathbf{v} \in T_p\Gamma_S$ we would then have

$$i_{\mathbf{v}}(u\,d\beta + \beta \wedge du) = 0,$$

contradicting the contact condition (4.8).

That contact condition also implies that Γ_S is non-empty, for otherwise we could write the contact structure as $\beta/u + dz = 0$, and by condition (4.8) we would obtain the exact area form $d(\beta/u)$ on the closed surface S.

Define

$$S_\pm = \{p \in S\colon\ \pm u(p) > 0\},$$

so that $S \setminus \Gamma_S = S_+ \sqcup S_-$. Observe that on the two open (not necessarily connected) sets $S_\pm \times \mathbb{R}$ we may write the contact structure ξ more conveniently as the kernel of the 1–form $\beta/|u| \pm dz$.

For the moment we write β for this 1–form $\beta/|u|$, bearing in mind that it is only defined on $S \setminus \Gamma_S$. The contact condition (4.8) on $(S \setminus \Gamma_S) \times \mathbb{R}$ simplifies further to

$$\pm d\beta > 0 \ \text{ on } \ S_\pm.$$

In particular, we may take $\pm d\beta$ as area form Ω_0 on $S_\pm$. With respect to this area form, the vector field X from Equation (4.4) defining the characteristic foliation satisfies $d(i_X\Omega_0) = d\beta = \pm\Omega_0$, that is, $\operatorname{div}_{\Omega_0}(X) = \pm 1$ on $S_\pm$.

We may therefore find a smooth vector field X on all of S that defines the characteristic foliation S_ξ, points out of S_+ and into S_- along Γ_S, and satisfies $\operatorname{div}_{\Omega_0}(X) = \pm 1$ on $S_\pm$ away from a small neighbourhood of Γ_S.

We claim that there is in fact an area form Ω on the whole of S such that the vector field X satisfies $\pm\operatorname{div}_\Omega(X) > 0$ on $S_\pm$ (and, by continuity,

$\mathrm{div}_\Omega(X) = 0$ along Γ_S). In general, of course, X and this new Ω will no longer be related by Equation (4.4).

Consider a component of Γ_S. Use the flow of X to identify a tubular neighbourhood of this component with an annulus $A = S^1 \times [-\varepsilon, \varepsilon]$, where the component of Γ_S is identified with $S^1 \times \{0\}$, and the transverse parametrisation is chosen such that $A \cap S_+ = S^1 \times (0, \varepsilon]$. We write s for the parameter in $(-\varepsilon, \varepsilon)$. Our choices entail that $X = -\partial_s$ on A. Write θ for the S^1–coordinate, oriented in such a way that the area form $d\theta \wedge ds$ defines the same orientation as Ω_0 on $A \cap (S \setminus \Gamma_S)$. Moreover, we may assume that the annulus A has been chosen so wide that X has divergence ± 1 with respect to Ω_0 near $S^1 \times \{\pm\varepsilon\}$.

On A we may easily write down an area form $\Omega_A = f(s)\, d\theta \wedge ds$ such that $\pm \mathrm{div}_{\Omega_A}(X) > 0$ on $A \cap S_\pm$. Indeed, one computes

$$\mathrm{div}_{\Omega_A}(-\partial_s) = -\frac{f'}{f},$$

so we simply have to choose a positive function $f(s)$ that is strictly increasing on $[-\varepsilon, 0)$ and strictly decreasing on $(0, \varepsilon]$.

We can patch the area forms Ω_0 and Ω_A together to an area form Ω on all of S such that $\Omega = \Omega_A$ on $S \times (-\varepsilon/3, \varepsilon/3)$, and such that X satisfies the divergence condition outside compact annuli of the form $A_- = S^1 \times [-2\varepsilon/3, -\varepsilon/3]$ and $A_+ = S^1 \times [\varepsilon/3, 2\varepsilon/3]$, say. If we change the area form Ω to $g\Omega$, where g is some smooth function $S \to \mathbb{R}^+$, the divergence of X becomes

$$\mathrm{div}_{g\Omega}(X) = \frac{X(g)}{g} + \mathrm{div}_\Omega(X); \tag{4.21}$$

the computation is analogous to that which lead to Equation (4.11).

Since the flow of X goes from one boundary component of $A_\pm$ to the other, and $|\mathrm{div}_\Omega(X)|$ is bounded on these annuli, one can solve the differential inequality

$$\pm\left(\frac{X(g)}{g} + \mathrm{div}_\Omega(X)\right) > 0 \text{ on } A_\pm$$

by having g increase (respectively, decrease) fast enough along the flow lines of $X = -\partial_s$ on A_+ (respectively, A_-). We may assume that g is a smooth function on all of S, with $g \equiv 1$ outside the annuli A and g identically equal to some large constant K on the annuli $S^1 \times (-\varepsilon/3, \varepsilon/3)$. Since there X satisfied the divergence condition with respect to $\Omega = \Omega_A$, it will also do so with respect to $g\Omega = K\Omega_A$.

These properties of the dividing set motivate the following definition.

Definition 4.8.3 Let $\mathfrak{F}$ be a singular 1–dimensional foliation on a closed surface S. A collection Γ of embedded circles is said to **divide** $\mathfrak{F}$ if the following conditions hold.

(i) Γ is transverse to $\mathfrak{F}$; in particular, it does not meet the singularities of $\mathfrak{F}$.

(ii) There is an area form Ω on S and a vector field X defining $\mathfrak{F}$ such that

- $\mathcal{L}_X\Omega \neq 0$ on $S \setminus \Gamma$, and
- with $S_\pm := \{p \in S\colon \pm \operatorname{div}_\Omega(X) > 0\}$, so that $S\setminus\Gamma = S_+ \sqcup S_-$, the vector field X points out of S_+ along Γ.

Our discussion above implies that this is a sensible definition, at least linguistically: the dividing set Γ_S of a convex surface S divides the characteristic foliation S_ξ on that surface. We shall see presently that the converse holds true: if the characteristic foliation S_ξ is divided by a collection Γ of embedded circles, then S is a convex surface.

Examples 4.8.4 (1) The characteristic foliation on $S^2 \subset (\mathbb{R}^3, \xi_2)$ from Example 2.5.19 is divided by the equator $\{z = 0\}$. Indeed, we may take the vector field

$$X = (xz - y)\,\partial_x + (yz + x)\,\partial_y - (x^2 + y^2)\,\partial_z$$

from that example and the area form

$$\Omega = x\,dy \wedge dz + y\,dz \wedge dx + z\,dx \wedge dy$$

on S^2. Then, computing in $\mathbb{R}^3$,

$$d(i_X\Omega) = y\,dy \wedge dz - x\,dz \wedge dx + 2\,dx \wedge dy.$$

Hence

$$(x\,dx + y\,dy + z\,dz) \wedge d(i_X\Omega) = 2z\,dx \wedge dy \wedge dz,$$

which implies that on S^2 we have $\operatorname{div}_\Omega(X)$ positive (respectively, negative) on the upper (respectively, lower) hemisphere, i.e. with $S_\pm := \{\pm z > 0\} \subset S^2$ the conditions of the foregoing definition are satisfied.

(2) Consider $M = T^2 \times \mathbb{R}$ with circle-valued coordinates φ, θ on the T^2–factor, and with z denoting the $\mathbb{R}$–coordinate. Let ξ be the contact structure $\ker(d\theta + z\,d\varphi)$ on M. Let $S \subset M$ be the 2–torus $T^2 \times \{0\}$. Then S_ξ is the linear foliation represented by the vector field $X = \partial_\varphi$ on S. Clearly, there can be no collection of circles that divides this non-singular foliation. This entails that S cannot be convex, which can be seen directly as follows.

Take $\Omega = d\varphi \wedge d\theta$ as area form on S, and set $\beta = i_X\Omega = d\theta$. This β is simply the restriction of the contact form $d\theta + z\,d\varphi$ to TS. Then $d\beta = 0$, so if S were a convex surface, then by the contact condition (4.8) we would find a function $u\colon S \to \mathbb{R}$ with $\beta \wedge du > 0$ (i.e. a positive multiple of Ω). This translates into the condition $du(X) < 0$, that is, u would have to decrease along the flow lines of X. This is clearly impossible. This line of reasoning is completely analogous to Example 4.6.21. In Example 4.8.10 below we exhibit explicit deformations of S into a convex surface.

(3) The characteristic foliation S_{ξ_i}, $i = 1, 2$, in Example 4.6.15 is divided by the set

$$\Gamma_i := \{(x, y) \in T^2 \colon \operatorname{div}_\Omega(X_i)(x, y) = 0\}.$$

This follows from the fact that $\operatorname{div}_\Omega(X_i)$ is strictly monotone along the flow lines of X_i, which can be verified by an explicit calculation. From the signs of the singular points we conclude that Γ_1 consists of a pair of embedded circles in the homotopy class of the circle $\{x = \text{const.}\}$, whereas Γ_2 is given by a homotopically trivial embedded circle around the positive elliptic point E_+. For Γ_1 this can also be seen easily from its defining equation $2\cos(\pi x) - \cos(\pi y) = 0$.

(4) Let $\xi = \ker\alpha$ be a contact structure supported by an open book decomposition (B, p) on a 3–manifold M (Defn. 4.4.7). Set

$$S = \overline{p^{-1}(0)} \cup_B \overline{p^{-1}(\pi)};$$

this is a smoothly embedded surface in M, and the closures of the two components of $M \setminus S$ are handlebodies,† i.e. S defines a Heegaard splitting of M. Choose an area form Ω on S that defines the same orientation as $d\alpha$ on $p^{-1}(0)$, and the opposite one on $p^{-1}(\pi)$. The characteristic foliation S_ξ is defined by the vector field X on S with $i_X\Omega = \alpha|_{TS}$. I claim that S_ξ is divided by $\Gamma := B$.

Condition (ii) of Definition 4.4.7 says that $\alpha|_{TB} \neq 0$. The defining equation for X then implies that X is non-zero along B, and because of $\alpha(X) = 0$ it must be transverse to B. This is condition (i) of Definition 4.8.3. Moreover, we have $\mathcal{L}_X\Omega = d(i_X\Omega) = (d\alpha)|_{TS}$, which is non-zero on $S \setminus B$ by condition (i) of Definition 4.4.7. By our choice of Ω, we have $S_+ = p^{-1}(0)$ and $S_- = p^{-1}(\pi)$, with $S_\pm$ as in Definition 4.8.3. Thus, condition (ii) of that definition is also satisfied.

With the next theorem, this last example shows that any contact structure

† For instance, one can describe a Morse function on either component of $M \setminus S$ with one critical point of index 0 and all other critical points of index 1: start with such a function on a page of the open book, and add the quadratic term $(\varphi - \pi/2)^2$ or $(\varphi - 3\pi/2)^2$, respectively, with φ as in the definition of an abstract open book in Section 4.4.2.

supported by an open book decomposition gives rise to a Heegaard splitting of the 3–manifold along a convex surface.

Theorem 4.8.5 *(a) A closed, oriented surface S in a 3–dimensional contact manifold (M, ξ) is convex if and only if the characteristic foliation S_ξ is divided by a collection Γ of embedded circles on S.*

(b) The dividing set of a convex surface as in (a) is determined by the characteristic foliation S_ξ, up to an isotopy via curves transverse to S_ξ.

Proof (a) One half of this statement has been proved in the preceding discussion: if S is convex, then S_ξ is divided by the dividing set Γ_S.

For the converse, assume that S_ξ is divided by some set Γ. We need to show that S is convex. Let Ω and X be the area form and vector field, respectively, from Definition 4.8.3. Set $\beta = i_X \Omega$ and consider the $\mathbb{R}$–invariant 1–form $\alpha = \beta + u\,dz$ on a neighbourhood $S \times \mathbb{R}$ of $S \equiv S \times \{0\}$ in M, with u a smooth function on S, and with z denoting the $\mathbb{R}$–coordinate. The singular foliation induced by α on S is precisely S_ξ. The condition for α to be a contact form is, as in Equation (4.10),

$$u\,\mathrm{div}_\Omega(X) - X(u) > 0.$$

On $S_\pm$ (in the notation of Definition 4.8.3), this condition will be satisfied for $u \equiv \pm 1$.

Now use the flow of X to identify a neighbourhood A of Γ in S, made up of a disjoint collection of annuli, with $\Gamma \times [-\varepsilon, \varepsilon]$, such that Γ becomes identified with $\Gamma \times \{0\}$. As before, the parameter $s \in [-\varepsilon, \varepsilon]$ is chosen such that $X = -\partial_s$, so that the sign of s accords with the sign in $S_\pm$.

Write points in $A = \Gamma \times [-\varepsilon, \varepsilon]$ in the form (p, s). Now set

$$h(p, s) = \exp\Big(-\int_0^s \mathrm{div}_\Omega(X)(p, t)\,dt\Big)$$

and make the ansatz

$$u(p, s) = g(p, s) \cdot h(p, s).$$

Then, observing that $X(u) = -\partial u/\partial s$, we find

$$u\,\mathrm{div}_\Omega(X) - X(u) = \frac{\partial g}{\partial s}\,h(p, s).$$

Therefore, α will be a contact form, coinciding away from Γ with $\beta \pm dz$, if we choose g to satisfy

$$\begin{cases} \partial g/\partial s > 0, \\ g(p, s) = \pm 1/h(p, s) \ \text{ near } \ s = \pm\varepsilon. \end{cases}$$

Such a smooth function $g(p,s)$ can indeed be found.†

We have thus constructed an $\mathbb{R}$–invariant contact structure $\ker\alpha$ on $S\times\mathbb{R}$, so S is a convex surface with respect to this new contact structure. Moreover, since $\ker\alpha$ induces the given characteristic foliation S_ξ, the germs of $\ker\alpha$ and ξ near S coincide by Theorem 2.5.22. This means that S is also a convex surface for ξ. We may additionally assume that $g(p,0)=0$. Then Γ will actually be the dividing set of S_ξ.

(b) Given two contact vector fields Y_0, Y_1 defined near and transverse to $S\subset(M,\xi)$, our aim is to show that the dividing sets

$$\Gamma_i := \{p\in S\colon\ Y_i(p)\in\xi_p\}$$

are isotopic via curves transverse to S_ξ. As in the proof of Lemma 4.6.19 we may assume that the vector fields Y_0, Y_1 are compactly supported near S, so that we may use their respective flows to identify a neighbourhood of S (in two different ways) with $S\times\mathbb{R}$. Under these identifications, the contact structure ξ gives us two vertically invariant contact structures ξ_0,ξ_1 on $S\times\mathbb{R}$, both inducing the characteristic foliation S_ξ on $S\equiv S\times\{0\}$.

We may write $\xi_i=\ker(\beta+u_i\,dz)$. Then Γ_i is the zero set of the function u_i. The contact condition (4.8), for fixed β, is convex in u, so we have the linear interpolation of vertically invariant contact forms $\beta+((1-t)u_0+tu_1)\,dz$, $t\in[0,1]$, on $S\times\mathbb{R}$. The proof of Theorem 2.5.22, applied to this family of contact forms, gives us an $\mathbb{R}$–equivariant isotopy $\psi_t=\varphi_t\times\mathrm{id}_{\mathbb{R}}$. Hence, the isotopy $\varphi_t\colon\ S\to S$ moves the zero set Γ_0 of u_0 to the zero set Γ_1 of u_1, via the dividing sets of S_ξ given by the zero set of $(1-t)u_0+tu_1$, $t\in[0,1]$, all of which constitute collections of curves transverse to S_ξ by the discussion after Definition 4.8.1. □

The contact condition from Equation (4.10) for an $\mathbb{R}$–invariant 1–form $\beta+u\,dz$ can be used to exclude certain foliations as characteristic foliations of convex surfaces.

Example 4.8.6 If there is a flow line of X (as in Definition 4.6.1) in S_ξ from a negative to a positive singularity (both of which would have to be hyperbolic), then S cannot be convex. This is seen as follows. Arguing by contradiction, assume that S were convex, so that we could write ξ (near S) as $\ker(\beta+u\,dz)$. At the negative (respectively, positive) singularity, u would

† For a fixed parameter p, this is very easy. To get smoothness in p, one needs a 'canonical' way of interpolating between the two functions $\pm 1/h(p,s)$. This is tedious, but not difficult. For instance, choose a linear interpolation, and then smooth this continuous function. Since p varies in a compact parameter space, this can be done such that the resulting function is strictly monotone in s.

have to be negative (respectively, positive). So there would be a point along the flow line where $u = 0$ and $X(u) > 0$, contradicting Equation (4.10).

The situation in this example is ruled out by the condition that S_ξ be of Morse–Smale type (Defn. 4.6.8). The Morse–Smale condition turns out to be sufficient for convexity.

Proposition 4.8.7 *If S_ξ is of Morse–Smale type, then S is convex.*

Proof By the preceding theorem it suffices to show that S_ξ admits dividing curves. We retain the vector field X defining S_ξ, but we are going to adjust the area form Ω on S.

Let $S^0_\pm \subset S$ be submanifolds with boundary made up of discs around the elliptic points (of the corresponding sign) of S_ξ, annuli around the repelling or attracting periodic orbits, respectively, and bands around the stable or unstable separatrices of the positive or negative hyperbolic points, respectively. We may assume that $S^0_+ \cap S^0_- = \emptyset$ and that X is transverse to the boundary of $S^0_\pm$, see Figure 4.41.

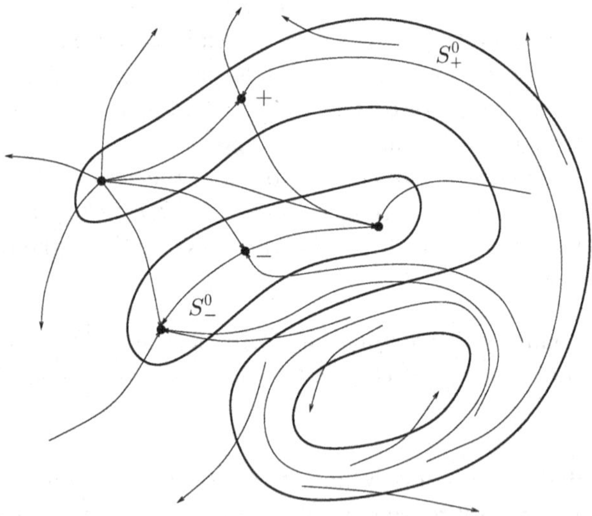

Fig. 4.41. The sets $S^0_\pm$.

Equation (4.21) shows that the divergence at the singular points will not change if we rescale the area form to $g\Omega$, with $g\colon S \to \mathbb{R}^+$. That same equation shows that we may achieve $\mathrm{div}_{g\Omega}(X) > 0$ on S^0_+ by having g grow fast enough along the flow lines of X (and equal to some constant near each singular point). Likewise, we can ensure that $\mathrm{div}_{g\Omega}(X) < 0$ on S^0_-.

On $A := S \setminus (S^0_+ \cup S^0_-)$ the characteristic foliation is non-singular, transverse to the boundary, and without closed leaves. Hence A is a collection of annuli, and we may choose g there as in the construction on page 231. Then the spines of these annuli divide S_ξ. □

Consequently, Proposition 4.6.11 can be rephrased in terms of convexity.

Proposition 4.8.8 *Let S be a closed, orientable surface in a contact 3–manifold $(M, \xi = \ker \alpha)$. Then there is a surface S' isotopic and C^∞–close to S such that S' is a convex surface for ξ.* □

Remark 4.8.9 The preceding proof does not require the full Morse–Smale property. The argument fails if there is a so-called **retrograde saddle–saddle connection**,† i.e. if the unstable separatrix of a negative hyperbolic point H_- coincides with the stable separatrix of a positive hyperbolic point H_+, so that there is a flow line from H_- to H_+. On the other hand, trajectories between hyperbolic points of the same sign or from a positive to a negative hyperbolic point do not invalidate the argument.

Example 4.8.10 We take another look at the non-convex surface S from Example 4.8.4 (2). The surface $S' := \{z = \varepsilon \sin\theta\}$ is isotopic and, for small ε, C^∞–close to S. Since S' is a graph over S, we may continue to use the coordinates φ, θ and the area form $\Omega = d\varphi \wedge d\theta$ on S'. Then the characteristic foliation S'_ξ is given by the vector field $X = \partial_\varphi - \varepsilon \sin\theta\, \partial_\theta$, see Figure 4.42 (left), and $\mathrm{div}_\Omega(X) = -\varepsilon\cos\theta$. It follows that S'_ξ is divided by the set Γ consisting of the two circles $\{\theta = \pi/2\}$ and $\{\theta = 3\pi/2\}$. In particular, S' is a convex surface.

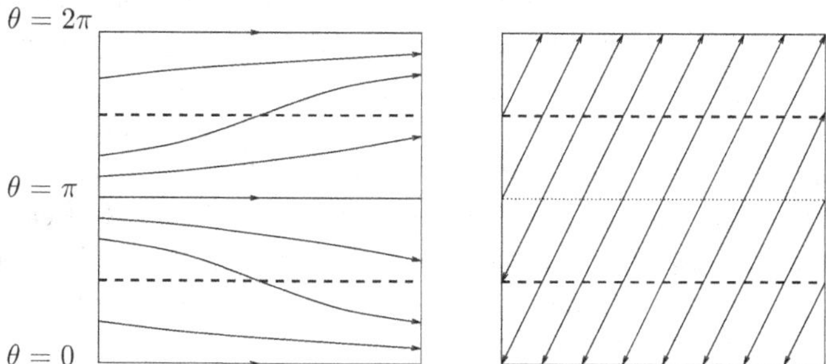

Fig. 4.42. Two characteristic foliations with the same dividing set.

The foliation $\mathfrak{F}$ in Figure 4.42 (right) is given by a vector field of the form

† In the words of Claudius, retrograde saddle–saddle connections are 'most retrograde to our desire.' [220]

$X = -\sin\theta(q\,\partial_\varphi + p\,\partial_\theta)$ with p a positive and q a non-negative integer. This foliation has singularities along the two circles $\{\theta = 0\}$ and $\{\theta = \pi\}$ parallel to the dividing curves. The divergence of X is $\mathrm{div}_\Omega(X) = -p\cos\theta$, so this second foliation $\mathfrak{F}$ is also divided by Γ. The slope p/q of the Legendrian leaves of $\mathfrak{F}$ (one may likewise realise irrational slopes) can take any value different from 0, the slope of the dividing curves.

This $\mathfrak{F}$, too, can be realised as the characteristic foliation on a suitably perturbed copy S' of S. Such a perturbation can be written down fairly explicitly, or see the following theorem.

The next theorem, read in conjunction with Theorem 2.5.22, gives a first glimpse of the power of convex surface theory. It asserts, in essence, that all the crucial information about the contact structure in a neighbourhood of a convex surface S is inherent in the dividing set Γ_S rather than a specific characteristic foliation.

Theorem 4.8.11 *Let S be a (closed, oriented) convex surface in a contact 3–manifold (M, ξ). Let $\mathfrak{F}$ be a singular 1–dimensional foliation on S divided by the dividing set Γ_S of the characteristic foliation S_ξ, and let $\mathcal{N}(S)$ be any neighbourhood of S in M. Then there is an isotopy $\psi_t \colon S \to \mathcal{N}(S)$, $t \in [0,1]$, of embeddings with the following properties.*

(i) ψ_0 is the inclusion $S \subset M$.
(ii) $\psi_t(S)$ is a convex surface with dividing set $\psi_t(\Gamma_S)$ for all $t \in [0,1]$.
(iii) The characteristic foliation $(\psi_1(S))_\xi$ coincides with $\psi_1(\mathfrak{F})$.

Proof By Lemma 4.6.19 (and the discussion about $\mathbb{R}$–invariant contact forms following it) we may identify a neighbourhood of S in (M, ξ) contactomorphically with $S \times \mathbb{R}$, equipped with a vertically invariant contact structure $\xi_0 = \ker\alpha_0$, with $\alpha_0 = \beta_0 + u_0\,dz$. Here $S \equiv S \times \{0\}$, and β_0, u_0 are a differential 1–form and smooth function, respectively, on S. Moreover, we may assume α_0 to be of the type described in the proof of Theorem 4.8.5.

Thus, as in the proof of that theorem we choose a collection A of closed annuli around the dividing curves in such a way that the flow lines of both S_{ξ_0} and $\mathfrak{F}$ foliate each annulus by line segments from one boundary component to the other (and transverse to the boundary).

By that same proof, we may assume that we have an area form Ω on S and a vector field X_0 defining S_{ξ_0} such that $\alpha_0 = i_{X_0}\Omega + u_0\,dz$ with $u_0 \equiv \pm 1$ on $S_\pm \setminus \mathrm{Int}\,(A)$ and $\pm\mathrm{div}_\Omega(X_0) > 0$ on $S_\pm$. As pointed out there, we may impose the additional condition that $u_0 = 0$ precisely along Γ_S, so that Γ_S still accords with Definition 4.8.1.

Since $\mathfrak{F}$ is also divided by Γ_S, we may assume that $\mathfrak{F}$ is defined by a vector field X_1' with $\pm\mathrm{div}_{g\Omega}(X_1') > 0$ on $S_\pm$ for some smooth function $g\colon S \to \mathbb{R}^+$. By comparing Equations (4.11) and (4.21) we see that

$$g\,\mathrm{div}_{g\Omega}(X) = \mathrm{div}_\Omega(gX).$$

So with $X_1 := gX_1'$ we have in fact $\pm\mathrm{div}_\Omega(X_1) > 0$ on $S_\pm$. We may assume that the annuli A around Γ_S have been chosen so small that the flow lines of the vector fields $X_t := (1-t)X_0 + tX_1$, $t \in [0,1]$, foliate A in the same way as those of X_0 and X_1. These vector fields will likewise satisfy the divergence condition $\pm\mathrm{div}_\Omega(X_t) > 0$ on $S_\pm$.

Notice that in the proof of Theorem 4.8.5 we used just this divergence condition in order to find a contact structure inducing the given singular foliation; only at the last step did we appeal to Giroux's theorem (Thm. 2.5.22) and the fact that we had started with the characteristic foliation induced by a contact structure to conclude that our construction realises the germ of the given contact structure. Therefore, we can find a whole family of $\mathbb{R}$–invariant contact forms $\alpha_t = i_{X_t}\Omega + u_t\,dz$, $t \in [0,1]$, with u_t satisfying the same constraints as u_0. With $\xi_t := \ker\alpha_t$ we have $\mathfrak{F} = S_{\xi_1}$.

In particular, we may ensure that $u_t = 0$ precisely along Γ_S, so that this is the dividing set for each contact structure in this 1–parametric family. Now apply the Moser trick from the proof of Theorem 2.2.2 to this family of contact forms. Since each member of this family is $\mathbb{R}$–invariant, so will be the time-dependent vector field Y_t defining the isotopy ψ_t with $T\psi_t(\xi_t) = \xi_0$.† This means that $\mathcal{L}_{Y_t}\partial_z = [Y_t, \partial_z] = \mathbf{0}$, so $T\psi_t(\partial_z) = \partial_z$.

This last condition implies that $\psi_t(S)$ is transverse to ∂_z for all $t \in [0,1]$, and hence a convex surface. From $T\psi_t(\xi_t) = \xi_0$ we have that $(\psi_t(S))_{\xi_0} = \psi_t(S_{\xi_t})$, which gives (iii). Finally, the dividing set of $\psi_t(S)$ consists of points $\psi_t(p)$ where ∂_z is contained in ξ_0, which is equivalent to $p \in S$ being a point where ∂_z is contained in ξ_t, i.e. a point p in the dividing set Γ_S of S_{ξ_t}; this is statement (ii). Condition (i) is obviously satisfied. □

Remark 4.8.12 By wiggling $\mathfrak{F}$ a little, we may assume that $X_0 = X_1$ along Γ_S, so that X_t does not depend on t along the dividing set. By construction, the same can then be said about u_t. Consequently, $\dot{\alpha}_t = 0$ along that set. In this case, by Remark 2.2.3, the isotopy ψ_t is stationary on the dividing set.

We end this chapter with a couple of simple applications of the rudiments of convex surface theory that we have just developed. The first is the easy part of *Giroux's criterion* for deciding whether a contact structure is tight near a convex surface.

† This would be ψ_t^{-1} in the notation of the Gray stability theorem.

Proposition 4.8.13 *Let S be a (closed, connected, oriented) convex surface in a contact 3–manifold (M, ξ). If the dividing set Γ_S contains a circle γ contructible in S, then any vertically invariant neighbourhood of S in (M, ξ) is overtwisted, unless $S = S^2$ and Γ_S is connected.*

Proof We may assume that γ has been chosen in such a way that it bounds a disc $\Delta \subset S$ not containing other components of Γ_S. We consider the case $\mathrm{Int}(\Delta) \subset S_-$; for $\mathrm{Int}(\Delta) \subset S_+$ the argument is analogous.

In order to show that ξ is overtwisted in any vertically invariant neighbourhood of S, by Theorem 4.8.11 it suffices to find a singular foliation $\mathfrak{F}$ on S that is divided by Γ_S and contains a circle γ_0 (without any singular points) bounding a disc Δ_0 in S.

We first want to find such a foliation $\mathfrak{F}$ in the case that Γ_S contains other components besides γ. Let γ_0, γ_+ be parallel copies of γ in S_+, see Figure 4.43. Our assumption implies that the component of $S \setminus \Gamma_S$ containing γ_+ has other boundary components apart from γ, along all of which S_ξ flows out transversely. Let S_0 be essentially this component, but with the annulus between γ and γ_+ removed, so that γ_+ is a boundary component of S_0.

On that annulus between γ and γ_+, define $\mathfrak{F}$ as shown. In particular, γ_0 is supposed to be a closed leaf of $\mathfrak{F}$. On the disc Δ_0 bounded by γ_0 we can choose $\mathfrak{F}$ in such a way that it is divided by γ, say with a single negative elliptic point in the interior of Δ.

On S_0 we want $\mathfrak{F}$ to be defined by a vector field of positive divergence, flowing in transversely along γ_+, and out along all other boundary components. Such a foliation can be constructed as the gradient flow of the 'height' function z on a copy of S_0† in $\mathbb{R}^3$, with γ_+ at the level $z = 0$, say, and the other boundary components at the level $z = 1$, and $S_0 \subset \mathbb{R}^3$ 'expanding' sufficiently as z increases. In particular, $\mathfrak{F}$ may be arranged to have only positive hyperbolic points as singularities. Near the boundary components of S_0 at $z = 1$ and everywhere else on S we may take $\mathfrak{F}$ to coincide with S_ξ. This concludes the construction of $\mathfrak{F}$ in the case of Γ_S being disconnected.

If $S = S^2$ and Γ_S is connected, then by Theorem 4.8.11 we may assume that the characteristic foliation S_ξ is the one from Example 4.8.4 (1). By Theorem 2.5.22, a neighbourhood of S in (M, ξ) looks like a neighbourhood of S^2 in the standard tight contact structure ξ_2 on $\mathbb{R}^3$.

If $S \neq S^2$ and Γ_S is connected, we want to show that S can be deformed inside a vertically invariant neighbourhood to a convex surface S' whose dividing set contains an additional circle. Then we conclude as before that ξ is overtwisted in a neighbourhood of S', and hence in a neighbourhood of S.

† The embedding $S_0 \subset \mathbb{R}^3$ can be chosen in such a way that the gradient flow is of Morse–Smale type.

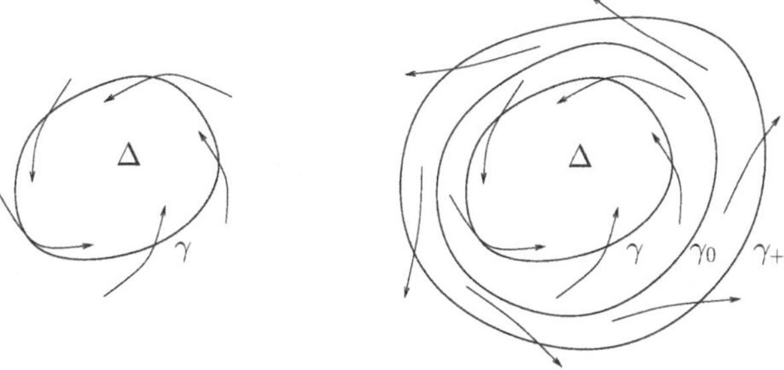

Fig. 4.43. The foliations S_ξ (left) and $\mathfrak{F}$ (right).

To do so, choose a non-separating embedded circle $\gamma_0 \subset S$ not intersecting γ (this is where we use $S \neq S^2$). As in the first part of the argument we find a singular foliation $\mathfrak{F}$ that has γ_0 as a repelling closed leaf. The role of S_0 is taken by the complement of $\mathrm{Int}\,(\Delta)$ with an open annulus around γ_0 removed. Since γ_0 is non-separating, S_0 will be connected. The singular foliation $\mathfrak{F}$ will be required to flow out of S_0 along γ, and into S_0 along the two boundary components coming from γ_0.

By Theorem 2.5.22 we may identify a neighbourhood of γ_0 in (M,ξ) with $S^1 \times (-\varepsilon,\varepsilon) \times \mathbb{R}$ with contact structure $\ker(dz - s\,d\theta + ds)$. Here θ, s, z are the coordinates in $S^1, (-\varepsilon,\varepsilon), \mathbb{R}$, respectively; γ_0 is identified with $S^1 \times \{0\} \times \{0\}$, and a neighbourhood of γ_0 in S with $S^1 \times (-\varepsilon,\varepsilon)$.

Now deform S to the surface S' as shown in Figure 4.44. (In the lower half of that figure, the vertical direction corresponds to the θ–coordinate, with the top and bottom of the rectangles identified.) One can verify by an explicit calculation that the characteristic foliation on S' has three parallel closed leaves, and a pair of dividing curves between these leaves. □

Theorem 4.8.11 greatly enhances the value of Theorem 2.5.22 about the characteristic foliation on a surface determining the germ of the contact structure near that surface. As emphasised earlier, this latter theorem requires the characteristic foliation to be known up to diffeomorphism. A simple picture of a given characteristic foliation, however, will only determine it up to homeomorphism. Often one can appeal to Theorem 4.8.11 by exhibiting a dividing set. Nonetheless, the following folklore result can be very useful for working directly with characteristic foliations. Ideas for proving it can be extracted from [64]; here I present a proof via convex surface theory.

Proposition 4.8.14 *Let $j_i\colon S \to (M_i,\xi_i)$, $i = 0,1$, be two embeddings of a (closed, connected, oriented) surface S into 3–dimensional contact manifolds*

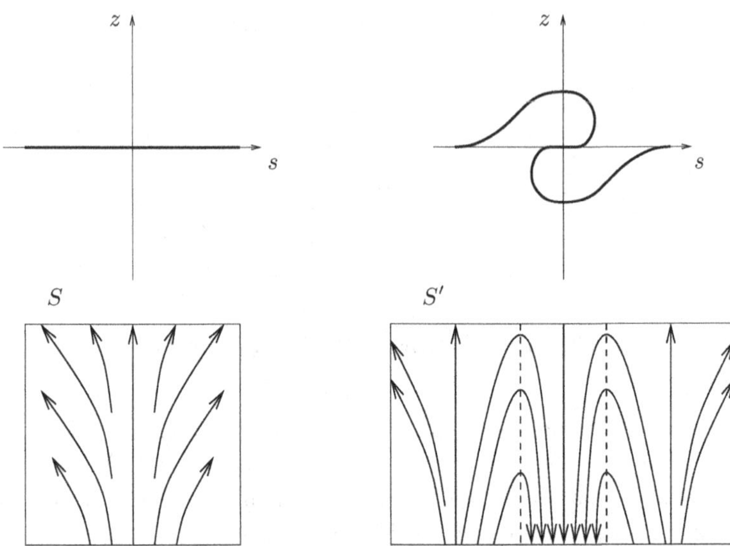

Fig. 4.44. Creating additional dividing curves.

such that the characteristic foliations $(j_i(S))_{\xi_i}$ are of Morse–Smale type and homeomorphic (as oriented singular foliations) via an orientation-preserving homeomorphism that respects the signs of the singular points (this is an extra condition for the hyperbolic points only). Then there is an embedding j_1', C^0-close to j_1, such that $(j_0(S))_{\xi_0}$ and $(j_1'(S))_{\xi_1}$ are diffeomorphic.

Proof We regard the given homeomorphism as a map $h\colon S \to S$, and we interpret $\mathfrak{F}_i := (j_i(S))_{\xi_i}$, $i = 0, 1$, as singular foliations on S. Since every element in the mapping class group of S has a smooth representative, we can write h as the composition of a diffeomorphism with a homeomorphism isotopic† to the identity and fixing the singular points. Therefore, without loss of generality, we may assume that h is a homeomorphism of that restricted type.

Construct a dividing set Γ_0 for $\mathfrak{F}_0$ as in the proof of Proposition 4.8.7. Qualitatively, the circles in the set $h(\Gamma_0)$ will look, relative to $\mathfrak{F}_1$, exactly as those in Γ_0 relative to $\mathfrak{F}_0$. That is, there are annulus-shaped regions in S, with the flow lines of $\mathfrak{F}_1$ going from one boundary component to the other, each intersecting a corresponding circle of the set $h(\Gamma_0)$ in a single point. Since h is only a homeomorphism, these intersections may fail to be trans-

† Recall that an isotopy of homeomorphisms is a homotopy h_t, $t \in [0, 1]$, such that each h_t is a homeomorphism. By a smooth isotopy we mean that each h_t is a diffeomorphism, and that the family (h_t) depends smoothly on t.

verse, but one can isotope $h(\Gamma_0)$ (inside these annuli and via circles having the same intersection property) to a set Γ_1 of smooth circles transversely intersecting the flow lines of $\mathfrak{F}_1$, so that Γ_1 is a dividing set for $\mathfrak{F}_1$.

Since h was assumed to be isotopic to the identity, the collection Γ_1 of circles is homotopic to Γ_0. By the proof of Baer's theorem [225, Thm. 6.2.5],† there is a smooth isotopy of S that moves Γ_0 to Γ_1.

Hence, after changing $\mathfrak{F}_1$ by a diffeomorphism, it can be assumed to have the same dividing set as $\mathfrak{F}_0$. Now apply Theorem 4.8.11. □

Finally, we use convex surface theory to give a simple alternative proof of the theorem on genus bounds in tight contact 3–manifolds.

Alternative proof of Theorem 4.6.30 After an isotopy we may assume (by Proposition 4.8.8) that S is convex. Write $S \setminus \Gamma_S = S_+ \sqcup S_-$ as before. Since the circle has zero Euler characteristic, we have

$$\chi(S) = \chi(S_+) + \chi(S_-).$$

All the positive or negative singularities of S_ξ are contained in S_+ or S_-, respectively, and S_ξ is transverse to the boundary of these sets. Hence, with an argument as in the proof of Proposition 4.6.14, we find

$$\langle e(\xi), [S] \rangle = \chi(S_+) - \chi(S_-).$$

If $S = S^2$, then tightness of ξ implies by Proposition 4.8.13 that both S_+ and S_- consist of a single disc, so that $\langle e(\xi), [S] \rangle = 0$.

If S is a surface of higher genus, then again thanks to Proposition 4.8.13 we know that no component of S_+ or S_- is a disc. Thus, each of these components is homeomorphic to a 2–sphere with at least two discs removed, or a surface of higher genus with at least one disc removed. It follows that $\chi(S_\pm) \leq 0$. Then

$$\langle e(\xi), [S] \rangle + \chi(S) = 2\chi(S_+) \leq 0$$

and

$$-\langle e(\xi), [S] \rangle + \chi(S) = 2\chi(S_-) \leq 0.$$

This gives the inequality as stated in the theorem. □

† Baer's theorem asserts that homotopic simple closed curves on an orientable surface are isotopic; the proof in [225] gives a smooth isotopy if we started with smooth curves. By the usual isotopy extension theorem from differential topology, this extends to a smooth isotopy of S.

4.9 Tomography

The key to the classification of tight contact structures on a host of closed 3–manifolds is an understanding of contact structures on product manifolds $S \times [-1,1]$, with S a closed, oriented surface.† There are two quite distinct approaches to this problem, albeit both based on convex surface theory. The first one is due to Giroux [105], where the key word is *tomography*. This means that a contact structure ξ on $S \times [-1,1]$ is studied via its 'film' of characteristic foliations $(S \times \{z\})_\xi$, $z \in [-1,1]$. The second approach, advanced by Honda [137], puts greater emphasis on dividing sets and their manipulation.

In the present section, the plan is to solve the classification problem for tight contact structures on $S^2 \times [-1,1]$ with the help of tomography. From that solution, the classification of tight contact structures on S^3, $S^2 \times S^1$ and the 3–ball D^3 will follow easily; this will be presented in Section 4.10.

Which of the two methods mentioned is the superior one may be a matter of debate. By restricting attention, right from the start, to the case $S = S^2$ and tight contact structures, many of the intricacies in either approach disappear. Here I give preference to tomography, because it requires fewer new concepts. All the results below are adaptations of statements in [105]; I retain the pregnant names Giroux has coined for them.

Throughout we assume that our contact structures ξ are cooriented and positive. As in Section 4.6 we write ξ as the kernel of a contact form

$$\alpha = \beta_z + u_z\, dz,$$

with the contact condition $\alpha \wedge d\alpha > 0$ given by Equation (4.3) on page 163.

In order to simplify notation, we are going to write S^z for $S \times \{z\}$. The following lemma says that a contact structure on $S \times [-1,1]$ can be reconstructed from its film of characteristic foliations.

Lemma 4.9.1 (Reconstruction Lemma) *Let ξ_0, ξ_1 be two contact structures on $S \times [-1,1]$ with the property that the characteristic foliations $S^z_{\xi_0}$ and $S^z_{\xi_1}$ coincide for each $z \in [-1,1]$. Then ξ_0 and ξ_1 are isotopic rel boundary.*

Proof The assumption on the characteristic foliations implies that we may choose contact forms α_i defining ξ_i of the form

$$\alpha_i = \beta_z + u^i_z\, dz,$$

that is, with β_z not depending on $i = 0, 1$. Given β_z, the functions u_z satisfy-

† A contact structure ξ on a manifold M with boundary is understood as a contact structure on M with an open collar attached to its boundary, i.e. $M \cup_\partial [0,\varepsilon) \times \partial M$, whose restriction to M equals ξ.

ing the contact condition (4.3) form a convex set. So the linear interpolation between α_0 and α_1 is via contact forms. It then follows from Gray stability that ξ_0 and ξ_1 are isotopic; an isotopy rel boundary can be constructed as in the proof of Theorem 2.5.23, since the homotopy of contact forms is stationary on $TS^{\pm 1}$. □

The next lemma is the analogous statement in the case that all the surfaces S^z are convex.

Lemma 4.9.2 (Uniqueness Lemma) *Let ξ_0, ξ_1 be two contact structures on $S \times [-1,1]$ with the following properties.*

- *The characteristic foliations $S^{\pm 1}_{\xi_i}$ on the boundary of $S \times [-1,1]$ coincide for $i = 0, 1$.*
- *Each surface S^z, $z \in [-1,1]$, is convex for either contact structure, and there is a smoothly varying family of multi-curves Γ_z dividing both $S^z_{\xi_0}$ and $S^z_{\xi_1}$.*

Then ξ_0 and ξ_1 are isotopic rel boundary.

Proof Choose contact forms

$$\alpha^i = \beta^i_z + u^i_z\, dz, \quad i = 0, 1,$$

defining $\xi_i = \ker \alpha^i$, with $\beta^0_{\pm 1} = \beta^1_{\pm 1}$. Recall the contact condition (4.3),

$$A^i := u^i_z\, d\beta^i_z + \beta^i_z \wedge (du^i_z - \dot{\beta}^i_z) > 0, \quad i = 0, 1,$$

with the dot denoting the derivative with respect to z.

The convexity assumption allows us to choose, for each $z \in [-1,1]$, functions v^0_z and v^1_z on S that vanish exactly on Γ_z (and only to first order) and that satisfy the $\mathbb{R}$–invariant contact condition (4.8),

$$B^i := v^i_z\, d\beta^i_z + \beta^i_z \wedge dv^i_z > 0, \quad i = 0, 1.$$

The construction in the proof of Theorem 4.8.5 gives us two such families of functions v^i_z varying smoothly with z, and with $v^0_{\pm 1} = v^1_{\pm 1}$. Notice that for each z the function v^0_z / v^1_z, defined *a priori* on $S \setminus \Gamma_z$ only, extends unambiguously to a positive function on all of S.

The computation

$$\begin{aligned} v^0\, d\Big(\frac{v^0}{v^1}\beta\Big) + \Big(\frac{v^0}{v^1}\beta\Big) \wedge dv^0 &= \Big(\frac{v^0}{v^1}\Big)^2 v^1\, d\beta + \frac{v^0}{v^1}\, dv^0 \wedge \beta \\ &\quad - \Big(\frac{v^0}{v^1}\Big)^2 dv^1 \wedge \beta + \frac{v^0}{v^1}\, \beta \wedge dv^0 \\ &= \Big(\frac{v^0}{v^1}\Big)^2 (v^1\, d\beta + \beta \wedge dv^1) \end{aligned}$$

shows that if we replace β_z^1 by $(v_z^0/v_z^1)\cdot\beta_z^1$, we may take $v_z^1 = v_z^0 =: v_z$. Observe that after this replacement we still have $\beta_{\pm 1}^0 = \beta_{\pm 1}^1$.

With $\lambda \in \mathbb{R}^+$ set

$$\alpha_t^i = \beta_z^i + \big((1-t)u_z^i + t\lambda v_z\big)\,dz,\ \ i = 0,1,\ \ t\in[0,1].$$

Then

$$\alpha_t^i \wedge d\alpha_t^i = \big((1-t)A^i - t\beta_z^i \wedge \dot{\beta}_z^i + t\lambda B^i\big)\wedge dz,$$

which is positive for all $t\in[0,1]$ if λ is chosen sufficiently large.

Finally, set

$$\alpha_s = (1-s)\alpha_1^0 + s\alpha_1^1 = (1-s)\beta_z^0 + s\beta_z^1 + \lambda v_z\,dz,\ \ s\in[0,1].$$

Then

$$\alpha_s\wedge d\alpha_s = \lambda\big((1-s)B^0 + sB^1\big)\wedge dz + O(1),$$

which is likewise positive for all $s\in[0,1]$, provided λ is sufficiently large.

Then

$$\begin{cases} \alpha_{3t}^0, & 0\le t\le 1/3,\\ \alpha_{3t-1}, & 1/3\le t\le 2/3,\\ \alpha_{3-3t}^1, & 2/3\le t\le 1,\end{cases}$$

defines a homotopy via contact forms from α^0 to α^1 that is stationary on $TS^{\pm 1}$. Now the proof concludes as in the reconstruction lemma. □

Remark 4.9.3 If in either of the two preceding lemmata the contact structures ξ_0,ξ_1 coincide near the boundary, the proofs give us isotopies that are stationary near the boundary.

From now on we specialise to the situation that $S = S^2$ and ξ be tight. Then the characteristic foliations S_ξ^z cannot contain any closed leaves by Proposition 4.6.28. Moreover, if S^z is convex, then its dividing set Γ_z consists of a single embedded circle by Proposition 4.8.13. This circle divides the sphere into two discs, each containing singular points of one particular sign only. We shall see that the condition in the foregoing proposition, *viz.*, that Γ_z vary smoothly with z, poses no additional difficulty.

The main result we want to prove in this section is the following.

Theorem 4.9.4 *A tight contact structure on $S^2\times[-1,1]$ is determined, up to isotopy rel boundary, by the characteristic foliation on the boundary.*

Starting from a tight contact structure ξ on $S\times[-1,1]$, $S = S^2$, such that $S^{\pm 1}$ are convex surfaces, we want to show how to manipulate the characteristic foliations S_ξ^z by an isotopy (rel boundary) of ξ until we arrive at a contact structure for which each S^z is a convex surface. Since any two embedded circles in S^2 are isotopic (by the easiest case of Baer's theorem), we can make the dividing sets of two such contact structures coincide by a level-preserving isotopy. The uniqueness lemma then implies that a tight contact structure on $S^2 \times [-1, 1]$ for which the boundary surfaces $S^2 \times \{\pm 1\}$ are convex is determined, up to isotopy rel boundary, by the characteristic foliation on the boundary. The convexity condition on the boundary is in fact superfluous. Indeed, given two contact structures on $S^2 \times [-1, 1]$ inducing the same characteristic foliation on the boundary, we may assume by Theorem 2.5.23 that after an isotopy fixing the boundary the contact structures coincide near the boundary. Proposition 4.8.8 allows us to find convex 2–spheres $S^2_{\pm}$ isotopic to $S^2 \times \{\pm 1\}$ inside these neighbourhoods where the two contact structures coincide. It therefore suffices to prove Theorem 4.9.4 under the additional assumption that the boundary spheres are convex for the given contact structure. This we shall now proceed to do.

We have seen in Proposition 4.6.11 that, given a surface in a contact 3–manifold, by perturbing the contact structure we can arrange the surface to have a characteristic foliation of Morse–Smale type, so that the surface is convex by Proposition 4.8.7. Now, however, we are dealing with a 1–parametric family of surfaces, and the Morse–Smale condition is not generic for 1–parametric families of singular foliations. For instance, the elimination process in Section 4.6.3 gives rise to a film of characteristic foliations where a couple of singular points in elimination position in S^z_ξ for $z < 0$, say, merge to a degenerate singular point of index 0 (and the same sign as the two previous singular points) in S^0_ξ, which then disappears in S^z_ξ for $z > 0$.

Nonetheless, there are some simple 'genericity' assumptions that we may impose. The key to this are the fact that the contact condition is open in the C^1–topology (on sections of the cotangent bundle, i.e. 1–forms), and Gray stability for contact structures. Therefore, any property that we may impose on a film of singular foliations by a C^1–small perturbation can be achieved by a corresponding isotopy of contact structures. By a relatively straightforward application of the usual transversality theorem in differential topology (and the stability of the various non-degeneracy phenomena in question), see [205, Lemma 2], we may assume that

- each S^z_ξ contains only finitely many singular points, and – on finitely many z–levels – at most one degenerate one;
- there are no trajectories from a negative to a positive singularity in S^z_ξ, except on a finite number of levels $z_1, \ldots, z_k$, where all critical points are non-degenerate and there is exactly one retrograde saddle–saddle connection;

- on these levels $z_1, \ldots, z_k$ there are no other saddle–saddle connections.

Let me briefly expand on these statements. That there can be no accumulation point (in terms of the parameter z) of retrograde saddle–saddle connections is in fact not simply a genericity result, but a consequence of the contact condition. Arguing by contradiction, assume that we had such an accumulation point at $z = 0$. Let z_ν, $\nu \in \mathbb{N}$, be a sequence of points with $\lim_{\nu\to\infty} z_\nu = 0$ such that each $S_\xi^{z_\nu}$ contains a retrograde saddle–saddle connection, accumulating at such a connection for $\nu \to \infty$. We may choose local coordinates in such a way that we are dealing with a contact structure on $\mathbb{R}^2 \times [-1, 1]$, with a positive (respectively, negative) hyperbolic point at $(\pm 1, 0, z_\nu)$, including $z_\infty = 0$, and with the line segment between these to points (for each ν) along the x–axis a retrograde saddle–saddle connection. On each of these slices S^{z_ν}, the vector field defining the characteristic foliation is a positive multiple of ∂_x along the connecting separatrix, and hence β_{z_ν} a positive multiple of dy (if we take $\Omega = dx \wedge dy$ as area form). It follows that $\dot{\beta}_z|_{z=0}$ is a multiple of dy along the separatrix γ_0 from $(-1, 0, 0)$ to $(1, 0, 0)$, and thus $(\beta_z \wedge \dot{\beta}_z)|_{z=0} = 0$ there. There has to be a point on γ_0 where $u_0 = 0$ and $du(\partial_x) \geq 0$, see Example 4.8.6. At such a point the contact condition (4.3) would be violated, which is the desired contradiction.

Non-retrograde saddle–saddle connections, however, may well appear in a 1–parametric family. By Remark 4.8.9 there is in fact a vertically invariant contact structure inducing such a phenomenon in the characteristic foliation. But by Peixoto's result this phenomenon is neither generic nor stable, so we may ensure that this happens only on isolated levels not containing a retrograde saddle–saddle connection. Stable phenomena, on the other hand, persist in neighbouring levels, so on pointed open neighbourhoods of the levels $z_1, \ldots, z_k$ we may assume not to have any saddle–saddle connections.

Remark 4.9.5 While these statements hold for any S, in the case $S = S^2$ we may in addition appeal to the Poincaré–Bendixson theorem (Thm. 4.7.8). Notice that closed orbits do not arise as α– or ω–limits of flow lines in the case that ξ is tight. Neither do poly-cycles made up of singular points and connecting orbits, because Lemma 4.6.33 rules out (on S^2) any poly-cycle containing singular points of one sign only.

Lemma 4.9.6 *Let S be an oriented 2–sphere embedded in a tight contact 3–manifold (M, ξ) such that S_ξ has finitely many non-degenerate singular points and there are no retrograde saddle–saddle connections. Then the graph G in S whose vertices are the positive singular points, and whose edges are the stable separatrices of the positive hyperbolic points, is a connected tree.*

Moreover, S is convex, and the dividing set is given, up to isotopy, by the boundary of a disc-like neighbourhood of G.

Proof † By the proof of Proposition 4.8.7, the absence of retrograde saddle–saddle connections guarantees that S is a convex surface with dividing set given by the boundaries of regions obtained by thickening G. If G were not a connected tree, it would give rise to a disconnected dividing set, in contradiction with Proposition 4.8.13. □

The next proposition is the main step in the proof of Theorem 4.9.4.

Proposition 4.9.7 *Let ξ be a tight contact structure on $S \times [-1, 1]$, where $S = S^2$, such that $S^{\pm 1}$ are convex surfaces. Then ξ is isotopic rel boundary to a contact structure for which each S^z, $z \in [-1, 1]$ is a convex surface.*

Proof By the discussion above, it suffices to find an isotopy that helps us to get rid of all retrograde saddle–saddle connections. First, we may make all the genericity assumptions discussed above. Then, by the preceding lemma, the surfaces S^z will be convex, except at finitely many levels $z_1, \ldots, z_k \in (-1, 1)$, where we have a single retrograde saddle–saddle connection. For ease of notation, we assume that $z = 0$ is such an exceptional level. We write $x_{\mathrm{h}}^{\pm}$ for the hyperbolic points on that level with a retrograde connection between them.

I claim that the stable separatrix γ of x_{h}^{+} not coming from x_{h}^{-} must emanate at a positive elliptic point x_{e}^{+}. Indeed, the Poincaré–Bendixson theorem (Thm. 4.7.8) tells us that the α–limit of γ has to be either a singular point (which would have to be a positive elliptic point, since there are no further saddle–saddle connections), a closed orbit (ruled out by tightness), or a poly-cycle. But only a poly-cycle whose vertices are hyperbolic points of the same sign can occur as such a limit set, so this last case is ruled out by our genericity assumption or by the arguments in the proof of the preceding lemma.

Thus, the points x_{e}^{+} and x_{h}^{+} are in elimination position. All the properties of the characteristic foliation S_{ξ}^{0}, except the retrograde saddle–saddle connection, are structurally stable, so they persist in all levels $z \in (-\varepsilon, \varepsilon)$ for some small $\varepsilon > 0$. We may choose local coordinates around γ in such a way

† The same proof gives the analogous statement for the graph of negative singularities and unstable separatrices. A 'classical' proof for either case can be given by appealing to Lemma 4.6.33, which forces G to be a (possibly disconnected) tree. Connectedness of G follows from the observation that, by the proof of Proposition 4.8.7, the complement of disc-like neighbourhoods around each of the components of G and the components of the 'negative' graph must be a collection of annuli.

that, on all levels $z \in (-\varepsilon, \varepsilon)$, the set-up is as in part (2) of the proof of the elimination lemma (Lemma 4.6.26). We may assume that $\|aX\|$ has a unique local maximum along γ, where X, as before, is the vector field defining the characteristic foliation, and a the coefficient function of the area form Ω as in the proof of the elimination lemma.

Now choose a compactly supported bump function $\varphi\colon (-\varepsilon, \varepsilon) \to [0,1]$, with $\varphi(0) = 1$. With g as in the said proof, define a compactly supported deformation of $S \times (-\varepsilon, \varepsilon)$ by

$$S \times (-\varepsilon, \varepsilon) \ni (p, z) \longmapsto (p, z + \varphi(z) g(p)).$$

In other words, we replace each slice S^z by the graph of $\varphi(z)g(p)$ over it. For this to make sense we need to ensure that

$$\frac{\partial}{\partial z}\left(z + \varphi(z) g(p)\right) = 1 + \dot{\varphi}(z) g(p)$$

is positive; this is possible, for the construction in the proof of the elimination lemma allows us to choose g as small as we like.

This deformation of $S \times [-1, 1]$ translates into an isotopy (supported in $S \times (-\varepsilon, \varepsilon)$) of contact structures, so we shall continue to speak of the spheres S^z_ξ, but we think of ξ as the perturbed contact structure.

Now recall the points (i) to (iv) in the proof of the elimination lemma, where we checked that the foliation becomes non-singular after the deformation. The estimates in (ii), (iii) and (iv) remain valid for the scaled perturbation function $\varphi(z)g$. The estimate in (i) and our assumption on the behaviour of aX along the separatrix γ imply that for the points $z \in (-\varepsilon, \varepsilon)$ with

$$\varphi(z) g_2'(0) = -\max_{|x|\leq 1}\big(a(x,0)\cdot X_1(x,0)\big)$$

we have – after the perturbation – a single degenerate singular point of index 0 and positive sign in S^z_ξ (in the neighbourhood of γ considered during the elimination process).† If φ is a bump function in the strict sense of the word, this happens for $z = \pm\delta$, say. For $z \in (-\delta, \delta)$, where $\varphi(z) > \varphi(\pm\delta)$, the pair of critical points has been eliminated; for $z \in (-\varepsilon, -\delta) \cup (\delta, \varepsilon)$ we always have the original pair of critical points, but at varying distance along the x–axis.

Observe that in the local model for this 1–parametric elimination process, the x–axis is always a flow line of the characteristic foliation, with at most

† The presence of such a singular point x_0^+ does not affect the convexity property. When we construct the set $(S^{\pm\delta})^0_+$ as in the proof of Proposition 4.8.7, a neighbourhood of x_0^+ and its stable manifold (a flow line whose ω–limit is x_0^+ and whose α–limit is some positive elliptic point) has to be included in that set.

two singular points on it. With the help of Lemma 4.6.23 we can ensure that the flow line in S^0_ξ, emanating at $x^+_{\rm e}$ in the negative x–direction ends at a negative elliptic point (because only finitely many trajectories emanating at $x^+_{\rm e}$ are separatrices of hyperbolic points). Thus, the elimination process also eliminates the retrograde saddle–saddle connection.

We need to check that we did not, by accident, introduce other retrograde saddle–saddle connections. We do this by showing that all stable separatrices of positive hyperbolic points in S^z_ξ – after the perturbation – come from positive elliptic points. For $x^+_{\rm h}$ (in the levels $z \in (-\varepsilon, -\delta) \cup (\delta, \varepsilon)$ where it is not eliminated) this holds because we did not change the direction of its stable separatrices during the elimination process. There might but be a problem with the stable separatrix of $x^+_{\rm h}$ not coming from $x^+_{\rm e}$ if, at some level $z \in (-\varepsilon, \varepsilon)$, it passed through the region where we perform the elimination at some other place besides the segment along the x–axis. The same goes for the stable separatrices of the other positive hyperbolic points. But here we are dealing with finitely many separatrices, and the deformation is only non-trivial inside a compact interval of parameters z. Thus, by performing the elimination inside a sufficiently small neighbourhood of γ, we are home and dry. □

This proof shows that by an isotopy rel boundary we may assume that on each slice S^z we have a situation as described in Lemma 4.9.6, except for finitely many z–levels where we have a single degenerate singular point. The last step in the proof of Theorem 4.9.4 is the following lemma.

Lemma 4.9.8 *Let ξ be a tight contact structure on $S \times [-1, 1]$, where $S = S^2$, with the properties just described. Then we can choose a dividing set Γ_z for S^z_ξ (consisting of a single circle) that varies smoothly with z.*

Proof The dividing set Γ_z can be defined as the boundary of a disc-like neighbourhood of the tree of positive or negative singularities and their stable or unstable manifolds, respectively, i.e. the collection of flow lines whose ω– or α–limit set, respectively, is one of the corresponding singular points. Any two such choices are isotopic by an isotopy via embedded circles transverse to the characteristic foliation S^z_ξ. As we vary z, without passing a level containing a degenerate singular point, the number of singular points in S^z_ξ does not change, but the trees of singularities may well change their geometry.† We can always ensure, however, that such a change occurs only in

† Here is a possible scenario. The α–limit $x^+_{\rm e}$ of the stable separatrix of a hyperbolic point $x^+_{\rm h}$ turns into a double-sun as in Figure 4.26, creating two additional singular points of positive sign, one elliptic and one hyperbolic. One of the unstable separatrices of that new hyperbolic point may then form a saddle–saddle connection with $x^+_{\rm h}$.

one of the two trees on any given level. Similarly, when a degenerate critical point of a certain sign appears, this will happen away from a neighbourhood of the tree of the opposite sign. We may therefore choose a smooth family of Γ_z that always stays close enough to the one of the two trees where no change in the topology of the characteristic foliation takes place. □

4.10 On the classification of tight contact structures

We are now going to use Theorem 4.9.4 for classifying the tight contact structures on S^3, $\mathbb{R}^3$, and $S^2 \times S^1$. The standard contact structure on S^3 and $\mathbb{R}^3$, respectively, has been described previously. By the standard contact structure on $S^2 \times S^1 \subset \mathbb{R}^3 \times S^1$ – with θ denoting the S^1–coordinate and (x, y, z) Cartesian coordinates on $\mathbb{R}^3$ – we mean the contact structure given by

$$z\,d\theta + x\,dy - y\,dx = 0.$$

We shall write ξ_{st} for each of these contact structures if the manifold in question is clear from the context. It will be shown in Corollary 6.5.10 that each of these ξ_{st} is tight. The following theorem says that these are the unique tight contact structures on the respective spaces. It is understood that we fix an orientation on each of these spaces and consider only cooriented positive contact structures.

Theorem 4.10.1 *(a) Each of the manifolds S^3, $\mathbb{R}^3$ and $S^2 \times S^1$ admits a unique tight contact structure up to isotopy.*

(b) Any two tight contact structures on the 3–ball D^3 that induce the same characteristic foliation on the boundary are isotopic rel boundary.

Proof (a – $S^2 \times S^1$) The vector field ∂_θ on $S^2 \times S^1$ is a contact vector field for the standard contact structure ξ_{st}. So the 2–sphere $S^2 \equiv S^2 \times \{0\}$, where we think of S^1 as $\mathbb{R}/2\pi\mathbb{Z}$, is a convex surface with respect to ξ_{st}. Its characteristic foliation is given by the vector field

$$xz\,\partial_x + yz\,\partial_y - (x^2 + y^2)\,\partial_z,$$

with elliptic points at the poles and each flow line a meridian. The dividing set is the equator $\{z = 0\}$. The same is true for any other 2–sphere $S^2 \times \{\theta\}$.

By Propositions 4.6.11 (or rather its proof) and 4.8.7, any given contact structure ξ on this manifold is isotopic to a contact structure ξ' for which S^2 is convex. If ξ was tight, so will be ξ'. Then, by Proposition 4.8.13, the dividing set of the characteristic foliation $S^2_{\xi'}$ is an embedded circle. By an isotopy vertically invariant in a neighbourhood of S^2, we may assume that the dividing set of $S^2_{\xi'}$ is the equator. Then the isotopy in Theorem 4.8.11 translates into an isotopy of ξ' to a contact structure ξ'' for which $S^2_{\xi''}$ coincides with $S^2_{\xi_{st}}$.

The proof of Theorem 2.5.22 yields an isotopy of ξ'' to a contact structure ξ''' that coincides with ξ_{st} on $S^2 \times [-\varepsilon, \varepsilon]$ for some small $\varepsilon > 0$.† On $S^2 \times [\varepsilon/2, 2\pi - \varepsilon/2]$, the contact structures ξ''' and ξ_{st} are isotopic, via an isotopy stationary near the boundary, by Theorem 4.9.4 and Remark 4.9.3. This concludes the proof that the given ξ is isotopic to the standard contact structure on $S^2 \times S^1$.

(b) The classification result for S^3 and $\mathbb{R}^3$ will follow from that of D^3, which we deal with now. Given two contact structures on D^3 that induce the same characteristic foliation on the boundary, they can be made to coincide near the boundary by an isotopy rel boundary, thanks to the proof of Theorem 2.5.23. Likewise, by the proof of Darboux's theorem there is a further isotopy that makes the two contact structures coincide near the origin $\mathbf{0}$, and such that a small standard 2–sphere S centred at $\mathbf{0}$ is convex.

Again by invoking Propositions 4.6.11 and 4.8.7 we find a convex 2–sphere S' (for either contact structure) inside the region near the boundary ∂D^3 where the two contact structures have been made to coincide.

Now the result follows by applying Theorem 4.9.4 to the region A between S and S'. We can ensure that the isotopy provided by that theorem glues smoothly with the identity map on the 3–ball inside S and the spherical shell between S' and ∂D^3 by thickening A slightly and appealing to Remark 4.9.3. (In the arguments below, smoothness is achieved by that same method, so I shall glue isotopies along stationary boundaries without further comment.)

(a – S^3) Fix an 'equator' S^2 that separates S^3 into two 3–balls $D^3_{\pm}$. Given any tight contact structure on S^3, the same arguments as before allow us to construct an isotopy to a contact structure that coincides with the standard contact structure near S^2. Now apply (b) to $D^3_{\pm}$.

(a – $\mathbb{R}^3$) Given any tight contact structure, we can construct an isotopy to a contact structure that coincides with the standard contact structure near the 2–spheres of radius n, where n ranges over $\mathbb{N}$, centred at the origin. Now apply (b) to the ball of radius 1 and Theorem 4.9.4 to the spherical shells between the 2–spheres of radius n and $n+1$, $n \in \mathbb{N}$. □

Remark 4.10.2 These classification results are due to Eliashberg. For S^3, $\mathbb{R}^3$ and D^3 they are proved, using different methods, in [69]. The proof for $S^2 \times S^1$ remained folklore for a long time; the proof given here was suggested in [105].

† Here we need the assumption that we are working with a positive contact structure.

By the discussion in Section 4.2, homotopy classes of tangent 2–plane fields on the 3–sphere are classified by $\pi_3(S^2) \cong \mathbb{Z}$. In order to fix an identification of homotopy classes of 2–plane fields with the integers, we need to specify a trivialisation of the tangent bundle TS^3.

With the trivialisation X_1, X_2, X_3 described on page 135, the tight standard contact structure ξ_{st} corresponds to $0 \in \pi_3(S^2)$, since ξ_{st} is spanned by X_2, X_3; and X_1 is the vector field orthogonal to ξ_{st} with respect to the standard metric. With the usual coorientation of ξ_{st}, given by writing it as

$$\xi_{\mathrm{st}} = \ker(x_0\, dx_1 - x_1\, dx_0 + x_2\, dx_3 - x_3\, dx_2),$$

the plane field ξ_{st} corresponds to the constant map $S^3 \to \{X_1\} \in S^2$.

With this identification understood, we can formulate the complete classification of (positive, cooriented) contact structures on S^3 as follows.

Theorem 4.10.3 *Up to isotopy, the class $0 \in \pi_3(S^2)$ contains two contact structures: the tight structure ξ_{st} and an overtwisted one. Every other class in $\pi_3(S^2)$ contains a unique (overtwisted) contact structure.* □

4.11 Proof of Cerf's theorem

In this section we are going to prove Cerf's theorem $\Gamma_4 = 0$, i.e. the result that every diffeomorphism of S^3 extends to a diffeomorphism of the 4–ball; see Section 1.7.1, where I gave an outline of the idea of the proof.

Lemma 4.11.1 *Any orientation-preserving diffeomorphism f of S^3 is isotopic to a diffeomorphism g preserving the standard contact structure ξ_{st}.*

Proof The contact structure $Tf(\xi_{\mathrm{st}})$ is positive and tight, and thus isotopic to ξ_{st} by Theorem 4.10.3. Let ψ_t, $t \in [0,1]$, be an isotopy with $\psi_0 = \mathrm{id}_{S^3}$ and $T\psi_1(Tf(\xi_{\mathrm{st}})) = \xi_{\mathrm{st}}$. Then $\psi_t \circ f$, $t \in [0,1]$, is the desired isotopy to a contactomorphism $g := \psi_1 \circ f$ of (S^3, ξ_{st}). □

Cerf's theorem is then a consequence of the following statement from Eliashberg's paper [69].

Proposition 4.11.2 *Any contactomorphism g of (S^3, ξ_{st}) extends to a diffeomorphism of D^4.*

The construction of such an extension is based on the technique of filling by holomorphic discs. The details of that technique are beyond the scope of this text, but here are some of the elementary aspects of it. To simplify notation, we shall write ξ instead of ξ_{st} in the sequel.

Regard the discs D^2 and D^4 as the unit discs in $\mathbb{C}$ and $\mathbb{C}^2$, respectively. Write $j\colon T\mathbb{C} \to T\mathbb{C}$ and $J\colon T\mathbb{C}^2 \to T\mathbb{C}^2$ for the complex bundle structures induced by multiplication with $\mathrm{i} = \sqrt{-1}$. A **holomorphic disc** in D^4 is a holomorphic map $\varphi\colon (D^2, S^1) \to (D^4, S^3)$, i.e. a smooth map satisfying the Cauchy–Riemann equation $T\varphi \circ j = J \circ T\varphi$.

Let $h\colon \mathbb{C}^2 \to \mathbb{R}$ be the function defined by $h(z_1, z_2) = |z_1|^2 + |z_2|^2$, so that $S^3 = h^{-1}(1)$. Define the 1–form J^*dh by

$$(J^*dh)(\mathbf{v}) = dh(J\mathbf{v}) \text{ for } \mathbf{v} \in T\mathbb{C}^2.$$

Then the standard contact form α defining ξ can be written as

$$\alpha = -\frac{1}{2} J^*dh.$$

As pointed out in Example 2.1.7, this implies in particular that the contact planes ξ are given by the J–invariant subspaces of $TS^3 \subset T\mathbb{C}^2|_{S^3}$.

Lemma 4.11.3 *(see [176]) Let $\varphi\colon (D^2, S^1) \to (D^4, S^3)$ be a holomorphic disc and $S \subset S^3$ an embedded surface.*

(a) The disc $\varphi(D^2)$ cannot be tangent to S^3 at a non-singular point, i.e.

$$T_p\varphi(T_pD^2) \not\subset T_{\varphi(p)}S^3 \quad \text{for all } p \in D^2 \text{ where } T_p\varphi \neq 0;$$

in particular, $\varphi(p) \notin S^3$ for any non-singular point $p \in \mathrm{Int}\,(D^2)$.

(b) If $\varphi|_{S^1}$ is an embedding of S^1 into the surface S, the circle $\varphi(S^1)$ is transverse to the characteristic foliation S_ξ.

Proof (a) Suppose $p \in D^2$ were a non-singular point of φ where we have $T_p\varphi(T_pD^2) \subset T_{\varphi(p)}S^3$. Since $T\varphi$ is j–J–equivariant, this is equivalent to saying that $T_p\varphi(T_pD^2) \subset \xi_{\varphi(p)}$.

Let $\Delta = \frac{\partial^2}{\partial x^2} + \frac{\partial^2}{\partial y^2}$ be the Laplace operator on $\mathbb{C} = \mathbb{R}^2$ with Cartesian coordinates (x, y). Then, for any smooth function $f\colon \mathbb{C} \to \mathbb{R}$ we have $\Delta f \cdot dx \wedge dy = -d(j^*df)$. We compute

$$\begin{aligned} d(j^*d(h \circ \varphi)) &= d(dh \circ T\varphi \circ j) = d(dh \circ J \circ T\varphi) = d(\varphi^*J^*dh) \\ &= \varphi^*(dJ^*dh). \end{aligned}$$

At the point p this would gives us

$$d(j^*d(h \circ \varphi)) = -2\varphi^*(d\alpha),$$

hence $\Delta_p(h \circ \varphi) > 0$. This implies that the function $h \circ \varphi$ would be strictly subharmonic in a neighbourhood of $p \in D^2$, and attain a maximum in p.

The strong maximum principle for subharmonic functions [89, Thm. 2.13] prevents the case $p \in \mathrm{Int}\,(D^2)$, the boundary point lemma [89, Lemma 2.12] also the case $p \in S^1$.

(b) If $\varphi(S^1)$ failed to be transverse to S_ξ at some point $\varphi(p)$, $p \in S^1$, either because $\varphi(p)$ were a singular point of S_ξ or because $T_p\varphi(T_pS^1)$ were tangent to the characteristic foliation, we would have $T_p\varphi(T_pS^1) \subset \xi_{\varphi(p)}$, and hence $T_p\varphi(T_pD^2) \subset \xi_{\varphi(p)}$ by the j–J–equivariance of $T\varphi$, contradicting (a). □

Proof of Proposition 4.11.2 Choose one of the (real) Cartesian coordinate functions, say y_2, on $\mathbb{C}^2$ and define the 2–sphere $S^t \in S^3 \subset \mathbb{C}^2$ by the equation $y_2 = t$, with t ranging over the open interval $(-1, 1)$. Set $\widetilde{S}^t = g(S^t)$.

By the contact disc theorem (Thm. 2.6.7), which holds equally for any finite number of embedded copies of the standard ball B_{st}, we may assume, after a contact isotopy, that g is the identity near the poles $\{y_2 = \pm 1\} \cap S^3$. Then $\widetilde{S}^t = S^t$ for t near ± 1. This means that the constructions in the sequel are only non-trivial on a compact interval of parameters contained in $(-1, 1)$.

As in Example 2.5.19 we see that the characteristic foliation S^t_ξ has two singular points $q^t_\pm$, both elliptic. Indeed, the contact form

$$x_1\,dy_1 - y_1\,dx_1 + x_2\,dy_2 - y_2\,dx_2$$

describing ξ restricts to the form

$$x_1\,dy_1 - y_1\,dx_1 - t\,dx_2$$

on TS^t, so S^t_ξ is described away from the poles

$$(x_1, y_1, x_2, y_2) = (0, 0, \pm\sqrt{1-t^2}, t) =: q^t_\pm$$

by the non-singular vector field

$$(x_1x_2 + ty_1)\,\partial_{x_1} + (y_1x_2 - tx_1)\,\partial_{y_1} - (x_1^2 + y_1^2)\,\partial_{x_2}.$$

The contactomorphism g maps this characteristic foliation diffeomorphically onto $\widetilde{S}^t_\xi$.

Any of the punctured spheres $S^t \setminus \{q^t_\pm\}$ admits a foliation $\mathfrak{C}_t$ by circles transverse to the characteristic foliation S^t_ξ. The circles in $\mathfrak{C}_t$ are given by the equation $x_2 = s$, with s ranging over the interval $(-\sqrt{1-t^2}, \sqrt{1-t^2})$. These circles bound obvious holomorphic discs

$$D^4 \cap \big(\mathbb{C} \times (x_2 = s, y_2 = t)\big).$$

The method of filling by holomorphic discs ([25], [118, 2.4.D], see also [63], [66]) – here in a relatively simple form, since we are dealing with an honest complex structure J – then allows us to find similar foliations $\widetilde{\mathfrak{C}}_t$ of the punctured spheres $\widetilde{S}^t \setminus \{g(q_\pm^t)\}$ by circles spanning holomorphic discs in the 4–ball D^4. The collection of these discs constitutes a smooth foliation of D^4 (with the $g(q_\pm^t)$ and the two poles of S^3 removed), and by Lemma 4.11.3 this foliation $\widetilde{\mathfrak{C}}_t$ is necessarily transverse to the characteristic foliation $\widetilde{S}^t_\xi$.

The foliation $g(\mathfrak{C}_t)$ of $\widetilde{S}^t \setminus \{g(q_\pm^t)\}$ shares this last property with $\widetilde{\mathfrak{C}}_t$. It follows that g is isotopic, by an isotopy along the leaves of $\widetilde{S}^t_\xi$, to a diffeomorphism (no longer, in general, a contactomorphism) of S^3 that maps $\mathfrak{C}_t$ to $\widetilde{\mathfrak{C}}_t$ for each $t \in (-1,1)$.

The curves $t \mapsto q_\pm^t$ and, correspondingly, $t \mapsto g(q_\pm^t)$ are transverse curves for ξ_{st}. The methods of the proof of Theorem 2.5.15 allow one to control the behaviour of g in a neighbourhood of these curves. This implies that, for each fixed t, we have only a compact interval of parameter values for s, contained in $(-\sqrt{1-t^2}, \sqrt{1-t^2})$, where there is no explicit control over the holomorphic discs.

We have now reduced the proof of the proposition to the following extension problem: given a smooth family $\phi_{s,t}$ of diffeomorphisms of S^1, with s,t ranging over a compact 'square' of parameters, and extensions to diffeomorphisms of D^2 over the boundary of that square, does that family extend to a smooth family of diffeomorphisms of D^2? The answer is yes, for the restriction homomorphism $\rho_2\colon \mathrm{Diff}\,(D^2) \to \mathrm{Diff}\,(S^1)$ is a Serre fibration – this is implicit in the usual proof that $\Gamma_2 = 0$ –, and the fibre of this Serre fibration, i.e. the group of diffeomorphisms of D^2 keeping S^1 pointwise fixed, is contractible by [222]. □

4.12 Prime decomposition of tight contact manifolds

Unless stated otherwise, all 3–manifolds in this section are assumed to be closed, connected and oriented. A 3–manifold is called **non-trivial** if it is not diffeomorphic to S^3. A non-trivial 3–manifold P is said to be **prime** if in every connected sum decomposition $P = P_0 \# P_1$ one of the summands P_0, P_1 is S^3. It is known that every non-trivial 3–manifold M admits a prime decomposition, i.e. M can be written as a connected sum of finitely many prime manifolds. The main step in the proof of this fact is due to Kneser [147], see also [128]. Moreover, as shown by Milnor [184], the summands in this prime decomposition are unique up to order and diffeomorphism.†

The purpose of the present section is to prove the analogous result for tight contact 3–manifolds. The basis for the argument is a connected sum construction for such manifolds, due to Colin [48] and reproved by Honda [138].

† There is also a prime decomposition theorem for non-orientable 3–manifolds, with some mild non-uniqueness, see [128].

Given a fixed connected sum decomposition $M = M_0 \# M_1$ of a 3–manifold M, Colin's result says that tight contact structures ξ_i on M_i, $i = 0, 1$, give rise to a tight contact structure $\xi_0 \# \xi_1$ on M, uniquely defined up to isotopy. Conversely, for any tight contact structure ξ on M there are – up to isotopy – unique tight contact structures ξ_i on M_i, $i = 0, 1$, such that $\xi_0 \# \xi_1$ is the given contact structure ξ. The prime decomposition theorem for tight contact 3–manifolds is an immediate consequence.

Colin's result is not quite strong enough to prove the *unique* decomposition theorem for tight contact 3–manifolds. This is due to the fact that the system of 2–spheres in a given manifold M defining the prime decomposition of M is not, in general, unique up to isotopy. The argument for the unique decomposition of tight contact 3–manifolds given here, which is based on joint work with Fan Ding [54], closely follows the variant of Milnor's argument given in Hempel's book [128].

We begin with a connected sum construction for tight contact 3–manifolds.

Lemma 4.12.1 *Let (M, ξ) be a (not necessarily connected) tight contact 3–manifold. Given embeddings $f_0, f_1 \colon S^2 \rightarrow M$, there is a tight contact structure η on $S^2 \times [0,1]$ such that the characteristic foliation $(S^2 \times \{i\})_\eta$ coincides with $S^2_{f_i^*\xi}$, $i = 0, 1$.*† *This contact structure η is unique up to isotopy rel boundary.*

Proof The uniqueness statement follows from Theorem 4.9.4.

(i) We first prove existence under the additional assumption that the spheres $f_i(S^2)$ are convex with respect to ξ. By Giroux's overtwistedness criterion (Prop. 4.8.13), the dividing set Γ_i of these characteristic foliations $S^2_{f_i^*\xi}$ consists of a single circle each.

Now consider $\mathbb{R}^3$ with its standard tight contact structure ξ_{st} given by $\alpha := dz + x\,dy - y\,dx = 0$. This contact structure admits the contact vector field $Y = x\,\partial_x + y\,\partial_y + 2z\,\partial_z$. Since $\alpha(Y) = 2z$, the spheres in $\mathbb{R}^3$ centred at the origin have the equator $\{z = 0\}$ as dividing set, see Example 4.8.4 (1).

Choose two orientation-preserving diffeomorphisms $S^2 \rightarrow S^2 \subset \mathbb{R}^3$ that map Γ_0 and Γ_1, respectively, to this equator. Using the flow of Y in $\mathbb{R}^3$ and the flexibility of characteristic foliations as expressed in Theorem 4.8.11, these diffeomorphisms (regarded as embeddings of S^2 into $\mathbb{R}^3$) can be deformed to embeddings g_i of $S^2 \equiv S^2 \times \{i\}$ that are transverse to Y and extend to an orientation-preserving embedding $g \colon S^2 \times [0,1] \rightarrow \mathbb{R}^3$ – since the space of orientation-preserving diffeomorphisms of S^2 is connected [222]

† Here $S^2_{f_i^*\xi}$ denotes the characteristic foliation induced by the embedding f_i, that is, the pull-back to S^2 via f_i of the characteristic foliation $(f_i(S^2))_\xi$.

– and induce the characteristic foliations $S^2_{g_i^*\xi_{st}} = S^2_{f_i^*\xi}$. Then $g^*\xi_{st}$ (i.e. $\ker(g^*\alpha)$) is the desired contact structure on $S^2 \times [0,1]$.†

(ii) If the $f_i(S^2)$ are not convex, with the help of Proposition 4.8.8 we first extend the f_i to embeddings

$$\tilde{f}_0\colon S^2 \times [-\varepsilon, \varepsilon] \longrightarrow M \text{ and } \tilde{f}_1\colon S^2 \times [1-\varepsilon, 1+\varepsilon] \longrightarrow M$$

for some small $\varepsilon > 0$ such that $\tilde{f}_0(S^2 \times \{\pm\varepsilon\})$ and $\tilde{f}_1(S^2 \times \{1 \pm \varepsilon\})$ are convex. According to the first part of the proof, and using the uniqueness statement from Theorem 4.9.4, we can find contactomorphic copies A_0, A_1 of $(\tilde{f}_0(S^2 \times [-\varepsilon,\varepsilon]), \xi)$ and $(\tilde{f}_1(S^2 \times [1-\varepsilon, 1+\varepsilon]), \xi)$, respectively, in $(\mathbb{R}^3, \xi_{st})$. The spherical shells A_0 and A_1 may be assumed to be disjoint and nested (i.e. A_0 is contained in the bounded component of $\mathbb{R}^3 \setminus A_1$). By construction, A_i then contains an embedded 2–sphere with the desired characteristic foliation, and such that these two 2–spheres bound a spherical shell $S^2 \times [0,1]$ in $\mathbb{R}^3$. □

We can now define surgery along a 0–sphere inside a given (not necessarily connected) tight contact 3–manifold (M, ξ) as follows; this includes the formation of a connected sum.‡ Notice that an orientation-preserving embedding of $S^0 \times D^3$ into M is the same as a pair of embeddings of D^3 with opposite orientation behaviour.

Equip the 3–disc D^3 with its standard orientation. Let $\phi_0, \phi_1\colon D^3 \to M$ be embeddings such that ϕ_0 reverses and ϕ_1 preserves orientation, and whose images $B_i := \phi_i(D^3) \subset M$ are disjoint. Let η be the contact structure on $S^2 \times [0,1]$, constructed in the preceding lemma, with the property that $(S^2 \times \{i\})_\eta = (\partial D^3)_{\phi_i^*\xi}$. Then set

$$(M', \xi') = (M \setminus \operatorname{Int}(B_0 \cup B_1), \xi) \cup_\partial (S^2 \times [0,1], \eta),$$

where $\cup_\partial$ denotes the obvious gluing along the boundary.

If $M = M_0 + M_1$ is the topological sum of two connected tight contact 3–manifolds (M_0, ξ_0), (M_1, ξ_1), and $B_i \subset M_i$, $i = 0, 1$, then M' is the connected sum $M_0 \# M_1$ of M_0 and M_1, and we write $\xi_0 \# \xi_1$ for the contact structure ξ' in this specific case. We also use the notation $(M_0, \xi_0)\#(M_1, \xi_1)$ for this connected sum of tight contact 3–manifolds. As in the topological case, this connected sum operation is commutative and associative, and (S^3, ξ_{st}) serves as neutral element; these are consequences of the discussion that follows.

† This argument, together with Theorem 4.9.4, shows that a tight contact structure on $S^2 \times [0,1]$ for which the boundary spheres are convex is always isotopic (rel boundary) to one that is invariant under the flow of a vector field transverse to all levels $S^2 \times \{t\}$, $t \in [0,1]$.

‡ General surgeries on arbitrary contact manifolds will be discussed in Chapter 6.

Lemma 4.12.2 *Up to contactomorphism, (M', ξ') only depends on the isotopy class of the embedding $\phi_0 + \phi_1\colon D^3 + D^3 \to M$.*

Proof (i) Let $\widetilde{\phi}_0 + \widetilde{\phi}_1\colon D^3 + D^3 \to M$ be an embedding isotopic to $\phi_0 + \phi_1$. Set $\widetilde{B}_i := \widetilde{\phi}_i(D^3)$. Suppose we are in the special situation that $\widetilde{B}_i \subset B_i$. Let $\widetilde{\eta}$ be the contact structure on $S^2 \times [0,1]$ determined by the requirement that $(S^2 \times \{i\})_{\widetilde{\eta}}$ coincide with $(\partial D^3)_{\widetilde{\phi}_i^*\xi}$. From the uniqueness statement in the preceding lemma it follows that $(S^2 \times [0,1], \eta)$ is contactomorphic rel boundary to

$$\big((B_0 \cup B_1) \setminus \operatorname{Int}(\widetilde{B}_0 \cup \widetilde{B}_1), \xi\big) \cup_\partial (S^2 \times [0,1], \widetilde{\eta}).$$

This proves the lemma in this particular case.

(ii) In the general case we appeal to the transitivity of the action of the contactomorphism group (Corollary 2.6.3). This allows us to find small embedded balls $B_i' \subset B_i$ and a contact isotopy ψ_t, $t \in [0,1]$, of M with $\psi_0 = \mathrm{id}_M$ and $\psi_1(B_i') \subset \widetilde{B}_i$.

The surgered manifolds with respect to the embedded balls $\psi_t(B_i')$ are all contactomorphic: on $S^2 \times [0,1]$ the contact structure stays fixed, and $M \setminus \operatorname{Int}(B_0' \cup B_1')$ is contactomorphic to $M \setminus \operatorname{Int}\big(\psi_t(B_0') \cup \psi_t(B_1')\big)$ via ψ_t. Now apply (i) to the balls $B_i' \subset B_i$ and $\psi_1(B_i') \subset \widetilde{B}_i$. □

Next we want to show that this surgery construction always yields a *tight* contact structure ξ'.

Lemma 4.12.3 *Let (M', ξ') be a contact 3–manifold and $f_t\colon S^2 \to M'$, $t \in [0,1]$, an isotopy of embeddings. If the spheres $S_i := f_i(S^2)$, $i = 0, 1$, are convex with respect to ξ', and $(M' \setminus S_0, \xi')$ is tight, then so is $(M' \setminus S_1, \xi')$.*

Proof (i) First assume that the 2–spheres S_0 and S_1 bound an embedded copy of $S^2 \times [0,1]$ in M'. Since S_0 is convex, any embedded disc in $S^2 \times [0,1] \subset M'$ can be pushed by a contact isotopy into the complement of $S_0 \equiv S \times \{0\}$, on which ξ' is tight by assumption. It follows that $(S^2 \times [0,1], \xi')$ is tight.

According to the footnote in the proof of Lemma 4.12.1, there is a contact flow on $S^2 \times [0,1]$ transverse to all levels $S^2 \times \{t\}$, $t \in [0,1]$. Given an embedded disc Δ in $M' \setminus S_1$, we can find a neighbourhood $S^2 \times [-\varepsilon, 1+\varepsilon]$ of $S^2 \times [0,1]$ in M' with $\Delta \cap S^2 \times [1, 1+\varepsilon] = \emptyset$. The mentioned contact flow may be assumed to exist on $S^2 \times [-\varepsilon, 1+\varepsilon]$, and by cutting off the corresponding Hamiltonian function with a bump function we obtain a global contact flow on M' that coincides with the original one in a neighbourhood of $S^2 \times [0,1]$. With its help we can push Δ into the tight contact manifold $M' \setminus S^2 \times [0,1] \subset M' \setminus S_0$. So Δ cannot have been an overtwisted disc.

(ii) In the general case, we can find for each $t \in [0,1]$ an embedding $g\colon S^2 \times [-1,1]$ with $g|_{S^2\times\{0\}} = f_t$, and with the sphere $f_s(S^2)$ contained in $g(S^2 \times [-1,1])$ for s sufficiently close to t. For any fixed t, we may assume after a C^∞–small perturbation of the original isotopy that the spheres $g(S^2 \times \{-1\})$ and $g(S^2\times\{0\}) = f_t(S^2)$ are convex. Thus, a simple compactness argument furnishes us with a subdivision $0 = t_0 < t_1 < \ldots < t_k = 1$ of the interval $[0,1]$, and embeddings $g_i\colon S^2 \times [-1,1] \to M'$ with the following properties:

- $g_i|_{S^2\times\{0\}} = f_{t_i}$,
- the embedded spheres $g_i(S^2 \times \{-1\})$ and $g_i(S^2 \times \{0\}) = f_{t_i}(S^2)$ are convex,
- $f_t(S^2)$ is contained in $g_i(S^2 \times [-1,1])$ for all $t \in [t_i, t_{i+1}]$.

Since $g_i(S^2 \times \{-1\})$ together with either of $f_{t_i}(S^2)$ and $f_{t_{i+1}}(S^2)$ bounds embedded copies of $S^2 \times [0,1]$ – to see this for $f_{t_{i+1}}$ simply apply the isotopy extension theorem inside $g_i(S^2 \times [0,1]) \cong S^2 \times [0,1]$ to the isotopy f_t, $t \in [t_i, t_{i+1}]$ – the argument reduces to an iterated application of the reasoning from (i). □

Proposition 4.12.4 *The manifold (M', ξ') obtained via surgery along a 0–sphere in a tight contact 3–manifold (M, ξ) is tight as well.*

Proof By Lemma 4.12.2 we may assume that the boundaries of the balls $B_i \subset M$, $i = 0, 1$, used in the formation of (M', ξ') are convex with respect to ξ. By construction, $(S^2 \times [0,1], \xi') \subset (M', \xi')$ is tight. Arguing as in the first part of the foregoing proof, we see that $(M' \backslash S_0, \xi')$, where $S_0 = S^2 \times \{0\}$, is tight.

Given an embedded disc Δ in M', we can find an isotopy of M' that moves S_0 to a convex sphere S_1 disjoint from Δ. By the preceding lemma, $(M' \setminus S_1, \xi')$ is tight, so Δ is not an overtwisted disc. □

We can now formulate the unique decomposition theorem for tight contact 3–manifolds.

Theorem 4.12.5 *Every non-trivial tight contact 3–manifold (M, ξ) is contactomorphic to a connected sum*

$$(M_1, \xi_1) \# \cdots \# (M_k, \xi_k)$$

of finitely many prime tight contact 3–manifolds. The summands are unique up to order and contactomorphism.

The proof of this theorem requires a few preparations. First of all, we observe that there is a well-defined procedure for capping off a compact tight contact 3–manifold with boundary consisting of a collection of 2–spheres. Indeed, suppose that (M, ξ) is a tight contact 3–manifold with boundary $\partial M = S_1 + \cdots + S_k$, where each S_i is diffeomorphic to S^2. Choose orientation-reversing diffeomorphisms $f_i\colon \partial D^3 \to S_i$. By a reasoning as in part (i) of the proof of Lemma 4.12.1, now applied to a single 2–sphere S_i, we find an orientation-preserving embedding $g_i\colon D^3 \to \mathbb{R}^3$ such that $S^2_{g_i^*\xi_{\mathrm{st}}} = S^2_{f_i^*\xi}$. The tight contact structure $\eta_i := g_i^*\xi_{\mathrm{st}}$ on D^3, which by Theorem 4.10.1 is uniquely determined by the characteristic foliation it induces in the boundary, can then be used to form the closed contact manifold

$$(\widehat{M}, \hat{\xi}) = (M, \xi) \cup_\partial \big((D^3, \eta_1) \cup \ldots \cup (D^3, \eta_k)\big),$$

where the gluing is defined by the embeddings f_i.

Theorem 4.10.1 (b) entails that we arrive at a contactomorphic manifold if instead of gluing discs along the S_i we first perturb the boundary spheres into convex spheres S_i' in the interior of (M, ξ), cut off the spherical shell between S_i and S_i', and then glue discs along the S_i'. Now we can use Lemma 4.12.3 as in the proof of Proposition 4.12.4 to conclude that $(\widehat{M}, \hat{\xi})$ is tight.

Given an embedded 2–sphere $S \subset \mathrm{Int}\,(M)$, we can find a product neighbourhood $S \times [-1, 1] \subset M$ of $S \equiv S \times \{0\}$. Set

$$M_S = M \setminus \big(S \times (-1, 1)\big).$$

Again by Theorem 4.10.1 (b), the contactomorphism type of $(\widehat{M_S}, \hat{\xi})$ is independent of the choice of this product neighbourhood; this follows by comparing the resulting manifolds using two given product neighbourhoods with a third manifold constructed from a product neighbourhood contained in the first two. In particular, this justifies our notation $(\widehat{M_S}, \hat{\xi})$.

Lemma 4.12.6 *If S_0 and S_1 are isotopic 2–spheres in* Int (M), *then the contact manifolds $(\widehat{M_{S_0}}, \hat{\xi})$ and $(\widehat{M_{S_1}}, \hat{\xi})$ are contactomorphic.*

Proof This is clear if S_1 is isotopic to S_0 inside a product neighbourhood $S_0 \times (-1, 1)$. The general case follows by an argument very similar to part (ii) of the proof of Lemma 4.12.3 (without any of the convexity issues). □

Given a connected sum decomposition $M = M_0 \# M_1$ of a closed, connected 3–manifold with a tight contact structure ξ, let $S \subset M$ be an embedded sphere defining this connected sum, i.e. $\widehat{M_S} = M_0 + M_1$. Our constructions imply that

$$(M, \xi) = (M_0, \hat{\xi}|_{M_0}) \# (M_1, \hat{\xi}|_{M_1}).$$

So the topological prime decomposition of M also gives us a decomposition of (M, ξ) into prime tight contact 3–manifolds. The only remaining issue is the uniqueness of this decomposition up to contactomorphism of the summands.

A 3–manifold M is said to be **irreducible** if every embedded 2–sphere bounds a 3–disc in M. Clearly, irreducible 3–manifolds (except S^3) are prime. There is but one orientable prime 3–manifold that is not irreducible, namely, $S^2 \times S^1$ [128, Lemma 3.13]. In a connected sum $M = M_0 \# S^2 \times S^1$ we obviously find an embedded non-separating 2–sphere S such that $\widehat{M}_S = M_0$; simply take S to be a fibre of $S^2 \times S^1$ not affected by the connected sum construction.

In the argument proving that the number of summands $S^2 \times S^1$ in a prime decomposition of M is uniquely determined by M, the crucial lemma is that for any two non-separating 2–spheres $S_0, S_1 \subset M$ there is a diffeomorphism of M sending S_0 to S_1. In the presence of a contact structure, this statement needs to be weakened slightly; the following is sufficient for our purposes.

Lemma 4.12.7 *Let (M, ξ) be a tight contact 3–manifold*† *and $S_0, S_1 \subset M$ two non-separating 2–spheres. Then $(\widehat{M}_{S_0}, \hat{\xi})$ and $(\widehat{M}_{S_1}, \hat{\xi})$ are contactomorphic.*

Proof By the preceding lemma we may assume that S_0 and S_1 are in general position with respect to each other, so that $S_0 \cap S_1$ consists of a finite number of embedded circles. We use induction on the number n of components of $S_0 \cap S_1$.

If $n = 0$, we find disjoint product neighbourhoods $S_i \times [-1, 1] \subset M$, $i = 0, 1$. In case $M \setminus (S_0 \cup S_1)$ is not connected, we may assume that the identifications of these neighbourhoods with a product have been chosen in such a way that $S_0 \times \{1\}$ and $S_1 \times \{1\}$ lie in the same component of $M \setminus (S_0 \cup S_1)$. As described above, we then obtain a well-defined tight contact manifold $(\widetilde{M}, \tilde{\xi})$ by capping off the boundary components $S_i \times \{\pm 1\}$ of

$$M \setminus \big(S_0 \times (-1, 1) \cup S_1 \times (-1, 1)\big)$$

with 3–discs $D_0^{\pm}, D_1^{\pm}$. Our assumptions imply that $D_0^- + D_0^+$ is isotopic to $D_1^- + D_1^+$ in $\widetilde{M}$. By performing surgery with respect to these embeddings of $S^0 \times D^3$, we obtain $(\widehat{M}_{S_1}, \hat{\xi})$ and $(\widehat{M}_{S_0}, \hat{\xi})$, respectively, so the result follows from Lemma 4.12.2.

If $n > 0$, then some component J of $S_0 \cap S_1$ bounds a 2–disc $D \subset S_1$ with

† Here connectedness of M is essential. The manifold M is allowed to have boundary, in which case we need to require that S_0 and S_1, as well as the product neighbourhoods used in the proof, are contained in $\mathrm{Int}\,(M)$.

$\operatorname{Int}(D) \cap S_0 = \emptyset$. Let E' and E'' be the 2–discs in S_0 bounded by J, and set $S_0' = D \cup E'$ and $S_0'' = D \cup E''$. At least one of S_0' and S_0'', say S_0', is non-separating.† Move S_0' slightly so that it becomes a smoothly embedded sphere disjoint from S_0 and intersecting S_1 in fewer than n circles. Then two applications of the inductive assumption prove the inductive step. □

Proof of Theorem 4.12.5 As indicated above, it only remains to prove the uniqueness statement. Thus, let

$$(M_1, \xi_1)\# \cdots \#(M_k, \xi_k)$$

and

$$(M_1^*, \xi_1^*)\# \cdots \#(M_l^*, \xi_l^*)$$

be two prime decompositions of a given tight contact 3–manifold (M, ξ). Without loss of generality we assume‡ $k \leq l$ and use induction on k. For $k = 1$ there is nothing to prove. Now assume $k > 1$ and the assumption to be proved for prime decompositions with fewer than k summands.

(i) Suppose some M_i (say M_k) is diffeomorphic to $S^2 \times S^1$. Then M contains a non-separating 2–sphere. By applying the argument in the footnote to the preceding proof to this non-separating 2–sphere and the 2–spheres defining the splitting of M into the connected sum of the M_j^*, one finds a non-separating 2–sphere in at least one of these summands, say M_l^*, which therefore must be a copy of $S^2 \times S^1$. By Theorem 4.10.1 (for $S^2 \times S^1$), (M_k, ξ_k) is contactomorphic to (M_l^*, ξ_l^*). Let S_0, S_1 be a fibre in M_k, M_l^*, respectively. From Theorem 4.10.1 (b) it follows that

$$(\widehat{M}_{S_0}, \hat{\xi}) = (M_1, \xi_1)\# \cdots \#(M_{k-1}, \xi_{k-1})$$

and

$$(\widehat{M}_{S_1}, \hat{\xi}) = (M_1^*, \xi_1^*)\# \cdots \#(M_{l-1}^*, \xi_{l-1}^*),$$

and by the preceding lemma these two manifolds are contactomorphic. So the conclusion of the theorem follows from the inductive assumption.

(ii) It remains to deal with the case where all the M_i are irreducible. Arguing as before (with the roles of the two connected sum decompositions reversed), we see that each M_j^* must be irreducible. Choose a separating

† Since S_0 is non-separating, there is a loop γ in M (in general position with respect to all spheres in question) that intersects S_0 in a single point, say one contained in the interior of E'. If S_0'' is separating, then γ intersects it in an even number of points. Since γ does not intersect E'', these points all lie in D. So γ intersects S_0' in an odd number of points, which means that S_0' is non-separating.

‡ Of course, from the topological prime decomposition theorem, one already knows that $k = l$, but this does not help to simplify the present proof.

2–sphere $S \subset M$ such that the closures U, V of the components of $M \setminus S$ satisfy

$$(\widehat{U}, \widehat{\xi|_U}) = (M_1, \xi_1)\# \cdots \#(M_{k-1}, \xi_{k-1})$$

and $(\widehat{V}, \widehat{\xi|_V}) = (M_k, \xi_k)$.†

Similarly, there exist pairwise disjoint 2–spheres $T_1, \ldots, T_{l-1}$ in M such that – with $W_1, \ldots, W_l$ denoting the closures of the components of the complement $M \setminus (T_1 \cup \ldots \cup T_{l-1})$, and ξ_j the restriction of ξ to W_j – we have $(\widehat{W}_j, \hat{\xi}_j) = (M_j^*, \xi_j^*)$, $j = 1, \ldots, l$.

Suppose that the system $T_1, \ldots, T_{l-1}$ of embedded spheres has been chosen in general position with respect to S and with $S \cap (T_1 \cup \ldots \cup T_{l-1})$ having the minimal number of components among all such systems.

Here we have to enter a caveat. The notation suggests that W_1 has boundary T_1, the W_j with $j \in \{2, \ldots, l-1\}$ have boundary $T_{j-1} \sqcup T_j$, and W_l has boundary T_{l-1}. In fact, some of the reasoning in the proof given in [128] appears to rely on such an assumption. However, under the minimality condition we have just described, it is perfectly feasible that some of the W_j have several boundary components (i.e. the connected sum looks like a tree rather than a chain). In particular, the numbering of the W_j is not meant to suggest any kind of order in which they are glued together.

We claim that the minimality assumption implies

$$S \cap (T_1 \cup \ldots \cup T_{l-1}) = \emptyset.$$

Assuming this claim, we have $S \subset W_j$ for some $j \in \{1, \ldots, l\}$. Since $\widehat{W}_j = M_j^*$ is irreducible, S bounds a 3–cell B in M_j^*. Thus, S cuts W_j into two pieces X and Y, where say $\widehat{Y} = S^3$. By the uniqueness of the tight contact structure on S^3 we have in fact $(\widehat{Y}, \widehat{\xi|_Y}) = (S^3, \xi_{\mathrm{st}})$. Moreover, $(\widehat{X}, \widehat{\xi|_X}) = (M_j^*, \xi_j^*)$ by Theorem 4.10.1 (b).

Of the 3–discs in M_j^* used for forming the connected sum with one or several of the other prime manifolds, at least one has to be contained in B, otherwise S would bound a disc in M. This means that of the closures U, V of the two components of $M \setminus S$, the one containing Y must contain at least one of $W_1, \ldots, W_{j-1}, W_{j+1}, \ldots, W_l$. Thus, in the case $Y \subset V$, the numbering (including that of W_j) can be chosen in such a way that $W_1, \ldots, W_{j-1}, X \subset U$ and $Y, W_{j+1}, \ldots, W_l \subset V$, with $j \leq l-1$. (The case with $X \subset V$ and $Y \subset U$ is analogous; here $j \geq 2$.) With Theorem 4.10.1, and in particular the fact that (S^3, ξ_{st}) is the neutral element for the connected

† The contact structure $\widehat{\xi|_U}$ is the same as the restriction of the contact structure $\hat{\xi}$ (defined on $\widehat{M}_S = \widehat{U} + \widehat{V}$) to $\widehat{U}$.

sum operation, we conclude that

$$\begin{aligned}(M_1,\xi_1)\#\cdots\#(M_{k-1},\xi_{k-1}) &= (\widehat{U},\widehat{\xi|_U})\\ &= (\widehat{W}_1,\hat{\xi}_1)\#\cdots\#(\widehat{W}_{j-1},\hat{\xi}_{j-1})\#(\widehat{X},\widehat{\xi|X})\\ &= (M_1^*,\xi_1^*)\#\cdots\#(M_j^*,\xi_j^*)\end{aligned}$$

and

$$\begin{aligned}(M_k,\xi_k) &= (\widehat{V},\widehat{\xi|_V})\\ &= (\widehat{Y},\widehat{\xi|_Y})\#(\widehat{W}_{j+1},\hat{\xi}_{j+1})\#\cdots\#(\widehat{W}_l,\hat{\xi}_l)\\ &= (M_{j+1}^*,\xi_{j+1}^*)\#\cdots\#(M_l^*,\xi_l^*).\end{aligned}$$

Since M_k is prime, we must have $j=l-1$, hence $(M_k,\xi_k)=(M_l^*,\xi_l^*)$. Once again, the theorem follows from the inductive assumption.

It remains to prove the claim. Arguing by contradiction, we assume that $S\cap(T_1\cup\ldots\cup T_{l-1})\neq\emptyset$. Then we find a 2–disc $D\subset S$ with $\partial D\subset T_i$ for some $i\in\{1,\ldots,l-1\}$, and $\text{Int}\,(D)\cap(T_1\cup\ldots\cup T_{l-1})=\emptyset$. This disc is contained in W_j for some $j\in\{1,\ldots,l\}$. For ease of notation we assume that $i=j=1$, and that W_2 is the other component adjacent to T_1.

Let E',E'' be the 2–discs in T_1 bounded by ∂D. Since $\widehat{W}_1$ is irreducible, the sets $D\cup E'$ and $D\cup E''$ (which are homeomorphic copies of S^2) bound 3–cells B',B'' in $\widehat{W}_1$. One of these must contain the other, otherwise it would follow that $\widehat{W}_1$ can be obtained by capping off the 3–cell $B'\cup_D B''$, and thus would be a 3–sphere.

So suppose that $B''\subset B'$. Then $D\cup E'$ can be deformed into a smooth 2–sphere T_1' that meets S in fewer components then T_1, see Figure 4.45. In the complement $M\setminus(T_1'\cup T_2\cup\ldots\cup T_{l-1})$ we still find $W_3,\ldots,W_l$, but W_1,W_2 have been changed to new components W_1', W_2'. Write ξ_1',ξ_2', respectively, for the restriction of ξ to these components. We are done if we can show that

$$(\widehat{W}_i',\hat{\xi}_i')=(\widehat{W}_i,\hat{\xi}_i),\quad i=1,2,$$

because then the new system of spheres $T_1',T_2,\ldots,T_{l-1}$ contradicts the minimality assumption on $T_1,T_2,\ldots,T_{l-1}$.

The 2–sphere T_1' is isotopic to T_1 in $\widehat{W}_1$: simply move $D\subset T_1'$ across the ball B'' to E''. But beware that T_1' need not be isotopic to T_1 in W_1 or M. However, B'' lies on the same side of T_1 as W_1, so T_1' is isotopic to T_1 in

$$\widehat{W_1\cup W_2}=\widehat{W_1'\cup W_2'}.$$

Cutting this latter manifold open along T_1 and then capping off with discs gives the disjoint union of $(\widehat{W}_1,\hat{\xi}_1)$ and $(\widehat{W}_2,\hat{\xi}_2)$; alternatively, cutting it open along T_1' and capping off yields the disjoint union of $(\widehat{W}_1',\hat{\xi}_1')$ and $(\widehat{W}_2',\hat{\xi}_2')$.

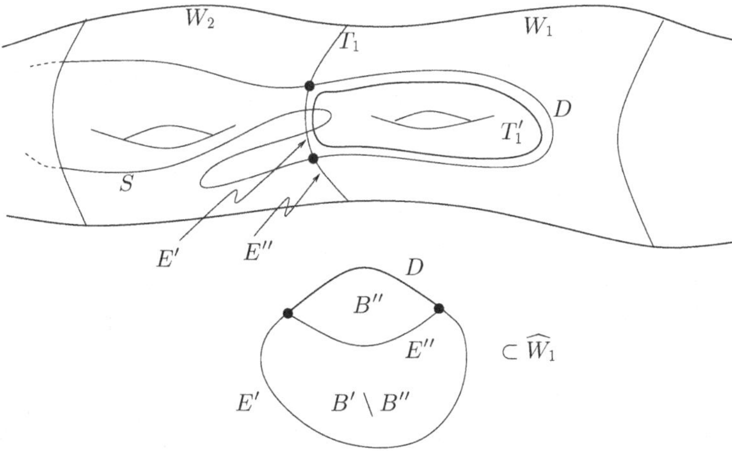

Fig. 4.45. Modification of the prime decomposition.

From Lemma 4.12.6 it follows that the results of either procedure are contactomorphic.

This was the last point we had to show in order to conclude the proof of the unique decomposition theorem. □

Remark 4.12.8 There is no unique decomposition theorem for overtwisted contact 3–manifolds. For instance, start with a connected sum of two distinct prime tight contact 3–manifolds. Now perform a full Lutz twist in one or the other summand, preserving the topology of the manifold and the homotopy class of the contact structure as a 2–plane field (Lemma 4.5.3). By Eliashberg's classification of overtwisted contact structures (Theorem 4.7.2), the resulting manifolds are contactomorphic, and we obviously have two distinct connected sum decompositions.

5

Symplectic fillings and convexity

> 'My wife was greatly impressed by the number of your "fillings".'
>
> Wyndham Lewis,
> *Letter dated 5 Dec. 1943*†

In Definition 1.7.14 we introduced the notion of weak and strong symplectic filling of a contact 3–manifold. In the first section of the present chapter we extend these definitions to contact manifolds of arbitrary dimension, and we discuss other notions of fillings and their interrelations.

If a contact manifold (M, ξ) admits a strong symplectic filling (W, ω), one also says that (M, ξ) is the *ω–convex* boundary of (W, ω). With the appropriate changes of signs in the definition, one arrives at the notion of *ω–concave* boundary. A symplectic manifold with a concave and a convex boundary component can then be interpreted as a *symplectic cobordism*. In particular, a strong filling of (M, ξ) may be interpreted as a symplectic cobordism from the empty set to (M, ξ). In Section 5.2 we discuss the gluing of symplectic cobordisms, which proves the transitivity of the cobordism relation.

With this discussion we prepare the ground for the following chapter. There we show how to perform surgery on a given contact manifold by regarding it as the convex boundary component of a symplectic manifold and then attaching symplectic handles to that boundary.

A word of warning is in order. In the literature on symplectic fillings, there are many subtle variations in terminology. One of the first systematic discussions of convexity in symplectic geometry can be found in a paper by Eliashberg and Gromov [76]. My only deviation from the terminology used there is that I call 'ω–convex' what they refer to as 'locally ω–convex'.

In Section 5.3 we show how one can define a notion of convexity in complex

† From the entry for 'filling' in the *Oxford English Dictionary.*

analysis in analogy with geometric convexity (in the naive sense), and in Section 5.4 we relate this so-called Levi pseudoconvexity to ω–convexity.

5.1 Weak versus strong fillings

The following definitions of weak and strong symplectic fillings of contact manifolds are the obvious modifications of those in dimension 3. In this general setting, it is useful to consider an intermediate notion of *domination*, see [79], [176].

Definition 5.1.1 Let M be a closed manifold of dimension $2n-1$ with cooriented contact structure $\xi = \ker\alpha$. Notice that the (positive) conformal class of the symplectic bundle structure $d\alpha|_\xi$ is independent of the choice of contact form defining ξ with given coorientation. We orient ξ by $(d\alpha)^{n-1}$.

(a) A compact symplectic manifold (W,ω) of dimension $2n$ is called a **weak (symplectic) filling** of (M,ξ) if $\partial W = M$ as oriented manifolds† and $\omega^{n-1}|_\xi > 0$.

(b) We say that a compact symplectic manifold (W,ω) **dominates** (M,ξ) if $\partial W = M$ as oriented manifolds and $\omega|_\xi$ is in the conformal class of $d\alpha|_\xi$.

(c) A compact symplectic manifold (W,ω) is called a **strong (symplectic) filling** of (M,ξ) if $\partial W = M$ and there is a Liouville vector field Y defined near ∂W, pointing outwards along ∂W, and satisfying $\xi = \ker(i_Y\omega|_{TM})$ (as cooriented contact structure). In this case we say that (M,ξ) is the **convex** (or more precisely: **ω–convex**) boundary of (W,ω).

Remark 5.1.2 In the terminology of Definition 1.4.5, the boundary of a strong filling is of *contact type.*

Example 5.1.3 By Examples 1.4.8 and 2.1.7, the disc D^{2n} with its standard symplectic form ω_{st} is a strong filling of the sphere S^{2n-1} with its standard contact structure ξ_{st}.

It turns out that two of these three notions coincide, but which two depends on the dimension. Before we formulate this theorem, we record some observations concerning strong fillings. Part (b) of the following lemma will be used in the proof of part (c), but it is also of more general interest. For instance, it will be instrumental in proving that one can attach symplectic handles to weak fillings of 3–dimensional contact manifolds, see Lemma 6.5.2.

† The orientations of M and W are given by the volume forms $\alpha\wedge(d\alpha)^{n-1}$ and ω^n, respectively.

Lemma 5.1.4 *(a) If (W,ω) is a strong filling of (M,ξ_0) and (M,ξ_1), then the contact structures ξ_0 and ξ_1 are isotopic.*

(b) (see also Exercise 3.36 in [177]) Let ω_0 and ω_1 be symplectic forms on a manifold W (inducing the same orientation) that coincide on the tangent bundle TM, where M is the compact boundary of W or, more generally, a compact, oriented hypersurface. Then there is a symplectomorphism† *$\psi\colon (\mathcal{N}(M),\omega_0) \to (\mathcal{N}'(M),\omega_1)$ of suitable neighbourhoods, fixing M pointwise.*

(c) (see also [240, Lemma 2]) Let (W,ω) be a compact symplectic manifold with boundary M. Write $j\colon M \to W$ for the inclusion of M in W. If $j^*\omega = d\alpha$ for some contact form α inducing the boundary orientation, then (W,ω) is a strong filling of $(M,\ker\alpha)$. (The converse of this statement has already been observed in Lemma/Definition 1.4.5.)

Proof (a) Let Y_0 and Y_1 be the Liouville vector fields near ∂W corresponding to the two contact structures ξ_0 and ξ_1, respectively. Then the vector field $Y_t := (1-t)Y_0 + tY_1$ is also a Liouville vector field for ω for all $t \in [0,1]$. Hence $i_{Y_t}\omega|_{TM}$ defines a homotopy of contact forms, and the result follows from Gray stability (Thm. 2.2.2).

(b) Observe that a skew-symmetric bilinear form ω on the vector space $\mathbb{R}\oplus\mathbb{R}^{2n-1}$ is symplectic if and only if

(i) $\omega|_{\mathbb{R}^{2n-1}}$ has rank $2n-2$, and thus a 1–dimensional kernel $\langle X\rangle \subset \mathbb{R}^{2n-1}$, and

(ii) for $0 \neq X' \in \mathbb{R}$ we have $\omega(X',X) \neq 0$.

Notice that $\ker(\omega|_{\mathbb{R}^{2n-1}})$ can be oriented canonically. In our given situation, let X be a non-zero section of the trivial line bundle $\ker(\omega_0|_{TM}) = \ker(\omega_1|_{TM})$, and X' a vector field on W along M, complementary to TM in $TW|_M$. Set $\omega_t = (1-t)\omega_0 + t\omega_1$, $t \in [0,1]$. Then $\omega_t|_{TM} = \omega_0|_{TM}$ for all $t \in [0,1]$. Our orientation assumptions imply that $\omega_0(X',X)$ and $\omega_1(X',X)$ have the same sign, hence $\omega_t(X',X) \neq 0$ for all $t \in [0,1]$. The linear algebraic criterion just mentioned then implies that ω_t is symplectic along M for all t. By the compactness of M, we can find a neighbourhood of M in W on which ω_t is symplectic for all $t \in [0,1]$.

Since $(\omega_1-\omega_0)|_{TM} = 0$, the generalised Poincaré Lemma (Cor. A.4) allows us to find a 1–form ζ on W with $\zeta|_{TW|_M} = 0$ and $d\zeta = \omega_1 - \omega_0$ in a neighbourhood of M.

Now we apply the Moser trick to the equation $\psi_t^*\omega_t = \omega_0$, where ψ_t is supposed to be the flow of some time-dependent vector field X_t. Differentiating

† A **symplectomorphism** is a diffeomorphism $\psi\colon (W,\omega) \to (W',\omega')$ between symplectic manifolds with $\psi^*\omega' = \omega$.

this equation gives

$$\psi_t^*\left(\dot{\omega}_t + \mathcal{L}_{X_t}\omega_t\right) = 0.$$

This translates into

$$d\zeta + d(i_{X_t}\omega_t) = 0.$$

Since ω_t is symplectic (in a neighbourhood of M), we may define the vector field X_t by

$$\zeta + i_{X_t}\omega_t = 0.$$

Then $X_t \equiv \mathbf{0}$ on M, so the flow ψ_t will be defined up to time 1 in a neighbourhood of M. The map $\psi := \psi_1$ is the desired symplectomorphism.

(c) Extend α to a 1–form α' on W. Then

$$j^*(d\alpha') = d\alpha = j^*\omega.$$

The generalised Poincaré Lemma, applied to $\omega - d\alpha'$, gives us a 1–form ζ in a neighbourhood of M in W with $d\zeta = \omega - d\alpha'$. In fact, Corollary A.4 tells us that we may even impose the condition that ζ vanish on M, but this is not needed here. However, if we do impose that condition, it is less abusive to write α again for the 1–form $\alpha' + \zeta$, with α now defined in a neighbourhood of M in W.

Define a vector field Y in that neighbourhood by $i_Y\omega = \alpha$. Then Y is a Liouville vector field for ω, and the calculation in the proof of Lemma/Definition 1.4.5, together with our orientation assumptions, shows that Y is pointing outwards along M. □

Theorem 5.1.5 *The following implications hold between the above definitions:*

$$\begin{array}{lllllll} \text{(a)} & \Longleftrightarrow & \text{(b)} & \Longleftarrow & \text{(c)} & \text{for } n = 2, \\ \text{(a)} & \Longleftarrow & \text{(b)} & \Longleftrightarrow & \text{(c)} & \text{for } n > 2. \end{array}$$

The implications not shown are not valid, in general. More precisely, in all dimensions (odd and at least 3*) there are contact manifolds with a weak filling that is not a strong filling; in dimension* 3 *there are even weakly fillable contact manifolds that do not admit any strong fillings.*

Proof The implications going to the left are straightforward. Given (c), we may take $\alpha := i_Y\omega|_{TM}$ as contact form defining ξ. Then, by the Liouville condition, $\omega|_\xi = d\alpha|_\xi$, so ω dominates ξ. That (b) implies (a) is equally straightforward.

The implication (a) ⇒ (b) for $n = 2$ is just as simple: if $\omega|_\xi > 0$, the 'area forms' $\omega|_\xi$ and $d\alpha|_\xi$ differ only by a conformal factor $f\colon M \to \mathbb{R}^+$.

The only implication left to be shown is (b) $\Rightarrow$ (c) for $n > 2$. This implication was first noticed in [176]. Thus, let (W, ω) be a compact symplectic manifold that dominates (M, ξ). Choose a defining 1–form α for ξ on M such that $\omega|_\xi = d\alpha|_\xi$. Let R be the Reeb vector field of α.

The ω–orthogonal complement $\xi^\perp$ of ξ in $TW|M$ is a symplectic vector bundle of rank 2. The vector field R can be projected along ξ to a non-vanishing vector field in $\xi^\perp$. This shows that one can choose a vector field $Y \in \xi^\perp$ with $\omega(Y, R) = 1$ and Y pointing outwards. Extend Y to a non-vanishing vector field on W near $\partial W = M$, and use the flow of $-Y$ to define a collar neighbourhood of M in W.

With the help of that flow we may regard α as a 1–form on this collar neighbourhood with $\alpha(Y) \equiv 0$ and $\mathcal{L}_Y \alpha \equiv 0$, hence also $i_Y d\alpha \equiv 0$ by Cartan's formula. Likewise we regard R as a vector field on this collar with $\mathcal{L}_Y R \equiv \mathbf{0}$.

With $\beta := i_R\omega$ it then follows that

$$\omega = d\alpha + \alpha \wedge \beta; \tag{5.1}$$

one simply has to verify this identity on pairs of vectors $(R_p, \mathbf{v})$, (R_p, Y_p) and $(\mathbf{v}, Y_p)$, with $\mathbf{v} \in \ker \alpha_p$.

We now find that

$$d\alpha \wedge \beta - \alpha \wedge d\beta = d\omega = 0. \tag{5.2}$$

Since both $i_R d\alpha$ and $\beta(R)$ vanish, we get

$$0 \equiv i_R(\alpha \wedge d\beta) = d\beta - \alpha \wedge i_R d\beta.$$

Wedging this equation with α we get $\alpha \wedge d\beta \equiv 0$. But then also $d\alpha \wedge \beta \equiv 0$ by (5.2).

We now want to deduce from this last identity that $\beta|_{TM} = 0$. Here the assumption $n > 2$ is used; in the case $n = 2$ the identity $(d\alpha \wedge \beta)|_\xi \equiv 0$ is satisfied for dimensional reasons and does not yield any non-trivial information. Given any point $p \in M$ and any tangent vector $\mathbf{v} \in T_pM$ at that point, we may choose (for $n > 2$) vectors $\mathbf{v}_1, \mathbf{v}_2 \in \xi_p$ that are $d\alpha$–orthogonal to $\mathbf{v}$ and satisfy $d\alpha(\mathbf{v}_1, \mathbf{v}_2) = 1$. Thus

$$0 = (d\alpha \wedge \beta)(\mathbf{v}_1, \mathbf{v}_2, \mathbf{v}) = d\alpha(\mathbf{v}_1, \mathbf{v}_2)\beta(\mathbf{v}) = \beta(\mathbf{v}),$$

hence $\beta|_{TM} = 0$. With Equation (5.1) we deduce $\omega|_{TM} = d\alpha$. Now apply Lemma 5.1.4 (c).

It remains to discuss the non-implications. For the 3–dimensional case see [71] and [50]. Higher-dimensional examples of weak fillings that are not strong were constructed in [76, 3.1.3] with the help of contact surgery. $\square$

The relation between symplectic convexity and the usual geometric convexity is anything but straightforward, as the following example shows.

Example 5.1.6 The unit circle S^1 in $\mathbb{R}^2$ with its standard symplectic structure $\omega_{\mathrm{st}} = r\,dr \wedge d\varphi$ (in polar coordinates) is simultaneously ω_{st}–convex and ω_{st}–concave. In other words, there are Liouville vector fields $X_\pm$ for ω_{st} defined near and transverse to S^1, with X_+ pointing out of the unit disc D^2 and X_- pointing into the unit disc. Such vector fields are given by

$$X_+ = \frac{1}{2} r\,\partial_r \quad\text{and}\quad X_- = -\frac{4-r^2}{2r}\,\partial_r .$$

The deeper reason behind this example is the fact that the punctured disc $\mathrm{Int}\,(D^2_R) \setminus \{\mathbf{0}\}$ of radius R can be turned 'inside out' symplectically by the map

$$(r,\varphi) \longmapsto (\sqrt{R^2-r^2}, -\varphi);$$

above I took $R=2$.

However, symplectic convexity is related to (strict) pseudoconvexity of (real) hypersurfaces in complex spaces, and this notion of convexity is in turn modelled on the definition of geometric convexity. I shall elaborate on this point in Section 5.3.

5.2 Symplectic cobordisms

There is a natural correspondence between surgery and cobordisms, which we are going to recall in detail in Section 6.1. When one wants to perform surgery on contact manifolds, this gives rise to a corresponding notion of symplectic cobordisms. With the following discussion of such cobordisms we thus pave the way to the next chapter.

Definition 5.2.1 *Let $(M_\pm, \xi_\pm)$ be closed contact manifolds of dimension $2n-1$ with cooriented contact structures, which induce the orientation of the respective manifold. A* **symplectic cobordism** *from (M_-, ξ_-) to (M_+, ξ_+) is a compact $2n$–dimensional symplectic manifold (W, ω), oriented by the volume form ω^n, such that the following conditions hold.*

- *The oriented boundary of W equals $\partial W = M_+ \sqcup \overline{M}_-$, where $\overline{M}_-$ stands for M_- with reversed orientation.*
- *In a neighbourhood of ∂W, there is a Liouville vector field Y for ω, transverse to the boundary and pointing outwards along M_+, inwards along M_-.*
- *The 1–form $\alpha := i_Y\omega$ restricts to $TM_\pm$ as a contact form for $\xi_\pm$.*

We call M_+ the **convex** *boundary of the cobordism W and M_- the* **concave** *boundary.*

Example 5.2.2 Every closed contact manifold $(M, \xi = \ker \alpha)$ is symplectically cobordant to itself via the trivial symplectic cobordism

$$\big([0,1] \times M,\ d(\mathrm{e}^t \alpha)\big),$$

with $Y = \partial_t$.

Remark 5.2.3 A strong symplectic filling of a contact manifold (M, ξ) is the same as a symplectic cobordism from the empty set to (M, ξ). If one wants to be more specific, this should be called a *convex* filling; later we shall also consider concave fillings (i.e. symplectic cobordisms from a given contact manifold to the empty set).

One could similarly define a notion of *weak* symplectic cobordisms in analogy with weak fillings. Although we shall have reason to mention such cobordisms briefly during the course of the proof of Theorem 1.7.15 in the next chapter, such cobordisms are less useful. The reason is that such cobordisms cannot be glued together, in general. For (strong) symplectic cobordisms, such a gluing construction exists, as we shall discuss now.

Let M be a hypersurface in the symplectic manifold (W, ω) transverse to a Liouville vector field Y that is defined and non-vanishing in a neighbourhood of M. Write j for the inclusion of M in W, and set $\alpha = j^*(i_Y \omega)$. In the following lemma we are going to consider two such set-ups, indexed by 0 and 1.

Lemma 5.2.4 *Given a strict contactomorphism $\phi\colon (M_0, \alpha_0) \to (M_1, \alpha_1)$, extend it to a diffeomorphism $\widetilde{\phi}$ of a cylindrical neighbourhood of M_0 in W_0 to a corresponding neighbourhood of M_1 in W_1 by sending the flow lines of Y_0 to those of Y_1. Then $\widetilde{\phi}$ is a symplectomorphism, i.e. $\widetilde{\phi}^* \omega_1 = \omega_0$.*

Proof Observe that $\widetilde{\phi}^* \omega_1$ is a symplectic form in a cylindrical neighbourhood of M_0 in W_0 for which Y_0 is a Liouville vector field. Moreover, we compute

$$j_0^*(i_{Y_0} \widetilde{\phi}^* \omega_1) = j_0^* \widetilde{\phi}^* (i_{Y_1} \omega_1) = \phi^* j_1^* (i_{Y_1} \omega_1) = \phi^* \alpha_1 = \alpha_0.$$

This implies that we may identify all corresponding objects (which we now write again without index), except for the two symplectic forms ω_0 and ω_1 (the latter replacing $\widetilde{\phi}^* \omega_1$). In order to prove $\omega_0 = \omega_1$, it suffices to show that $\omega := \omega_0$ is completely determined by α and Y.

Set $\widetilde{\alpha} = i_Y \omega$. Since $\widetilde{\alpha}(Y) = 0$ and, by the Cartan formula,

$$\mathcal{L}_Y \widetilde{\alpha} = i_Y d(i_Y \omega) = i_Y \mathcal{L}_Y \omega = i_Y \omega = \widetilde{\alpha},$$

the 1–form $\widetilde{\alpha}$ on the cylindrical neighbourhood of M in W is completely determined by its restriction α to TM. With t denoting the parameter of the flow lines of Y, passing through M at $t = 0$, we have in fact $\widetilde{\alpha} = e^t \alpha$. (Strictly speaking, instead of α we have to write its pull-back under the projection of the cylindrical neighbourhood to M along the flow lines.)

It follows that $\omega = d\widetilde{\alpha}$ is also determined by α and Y. □

We can now formulate the result about the gluing of symplectic cobordisms. This entails that the relation of being symplectically cobordant is *transitive*, see also [5].

Proposition 5.2.5 *Let (W_-, ω_-) be a symplectic cobordism from the contact manifold (M_-, ξ_-) to (M, ξ), and (W_+, ω_+) a symplectic cobordisms from (M, ξ) to (M_+, ξ_+). Then there is a symplectic cobordism from (M_-, ξ_-) to (M_+, ξ_+), which is topologically given by gluing W_- and W_+ along M.*

Proof Write $j_\pm \colon M \to W_\pm$ for the inclusion maps and $Y_\pm$ for the Liouville vector fields on $(W_\pm, \omega_\pm)$. Set $\alpha_\pm := j_\pm^*(i_{Y_\pm} \omega_\pm)$. Since both α_- and α_+ are contact forms for the cooriented contact structure ξ, we find a function $f \colon M \to \mathbb{R}$ such that $\alpha_+ = e^f \alpha_-$. After multiplying ω_+ (and hence α_+) by a large positive constant (while keeping Y_+ as it is), we may assume that f takes values in $\mathbb{R}^+$.

By the proof of Lemma 5.2.4, there is an $\varepsilon > 0$ and a collar neighbourhood of M in (W_-, ω_-) of the form

$$\big((-\varepsilon, 0] \times M,\ \omega_- = d(e^t \alpha_-)\big),$$

and a collar neighbourhood of M in (W_+, ω_+) of the form

$$\big([0, \varepsilon) \times M,\ \omega_+ = d(e^t \alpha_+) = d(e^{t+f} \alpha_-)\big).$$

Set

$$W_0 := \{(t, x) \in \mathbb{R} \times M \colon 0 \leq t \leq f(x)\},$$

with symplectic form $d(e^t \alpha_-)$. Then $W_- \cup_M W_0 \cup_M W_+$ is a symplectic cobordism from (M_-, ξ_-) to (M_+, ξ_+), where the identification along the boundaries is given, in terms of the collar description, by

$$W_- \ni (0, x) \sim (0, x) \in W_0$$

and

$$W_0 \ni (f(x), x) \sim (0, x) \in W_+.$$

This concludes the proof. □

Remark 5.2.6 Example 5.2.2 shows that the notion of being symplectically cobordant is also *reflexive*. In contrast with topological cobordisms, however, the notion is *not symmetric*. For instance, as we shall see in Section 6.5, there is a symplectic cobordism from any contact 3–manifold to the empty set, but a symplectic cobordism from the empty set to the given contact 3–manifold imposes strong conditions on that contact manifold, e.g. tightness of the contact structure.

5.3 Convexity and Levi pseudoconvexity

In this section I give a very brief introduction to Levi pseudoconvexity, one of several important notions of convexity studied in complex analysis. For more details and other notions of convexity in complex analysis see the well-written account in Chapter II of the textbook by Range [211]. Another standard reference is Chapter IX in the classic monograph [120] by Gunning and Rossi. The definition of Levi pseudoconvexity is modelled on the definition of geometric convexity, so we begin with the latter.

A region $G \subset \mathbb{R}^m$ (i.e. an open connected subset) has **differentiable boundary** ∂G if there is a smooth function $\rho\colon \mathbb{R}^m \to \mathbb{R}$ such that

$$G = \{x \in \mathbb{R}^m :\ \rho(x) < 0\}$$

and

$$d\rho_x \neq 0 \ \text{ in a neighbourhood of } \ \partial G,$$

so that ∂G is a submanifold of $\mathbb{R}^m$ of codimension 1, given by the equation $\rho = 0$, with 0 a regular value of ρ. For the following discussion it would be enough to assume that ρ is of class C^2.

The tangent space $T_p(\partial G)$ to ∂G at a point $p \in \partial G$ can be described as

$$T_p(\partial G) = \{X \in \mathbb{R}^m :\ d\rho_p(X) = 0\}.$$

Given two defining functions ρ_1, ρ_2 for G and a point $p \in \partial G$, choose local coordinates $(x_1, \ldots, x_m)$ near p such that $p = \mathbf{0}$ and

$$\rho_2(x_1, \ldots, x_m) = x_m.$$

With $x' := (x_1, \ldots, x_{m-1})$ we then have $\rho_1(x', 0) = 0$ for x' near $\mathbf{0} \in \mathbb{R}^{m-1}$. The fundamental theorem of calculus applied to the function $t \mapsto \rho_1(x', tx_m)$ yields

$$\rho_1(x', x_m) = \rho_1(x', x_m) - \rho_1(x', 0) = x_m \int_0^1 \frac{\partial \rho_1}{\partial x_m}(x', tx_m)\, dt.$$

In other words, we have found a smooth function h such that $\rho_1 = h\rho_2$ on $\mathbb{R}^m$. Then $d\rho_1 = h\,d\rho_2$ on ∂G. These last two equations determine h, and in particular we see that h is everywhere positive.

We now want to describe the (geometric) convexity of G in terms of the function ρ.

Lemma 5.3.1 *If G is convex near $p \in \partial G$, then*

$$\sum_{j,k=1}^{m} \frac{\partial^2 \rho}{\partial x_j\,\partial x_k}(p)\,X_j X_k \geq 0 \text{ for all } X = (X_1, \ldots, X_m) \in T_p(\partial G). \tag{5.3}$$

Conversely, if Equation (5.3) holds with strict inequality for all $\mathbf{0} \neq X \in T_p(\partial G)$, in which case one says that G is **strictly convex** *at p, then G is convex in a neighbourhood of p in $\mathbb{R}^m$.*

Proof First of all we notice that the statements in this lemma are independent of the choice of defining function ρ and invariant under *linear* coordinate changes in $\mathbb{R}^m$. Indeed, given two defining functions ρ_1 and ρ_2, with $\rho_1 = h\rho_2$ as above, we have

$$\frac{\partial^2 \rho_1}{\partial x_j\,\partial x_k} = \frac{\partial^2 h}{\partial x_j\,\partial x_k}\,\rho_2 + \frac{\partial h}{\partial x_j}\cdot\frac{\partial \rho_2}{\partial x_k} + \frac{\partial h}{\partial x_k}\cdot\frac{\partial \rho_2}{\partial x_j} + h\,\frac{\partial^2 \rho_2}{\partial x_j\,\partial x_k}.$$

For $p \in \partial G$ and $X \in T_p(\partial G)$ we have

$$\sum_{j=1}^{m} \frac{\partial \rho_2}{\partial x_j}(p)\,X_j = 0,$$

and hence

$$\sum_{j,k=1}^{m} \frac{\partial^2 \rho_1}{\partial x_j\,\partial x_k}(p)\,X_j X_k = h(p) \sum_{j,k=1}^{m} \frac{\partial^2 \rho}{\partial x_j\,\partial x_k}(p)\,X_j X_k.$$

Moreover, if $(x_1, \ldots, x_m)$ and $(y_1, \ldots, y_m)$ are two local systems of coordinates, then the corresponding coordinate descriptions $(X_1, \ldots, X_m)$ and $(Y_1, \ldots, Y_m)$ of a tangent vector are related by

$$Y_r = \sum_{j=1}^{m} \frac{\partial y_r}{\partial x_j}\,X_j.$$

We compute

$$\frac{\partial^2 \rho}{\partial x_j\,\partial x_k} = \sum_{r,s=1}^{m} \frac{\partial^2 \rho}{\partial y_r\,\partial y_s}\cdot\frac{\partial y_r}{\partial x_j}\cdot\frac{\partial y_s}{\partial x_k} + \sum_{s=1}^{m} \frac{\partial \rho}{\partial y_s}\cdot\frac{\partial^2 y_s}{\partial x_j\,\partial x_k}.$$

The second summand disappears for a linear coordinate change, though not for a general one, and we get

$$\sum_{j,k=1}^{m} \frac{\partial^2 \rho}{\partial x_j\, \partial x_k}(p)\, X_j X_k = \sum_{r,s=1}^{m} \frac{\partial^2 \rho}{\partial y_r\, \partial y_s}(p)\, Y_r Y_s.$$

These preliminary considerations show that we may assume without loss of generality that $p = \mathbf{0}$ and that $T_p(\partial G)$ is the linear subspace of $\mathbb{R}^m$ corresponding to the first $m-1$ coordinates. Moreover, we may then take the function ρ near $p = \mathbf{0}$ to be of the form

$$\rho(x_1, \ldots, x_m) = f(x_1, \ldots, x_{m-1}) - x_m$$

with $df_0 = 0$. Convexity of ρ at p then translates into saying that the function $t \mapsto f(tX_1, \ldots, tX_{m-1})$ is convex at $t = 0$ for any non-zero vector $(X_1, \ldots, X_{m-1})$. The claims follow. □

Now suppose $m = 2n$ and identify $\mathbb{R}^{2n}$ with $\mathbb{C}^n$. Write J for the complex bundle structure on the tangent bundle $T\mathbb{R}^{2n}$ induced by multiplication by $\mathrm{i} = \sqrt{-1}$ on $\mathbb{C}^n$. Write $\partial_{x_j}, \partial_{y_j}$, $j = 1, \ldots, n$, for the (pointwise) standard basis for $T\mathbb{R}^{2n}$ with $J(\partial_{x_j}) = \partial_{y_j}$. Set

$$\partial_{z_j} = \frac{1}{2}(\partial_{x_j} - \mathrm{i}\, \partial_{y_j}), \quad \partial_{\overline{z}_j} = \frac{1}{2}(\partial_{x_j} + \mathrm{i}\, \partial_{y_j}), \quad j = 1, \ldots, n.$$

These are vector fields on the complexified bundle $T\mathbb{R}^{2n} \otimes_{\mathbb{R}} \mathbb{C}$. The $\mathbb{C}$–linear extension of J (still satisfying $J^2 = -\mathrm{id}$) to this bundle is characterised by

$$J(\partial_{z_j}) = \mathrm{i}\, \partial_{z_j}, \quad J(\partial_{\overline{z}_j}) = -\mathrm{i}\, \partial_{\overline{z}_j}.$$

Observe that a real vector field X on $\mathbb{R}^{2n}$ can be written as

$$X = \sum_{j=1}^{n} (X_j\, \partial_{x_j} + Y_j\, \partial_{y_j}) = \sum_{j=1}^{n} (Z_j\, \partial_{z_j} + \overline{Z}_j\, \partial_{\overline{z}_j}) \quad \text{with } Z_j := X_j + \mathrm{i} Y_j,$$

that is $X = 2\,\mathrm{Re}(Z)$ with $Z := \sum_j Z_j\, \partial_{z_j}$.

We usually identify $T_p\mathbb{R}^{2n}$ with $\mathbb{R}^{2n}$ in a canonical fashion. Likewise, we identify the *complex* tangent space $T_p^{\mathbb{C}}\mathbb{C}^n \subset T_p\mathbb{R}^{2n} \otimes_{\mathbb{R}} \mathbb{C}$ (spanned by $\partial_{z_1}, \ldots, \partial_{z_n}$) with $\mathbb{C}^n$, that is, we identify the vector $\sum_j Z_j\, \partial_{z_j} \in T_p^{\mathbb{C}}\mathbb{C}^n$ with $(Z_1, \ldots, Z_n) \in \mathbb{C}^n$. Then the natural identification of $\mathbb{R}^{2n}$ with $\mathbb{C}^n$, mapping

$$(X_1, Y_1, \ldots, X_n, Y_n) \longmapsto (Z_1, \ldots, Z_n) := (X_1 + \mathrm{i} Y_1, \ldots, X_n + \mathrm{i} Y_n),$$

defines an isomorphism

$$\begin{array}{ccc} T_p\mathbb{R}^{2n} & \longrightarrow & T_p^{\mathbb{C}}\mathbb{C}^n \\ X & \longmapsto & Z = \frac{1}{2}(X - \mathrm{i} JX). \end{array}$$

Observe that the complex tangent space $T_p^{\mathbb{C}}\mathbb{C}^n \subset T_p\mathbb{R}^{2n} \otimes_{\mathbb{R}} \mathbb{C}$ can also be characterised as the eigenspace of the endomorphism J corresponding to the eigenvalue $+\mathrm{i}$.

Definition 5.3.2 *Given G as before, we define the* **complex tangent space** *to ∂G at $p \in \partial G$ as*

$$T_p^{\mathbb{C}}(\partial G) := T_p(\partial G) \cap J(T_p(\partial G)).$$

This is the largest J–invariant subspace of $T_p(\partial G) \subset T_p\mathbb{R}^{2n}$, of real dimension $2n-2$.

In terms of standard complex Cartesian coordinates $(z_1, \ldots, z_n)$ on $\mathbb{C}^n$ we define complex-valued 1–forms on $\mathbb{C}^n$ by

$$\partial\rho = \sum_{j=1}^{n} \frac{\partial\rho}{\partial z_j}\, dz_j \quad \text{and} \quad \overline{\partial}\rho = \sum_{j=1}^{n} \frac{\partial\rho}{\partial \overline{z}_j}\, d\overline{z}_j,$$

so that

$$d\rho = \partial\rho + \overline{\partial}\rho.$$

One should really regard these as sections of the bundle $T^*\mathbb{R}^{2n} \otimes_{\mathbb{R}} \mathbb{C}$. Notice that $\partial\rho$ is $\mathbb{C}$–linear and $\overline{\partial}\rho$ is $\mathbb{C}$–antilinear.

Lemma 5.3.3 *The complex tangent space $T_p^{\mathbb{C}}(\partial G)$ is given by*

$$T_p^{\mathbb{C}}(\partial G) = \{X \in T_p\mathbb{R}^{2n} : \partial\rho_p(X) = 0\} = \{Z \in T_p^{\mathbb{C}}\mathbb{C}^n : \partial\rho_p(Z) = 0\}.$$

Proof Since ρ is real-valued we have $\overline{\partial}\rho = \overline{\partial\rho}$. This implies that $d\rho(X) = 2\,\mathrm{Re}(\partial\rho(X))$. Moreover, the $\mathbb{C}$–linearity of the 1–form $\partial\rho$ gives

$$\mathrm{Re}(\partial\rho(JX)) = \mathrm{Re}(\mathrm{i}\partial\rho(X)) = -\mathrm{Im}(\partial\rho(X)).$$

Since $T_p^{\mathbb{C}}(\partial G)$ can be characterised as the vector space of those $X \in \mathbb{R}^{2n}$ with

$$d\rho_p(X) = d\rho_p(JX) = 0,$$

the first description of $T_p^{\mathbb{C}}(\partial G)$ follows. The second description follows by writing X in the form $X = \sum_j (Z_j\, \partial_{z_j} + \overline{Z}_j\, \partial_{\overline{z}_j})$. □

Remark 5.3.4 When below we write $Z \in T_p^{\mathbb{C}}(\partial G)$, where we regard Z as an element of $\mathbb{C}^n$, this is always meant to correspond to $X = \sum_j (Z_j\, \partial_{z_j} + \overline{Z}_j\, \partial_{\overline{z}_j})$ in the real description of $T_p^{\mathbb{C}}(\partial G)$. This is the natural identification of the J–invariant subspace $T_p^{\mathbb{C}}(\partial G)$ of $T_p\mathbb{R}^{2n}$ with a complex linear subspace of $\mathbb{C}^n = T_p^{\mathbb{C}}\mathbb{C}^n$.

We would now like to find a complex analogue of the convexity condition (5.3). In order for it to have any chance to bear a relation to other notions of convexity in complex analysis, which are typically defined in terms of the existence of certain holomorphic functions (see [211]), it should be a condition that is invariant under biholomorphisms.

The quadratic term in the Taylor expansion of $\rho(p+Z)$ at $p \in \partial G$, with $Z = (Z_1, \ldots, Z_n) \in \mathbb{C}^n$, can be written as

$$\operatorname{Re} \sum_{j,k=1}^{n} \frac{\partial^2 \rho}{\partial z_j\, \partial z_k}(p)\, Z_j Z_k + \sum_{j,k=1}^{n} \frac{\partial^2 \rho}{\partial z_j\, \partial \overline{z}_k}(p)\, Z_j \overline{Z}_k =: Q_p\rho(Z) + L_p\rho(Z).$$

Observe that $Q_p\rho(\mathrm{i}Z) = -Q_p\rho(Z)$ and $L_p\rho(\mathrm{i}Z) = L_p\rho(Z)$.

If the geometric convexity condition (5.3) holds, then

$$Q_p\rho(Z) + L_p\rho(Z) \geq 0 \text{ for all } Z = \frac{1}{2}(X - \mathrm{i}JX) \in T_p(\partial G).$$

For $Z \in T_p^{\mathbb{C}}(G)$, this equation holds both for Z and $JZ = \mathrm{i}Z$; summing these two equations we obtain

$$L_p\rho(Z) \geq 0 \text{ for all } Z \in T_p^{\mathbb{C}}(\partial G). \tag{5.4}$$

Definition 5.3.5 *The Hermitian form*

$$L_p\rho(Z) = \sum_{j,k=1}^{n} \frac{\partial^2 \rho}{\partial z_j\, \partial \overline{z}_k}(p)\, Z_j \overline{Z}_k$$

is called the **Levi form** *(or* **complex Hessian***) of ρ at $p \in \partial G$. One says that G is* **(strictly) Levi pseudoconvex** *if its boundary ∂G is smooth and condition (5.4) holds for all $p \in \partial G$ (with strict inequality for $Z \neq \mathbf{0}$).*

From this definition and the preceding considerations it follows immediately that any (strictly) convex region with smooth boundary is (strictly) Levi pseudoconvex. The converse is false, of course. For instance, for $n = 1$ condition (5.4) is empty, so any region in $\mathbb{C}$ with smooth boundary is strictly Levi pseudoconvex.

Here is the verification that Levi pseudoconvexity is indeed a condition invariant under biholomorphisms. Let ϕ^{-1} be a biholomorphism defined in a neighbourhood of $p \in \partial G$. To keep notation simple, I am not going to specify that neighbourhood and write things as if ϕ was globally defined. The region $\phi^{-1}(G)$ is defined by the condition $\rho \circ \phi(z) < 0$, and with ϕ' denoting the holomorphic Jacobian of ϕ we have

$$\partial(\rho \circ \phi)_p(Z) = \partial\rho_{\phi(p)}(\phi'(p)Z).$$

Given the characterisation of the complex tangent space in Lemma 5.3.3, we see that multiplication by $\phi'(p)$ defines an isomorphism from $T_p^{\mathbb{C}}(\partial G)$ to $T_{\phi(p)}^{\mathbb{C}}(\partial(\phi(G))$. Then the equation

$$L_p(\rho \circ \phi)(Z) = L_{\phi(p)}\rho(\phi'(p)Z)$$

shows the invariance of the Levi condition (5.4). That last equation is derived as in the real case. The essential difference is that the term $\partial^2 w_k/(\partial z_j\, \partial \overline{z}_k)$ in the formula relating the Levi form in terms of coordinates $(z_1, \ldots, z_n)$ to that in terms of $(w_1, \ldots, w_n)$, see the computations in the proof of Lemma 5.3.1, disappears for all holomorphic coordinate changes, not only linear ones.

5.4 Levi pseudoconvexity and ω–convexity

In this section we are going to relate ω–convexity in the sense of Definition 5.1.1 to (strict) Levi pseudoconvexity. Before we address the issue of fillings, we show that strictly Levi pseudoconvex hypersurfaces carry a natural contact structure.

Given a region $G \subset \mathbb{R}^{2n} \equiv \mathbb{C}^n$ as in the previous section, define a real 1–form α on $T(\partial G)$ by

$$\alpha := -d\rho \circ J|_{T(\partial G)}.$$

Since $dz_j \circ J = \mathrm{i}\, dz_j$ and $d\overline{z}_j \circ J = -\mathrm{i}\, d\overline{z}_j$, this can be written as

$$\alpha = \mathrm{i} \sum_{j=1}^{n} \left(\frac{\partial \rho}{\partial \overline{z}_j}\, d\overline{z}_j - \frac{\partial \rho}{\partial z_j}\, dz_j \right),$$

where restriction to $T(\partial G)$ is from now on understood. We then compute

$$d\alpha = 2\mathrm{i} \sum_{j,k=1}^{n} \frac{\partial^2 \rho}{\partial z_j\, \partial \overline{z}_k}\, dz_j \wedge d\overline{z}_k;$$

the purely holomorphic or purely antiholomorphic terms disappear by the symmetry of the second derivatives of ρ and the antisymmetry of the wedge product.

Observe that $\alpha = \mathrm{i}(\overline{\partial}\rho - \partial\rho) = 2\,\mathrm{Im}(\partial\rho)$. Arguing as in the proof of Lemma 5.3.3, we see that $\ker \alpha_p = T_p^{\mathbb{C}}(\partial G)$.

Lemma 5.4.1 *For $X \in T_p^{\mathbb{C}}(\partial G)$ and with $Z = (X - \mathrm{i}JX)/2$ as before, we have*

$$d\alpha_p(X, JX) = 4\, L_p\rho(Z).$$

Proof Write $X = Z + \overline{Z}$ with $Z = \sum_j Z_j\, \partial_{z_j}$ and $\overline{Z} = \sum_j \overline{Z}_j\, \partial_{\overline{z}_j}$. Then $JX = \mathrm{i}Z - \mathrm{i}\overline{Z}$. Hence

$$\begin{aligned} d\alpha_p(X, JX) &= 2\mathrm{i} \sum_{j,k=1}^{n} \frac{\partial^2 \rho}{\partial z_j\, \partial \overline{z}_k}(p)\, dz_j \wedge d\overline{z}_k (Z + \overline{Z}, \mathrm{i}Z - \mathrm{i}\overline{Z}) \\ &= 4 \sum_{j,k=1}^{n} \frac{\partial^2 \rho}{\partial z_j\, \partial \overline{z}_k}(p)\, Z_j \overline{Z}_k \\ &= 4\, L_p \rho(Z), \end{aligned}$$

which is the claimed identity. □

Proposition 5.4.2 *Let $G \subset \mathbb{C}^n$ be a strictly Levi pseudoconvex region. Then the hyperplane distribution ξ on ∂G defined by $\xi_p = T_p^{\mathbb{C}}(\partial G)$, $p \in \partial G$, is a contact structure.*

Proof Strict Levi pseudoconvexity means, by the foregoing lemma, that

$$d\alpha_p(X, JX) > 0 \quad \text{for all } \mathbf{0} \neq X \in \xi_p = T_p^{\mathbb{C}}(\partial G) = \ker \alpha_p.$$

So $d\alpha|_\xi$ is symplectic, and hence ξ a contact structure. □

Remark 5.4.3 More generally, with $Y = W + \overline{W}$, we have

$$\begin{aligned} d\alpha_p(X, Y) &= 2\mathrm{i} \sum_{j,k=1}^{n} \frac{\partial^2 \rho}{\partial z_j\, \partial \overline{z}_k}(p)\, dz_j \wedge d\overline{z}_k (Z + \overline{Z}, W + \overline{W}) \\ &= 2\mathrm{i} \sum_{j,k=1}^{n} \frac{\partial^2 \rho}{\partial z_j\, \partial \overline{z}_k}(p)\, (Z_j \overline{W}_k - W_j \overline{Z}_k), \end{aligned}$$

from which one sees that $d\alpha(JX, JY) = d\alpha(X, Y)$, i.e. $J|_\xi$ is compatible with the symplectic structure $d\alpha|_\xi$.

Example 5.4.4 (see also Example 2.1.7) With $\rho\colon \mathbb{C}^n \to \mathbb{R}$ given by

$$\rho(\mathbf{z}) := \frac{\|\mathbf{z}\|^2 - 1}{4},$$

so that the corresponding domain $G = \{\rho(\mathbf{z}) < 0\} \subset \mathbb{C}^n$ is the open unit ball, we have

$$d\alpha = \frac{\mathrm{i}}{2} \sum_{j=1}^{n} dz_j \wedge d\overline{z}_j = \sum_{j=1}^{n} dx_j \wedge dy_j,$$

i.e. the standard symplectic form on $\mathbb{C}^n$. So the unit ball has a strictly Levi pseudoconvex boundary, and the complex tangencies to this boundary (the unit sphere in $\mathbb{C}^n$) define a contact structure.

Since (strict) Levi pseudoconvexity is a notion invariant under biholomorphisms, one can sensibly speak of pseudoconvex hypersurfaces in arbitrary complex manifolds, or pseudoconvex boundaries of complex manifolds. This gives rise to a further type of filling. Concerning compatibility of orientations, observe that the condition

$$d\alpha(X, JX) > 0 \quad \text{for all } \mathbf{0} \neq X \in \xi_p = T_p^{\mathbb{C}}(\partial G)$$

says that the orientations on $T_p^{\mathbb{C}}(\partial G)$ induced by the complex structure J on the one hand, and the symplectic structure $d\alpha$ on the other, coincide. If Y is a vector orthonormal to ∂G (with respect to the standard inner product on $\mathbb{R}^{2n}$) and pointing outwards, then JY is tangent to ∂G and

$$\alpha(JY) = -d\rho(J^2 Y) = d\rho(Y) > 0.$$

This implies that the orientation on ∂G induced by the cooriented contact structure $\xi = \ker \alpha$ coincides with the boundary orientation, with G being naturally oriented as open subset of $\mathbb{C}^n$.

In the following definition, this choice of coorientation of ξ (and hence induced orientation of M) is understood.

Definition 5.4.5 *A compact complex manifold W is called a* **holomorphic filling** *of the contact manifold (M, ξ) if W has strictly Levi pseudoconvex boundary ∂W, equal to M as oriented manifold, and the contact structure ξ is given by $\xi_p = T_p^{\mathbb{C}}(\partial W)$, $p \in \partial W$.*

The following theorem relates this notion of filling to those encountered earlier, at least in dimension 3.

Theorem 5.4.6 *Let (M, ξ) be a (closed) 3-dimensional contact manifold. If (M, ξ) admits a holomorphic filling, then it is also strongly symplectically fillable.*

The proof of this theorem relies on the following notions.

Definition 5.4.7 *(a) A* **Stein manifold** *is an affine complex manifold, i.e. a complex manifold that admits a proper holomorphic embedding into $\mathbb{C}^N$ for some large integer N.*

(b) A smooth function $\rho\colon W \to \mathbb{R}$ on a complex manifold W is called **strictly plurisubharmonic** *if the Levi form $L_p\rho$ is positive definite for all $p \in W$.*

(c) A function ρ as in (b) is called an **exhausting function** *if it is proper and bounded below.*

The definition of Stein manifold given here is not the original one, which is in terms of holomorphic convexity and the existence of 'sufficiently many' (which I am not going to elaborate) holomorphic functions, see [120].

The following theorem relates these notions to each other.

Theorem 5.4.8 (Grauert [116]) *A complex manifold is Stein if and only if it admits an exhausting plurisubharmonic function.* □

One direction of this theorem is simple. The function $\rho = \sum_{j=1}^{N} |z_j|^2$ on $\mathbb{C}^N$ restricts to an exhausting plurisubharmonic function on any Stein manifold $W \subset \mathbb{C}^N$.

Given a strictly plurisubharmonic function $\rho\colon W \to \mathbb{R}$ on a complex manifold (W, J) and a regular value c of ρ, the complex tangencies define a contact structure on the level set $M_c := \{p \in W\colon \rho(p) = c\}$ as in Proposition 5.4.2.

Proposition 5.4.9 *Let (W, J) be a Stein manifold and $\rho\colon W \to \mathbb{R}$ an exhausting plurisubharmonic function on W. For any regular value c of ρ, the manifold $W_c := \{p \in W\colon \rho(p) \leq c\}$ is in a natural way a strong symplectic filling of the contact manifold M_c.*

Proof The manifold W_c is compact, since ρ is an exhausting function.

As we have seen above, the contact structure on M_c is defined by the restriction of the 1–form $\alpha := -d\rho \circ J$ to TM_c. Our local calculations at the beginning of this section show that $\omega := d\alpha$ is a symplectic form on W_c, and that

$$g(X, Y) := \omega(X, JY)$$

defines a Riemannian metric on W_c.

Let $\nabla\rho$ be the gradient of ρ with respect to this Riemannian metric g. Then

$$\omega(\nabla\rho, -) = -g(\nabla\rho, J(-)) = -d\rho \circ J = \alpha.$$

This implies that $\nabla\rho$ is a Liouville vector field for ω, since

$$\mathcal{L}_{\nabla\rho}\omega = d(i_{\nabla\rho}\omega) = d\alpha = \omega.$$

This proves the proposition. □

If we call W_c a **Stein filling** of M_c, then the proposition can be phrased more succinctly as 'Every Stein filling is a strong symplectic filling'. In fact, as shown by Hill and Nacinovich [131], it is enough to require that $\mathrm{Int}\,(W_c)$ be a Stein manifold (with no assumption on ρ other than c being a regular value and W_c compact), and that the complex structure on $\mathrm{Int}\,(W_c)$

extend smoothly (as an almost complex structure) to the boundary M_c (there defining the contact structure), in order to find a Stein filling in our sense.

The converse of the proposition is false. As shown by Ghiggini [100], there are contact 3–manifolds that admit a strong symplectic filling, but no Stein filling.

Proof of Theorem 5.4.6 If (M, ξ) has a holomorphic filling, then by [120, Thm. IX.C.4] one may regard M as the level set of a strictly plurisubharmonic function on a Stein surface with singularities not in M. As shown in [28], one can then find a Stein filling of M as follows. First resolve the singularities; this preserves M and the complex structure near it. Then the main theorem of [28] says that a small deformation of this desingularised surface will be Stein, and that M remains a level set of a strictly plurisubharmonic function during the deformation. By Gray stability, the contact structures on M defined by the complex tangencies (for each complex structure in the deformation family) are isotopic. □

For more details on some of the topics discussed in this section see the notes taken by Abreu of lectures by Eliashberg [72].

6
Contact surgery

> 'The afternoon he came to say goodbye there was a positively surgical atmosphere in the flat.'
>
> Christopher Isherwood,
> *Goodbye to Berlin*

The proof of the Lutz–Martinet theorem in Chapter 4 was based on Dehn surgery along transverse knots in a given contact 3–manifold. This construction does not admit any direct extension to higher dimensions. In 1982, Meckert [178] developed a connected sum construction for contact manifolds. Now, forming the connected sum of two manifolds is the same as performing a surgery along a 0–sphere (i.e. two points, one in each of the two manifolds we want to connect). Since a point in a contact manifold is the simplest example of an isotropic submanifold, this intimated that there might be a more general form of 'contact surgery' along isotropic submanifolds. On the other hand, Meckert's construction is so complex that such a generalisation did not immediately suggest itself.

Then, in 1990, Eliashberg [65] did indeed find such a general form of contact surgery. In fact, he solved a much more intricate problem about the topology of Stein manifolds, involving the construction of complex structures on certain handlebodies such that their boundaries are strictly pseudoconvex (and hence inherit a contact structure), see Example 2.1.7 and Section 5.3. A little later it was realised by Weinstein [241] that the symplectic and contact geometric part of Eliashberg's result can be described in a considerably simplified form.

In Section 6.1 we recall the notion of surgery on differential manifolds and the equivalent description via attaching of handles to manifolds with boundary. A special instance of this has already been discussed in Section 1.7.2.

In Section 6.2 we give Weinstein's description how this topological construction can be performed with the additional structures we are interested in, that is, contact structures on surgered manifolds or symplectic structures on cobordisms. In passing, we discuss Eliashberg's more difficult construction of Stein manifolds with a given topology.

When trying to apply contact surgery to the construction of contact manifolds, special attention has to be paid to the framings of the surgeries in question (i.e. the framings of the spheres along which surgery is performed). This question was addressed in Eliashberg's paper on Stein manifolds, but not by Weinstein. In Section 6.3 we deal with that part of Eliashberg's general discussion that is directly relevant to contact geometry. Essentially, it will be shown that all framings that satisfy an obvious topological condition (extendibility of an almost complex structure over the corresponding handle) can be used to perform contact surgery. In this argument we appeal to various so-called h–principles; Section 6.3.1 is devoted to a discussion of the h–principle for isotropic immersions. As a by-product, we obtain in Section 6.3.2 a contact geometric proof of the Whitney–Graustein theorem, stating that immersions of S^1 in $\mathbb{R}^2$ are classified by their rotation number.

There are special problems related to framings in the case of 3–dimensional contact manifolds; these will be discussed in Section 6.4. We then apply these 3–dimensional results in Section 6.5 to questions concerning symplectic fillings of contact 3–manifolds. In particular, we provide a proof of Theorem 1.7.15, based on a surgery description of contact 3–manifolds found by Fan Ding and this author [51].

6.1 Topological surgery

Basic references on surgery are the textbooks by Kosinski [153] and Ranicki [213]. Milnor's beautiful article [183] remains one of the best introductions to surgery; the survey [216] by Rosenberg is also very useful. In the topological setting, it is usually not vital to have an explicit coordinate description for surgeries or the corresponding handle attachments. This situation changes, however, once we want to construct geometric structures on the surgered manifolds or the corresponding cobordisms. Such explicit coordinate descriptions are given in Milnor's article, and, following Weinstein, we are going to adapt them to the contact geometric or symplectic setting, respectively. In the present section we first give the topological picture.

Let D^{k+1} denote the unit disc in the Euclidean space $\mathbb{R}^{k+1}$, with boundary $\partial D^{k+1} = S^k$ the unit k–sphere. The product manifold $S^k \times S^{n-k-1}$ may be regarded as the boundary either of $S^k \times D^{n-k}$ or of $D^{k+1} \times S^{n-k-1}$.

Now let M be an n–dimensional differential manifold, and suppose that we have an embedded k–sphere $S^k \subset M$ with trivial normal bundle. This implies that we can find an embedded copy of $S^k \times D^{n-k}$ in M.

Definition 6.1.1 Given $S^k \times D^{n-k} \subset M$, one can form a new manifold

$$M' := \big(M \setminus S^k \times \mathrm{Int}\,(D^{n-k})\big) \cup_{S^k \times S^{n-k-1}} (D^{k+1} \times S^{n-k-1})$$

by making the obvious identification along

$$S^k \times S^{n-k-1} = \partial\big(M \setminus S^k \times \mathrm{Int}\,(D^{n-k})\big) = \partial(D^{k+1} \times S^{n-k-1}).$$

(The fact that M' is indeed a manifold is a simple consequence of the collaring theorem for manifolds with boundary, see [38], [132] or [153].) The procedure of constructing M' from M is called a **surgery along** $S^k \subset M$.

Observe that M may be recovered from M' by a surgery along S^{n-k-1}.

Remark 6.1.2 The result M' of a surgery along $S^k \subset M$ is, up to diffeomorphism, determined by the choice of embedding $S^k \times D^{n-k} \to M$, which amounts to a choice of trivialisation of the normal bundle of $S^k \subset M$, i.e. a choice of **framing**. In fact, the diffeomorphism type of M' depends only on the isotopy class of the embedding $S^k \times D^{n-k} \to M$. This is a consequence of the standard isotopy extension theorem in differential topology.

Examples 6.1.3 (1) Suppose that M consists of two connected components M_+ and M_-. Let $S^0 \to M$ be an embedding of $S^0 = \{\pm 1\} \subset \mathbb{R}$ into M sending ± 1 to $M_\pm$, respectively. Then surgery along S^0 is the same as taking the connected sum of M_+ and M_-. In the oriented case, it is understood that the two embeddings $\{\pm 1\} \times D^n \to M_\pm$ are chosen with opposite orientation behaviour. This amounts to the same as an orientation-preserving embedding of $S^0 \times D^n$, oriented as (part of) the boundary of $D^1 \times D^n$. See also the third example below.

(2) The unit n–sphere $S^n \subset \mathbb{R}^{n+1}$ may be viewed as a union

$$S^n = (D^{k+1} \times S^{n-k-1}) \cup (S^k \times D^{n-k}),$$

with

$$S^k \times D^{n-k} := \big\{(x_0, \ldots, x_n) \in S^n : \, x_{k+1}^2 + \cdots + x_n^2 \leq 1/2\big\}.$$

The surgery along $S^k \subset S^n$ defined by this embedding of $S^k \times D^{n-k}$ produces

$$(D^{k+1} \times S^{n-k-1}) \cup (D^{k+1} \times S^{n-k-1}) = S^{k+1} \times S^{n-k-1}.$$

(3) Fix $S^k \times D^{n-k} \subset S^n$ as in the preceding example. Up to homotopy, any other choice of normal framing of $S^k \subset S^n$ (for $k \geq 1$) is realised by an embedding

$$\begin{array}{rccccccl} \phi\colon & S^k & \times & D^{n-k} & \longrightarrow & S^k & \times & D^{n-k} \quad \subset S^n \\ & (\mathbf{x} & , & \mathbf{y}) & \longmapsto & (\mathbf{x} & , & \alpha(\mathbf{x})(\mathbf{y})) \end{array}$$

for some $\alpha \in \pi_k(\mathrm{SO}(n-k))$.† The corresponding surgered manifold

$$M' = (D_-^{k+1} \times S^{n-k-1}) \cup (D_+^{k+1} \times S^{n-k-1}),$$

where the subscripts $\pm$ merely serve to distinguish the two copies of the manifold $D^{k+1} \times S^{n-k-1}$, is obtained by identifying

$$(\mathbf{x}, \mathbf{y}) \in S^k \times S^{n-k-1} = \partial(D_+^{k+1} \times S^{n-k-1})$$

with

$$\phi(\mathbf{x}, \mathbf{y}) = \big(\mathbf{x}, \alpha(\mathbf{x})(\mathbf{y})\big) \in S^k \times S^{n-k-1} = \partial(D_-^{k+1} \times S^{n-k-1}).$$

The projection mapping $(\mathbf{x}, \mathbf{y}) \mapsto \mathbf{x}$ is compatible with this gluing, so M' inherits the structure of an S^{n-k-1}–bundle over the $(k+1)$–sphere $S^{k+1} := D_+^{k+1} \cup D_-^{k+1}$. The map $\alpha \in \pi_k(\mathrm{SO}(n-k))$ is the *characteristic map* of that bundle; it describes the gluing along the equator $S^k = D_+^{k+1} \cap D_-^{k+1}$ of two trivial bundles over the hemispheres $D_\pm^{k+1}$.

Conversely, given any S^{n-k-1}–bundle over $S^{k+1} = D_+^{k+1} \cup D_-^{k+1}$, it restricts to trivial bundles $D_\pm^{k+1} \times S^{n-k-1}$ over the hemispheres, and – up to homotopy – the gluing along the equator can be described by a characteristic map $\alpha \in \pi_k(\mathrm{SO}(n-k))$ as above.

In the case $k = 0$ we have the same one-to-one correspondence between surgeries on S^n along $S^0 \subset S^n$ and S^{n-1}–bundles over S^1, except that here we may take $\alpha \in \pi_0(\mathrm{O}(n))$, i.e. there are two choices.

We now give the alternative description of surgery in terms of handle attaching. Start with the cylinder $[-1,1] \times M$ over the n–manifold M. Given an embedded copy of $S^k \times D^{n-k} \subset M \equiv \{1\} \times M$, we can form the manifold

$$W = ([-1,1] \times M) \cup_{S^k \times D^{n-k}} (D^{k+1} \times D^{n-k}),$$

see Figure 1.6, but beware that there the notation W is used differently. One says that W is obtained from $[-1,1] \times M$ by attaching a $(k+1)$–**handle** (or **handle of index** $k+1$) $D^{k+1} \times D^{n-k-1}$ to the boundary component $\{1\} \times M$. As it stands, W is not actually a manifold – one first needs to 'smooth the corners'‡ at $S^k \times S^{n-k-1}$. This being understood, the boundary of W is given by the disjoint union of $M \equiv \{-1\} \times M$ (or $\overline{M}$ in the oriented case) and the result M' of performing surgery on M defined by $S^k \times D^{n-k} \subset M$. In other words, W is a cobordism between M and M'.

† More precisely: α should be a smooth representative of an element in the homotopy group $\pi_k(\mathrm{SO}(n-k))$.

‡ Kirby writes in [146]: 'The phrase "corners can be smoothed" has been a phrase that I have heard for 30 years, and this is not the place to explain it.' Neither is this, but see [38, (13.12)] and Section 7.6. An alternative expression for 'smoothing corners' is 'straightening the angle'.

The process of smoothing corners may be rather awkward in the presence of additional geometric structures. The following description of the cobordism W given in [183] uses only identification over open subsets. It thus gives the manifold structure for free (only the Hausdorff property needs to be checked), and we shall see later how it can be modified so as to yield a symplectic structure on the cobordism with boundaries of contact type, see Definition 1.4.5.

Let H (for 'handle') denote the locus of points $(\mathbf{x}, \mathbf{y}) \in \mathbb{R}^{k+1} \times \mathbb{R}^{n-k}$ satisfying the inequalities

$$-1 \leq \|\mathbf{y}\|^2 - \|\mathbf{x}\|^2 \leq 1$$

and

$$\|\mathbf{x}\| \cdot \|\mathbf{y}\| < \sinh 1 \cdot \cosh 1,$$

see Figure 6.1.

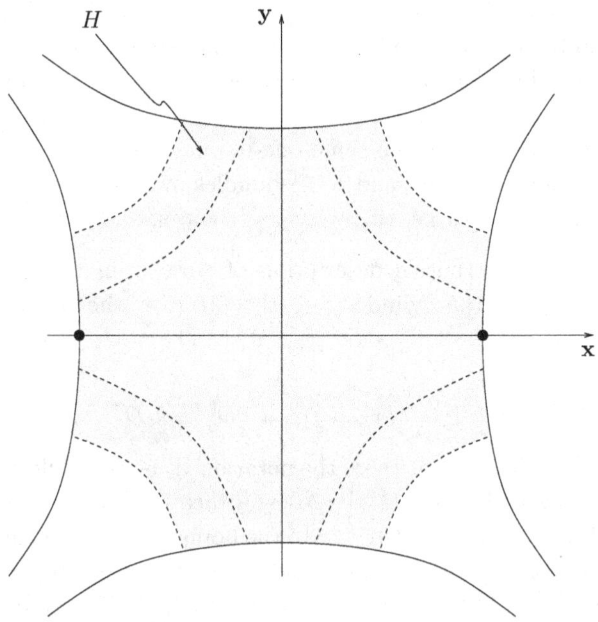

Fig. 6.1. The model handle H.

This manifold H is not diffeomorphic to $D^{k+1} \times D^{n-k}$, i.e. not a handle in the strict sense, but rather a copy of $D^{k+1} \times D^{n-k}$ with the corners cut off. We call the subset of H described by $\|\mathbf{y}\|^2 - \|\mathbf{x}\|^2 = -1$ its 'lower' boundary $\partial^- H$ (in Figure 6.1 it is actually on the left and right); the subset described by $\|\mathbf{y}\|^2 - \|\mathbf{x}\|^2 = 1$ will be referred to as the 'upper' boundary $\partial^+ H$ of H (top and bottom in our figure). The lower boundary is connected for $k \geq 1$, and it has two components (as in the figure) if $k = 0$. Analogous statements hold for the upper boundary. The k–sphere S^k_H in Figure 6.1 in the lower boundary, visualised by the two points on the $\mathbf{x}$–axis, will correspond to the k–sphere along which we want to perform surgery. The $(k+1)$–disc $D^{k+1} \times \{\mathbf{0}\} \subset H$ is called the **core** of the handle; the $(n-k)$–disc $\{\mathbf{0}\} \times D^{n-k} \subset H$, the **belt disc** or **cocore**; the $(n-k-1)$–sphere $\{\mathbf{0}\} \times \partial D^{n-k} \subset \partial H$, the **belt sphere**.

The lower boundary is diffeomorphic to $S^k \times \mathrm{Int}\,(D^{n-k})$ under the correspondence

$$(\mathbf{u}\cosh r, \mathbf{v}\sinh r) \longleftrightarrow (\mathbf{u}, r\mathbf{v}), \quad \mathbf{u} \in S^k,\ \mathbf{v} \in S^{n-k-1},\ 0 \leq r < 1.$$

Similarly, the upper boundary is diffeomorphic to $\mathrm{Int}\,(D^{k+1}) \times S^{n-k-1}$ under the correspondence

$$(\mathbf{u}\sinh r, \mathbf{v}\cosh r) \longleftrightarrow (r\mathbf{u}, \mathbf{v}), \quad \mathbf{u} \in S^k,\ \mathbf{v} \in S^{n-k-1},\ 0 \leq r < 1.$$

For given $(\mathbf{x}_0, \mathbf{y}_0) \in \mathbb{R}^{k+1} \times \mathbb{R}^{n-k}$, the velocity vector of the curve

$$\gamma\colon\ t \longmapsto (t^{-1}\mathbf{x}_0, t\mathbf{y}_0), \quad t \in \mathbb{R}^+,$$

equals the gradient of the function $(\mathbf{x}, \mathbf{y}) \mapsto \|\mathbf{y}\|^2 - \|\mathbf{x}\|^2$ up to positive scaling, so γ is orthogonal to the level sets of that function. Notice that $\|\mathbf{x}\| \cdot \|\mathbf{y}\| \equiv \text{const.}$ along γ. If one of $\mathbf{x}_0$ or $\mathbf{y}_0$ equals zero, the curve γ is a straight line through the origin. If both $\mathbf{x}_0$ and $\mathbf{y}_0$ are different from zero, the curve γ is a hyperbola that intersects the upper and lower boundary of H in one point each. Some of these hyperbolic segments joining the two boundaries are indicated in Figure 6.1.

Now we use this handle H to describe the cobordism corresponding to a surgery on the n–manifold M defined by an embedding $\phi\colon\ S^k \times D^{n-k} \to M$.† Start with the topological sum

$$[-1,1] \times \bigl(M \setminus \phi(S^k \times \{\mathbf{0}\})\bigr) + H,$$

and define W as the quotient space obtained by the following identification. For each $\mathbf{u} \in S^k$, $\mathbf{v} \in S^{n-k-1}$, $0 < r < 1$ and $c \in [-1,1]$, identify

$$\bigl(c, \phi(\mathbf{u}, r\mathbf{v})\bigr) \in [-1,1] \times \bigl(M \setminus \phi(S^k \times \{\mathbf{0}\})\bigr)$$

with the unique point $(\mathbf{x}, \mathbf{y}) \in H$ defined by the conditions

- $\|\mathbf{y}\|^2 - \|\mathbf{x}\|^2 = c$,
- $(\mathbf{x}, \mathbf{y})$ lies on the trajectory γ through the point $(\mathbf{u}\cosh r, \mathbf{v}\sinh r)$.

† The sphere $\phi(S^k \times \{\mathbf{0}\}) \subset M$ is the **attaching sphere** for the handle; occasionally I also refer to $S^k_H := S^k \times \{\mathbf{0}\} \subset \partial^- H$ as the attaching sphere.

This correspondence defines a diffeomorphism

$$[-1,1] \times \phi\big(S^k \times \operatorname{Int}(D^{n-k} \setminus \{\mathbf{0}\})\big) \longrightarrow H \cap (\mathbb{R}^{k+1} \setminus \{\mathbf{0}\}) \times (\mathbb{R}^{n-k} \setminus \{\mathbf{0}\})$$

of open subsets. So the resulting quotient space W, which is easily verified to be a Hausdorff space, is again a differential manifold, see Figure 6.2.

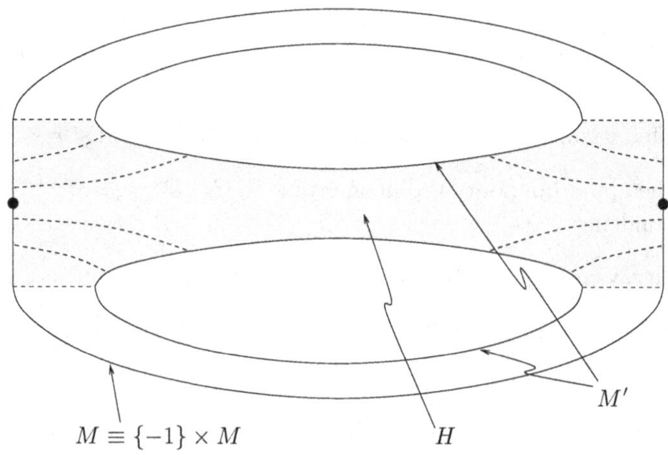

Fig. 6.2. Cobordism W corresponding to surgery.

The boundary component of W corresponding to $c = -1$ can be identified with the original manifold M by sending $p \in M$ to

$$\begin{cases} (-1,p) \in [-1,1] \times \big(M \setminus \phi(S^k \times \{\mathbf{0}\})\big) & \text{for } p \notin \phi(S^k \times \{\mathbf{0}\}), \\ (\mathbf{u}\cosh r, \mathbf{v}\sinh r) \in H & \text{for } p = \phi(\mathbf{u}, r\mathbf{v}),\ r < 1. \end{cases}$$

This is well defined, since for $p \in \phi\big(S^k \times \operatorname{Int}(D^{n-k} \setminus \{\mathbf{0}\})\big)$ the two definitions agree.

The surgered manifold M' can be described as a quotient space of the topological sum

$$\big(M \setminus \phi(S^k \times \{\mathbf{0}\})\big) + (D^{k+1} \times S^{n-k-1}),$$

where $(r\mathbf{u}, \mathbf{v}) \in D^{k+1} \times S^{n-k-1}$ is identified with $\phi(\mathbf{u}, r\mathbf{v}) \in M$ for $0 < r \leq 1$. We see that M' can be identified with the upper (or, in Figure 6.2, inner) boundary of W by sending $p \in \big(M \setminus \phi(S^k \times \{\mathbf{0}\})\big)$ to $(1,p)$, and $(r\mathbf{u}, \mathbf{v}) \in D^{k+1} \times S^{n-k-1}$ to $(\mathbf{u}\sinh r, \mathbf{v}\cosh r)$. That this is well defined follows from the following commutative diagram (for $0 < r \leq 1$).

$$\begin{array}{ccccc} M \ni & \phi(\mathbf{u}, r\mathbf{v}) & \stackrel{\sim}{\longleftrightarrow} & (r\mathbf{u}, \mathbf{v}) & \in D^{k+1} \times S^{n-k-1} \\ & \downarrow & & \downarrow & \\ \{1\} \times M \ni & \big(1, \phi(\mathbf{u}, r\mathbf{v})\big) & \stackrel{\sim}{\longleftrightarrow} & (\mathbf{u} \sinh r, \mathbf{v} \cosh r) & \in H \end{array}$$

For the identification at the bottom of this diagram, observe that the point $(\mathbf{x}, \mathbf{y}) = (\mathbf{u} \sinh r, \mathbf{v} \cosh r)$ is the unique point in H satisfying $\|\mathbf{y}\|^2 - \|\mathbf{x}\|^2 = 1$ and lying on the trajectory $t \mapsto (t^{-1}\mathbf{u} \cosh r, t\mathbf{v} \sinh r)$, *viz.*, corresponding to the value $t = \cosh r / \sinh r$ of the curve parameter.

6.2 Contact surgery and symplectic cobordisms

In the preceding section, the key point in the construction of the cobordism corresponding to a surgery was the following. Given $S^k \subset M$ with trivial normal bundle, it has a neighbourhood in M diffeomorphic to a neighbourhood of the k–sphere given by $\{\mathbf{y} = \mathbf{0}\}$ in the lower boundary of H. This diffeomorphism extends to a diffeomorphism of the punctured neighbourhood $S^k \times (D^{n-k} \setminus \{\mathbf{0}\})$ crossed with the interval $[-1, 1]$ on the one side, and H with the coordinate planes removed on the other, by mapping the obvious foliation by line segments in one space to the foliation by gradient flow lines γ on the other. This allowed the gluing that gave us the desired cobordism.

We now want to show that one can perform surgery along an isotropic sphere with trivial conformal symplectic normal bundle in a given contact manifold such that the resulting manifold carries again a contact structure (which coincides with the old one outside the neighbourhood where surgery takes place). We shall call this a **contact surgery**. We are going to perform this surgery by constructing the corresponding *symplectic* cobordism. Thus, the task becomes obvious. First of all, we need a neighbourhood theorem for isotropic submanifolds in a contact manifold. This will enable us to identify (by a contactomorphism) a neighbourhood of an isotropic sphere in a given contact manifold with a neighbourhood of an isotropic sphere in a model handle. Such a neighbourhood theorem has already been proved (Thm. 2.5.8). As we shall see, the isotropic spheres in the model handle will have trivial conformal symplectic normal bundle, so the same condition will have to be imposed on any sphere along which we want to perform contact surgery.

Secondly, we need to work with a suitable flow, replacing the gradient flow in the topological picture, that will yield a symplectomorphism for the gluing construction. Observe that if $(M, \xi = \ker \alpha)$ is a contact manifold, then a natural symplectic structure ω on $[-1,1] \times M$ is obtained by symplectisation (see Example 1.4.7), that is, $\omega := d(e^t \alpha)$. Here the obvious foliation by line segments is given by the flow of the Liouville vector field ∂_t. Correspondingly, the gradient flow in our topological handle H will have to be replaced by a Liouville flow on a symplectic handle. Moreover, the neighbourhood theorem for isotropic submanifolds will have to be strong enough so that the identification of flow lines in $[-1,1] \times M$ with those in H actually yields a symplectomorphism.

For that last point, we shall need a version of Theorem 2.5.8 for strict contact manifolds, see Definition 2.1.2. In the process, we also provide the argument hinted at in Remark 2.5.12. For the following definition recall Lemma 1.3.3 and Definition 2.5.3.

Definition 6.2.1 Let L be an isotropic submanifold of a strict contact manifold (M, α). The quotient bundle

$$\mathrm{SN}_M(L) := (TL)^\perp / TL$$

with the symplectic bundle structure induced by $d\alpha$ is called the **symplectic normal bundle** of L in M. Here $(TL)^\perp \subset \ker \alpha$ denotes, as in Chapter 2, the symplectic orthogonal complement.

Theorem 6.2.2 *Let (M_i, α_i), $i = 0, 1$, be strict contact manifolds with closed isotropic submanifolds L_i. Suppose there is an isomorphism of symplectic normal bundles $\Phi\colon \mathrm{SN}_{M_0}(L_0) \to \mathrm{SN}_{M_1}(L_1)$ that covers a diffeomorphism $\phi\colon L_0 \to L_1$. Then this diffeomorphism ϕ extends to a strict contactomorphism $\psi\colon \mathcal{N}(L_0) \to \mathcal{N}(L_1)$ of suitable neighbourhoods $\mathcal{N}(L_i)$ of L_i such that $T\psi|_{\mathrm{SN}_{M_0}(L_0)} = \Phi$.*

Proof By the proof of Theorem 2.5.8 we may assume that

- $M_0 = M_1 =: M$, $L_0 = L_1 =: L$, $\phi = \mathrm{id}_L$,
- $\alpha_0 = \alpha_1$ and $d\alpha_0 = d\alpha_1$ on $TM|_L$, $\Phi = \mathrm{id}|_{\mathrm{SN}_M(L)}$ (notice that $\mathrm{SN}_M(L)$ is the same for both contact forms).

Observe that these conditions entail that the Reeb vector fields R_0 and R_1 also coincide on L. Now choose an open neighbourhood $\mathcal{N}(L)$ of L in M and a hypersurface Σ in $\mathcal{N}(L)$ with the following properties:

- $L \subset \Sigma$,
- Σ is transverse to R_0 and R_1, and each integral curve of R_0 or R_1, respectively, in $\mathcal{N}(L)$ intersects Σ exactly once,

- $\xi|_L = T\Sigma|_L$.

These conditions imply that both $d\alpha_0$ and $d\alpha_1$ induce symplectic forms on Σ, with $d\alpha_0 = d\alpha_1$ on $T\Sigma|_L$. With $j\colon \Sigma \to M$ denoting the inclusion and $\alpha_t := (1-t)\alpha_0 + t\,\alpha_1$ it follows (by shrinking $\mathcal{N}(L)$ and Σ, if necessary) that the 2–form $\omega_t := j^*d\alpha_t$ is a symplectic form on Σ for all $t \in [0,1]$.

We now apply the symplectic Moser trick to this family of symplectic forms. Our aim is to find an isotopy ψ_t of a neighbourhood of $L \subset \Sigma$ with $\psi_t^*\omega_t = \omega_0$, and furthermore $\psi_t|_L = \mathrm{id}_L$ and $T\psi_t = \mathrm{id}$ on $T\Sigma|_L$.

The generalised Poincaré Lemma (Cor. A.4) applied to the 2–form

$$\eta := \omega_1 - \omega_0 = \dot{\omega}_t$$

allows us to find a 1–form ζ in a neighbourhood of $L \subset \Sigma$, vanishing to second order on L, such that $\eta = d\zeta$.

Assuming that ψ_t is the flow of a time-dependent vector field X_t, we obtain from $\psi_t^*\omega_t = \omega_0$ by differentiation the equation

$$\psi_t^*\big(d\zeta + d(i_{X_t}\omega_t)\big) = 0.$$

In order to find a solution X_t to this equation, we need only solve

$$\zeta + i_{X_t}\omega_t = 0,$$

which is possible (in a unique way) because ω_t is a symplectic form in a neighbourhood of $L \subset \Sigma$ for all $t \in [0,1]$. Since ζ vanishes to second order on L, the same is true for the vector field X_t. In particular, the (local) flow of X_t fixes L, and therefore it can be defined up to time 1 in a neighbourhood of L (see the argument in the proof of Theorem 2.5.1). Moreover, we have $T\psi_t = \mathrm{id}$ on $T\Sigma|_L$ for all $t \in [0,1]$, since this is true for $\psi_0 = \mathrm{id}_\Sigma$, and – for any vector field Y on Σ and with ψ_t^* defined as $T(\psi_t^{-1})$ on vector fields –

$$\frac{d}{dt}(\psi_t^*Y) = \psi_t^*(\mathcal{L}_{X_t}Y) = \psi_t^*([X_t,Y]) = \mathbf{0} \text{ on } L.$$

The formula for the time derivative of ψ_t^*Y used here follows by dualising the argument in the proof of Lemma 2.2.1.

Now extend ψ_1 to a diffeomorphism $\psi\colon \mathcal{N}(L) \to \mathcal{N}_1(L)$ of the neighbourhood $\mathcal{N}(L)$ of $L \subset M$ (again after shrinking $\mathcal{N}(L)$ and Σ, if necessary) to another such neighbourhood $\mathcal{N}_1(L)$ by requiring that ψ send flow lines of the Reeb vector field R_0 to those of R_1. Since $R_0 = R_1$ along L, this gives $T\psi = \mathrm{id}$ on $TM|_L$. This also implies $\psi^*\alpha_1 = \alpha_0$ on $TM|_L$. Moreover, since the Reeb vector field R of a contact form α satisfies $i_R d\alpha \equiv 0$ and $\mathcal{L}_R d\alpha \equiv 0$, this extension guarantees that $\psi^*d\alpha_1 = d\alpha_0$ on $\mathcal{N}_1(L)$.

Thus, changing notation in the obvious way, we may now assume that

- $\alpha_0 = \alpha_1$ on $TM|_L$, $d\alpha_0 = d\alpha_1$ on $\mathcal{N}_1(L)$, and $\Phi = \mathrm{id}|_{\mathrm{SN}_M(L)}$.

The 1–form α_t, defined as before, will then be a contact form on $\mathcal{N}(L)$ for all $t \in [0,1]$, provided this neighbourhood has been chosen small enough. As in the proof of Darboux's theorem for contact forms (Thm. 2.5.1), we now apply the Moser trick to the equation $\psi_t^* \alpha_t = \alpha_0$. As in that proof, ψ_t is found as the flow of the vector field X_t satisfying the equation

$$\dot{\alpha}_t + d(\alpha_t(X_t)) + i_{X_t} d\alpha_t = 0,$$

see Equation (2.4). By the generalised Poincaré Lemma, applied to the 1–form

$$\eta := \alpha_1 - \alpha_0 = \dot{\alpha}_t,$$

we find a function f on $\mathcal{N}(L)$ – after shrinking that neighbourhood, if necessary – vanishing to second order along L, such that $\eta = df$ (which means that Equation (2.5) is satisfied for $H_t = -f$). The unique solution X_t (with $\alpha_t(X_t) = -f$) to its defining equation above is then given by $X_t = -fR_t$.

By the same argument as that used in the first part of this proof, we deduce from the vanishing of X_t to second order on L that the flow of X_t exists up to time 1 in a sufficiently small neighbourhood of L, and that $T\psi_t = \mathrm{id}$ on $TM|_L$ for all $t \in [0,1]$, in particular for $\psi := \psi_1$. □

Remark 6.2.3 The foregoing proof contains the general principle behind Example 2.5.13.

Next we describe the symplectic model handle. For notational convenience we consider the handle corresponding to an isotropic $(k-1)$–sphere L in a $(2n-1)$–dimensional contact manifold M, where $1 \leq k \leq n$. The picture is essentially the same as in Figure 6.1, except that we replace $\mathbf{x}$ by

$$(q_1, \ldots, q_k) \in \mathbb{R}^k,$$

and $\mathbf{y}$ by

$$(q_{k+1}, \ldots, q_n, p_1, \ldots, p_n) \in \mathbb{R}^{2n-k}.$$

On $\mathbb{R}^{2n} = \mathbb{R}^k \times \mathbb{R}^{2n-k}$ we have the standard symplectic form

$$\omega_0 = \sum_{j=1}^{n} dp_j \wedge dq_j .$$

A Liouville vector field Y for ω_0 of the same qualitative behaviour as the gradient vector field of the function $(\mathbf{x}, \mathbf{y}) \mapsto \|\mathbf{y}\|^2 - \|\mathbf{x}\|^2$ in Figure 6.1 is given by

$$Y := \sum_{j=1}^{k} (-q_j\,\partial_{q_j} + 2p_j\,\partial_{p_j}) + \frac{1}{2} \sum_{j=k+1}^{n} (q_j\,\partial_{q_j} + p_j\,\partial_{p_j}).$$

Notice that Y is the gradient vector field, with respect to the standard Euclidean metric on $\mathbb{R}^{2n}$, of the function

$$g\colon (\mathbf{q},\mathbf{p}) \longmapsto \sum_{j=1}^{k}(-\frac{1}{2}q_j^2 + p_j^2) + \frac{1}{4}\sum_{j=k+1}^{n}(q_j^2+p_j^2).$$

Let $\mathcal{N}_H \cong S^{k-1}\times \mathrm{Int}\,(D^{2n-k})$ be an open neighbourhood in the hypersurface $g^{-1}(-1)\subset \mathbb{R}^{2n}$ of the $(k-1)$–sphere

$$S_H^{k-1} := \{\sum_{j=1}^{k} q_j^2 = 2,\ q_{k+1} = \cdots = q_n = p_1 = \cdots = p_n = 0\}.$$

This $\mathcal{N}_H$ is going to play the role of the lower boundary. We now define the **symplectic handle** H as the locus of points $(\mathbf{q},\mathbf{p}) \in (\mathbb{R}^{2n},\omega_0)$ satisfying the inequality

$$-1 \leq g(\mathbf{q},\mathbf{p}) \leq 1$$

and lying on a gradient flow line of g through a point of $\mathcal{N}_H$.

Since the Liouville vector field Y is obviously transverse to the level sets of g, the 1–form

$$\alpha_0 := i_Y\omega_0 = \sum_{j=1}^{k}(q_j\,dp_j + 2p_j\,dq_j) + \frac{1}{2}\sum_{j=k+1}^{n}(-q_j\,dp_j + p_j\,dq_j)$$

induces a contact form on the lower and upper boundary of H, see Lemma 1.4.5. With respect to this contact form, S_H^{k-1} is an isotropic sphere in the lower boundary. Its symplectic normal bundle $\mathrm{SN}_{\partial H}(S_H^{k-1})$ is trivialised by the vector fields $\partial_{q_j}, \partial_{p_j}$, $j = k+1,\ldots,n$, see Example 2.5.6.

Remark 6.2.4 The inequality $\|\mathbf{x}\|\cdot\|\mathbf{y}\| < \sinh 1\cdot\cosh 1$ in the topological picture merely served to make a specific choice for $\mathcal{N}_H$. Given an isotropic sphere S^{k-1} in a contact manifold (M,α) of dimension $2n-1$, the preceding theorem allows us to identify a small neighbourhood $\mathcal{N}(S^{k-1})$ of $S^{k-1}\subset M$ with a small neighbourhood of $S_H^{k-1}\subset \partial H$ via a strict contactomorphism. Since there is no *a priori* estimate on the size of that last neighbourhood, we simply take that to be $\mathcal{N}_H$ and define H accordingly.

Lemma 5.2.4 now allows us to glue this model handle H symplectomorphically to $[-1,1]\times M$ so as to obtain a symplectic structure on the cobordism corresponding to the surgery.

One last issue that we have to address before formulating the contact surgery theorem is the framing of the surgery corresponding to the cobordism constructed by using the symplectic handle H. Recall the observation following Definition 2.5.3 that the normal bundle NL of an isotropic submanifold L in a (strict) contact manifold (M, α) splits as

$$NL \cong \langle R \rangle \oplus J(TL) \oplus \mathrm{SN}_M(L),$$

where $\langle R \rangle$ is the trivial line bundle spanned by the Reeb vector field R of α, and J is a complex bundle structure on $\ker \alpha$ compatible with $d\alpha$.

Now suppose that L is a $(k-1)$–dimensional sphere, and fix an identification of L with $S^{k-1} \subset \mathbb{R}^k$. The stabilised tangent bundle $TS^{k-1} \oplus \epsilon$ has a natural trivialisation, corresponding to the identification of the trivial line bundle ϵ with the normal bundle of $S^{k-1} \subset \mathbb{R}^k$. This induces a natural trivialisation of $\langle R \rangle \oplus J(TL)$.† Therefore, in order to specify a trivialisation of NS^{k-1}, it suffices to do so for $\mathrm{SN}_M(S^{k-1})$.

The trivialisation of the normal bundle $S_H^{k-1} \subset \partial H$ is given by the vector fields $\partial_{p_1}, \ldots, \partial_{p_k}$ and the symplectic trivialisation of $SN_{\partial H}(S_H^{k-1})$ described earlier. Consider the standard complex structure J_0 on $\mathbb{R}^{2n}$. (This is given by $J_0(\partial_{p_j}) = \partial_{q_j}$, $j = 1, \ldots, n$. It preserves the contact hyperplanes along S_H^{k-1} and is compatible with $d\alpha_0$.) Under J_0 the Liouville vector field Y along S_H^{k-1} is mapped to the Reeb vector field. The trivialisation $\partial_{p_1}, \ldots, \partial_{p_k}$ is mapped to the standard trivialisation $\partial_{q_1}, \ldots, \partial_{q_k}$ of $T\mathbb{R}^k|_{S_H^{k-1}}$, where that last bundle should be thought of as $\langle -Y \rangle \oplus TS_H^{k-1}$ if S_H^{k-1} is oriented as the boundary of the unit disc in $\mathbb{R}^k$.

Thus, when we perform a surgery on an isotropic sphere $S^{k-1} \subset M$ with the help of the symplectic handle H, this corresponds to the framing given by the natural trivialisation of $\langle -R \rangle \oplus J(TS^{k-1})$ and an arbitrary choice of symplectic trivialisation of $\mathrm{SN}_M(S^{k-1})$, which then determines a bundle isomorphism $\phi\colon \mathrm{SN}_M(S^{k-1}) \to \mathrm{SN}_{\partial H}(S_H^{k-1})$. We call this framing of S^{k-1} the **natural framing** determined by the chosen trivialisation of the (conformal) symplectic normal bundle.

With this discussion we have proved the following contact surgery theorem.

Theorem 6.2.5 *Let S^{k-1} be an isotropic sphere in a contact manifold $(M, \xi = \ker \alpha)$ with a trivialisation of the conformal symplectic normal bundle $\mathrm{CSN}_M(S^{k-1})$. Then there is a symplectic cobordism from (M, ξ) to the manifold M' obtained from M by surgery along S^{k-1} using the natural framing. In particular, the surgered manifold M' carries a contact structure that coincides with the one on M away from the surgery region.* □

Remarks 6.2.6 (1) The Liouville vector field on the symplectic cobordism

† There is a subtle point here concerning the orientation of the normal bundle, as the subsequent discussion will show. We shall deal with this in an *ad hoc* fashion in the applications, see Example 6.2.7 and Section 8.3.

corresponding to a single surgery is globally defined, and it vanishes only in the single point $\mathbf{0} \in H$.

(2) The isotopy extension and neighbourhood theorems from Chapter 2 imply that the resulting contact structure on M' is uniquely determined up to isotopy by the isotropic isotopy class of S^{k-1} and the homotopy class of the trivialisation of $\mathrm{CSN}_M(S^{k-1})$.

(3) For $k=1$ the assumptions of the theorem are trivially satisfied. In particular, one can always form the connected sum of equidimensional contact manifolds.

(4) The rank of $\mathrm{CSN}_M(S^{k-1})$ is $2(n-k)$ (for $\dim M = 2n-1$). So contact surgery is always possible along a Legendrian sphere $S^{n-1} \subset M$, and the framing for the contact surgery is completely determined by the embedding of that sphere.

Example 6.2.7 The purpose of this example is to show that the natural framing of contact surgery along a Legendrian knot in a contact 3–manifold differs from the contact framing by one negative (i.e. left) twist. The computations are quite similar to those in Example 3.5.7.

Thus, consider the case $k=n=2$. Here

$$(\alpha_0 \wedge d\alpha_0)|_{S^1_H} = q_1\, dp_1 \wedge dp_2 \wedge dq_2 + q_2\, dp_2 \wedge dp_1 \wedge dq_1.$$

Orient S^1_H by the vector field $q_1\,\partial_{q_2} - q_2\,\partial_{q_1}$. Then the normal orientation of S^1_H is given by $dp_1 \wedge dp_2$, so the oriented framing of the surgery is given by the ordered basis $\{\partial_{p_1}, \partial_{p_2}\}$.

On the other hand, the contact framing of the Legendrian sphere S^1_H is given by the vector field $q_2\,\partial_{p_1} - q_1\,\partial_{p_2} \in (\ker \alpha_0)|_{S^1_H}$. As we go once along S^1_H, this vector field makes one positive (right-handed or counter-clockwise) twist relative to the basis $\{\partial_{p_1}, \partial_{p_2}\}$. Hence, conversely, the natural framing of the contact surgery makes one negative twist relative to the contact framing.

Given closed manifolds M_-, M, M_+ and (topological) cobordisms W_- from M_- to M and W_+ from M to M_+, the collaring theorem for manifolds with boundary says that there is a neighbourhood of M in W_- diffeomorphic to $(-\varepsilon, 0] \times M$, and a neighbourhood of M in W_+ diffeomorphic to $[0,\varepsilon) \times M$. Since the gluing

$$\big((-\varepsilon, 0] \times M\big) \cup_{\{0\}\times M} \big([0,\varepsilon) \times M\big)$$

clearly produces the smooth manifold $(-\varepsilon,\varepsilon) \times M$, one can likewise form a smooth cobordism $W_- \cup_M W_+$ from M_- to M_+. In particular, a sequence of surgeries gives rise to a cobordism from the original manifold to the surgered one. If the original manifold is a sphere, one may further attach a ball (i.e. a cobordism from the empty set to the sphere), so as to obtain a manifold with only the surgered manifold as boundary. Such a manifold, obtained from a ball by attaching handles, is called a **handlebody**.

As we have seen in Proposition 5.2.5, gluing of cobordisms is also possible in the symplectic category. This entails the symplectic analogues of the preceding statements.

Example 6.2.8 Consider $S^{2n-1} \subset \mathbb{R}^{2n}$ with its standard contact structure as in Example 2.1.7, admitting a strong symplectic filling by $(D^{2n}, \omega_{\mathrm{st}})$. What is the effect of performing contact surgery along the Legendrian sphere $S^{n-1} \subset S^{2n-1}$ given by

$$\{x_1^2 + \cdots + x_n^2 = 1,\ y_1 = \cdots = y_n = 0\}?$$

From the Examples 6.2.7 and 3.5.7 we deduce that, for $n = 2$, contact surgery along $S^1 \subset S^3$ corresponds to a surgery with framing -2 relative to the *surface* framing. So the handlebody corresponding to this surgery will be a D^2–bundle over S^2 with Euler number -2, see [111, Example 4.4.2] or Lemma 6.5.7 below. Topologically this corresponds to the cotangent disc bundle of S^2. (The zero section of the tangent bundle TS^2 has self-intersection 2 (so this is the Euler number of the bundle); the Euler number of T^*S^2 is minus that of TS^2; see [31, pp. 125/6, 267, 273].) So the result of the contact surgery is, at least topologically, the unit cotangent bundle ST^*S^2, which is a copy of $\mathbb{R}P^3$.

I claim that the same is true in general, i.e. contact surgery along $S^{n-1} \subset S^{2n-1}$ produces ST^*S^n. We begin with the observation that S^{n-1} bounds a Lagrangian† disc

$$D^n = \{x_1^2 + \cdots + x_n^2 \leq 1,\ y_1 = \cdots = y_n = 0\}$$

in $(D^{2n}, \omega_{\mathrm{st}})$. The core of the symplectic model handle is likewise a Lagrangian disc. So the symplectic handlebody resulting from the surgery is a D^n–bundle over S^n with Lagrangian zero section. The neighbourhood theorem for Lagrangian submanifolds ([239], [177, Thm. 3.33]) implies that a neighbourhood of that zero section looks (as a *symplectic* manifold) like a neighbourhood of the zero section in $(T^*S^n, d\lambda)$, with λ the Liouville form from Section 1.4. This is sufficient for concluding that the handlebody is *diffeomorphic* to the cotangent disc bundle, and hence the surgered manifold diffeomorphic to ST^*S^n. The Liouville vector fields that induce the contact structure $\ker \lambda$ on ST^*S^n on the one hand, and the contact structure on the

† A **Lagrangian submanifold** L in a symplectic manifold (W, ω) is a submanifold with the property that T_qL is a Lagrangian subspace of (T_qW, ω_q) for each point $q \in L$.

surgered manifold on the other, can be described explicitly. It is then not very difficult to see that the result of contact surgery is actually *contactomorphic* to $(ST^*S^n, \ker\lambda)$, see also the arguments in the proof of Theorem 7.6.2.

If one is merely interested in obtaining a contact structure on the surgered manifold, and not a symplectic cobordism, the above picture can be simplified. To keep things as concrete as possible, I only discuss the forming of the connected sum of two (connected) contact 3–manifolds. Let $(M_\pm, \xi_\pm)$ be the two contact 3–manifolds that we want to connect. Choose points $p_\pm \in M_\pm$. By Darboux's theorem (modified in the obvious way) we can find local coordinates in a neighbourhood of $p_\pm$ such that

$$\xi_\pm = \ker\Big(\pm dz + \frac{1}{2}x\,dy - \frac{1}{2}y\,dx\Big).$$

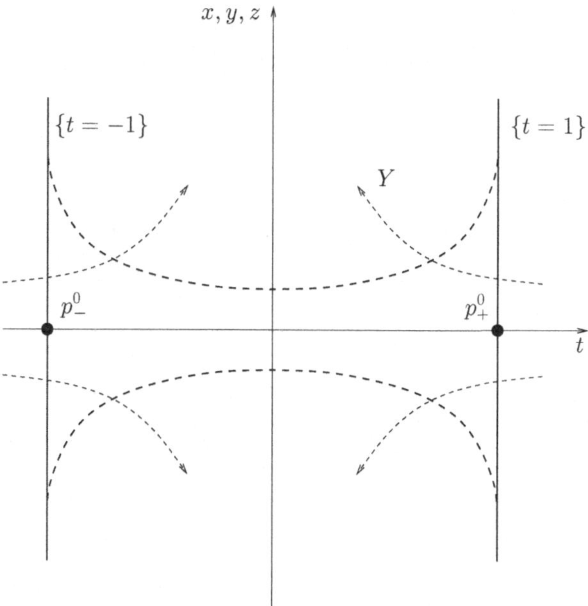

Fig. 6.3. Contact surgery.

Now consider $\mathbb{R}^4$ with coordinates x, y, z, t and standard symplectic form $\omega = dx \wedge dy + dz \wedge dt$. The vector field

$$Y := \frac{1}{2}x\,\partial_x + \frac{1}{2}y\,\partial_y + 2z\,\partial_z - t\,\partial_t$$

is a Liouville vector field for ω. It is transverse to the hypersurfaces $\{t = \pm 1\}$, where it induces the contact forms

$$(i_Y\omega)|_{T(\{t=\pm 1\})} = \pm dz + \frac{1}{2}x\,dy - \frac{1}{2}y\,dx,$$

see Figure 6.3.

The cylindrical tube in that picture, smoothly connecting the two hyperplanes $\{t = \pm 1\}$, is everywhere transverse to Y and hence also inherits a contact structure. This means that we can form the contact connected sum of these two hyperplanes. The Darboux theorem allows us to identify neighbourhoods of $p_\pm \in (M_\pm, \xi_\pm)$ with neighbourhoods of the points $p_\pm^0 \in \{t = \pm 1\}$ (see Figure 6.3) via a contactomorphism. If we choose the cylindrical tube so narrow that it connects to the two hyperplanes inside those neighbourhoods, we can glue that tube to $M_\pm$ to give us the contact connected sum of the two manifolds.

The picture for higher-dimensional surgeries is completely analogous.

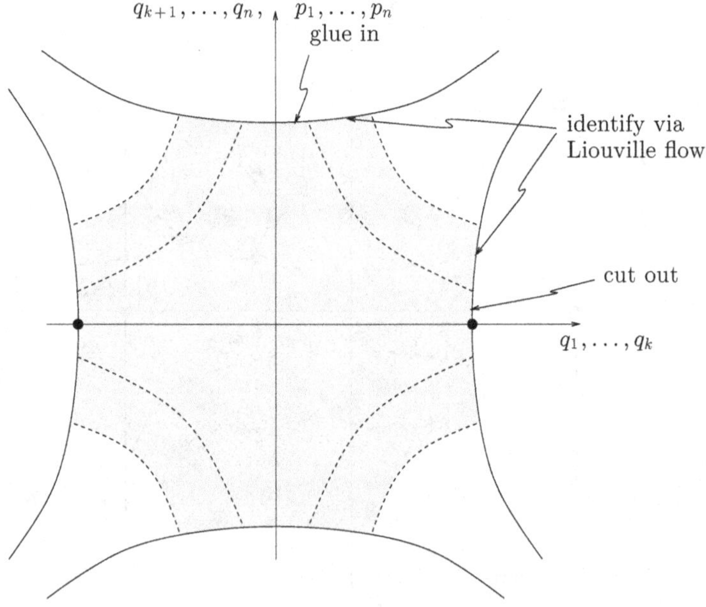

Fig. 6.4. Contact surgery via Liouville flow.

The following 'intermediate' point of view will prove useful in Section 6.4. Again, we are only interested in obtaining a contact structure on the surgered manifold. We use the symplectic model handle to do so, but not by forming a symplectic cobordism. Rather, we perform the construction indicated in Figure 6.4. First we remove a neighbourhood $S^{k-1} \times D^{2n-k}$ of the isotropic sphere S^{k-1} in (M, ξ). This corresponds (contactomorphically) to a neighbourhood $S_H^{k-1} \times D^{2n-k}$ of S_H^{k-1} in $g^{-1}(-1)$ in Figure 6.4 labelled 'cut out'. We choose these neighbourhoods so small that we also have a contactomorphisms of collars of the resulting manifolds with boundary (that is, the manifolds $M \setminus \mathrm{Int}\,(S^{k-1} \times D^{2n-k})$ and $g^{-1}(-1) \setminus \mathrm{Int}\,(S_H^{k-1} \times D^{2n-k})$). Using the Liouville flow, see Lemma 2.1.5, we can identify this collar neighbourhood with a subset of $g^{-1}(1)$ as shown in the figure, which then enables us to glue in a copy of $D^k \times S^{2n-k-1} \subset g^{-1}(1)$ so as to obtain the surgered contact manifold.

6.3 Framings in contact surgery

According to Theorem 6.2.5, the framing for a contact surgery along an isotropic sphere with trivial conformal symplectic normal bundle (CSN) is determined by a choice of trivialisation of that bundle.

For topological applications of this theorem, the following problems need to be discussed.

(1) Given an embedding f of a $(k-1)$–dimensional sphere in a contact manifold of dimension $2n-1$, where $1 \leq k \leq n$, is it possible to find an isotropic embedding (with trivial CSN) topologically isotopic to f?

(2) For a given topological isotopy class of embeddings f, is it possible to realise different topological framings as natural framings in the sense of Theorem 6.2.5 (beyond the freedom in trivialising CSN) by choosing different isotropic embeddings in that topological isotopy class?

The answer to the first question is provided by an h–principle. For a gentle introduction to the method of h–principles see [95]; all the results quoted here can be found in [77]; the brave-hearted should also consult [119]. The relevant results will be described in Section 6.3.1.

The answer to the second question, which will be presented in detail below, can be conveniently formulated in terms of the extension of an almost complex structure over the handle corresponding to the surgery.

Since, in this section, we are dealing only with homotopical questions, we may revert to the original more simple-minded handle attaching picture of Section 6.1, except that we want to be more explicit about the embedding of $S^{k-1} \times D^{2n-k}$ used for the surgery or the handle attachment. Thus, we define a k–handle H (of dimension $2n$) as

$$H = D^k \times D^{2n-k}.$$

The lower boundary of H is

$$\partial^- H := \partial D^k \times D^{2n-k}.$$

We can attach H to a boundary component M of a $2n$–dimensional manifold (e.g. $[-1,1] \times M$) with the help of an embedding ('attaching map')

$$f\colon\ \partial^- H \longrightarrow M.$$

Now assume that M carries a contact structure $\xi = \ker\alpha$. Let (W,ω) be a symplectic manifold which has (M,ξ) as a convex boundary component, e.g. $W = [-1,1] \times M$ with the symplectic form $\omega = d(\mathrm{e}^t\alpha)$ (and $M \equiv \{1\} \times M$). By Proposition 2.4.5, this determines a homotopically unique ω–compatible almost complex structure J on W. Along M this J may be described as the homotopically unique complex bundle structure on $TW|_M$ that is ξ–compatible and sends the outer normal vector field (or the Liouville vector field) to a vector field positively transverse to ξ (e.g. the Reeb vector field of α). In particular, along M the almost complex structure J does not depend (up to homotopy) on the specific choice of (W,ω).

Here is the main theorem of the present section. Observe that it only applies to contact manifolds of dimension at least 5. The 3–dimensional case will have to be dealt with separately.†

Theorem 6.3.1 *Let (M,ξ) be a $(2n-1)$–dimensional contact manifold and assume that it contains a $(k-1)$–dimensional embedded sphere with trivial (topological) normal bundle, where $n > 2$ and $1 \leq k \leq n$. Describe the sphere and a trivialisation of its normal bundle by an embedding*

$$f\colon\ S^{k-1} \times D^{2n-k} \longrightarrow M.$$

Let (W,ω) be a symplectic manifold with (M,ξ) as a convex boundary component. Let $W' = W \cup_f H$ be the manifold obtained from W by attaching a k–handle H along its lower boundary $\partial^- H = S^{k-1} \times D^{2n-k}$ with the help of f, and call M' the new boundary component (the result of surgery on M). If the almost complex structure J described above extends over H to an almost complex structure on all of W', then W' carries a symplectic form ω' and an ω'–compatible almost complex structure homotopic to J such that M' is a convex boundary component of (W',ω'). The induced contact structure ξ' on M' is the result of performing contact surgery along an isotropic sphere topologically isotopic to $f(S^{k-1} \times \{\mathbf{0}\})$.

† In different terms, the essential part of the corresponding result in dimension 3 has already been formulated in Example 6.2.7.

Remarks 6.3.2 (1) If, conversely, (M', ξ') is the result of performing a contact surgery on (M, ξ), then the corresponding symplectic cobordism gives an extension of J over H. Thus, the theorem gives a complete description of the necessary and sufficient condition for being able to perform contact surgery in a given topological type (i.e. given a topological isotopy class of embeddings $S^{k-1} \to M$ and a choice of normal framing).

(2) The Liouville vector fields for ω' defined near M' and pointing outwards along M' form a convex set, hence so do the induced contact forms on M'. Since the modification of M to M' occurs in a compact set, we can sensibly speak (by Gray stability) of *the* contact structure ξ' induced by ω'.

(3) For $n = 2$ and $k = 1$ (i.e. connected sums or more general surgeries of index 1 of contact 3–manifolds), the theorem remains true without changes; the homotopical condition (extension of the almost complex structure over the 1–handle) is trivially satisfied. For $n = k = 2$, the theorem remains true provided ξ is an overtwisted contact structure (and W necessarily non-compact or with additional non-convex boundary components).

By attaching handles along the convex boundary of the disc D^{2n} with its standard symplectic form, we obtain the following corollary.

Corollary 6.3.3 *Let W be a manifold with boundary obtained from the disc D^{2n}, $n > 2$, by attaching handles of index between* 1 *and n. If W admits an almost complex structure, then it admits a symplectic form for which the boundary ∂W is convex; in particular, ∂W admits a contact structure.* □

Remark 6.3.4 Eliashberg [65] proved a stronger version of this corollary in the holomorphic setting. Under the same assumptions on W, one can actually homotope the almost complex structure to a genuine complex structure such that the boundary ∂W is strictly pseudoconvex, see Example 2.1.7 and Section 5.3. Moreover, one can achieve that the interior of W with that complex structure is a Stein manifold. By a famous theorem of Lefschetz [156] and Andreotti–Frankel [11], see [185, Thm. 7.2], a Stein manifold of real dimension $2n$ has the homotopy type of an n–dimensional CW–complex, which implies that it can be constructed as a handlebody of the type described in the corollary. Thus, the corollary (in Eliashberg's version) establishes that in dimension $2n > 4$ there are no further obstructions to the existence of a Stein structure beyond that topological restriction and the existence of an almost complex structure.

In dimension $2n = 4$ there is a stronger restriction on the framings that can be used to attach symplectic 2–handles. This will be discussed below. What remains true, however, is that any symplectic handlebody obtained by attaching symplectic 1– and 2–handles (with the framing restriction) to D^4 likewise carries a Stein structure. See also [110] and [111].

In the following subsections we prepare the ground for the proof of Theorem 6.3.1. Several of the results presented there are of some independent interest. In all the applications of this theorem discussed in this book (notably in Chapter 8), we can deal with the framing issue in an *ad hoc* fashion. Therefore, I refrain from giving a proof of Theorem 6.3.1 in its full generality, for this would require us to delve deep into such matters as Whitney's classification of half-dimensional immersions via the self-intersection index. Besides, a readable account of such a general proof is going to be available in [47]. Instead, I only present a complete proof for the cases $k < n$ and, *mutatis mutandis*, $k = n = 2$. In spite of the special features of that last case, it is quite suited to illustrate the result for $n = k > 2$.

6.3.1 The h–principle for isotropic immersions

The first task in the proof of Theorem 6.3.1 will be to find an isotropic embedding of S^{k-1} into M, isotopic to the given embedding $f|_{S^{k-1}\times\{0\}}$. This will be done via the h–principle for isotropic immersions. Given a differentiable map $g\colon L \to M$ with $Tg(TL) \subset \xi$, we can form the **complexification** $T^{\mathbb{C}}g$ of the differential Tg (with respect to a ξ–compatible complex bundle structure J on ξ) by setting

$$T_p^{\mathbb{C}}g\big((a+ib)\mathbf{v}\big) := aT_pg(\mathbf{v}) + bJ(T_pg(\mathbf{v})) \;\text{ for }\; p \in L,\, \mathbf{v} \in T_pL,\, a, b \in \mathbb{R}.$$

For g being an isotropic *immersion*, this defines a fibrewise injective complex bundle map $TL \otimes \mathbb{C} \to \xi$ over g. The h–principle says, in essence, that this homotopical condition, i.e. the existence of such a bundle map, is sufficient for the existence of an isotropic immersion.

Theorem 6.3.5 *Let L be a $(k-1)$–dimensional manifold and $f_0\colon L \to M$ a map covered by a fibrewise injective complex bundle map $F_0\colon TL\otimes\mathbb{C} \to \xi$, where (M, ξ) is a $(2n-1)$–dimensional contact manifold (the existence of such a bundle map forces $k \leq n$). Then f_0 is homotopic to an isotropic immersion $f_1\colon L \to M$ homotopic to f_0 and with $T^{\mathbb{C}}f_1$ homotopic to F_0 via fibrewise injective complex bundle maps. Moreover, the map f_1 and the homotopy between f_0 and f_1 can be chosen arbitrarily C^0–close to f_0.*

Proof See [77, Chapter 16]. For the special case $L = S^1$ and $\dim M = 3$ (where the bundle condition is trivially satisfied, because any orientable vector bundle over S^1 is trivial), this is essentially Theorem 3.3.1. That theorem was only stated for f_0 an embedding, but the proof equally applies to an arbitrary (continuous) map f_0. □

The following is the extension of this result to embeddings.

Proposition 6.3.6 *Under the assumptions of the preceding theorem, if f_0 is an embedding and L is closed, one may further achieve that f_1 is likewise an embedding isotopic to f_0 (again in any C^0–small neighbourhood of f_0).*

Proof This follows from general position arguments. Consider the Legendrian case (i.e. $k = n$) – the argument will go through *a fortiori* in the isotropic case with $k < n$. By applying the neighbourhood theorem for isotropic submanifolds (Thm. 2.5.8) locally on $f_1(L)$, we can find finitely many points $p_\nu \in f_1(L)$, $\nu = 1, \ldots, m$, and Darboux charts $U_\nu \subset M$, with the contact structure ξ on each U_ν given by

$$\xi = \ker\Big(dz + \sum_{j=1}^{n-1} x_j\, dy_j\Big),$$

having the following properties (for each ν).

- $f_1^{-1}(p_\nu) \subset L$ consists of a single point.
- $f_1^{-1}(U_\nu)$ consists of finitely many connected components (use compactness of L).
- The component of $f_1(L) \cap U_\nu$ containing $f_1(p_\nu)$ is given in the Darboux coordinates by the set

  $$\{z = x_1 = \cdots = x_{n-1} = 0,\ |y_j| < 3,\ j = 1, \ldots, n-1\}.$$

  By mild abuse of notation we write this set as $C(3)_\nu$ and regard it as a subset of L (rather than $f_1(L)$).
- The subsets

  $$\{z = x_1 = \cdots = x_{n-1} = 0,\ |y_j| < 1,\ j = 1, \ldots, n-1\}$$

  from all m Darboux charts still cover L. We label these sets as $C(1)_\nu$. Correspondingly, we define sets $C(2)_\nu$.

We are now going to homotope the Legendrian immersion f_1 to a Legendrian embedding on one of the sets $C(1)_\nu$ at a time. Thus, assume that for some $i \in \{1, \ldots, m\}$, the map f_1 is already an embedding on $\cup_{\nu=1}^{i-1} C(1)_\nu$. We now want to turn f_1 into an embedding on $\cup_{\nu=1}^{i} C(1)_\nu$ by changing f_1 only in $C(2)_i$. That is, we want to remove intersection points of $f_1(C(1)_i)$ with any other part of the image $f_1(L)$, without creating intersection points of the image of $C(2)_i$ with that of $\cup_{\nu=1}^{i-1} \overline{C(1)}_\nu$.

In the sequel, we drop the subscript i. Represent $f_1(C(3))$ by its image under the front projection $\pi_F\colon \mathbb{R}^{2n-1} \to \mathbb{R}^n$, defined by

$$\pi_F(z, x_1, \ldots, x_{n-1}, y_1, \ldots, y_{n-1}) = (z, y_1, \ldots, y_{n-1}),$$

see Definition 3.2.2. This image may be regarded as the graph of the identically zero function on

$$C_{\mathbf{y}}(3) := \{\mathbf{y} \in \mathbb{R}^{n-1}\colon |y_j| < 3, \, j = 1, \ldots, n-1\}.$$

Observe that for any smooth function φ on $C_{\mathbf{y}}(3)$, the graph

$$\{(\mathbf{y}, z) \in \mathbb{R}^n\colon \mathbf{y} \in C_{\mathbf{y}}(3), z = \varphi(\mathbf{y})\}$$

lifts in a unique way to a Legendrian submanifold in $(\mathbb{R}^{2n-1}, \xi)$ by the requirement $x_j = -\partial\varphi/\partial y_j$, $j = 1, \ldots, n-1$, see Lemma 3.2.3. In particular, two graphs intersecting transversely lift to disjoint Legendrian submanifolds.

Now consider the connected components of the open subset $f^{-1}(U_i)$ of L, one of which is $C(3) = C(3)_i$. Regard the composition of f, restricted to any of these components, with the front projection as a collection of maps of open subsets of L to $\mathbb{R}^n$. The map on the component $C(3)$ may be identified with the map $\mathbf{y} \mapsto (\mathbf{y}, 0)$ on $C_{\mathbf{y}}(3)$. The usual transversality theorem for sections (see [153, Chapter IV], for instance) allows us to find a function φ_1, identically zero on $C_{\mathbf{y}}(3) \setminus \overline{C_{\mathbf{y}}(2)}$, whose graph is transverse to the other maps on $C_{\mathbf{y}}(1)$, and everywhere remains transverse to the restriction of those other maps to $\cup_{\nu=1}^{i-1} \overline{C(1)}_\nu$. Indeed, by Sard's theorem there is a dense set of values $c \in \mathbb{R}$ such that the constant function $\varphi \equiv c$ satisfies the transversality condition; the function φ_1 can now be defined as a smooth interpolation between $\varphi_1 \equiv c$ on $C_{\mathbf{y}}(1)$ and $\varphi_1 \equiv 0$ on $C_{\mathbf{y}}(3) \setminus \overline{C_{\mathbf{y}}(2)}$. For c sufficiently small, this map will have the desired properties (see Figure 6.5).

Thus, we have now achieved that the (perturbed) f_1 is a Legendrian embedding homotopic and C^0–close to f_0. We may regard f_1 as a map from L into a tubular neighbourhood of $f_0(L)$, inducing isomorphisms on all homotopy groups. By a result of Haefliger quoted below (Thm. 8.2.6) – for $k < n$ this is a simple general position result – f_1 is isotopic to f_0 (inside that tubular neighbourhood) provided $k < n$ or $n \geq 3$. The only case where this general argument does not apply is $k = n = 2$; for that case, however, we provided an *ad hoc* argument in Theorem 3.3.1. □

Example 6.3.7 It is well known that every finitely presentable group π is the fundamental group of some closed manifold M of dimension $m \geq 4$. Given a presentation of π with generators $g_1, \ldots, g_k$ and relations $r_1, \ldots, r_l$, the desired manifold is obtained as follows. Start with the connected sum of k copies of $S^1 \times S^{m-1}$. The fundamental group of this connected sum is

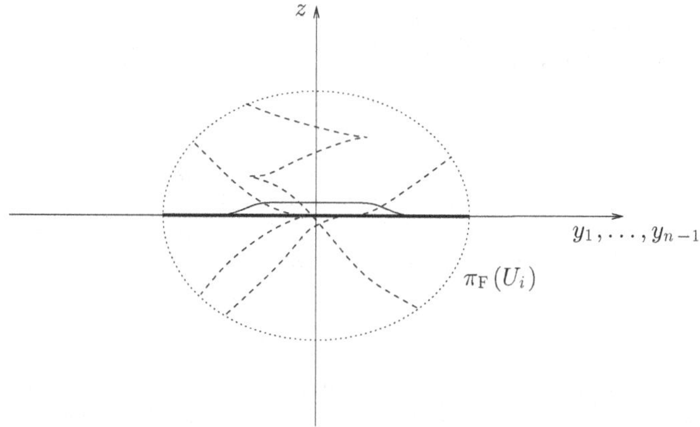

Fig. 6.5. Achieving transversality.

the free group on k obvious generators, which we identify with $g_1, \ldots, g_k$. Now perform surgery along disjoint embedded circles representing the words $r_1, \ldots, r_l$ in these generators. With each surgery, a copy of $S^1 \times D^{m-1}$ is replaced by $D^2 \times S^{m-2}$. Because of $m \geq 4$, this has – by Seifert–van Kampen – the effect of killing the corresponding word.

It follows immediately that every finitely presentable group is the fundamental group of a contact manifold of dimension $2m - 1 \geq 7$: simply take the unit cotangent bundle of the manifold constructed above.

Here is the argument for proving the same statement in dimension 5. This was first observed by A'Campo and Kotschick [4]. By surgery along k copies of S^0 in (S^5, ξ_{st}) we obtain a contact structure ξ on the connected sum of k copies of $S^1 \times S^4$. Over the embedded circles representing the relations, the U(2)–bundle ξ is trivial, so the h–principle for isotropic embeddings applies. Alternatively, see [4] for an elementary *ad hoc* argument that the embedded circles can be made isotropic. Contact surgery along these circles produces a contact 5–manifold with the desired fundamental group.

6.3.2 Interlude: the Whitney–Graustein theorem

In Theorem 6.3.5 we formulated the h–principle for isotropic immersions in its *non-parametric* form only. In this interlude we briefly discuss the *parametric* version. We prove the one-parametric version via the Whitney–Graustein theorem, i.e. the classification of immersions of S^1 in $\mathbb{R}^2$. We then give an alternative direct proof, and in turn deduce the Whitney–Graustein theorem. For a more elementary account of this contact geometric proof of the Whitney–Graustein theorem see [99].

Let (M, ξ) be a contact manifold and L an arbitrary manifold. Fix a complex bundle structure J on ξ. Write $\mathrm{Imm}_{\xi}(L, M)$ for the space of isotropic C^1–immersions of L in (M, ξ), equipped with the weak C^1–topology. Write $\mathrm{Mon}\,(TL \otimes \mathbb{C}, \xi)$ for the space of continuous and fibrewise injective complex bundle maps $TL \otimes \mathbb{C} \to \xi$ with the weak C^0–topology. Then the parametric h–principle for isotropic immersions says that the map

$$\begin{array}{ccc} \mathrm{Imm}_{\xi}(L, M) & \longrightarrow & \mathrm{Mon}\,(TL \otimes \mathbb{C}, \xi) \\ f & \longmapsto & T^{\mathbb{C}} f \end{array}$$

is a weak homotopy equivalence. We now want to prove this on the level of π_0 (i.e. the one-parametric h–principle) for the special case $L = S^1$ and $(M, \xi) = (\mathbb{R}^3, \xi_{\mathrm{st}})$. Observe that for two isotropic immersions to be in the same connected component of $\mathrm{Imm}_{\xi}(L, M)$ is the same as saying that they are C^1–homotopic via isotropic immersions (any continuous path in $\mathrm{Imm}_{\xi}(L, M)$ can be homotoped to a C^1–path). A C^1–homotopy via immersions is called a **regular homotopy**.

Remark 6.3.8 If we write $\xi = \ker \alpha$ (and choose J compatible with $d\alpha$), the space $\mathrm{Mon}\,(TL \otimes \mathbb{C}, \xi)$ may be replaced by the space of fibrewise injective bundle maps $F \colon TL \to \xi$ satisfying $F^*(d\alpha) \equiv 0$. The map $F \mapsto F^{\mathbb{C}}$ sending F to its complex linear extension $F^{\mathbb{C}}$ defines a homotopy equivalence between these two spaces.

Indeed, suppose we start with a bundle monomorphism $F \colon TL \to \xi$ satisfying $F^*(d\alpha) \equiv 0$. Assume $F^{\mathbb{C}}(X + \mathrm{i}Y) = \mathbf{0}$, that is, $F(X) + JF(Y) = \mathbf{0}$. Then

$$0 = F^*(d\alpha)(X, Y) = d\alpha(F(X), F(Y)) = d\alpha(F(X), JF(X)).$$

This implies $F(X) = \mathbf{0}$, hence $X = \mathbf{0}$, and then likewise $Y = \mathbf{0}$.

Conversely, starting from an element $\widetilde{F} \in \mathrm{Mon}\,(TL \otimes \mathbb{C}, \xi)$, there is a canonical procedure (depending only on a choice of a suitable covering of M by open sets) for constructing an F as above with $F^{\mathbb{C}}$ homotopic to $\widetilde{F}$, via local orthonormalisation of frames $\widetilde{F}(X_1), \ldots, \widetilde{F}(X_{k-1})$ (where $k - 1 = \dim L$) with respect to a Hermitian structure associated with $d\alpha$ and J. See [59, Lemma 2.5] for details.

Given a Legendrian immersion $\gamma \colon S^1 \to (\mathbb{R}^3, \xi_{\mathrm{st}})$, we write $\mathtt{rot}(\gamma)$ for its rotation number (see Remark 3.5.22).

Lemma 6.3.9 *Let γ_0, γ_1 be two Legendrian immersions of the circle S^1 in $(\mathbb{R}^3, \xi_{\mathrm{st}})$. Then $\mathtt{rot}(\gamma_0) = \mathtt{rot}(\gamma_1)$ if and only if $T^{\mathbb{C}}\gamma_0$ and $T^{\mathbb{C}}\gamma_1$ are homotopic as complex bundle isomorphisms $TS^1 \otimes \mathbb{C} \to \xi_{\mathrm{st}}$.*

Proof Choose a homotopy H between γ_0 and γ_1, i.e. a continuous map $H\colon S^1 \times [0,1] \to \mathbb{R}^3$ with $H(\theta, i) = \gamma_i(\theta)$ for $i = 0,1$ and all $\theta \in S^1$. We do not need to assume this to be a regular homotopy, although it would be possible to do so.

The bundle $\xi := \xi_{\mathrm{st}}$ is trivialised by the vector fields ∂_x and $\partial_y - x\,\partial_z$, and the rotation number measures rotations relative to this trivialisation. Then $H^*\xi$ is a trivialised 2–plane bundle over $S^1 \times [0,1]$. Notice that we can interpret the vector field γ_i' as a section of $H^*\xi$ over $S^1 \times \{i\}$, $i = 0,1$.

To argue in one direction, assume that $\mathtt{rot}(\gamma_0) = \mathtt{rot}(\gamma_1)$. This means that $\gamma_0'(\theta)$ and $\gamma_1'(\theta)$ make the same number of rotations with respect to the trivialisation of ξ as θ ranges over S^1. This implies that there is a global non-vanishing section of $H^*\xi$ that restricts to γ_i' over $S^1 \times \{i\}$. The map from $TS^1 \times [0,1] \subset T(S^1 \times [0,1])$ to $H^*\xi$ sending ∂_θ to that section defines a homotopy between $T\gamma_0$ and $T\gamma_1$ as injective bundle homomorphisms into ξ. In this dimension this is equivalent to saying that $T^{\mathbb{C}}\gamma_0$ and $T^{\mathbb{C}}\gamma_1$ are homotopic as complex bundle isomorphisms to ξ. (For isotropic immersions f in higher dimensions, the condition that $T^{\mathbb{C}}f$ be an injective complex bundle map is stronger than just Tf being injective, as is clear from the remark preceding this lemma.)

Conversely, assume that $T^{\mathbb{C}}\gamma_0$ and $T^{\mathbb{C}}\gamma_1$ are homotopic as complex bundle isomorphisms to ξ. Then $T\gamma_0$ and $T\gamma_1$ are homotopic as injective bundle maps. Let $\widetilde{H}\colon TS^1 \times [0,1] \to \xi$ be such a homotopy. Given the trivialisation of ξ, we can regard this as a homotopy of maps $S^1 \to \mathbb{R}^2 \setminus \{\mathbf{0}\}$. The degree of these maps is a homotopy invariant, and at $t = 0,1$ this degree equals $\mathtt{rot}(\gamma_0)$ and $\mathtt{rot}(\gamma_1)$, respectively. □

With this lemma, the one-parametric h–principle for Legendrian immersions $S^1 \to (\mathbb{R}^3, \xi_{\mathrm{st}})$ now takes the following form. Here a **Legendrian regular homotopy** means a C^1–homotopy via Legendrian immersions.

Theorem 6.3.10 *Legendrian regular homotopy classes of Legendrian immersions $\gamma\colon S^1 \to (\mathbb{R}^3, \xi_{\mathrm{st}})$ are in one-to-one correspondence with the integers, the correspondence being given by $[\gamma] \mapsto \mathtt{rot}(\gamma)$.*

We are going to relate this h–principle to that for immersions of S^1 in $\mathbb{R}^2$, known as the Whitney–Graustein† theorem. Recall the definition of the rotation number of such an immersion (Defn. 3.5.20).

Theorem 6.3.11 (Whitney–Graustein) *Regular homotopy classes of im-*

† The theorem is proved in a paper by Whitney [243], who writes: 'This theorem, together with its proof, was suggested to me by W. C. Graustein.'

mersions $\overline{\gamma}\colon S^1 \to \mathbb{R}^2$ are in one-to-one correspondence with the integers, the correspondence being given by $[\overline{\gamma}] \mapsto \mathtt{rot}(\overline{\gamma})$.

For a direct proof of this proposition see [95, Prop. 4.5].

Proof of Theorem 6.3.10 from Theorem 6.3.11 It is easy to see that any integer can be realised as the rotation number of a Legendrian immersion $S^1 \to \mathbb{R}^3$. What we need to show is that two such immersions γ_0, γ_1 with $\mathtt{rot}(\gamma_0) = \mathtt{rot}(\gamma_1)$ are Legendrian regularly homotopic.

By the discussion in Chapter 3, the Lagrangian projections $\gamma_{i,\mathrm{L}} = \pi_{\mathrm{L}} \circ \gamma_i$, $i = 0, 1$, are immersions $S^1 \to \mathbb{R}^2$ with $\mathtt{rot}(\gamma_{i,\mathrm{L}}) = \mathtt{rot}(\gamma_i)$, satisfying the area condition $\oint_{\gamma_{i,\mathrm{L}}} x\,dy = 0$. By the Whitney–Graustein theorem, $\gamma_{0,\mathrm{L}}$ and $\gamma_{1,\mathrm{L}}$ are regularly homotopic. The proof of that theorem can easily be adapted to yield a homotopy via immersions satisfying the area condition $\oint x\,dy = 0$. Then this homotopy lifts to a Legendrian regular homotopy between γ_0 and γ_1. □

Proof of Theorem 6.3.11 from Theorem 6.3.10 Again we only have to show that two immersions $\overline{\gamma}_0, \overline{\gamma}_1\colon S^1 \to \mathbb{R}^2$ (where we think of $\mathbb{R}^2$ as the (x,y)–plane) with $\mathtt{rot}(\overline{\gamma}_0) = \mathtt{rot}(\overline{\gamma}_1)$ are regularly homotopic.

After a regular homotopy we may assume that the $\overline{\gamma}_i$ satisfy the area condition $\oint_{\overline{\gamma}_i} x\,dy = 0$ and thus lift to Legendrian immersions $\gamma_i\colon S^1 \to (\mathbb{R}^3, \xi_{\mathrm{st}})$ with $\mathtt{rot}(\gamma_i) = \mathtt{rot}(\overline{\gamma}_i)$. By Theorem 6.3.10, γ_0 and γ_1 are Legendrian regularly homotopic. The Lagrangian projection of this homotopy gives a regular homotopy between $\overline{\gamma}_0$ and $\overline{\gamma}_1$. □

For this last argument to be of any interest, we do of course need to supply a direct proof of Theorem 6.3.10. In order to furnish such a proof, we revert to the front projection picture.

Proof of Theorem 6.3.10 Again we need only show that two Legendrian immersions $S^1 \to \mathbb{R}^3$ with the same rotation number are Legendrian regularly homotopic.

In the front projection of the Legendrian immersion γ, left and right cusps alternate. We label the up-cusps with $+$ and the down-cusps with $-$. Up to Legendrian regular homotopy, γ is completely determined by the sequence of these labels, starting at a right-cusp, say, and going once around S^1. (This can be seen by homotoping γ_{F} so that all left cusps come to lie on the line $\{y = 0\}$ and all right cusps on the line $\{y = 1\}$, say. The cusps on either line can be shuffled by further homotopies.) Moreover, a pair $+-$ or $-+$ can be cancelled from this sequence by a Legendrian regular homotopy; locally this is in fact achieved by a Legendrian isotopy, *viz.*, the first Legendrian

Reidemeister move (see Figure 6.6; there is an analogous move with the picture rotated by 180°).

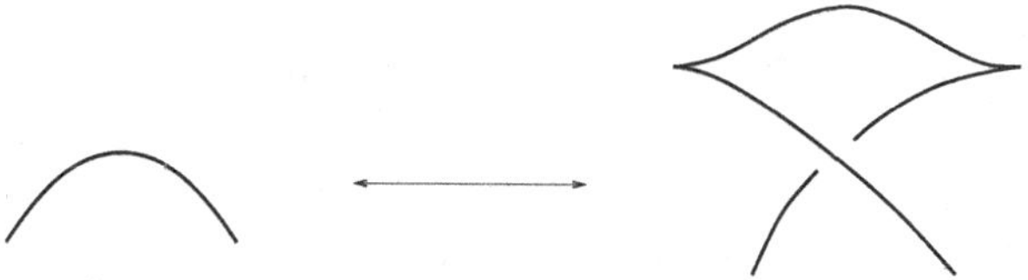

Fig. 6.6. The first Legendrian Reidemeister move.

Therefore, this sequence of labels can be reduced to a sequence containing only plus or only minus signs, or to one of the sequences $(+,-)$, $(-,+)$; see Figure 6.7 for an example. The formula $\texttt{rot}(\gamma) = (c_- - c_+)/2$ from Proposition 3.5.19 shows that there are the following possibilities: if $\texttt{rot}(\gamma)$ is positive (or negative, respectively), we must have a sequence of $2\,\texttt{rot}(\gamma)$ minus signs (or plus signs, respectively); if $\texttt{rot}(\gamma) = 0$, we must have the sequence $(+,-)$ or $(-,+)$. The proof is completed by observing that these last two sequences correspond to Legendrian isotopic knots: use a first Reidemeister move as in Figure 6.6, followed by the inverse of the rotated move. □

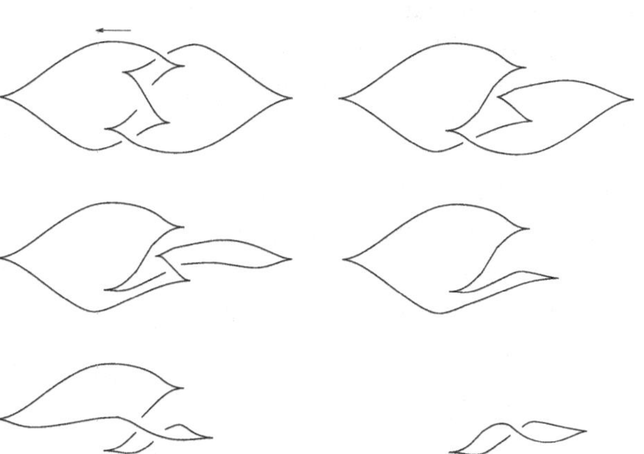

Fig. 6.7. An example of a Legendrian regular homotopy.

Remark 6.3.12 Self-tangencies in the front projection γ_F correspond to self-intersections of the Legendrian curve γ, since the negative slope of γ_F gives the x–component of γ. Therefore, as we pass such a self-tangency in the moves of Figure 6.7, we effect a crossing change in the Legendrian curve. With the orientation indicated in the figure, this example has $\mathtt{rot}(\gamma) = -1$.

6.3.3 Proof of the framing theorem

We continue with the proof of Theorem 6.3.1. The following lemma shows that the extendibility of the almost complex structure over the handle, as described in that theorem, allows us to apply the above h–principle so as to obtain a handle attached along an isotropic sphere with trivial conformal symplectic normal bundle. Write $\epsilon_{\mathbb{C}}$ for a trivial complex line bundle (over a base that will be clear from the context).

Lemma 6.3.13 *Let f_0 be the restriction to $S^{k-1} \equiv S^{k-1} \times \{\mathbf{0}\}$ of the attaching map*

$$f\colon\, S^{k-1} \times D^{2n-k} \longrightarrow M = \partial W$$

of a k–handle H as in Theorem 6.3.1. If the almost complex structure J on W extends over H, then f_0 is covered by a fibrewise isomorphic complex bundle map

$$\widetilde{F}_0\colon\, (TS^{k-1} \otimes \mathbb{C}) \oplus \epsilon_{\mathbb{C}}^{n-k} \longrightarrow \xi.$$

Proof Use Cartesian coordinates $(x_1, \ldots, x_n, y_1, \ldots, y_n)$ on $\mathbb{R}^{2n}$ in such a way that in the product $H = D^k \times D^{2n-k}$ the D^k–factor belongs to the coordinates $(x_1, \ldots, x_k)$, and correspondingly D^{2n-k} to the coordinates $(x_{k+1}, \ldots, x_n, y_1, \ldots, y_n)$. Write J_0 for the standard complex structure on H given by $J_0(\partial_{x_j}) = \partial_{y_j}$, $j = 1, \ldots, n$.

We write J also for the extension of the almost complex structure over H postulated in the lemma. Since the space of complex vector space structures on $\mathbb{R}^{2n}$ compatible with a given orientation is connected, see Lemma 8.1.7 below, we may assume that J coincides with J_0 on $D^k_{1/2} \times D^{2n-k}$, say. Here D_r denotes the closed disc of radius r.

Set $S^{k-1}_r = \partial D^k_r$, which we identify with $S^{k-1}_r \times \{\mathbf{0}\} \subset H$. Let ξ_r be the complex bundle of rank $n-1$ over S^{k-1}_r given by the maximal complex tangency of $\partial D^k_r \times D^{2n-k}$, that is,

$$\xi_r = \big(T(\partial D^k_r \times D^{2n-k}) \cap J(T(\partial D^k_r \times D^{2n-k}))\big)|_{S^{k-1}_r}.$$

Notice that for $r = 1/2$, where J is assumed to coincide with J_0, we have $\xi_{1/2} \cong (TS^{k-1} \otimes \mathbb{C}) \oplus \epsilon_{\mathbb{C}}^{n-k}$. By the homotopy properties of vector bundles, see [140, Chapter 3], we have the same bundle isomorphism for ξ_1. Moreover, since J is the extension over the handle H of the almost complex structure given on W, the differential of the attaching map of H sends the maximal complex tangency in $\partial^- H$ isomorphically to the maximal complex tangency in ∂W, which is the given contact structure ξ. Thus, under the described bundle isomorphisms, the differential of the attaching map gives the desired bundle map $\widetilde{F}_0$. □

The reader should be warned that the framing theorem is anything but a straightforward consequence of the discussion so far. The trouble is that the isotropic embedding we find with the help of the h–principle will not, in general, give us the same normal framing as the original attaching map. To illustrate this point, consider the case $k = n = 2$. The almost complex structure on the handle determines the homotopy class of the bundle map $\widetilde{F}_0 = F_0$ in Theorem 6.3.5, and hence – essentially by Lemma 6.3.9, assuming for simplicity that $M = S^3$ – the rotation number `rot` of the Legendrian knot along which we want to perform surgery. The normal framing of this Legendrian knot, however, is given by its Thurston–Bennequin invariant `tb`, and only the parity of `tb` is determined by `rot`, see Proposition 3.5.23.

As shown in Example 6.2.7, the natural framing (as defined in Section 6.2) of the contact surgery along a Legendrian knot K in (S^3, ξ_{st}) differs from its contact framing by one negative twist.

In the case $k < n$, the proof of the framing theorem concludes as follows. Identify L with the zero section in the 1–jet space $J^1(L)$ with its natural contact structure (see Example 2.5.11) and consider contact embeddings of an open neighbourhood of L in $J^1(L)$ into (M, ξ). Here we have a similar h–principle, which now allows us to control the normal framing. For details see [47]. Moreover, for $k < n$ one can apply a general position argument to the homotopy of f_0 to an isotropic embedding f_1, and turn this into an isotopy while fixing the 'endpoints' f_0 and f_1.

The following proposition concludes the proof of the framing theorem for $k = n = 2$ and ξ an overtwisted contact structure.

Proposition 6.3.14 *Let $\gamma_0 \colon S^1 \to (M, \xi)$ be a Legendrian embedding into a contact 3–manifold. Then there is a Legendrian embedding $\gamma_1 \colon S^1 \to (M, \xi)$ with the following properties.*

- *There is a Legendrian regular homotopy of γ_0 to γ_1; equivalently: the complexified differential $T^{\mathbb{C}}\gamma_0$ is homotopic to $T^{\mathbb{C}}\gamma_1$ via complex bundle isomorphisms $TS^1 \otimes \mathbb{C} \to \xi$.*
- *There is an isotopy of topological embeddings from γ_0 to γ_1.*
- *The contact framing of $\gamma_1(S^1)$ differs from that of $\gamma_0(S^1)$ by two negative twists.*

If ξ is overtwisted, one can also achieve that the contact framing changes by two positive twists.

Notice that by the parity condition of Proposition 3.5.23 (and Remark 4.6.35), fixing the homotopy class of $T^{\mathbb{C}}\gamma$ entails that the contact framing can at best be changed by an even number of twists – at least for homologically trivial knots and with an adequate (and simple) generalisation of Lemma 6.3.9. But this observation holds true in general, for the following reason (which in turn gives an alternative argument for the parity condition in the previous setting).

Given γ_0, γ_1 satisfying the first two conditions in the proposition, we have an isotopy of embeddings $\gamma_t\colon S^1 \to M$, covered by a homotopy of bundle isomorphisms $\phi_t\colon TS^1 \otimes \mathbb{C} \to \xi|_{\gamma_t(S^1)}$, with $\phi_i = T^{\mathbb{C}}\gamma_i$ for $i = 0, 1$. This extends to a homotopy of bundle isomorphisms

$$\widetilde{\phi}_t\colon (TS^1 \otimes \mathbb{C}) \oplus \epsilon \to TM|_{\gamma_t(S^1)}.$$

The contact framings of the Legendrian knots $\gamma_i(S^1)$ are given by the image under $\widetilde{\phi}_i$ of the obvious trivialisation of $\mathrm{i}TS^1 \oplus \epsilon$, $i = 0, 1$. These two framings differ by an element of $\pi_1(\mathrm{SO}(2))$. The existence of the homotopy $\widetilde{\phi}_t$ implies that this element has to lie in the kernel of the stabilising map $\mathbb{Z} \cong \pi_1(\mathrm{SO}(2)) \to \pi_1(\mathrm{SO}(3)) = \mathbb{Z}_2$.

Proof of Proposition 6.3.14 The construction is local, so we may work with the front projection of a Legendrian knot in $(\mathbb{R}^3, \xi_{\mathrm{st}})$. Figure 6.8 (taken from [62]) shows a Legendrian regular homotopy that by Proposition 3.5.9 adds two negative twists to the contact framing. Moreover, the final knot in this sequence is clearly topologically isotopic to the initial one.

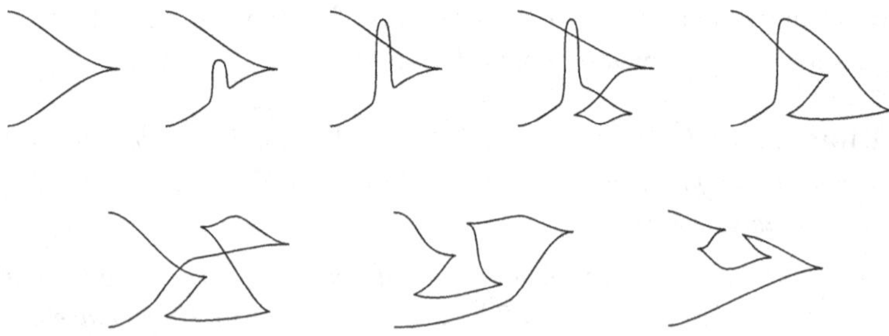

Fig. 6.8. A Legendrian regular homotopy decreasing `tb` by 2.

If ξ is overtwisted, we can find a Legendrian knot K_1 (*viz.*, the boundary of an overtwisted disc) whose surface framing coincides with its contact framing, and whose rotation number (relative to the overtwisted disc) is ± 1, depending on its orientation. The connected sum, as shown in Figure 6.9,† of two copies $K_1^{\pm}$ of K_1 with opposite orientations yields a topologically trivial Legendrian knot K_2 that, by Proposition 3.5.9 has

$$\mathtt{tb}(K_2) = \mathtt{tb}(K_1^+) + \mathtt{tb}(K_1^-) + 1 = 1$$

and, by Proposition 3.5.19,

$$\mathtt{rot}(K_2) = \mathtt{rot}(K_1^+) + \mathtt{rot}(K_1^-) = 0.$$

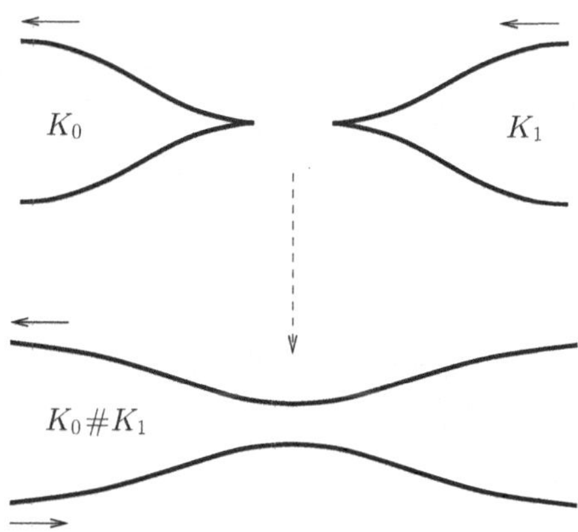

Fig. 6.9. Connected sum of two knots.

By a *topological* isotopy we can isotope the embedding γ_0 to an embedding $\widetilde{\gamma}_0$ disjoint from a disc bounded by K_2. By Theorem 3.3.1 we may assume that $\widetilde{\gamma}_0$, too, is a Legendrian embedding. Adding 'zigzags' as in Figure 6.10 allows one to change the rotation number at will, so we can use this to obtain a Legendrian embedding $\widetilde{\gamma}_1$ with $T^{\mathbb{C}}\widetilde{\gamma}_1 = T^{\mathbb{C}}\gamma_0$. The contact framing of $\widetilde{\gamma}_1$ can now be changed by any even number. The addition of negative twists has already been described. To add positive twists, we perform the connected sum of the oriented Legendrian knot $\widetilde{\gamma}_1(S^1)$ with K_2. This does not change the topological isotopy class of $\widetilde{\gamma}_1$, the homotopy class of $T^{\mathbb{C}}\widetilde{\gamma}_1$ remains unchanged by Proposition 3.5.19 and Lemma 6.3.9, and the contact framing receives two positive twists by Proposition 3.5.9. □

Remarks 6.3.15 (1) In tight contact 3–manifolds, a (homologically trivial) Legendrian knot satisfies the Bennequin inequality (Thm. 4.6.36). This implies that it is not possible, in general, to add positive twists to its contact

† Of course, the $K_1^{\pm}$ do not sit in $(\mathbb{R}^3, \xi_{\mathrm{st}})$, but the connected sum may be taken in such a Darboux chart, and the computations of the change in tb and rot can be done in this chart.

framing by a Legendrian regular homotopy. Therefore, the framing theorem fails for $k = n = 2$ and ξ a tight contact structure.

(2) When we are dealing with framings of Legendrian knots γ in contact 3–manifolds, it is usually not essential to keep track of the homotopy class of the bundle map $T^{\mathbb{C}}\gamma$. So the preceding proposition may, alternatively, be phrased as follows.

- It is always possible to add a negative twist to the contact framing of a Legendrian knot by adding (in a Darboux chart) a 'zigzag' (Figure 6.10) to the front projection of the given knot. The rotation number increases or decreases by one, depending on the two extra cusps being oriented downwards or upwards. This is called a (**positive** or **negative**) **stabilisation** of the Legendrian knot, and up to Legendrian isotopy this process is independent of where we put the zigzag, see [82].

Fig. 6.10. Legendrian 'zigzags'.

- Given a Legendrian knot K in an overtwisted contact manifold (M, ξ), one can find a Legendrian knot K' topologically isotopic to K whose contact framing differs from that of K by one positive twist. (The topological isotopy between K and K' allows us to compare framings.)

A word of warning is in order. If $\xi|_{M\setminus K}$ is overtwisted (a knot K having this property is called **loose** [74]), one can directly form the connected sum of K with $K_1^{\pm}$ as above. The resulting K' is then a **destabilisation** of K, i.e. a stabilisation of K' of the appropriate sign brings us back to K. It may well happen, however, that ξ is an overtwisted contact structure on M that becomes tight when restricted to $M \setminus K$; see [60]. Then we first have to separate K from the overtwisted disc bounded by $K_1^{\pm}$ – by a *topological* isotopy – before we can form the connected sum without changing the topological type of K. In this case, although we can still produce a knot K' whose classical invariants look like those of a destabilisation of K, it is not clear, in general, whether K' actually is such a destabilisation.

Here is an alternative view of the phenomenon described in this last remark, suggested to me by Paolo Lisca. Consider a *non-loose* knot K in an overtwisted contact 3–manifold (M, ξ_0), e.g. the one from [60]. Choose an overtwisted contact structure ξ' on S^3 that is homotopic (as a 2–plane field) to the standard contact structure ξ_{st}. Then the contact structure $\xi_1 := \xi_0 \# \xi'$ on $M \# S^3 = M$ – where we form the connected sum in the complement of K – is homotopic (as a 2–plane field) to ξ_0. By Eliashberg's classification of overtwisted contact structures (Thm. 4.7.2), ξ_1 is actually isotopic as a contact structure to ξ_0. However, $K \subset (M \# S^3, \xi_1 = \xi_0 \# \xi')$ is obviously loose.

Write K_1 for the knot K when regarded as a Legendrian knot in (M, ξ_1). Let ψ_t, $t \in [0,1]$, be an isotopy of M with $T\psi_1(\xi_0) = \xi_1$. Then set $K_0 := \psi_1^{-1}(K_1)$; this is a *loose* Legendrian knot in (M, ξ_0). So the Legendrian knots K and K_0 in (M, ξ_0) cannot be Legendrian isotopic (by Theorem 2.6.2).

However, we have a continuous family of Legendrian knots $\psi_t^{-1}(K_1)$ in $(M, T\psi_t^{-1}(\xi_1))$, starting at $K_1 \subset (M, \xi_1)$ and ending at $K_0 \subset (M, \xi_0)$. The contact framing of $K_1 \subset (M, \xi_1)$ is the same as that of $K \subset (M, \xi_0)$, so K and K_0 are isotopic as framed knots (in particular, they have the same Thurston–Bennequin invariant, provided they are homologically trivial).

Finally, we want to convince ourselves that there is also a Legendrian regular homotopy between the Legendrian immersions $\gamma_0\colon S^1 \xrightarrow{\cong} K_0$ and $\gamma\colon S^1 \xrightarrow{\cong} K$, where we may assume that $\gamma = \psi_1 \circ \gamma_0$. By the parametric h–principle for Legendrian immersions, this is equivalent to saying that $T^{\mathbb{C}}\gamma_0$ and $T^{\mathbb{C}}\gamma$ are homotopic as complex bundle isomorphisms $TS^1 \otimes \mathbb{C} \rightarrow \xi_0$.

By the construction of the contact structure ξ_1, there is a family of 2–plane fields η_t, $t \in [0,1]$, with $\eta_i = \xi_i$ for $i = 0, 1$, and $\eta_t = \xi_0$ near K for all $t \in [0,1]$. By Remark 4.7.25 (1) we may assume that the path of contact structures from ξ_0 to ξ_1 given by $\xi_t := T\psi_t(\xi_0)$, $t \in [0,1]$, is homotopic rel $\{0,1\}$ – as a path of 2–plane fields – to $(\eta_t)_{t \in [0,1]}$. Then, by the homotopy properties of vector bundles (see in particular the proof of [140, Thm. 3.4.3]), one can construct a family of complex bundle isomorphisms $F_t\colon \xi_0 \rightarrow \xi_t$, with $F_0 = \mathrm{id}_{\xi_0}$ and F_1 equal to the identity over a neighbourhood of K. Then

$$F_t^{-1} \circ T^{\mathbb{C}}(\psi_t \circ \gamma_0)\colon TS^1 \otimes \mathbb{C} \longrightarrow \xi_0, \;\; t \in [0,1],$$

is a homotopy of bundle isomorphisms as required; indeed, it starts at $T^{\mathbb{C}}\gamma_0$ and ends at $T^{\mathbb{C}}(\psi_1 \circ \gamma_0) = T^{\mathbb{C}}\gamma$.

It turns out that the two negative twists that the move in Figure 6.8 adds to the contact framing can be 'read off' from the fact that the part of the lower strand that pushes across the upper strand in the third picture describing that move has Euler characteristic +1, and only positive Euler characteristics can be realised by a disjoint union of 1–dimensional manifolds with boundary.

For $k = n > 2$, one has a similar front projection picture. Again one works in a neighbourhood of a cusp, but now one pushes part of a lower 'strand' of $\mathbb{R}^n$ across a second copy of $\mathbb{R}^n$. By pushing up a neighbourhood of a ball, one realises Euler characteristic $+1$; by pushing up a tubular neighbourhood of a figure eight, one can realise Euler characteristic -1. This is the key to proving the framing theorem in that case.

6.4 Contact Dehn surgery

Recall from Section 1.7.2 that Dehn surgery with coefficient $p/q \in \mathbb{Q} \cup \{\infty\}$ along a knot K in S^3 means cutting out a tubular neighbourhood νK and gluing in a solid torus by sending its meridian to $p\mu + q\lambda$, where μ is the meridian of $\partial(\nu K)$, and λ its preferred longitude determining the surface framing. Also recall that surgery in the sense of the present chapter is the same as Dehn surgery with integer coefficients.

As shown in Example 6.2.7, the natural framing of contact surgery along a Legendrian knot in a contact 3–manifold (M, ξ) is obtained by adding one negative twist to the contact framing. Given an arbitrary knot K in (M, ξ), we may isotope it by Theorem 3.3.1 to a Legendrian knot. By Remarks 6.3.15 we can therefore perform any integer Dehn surgery along K as a contact surgery, provided ξ is an overtwisted contact structure. If ξ is tight, the surgery coefficient needs to be a sufficiently small integer (because of the Bennequin inequality).

When talking about Dehn surgery along a Legendrian knot K in (M, ξ), it is convenient to express the surgery coefficient in terms of the longitude determined by the contact framing; this makes sense even if K is not homologically trivial. With this convention, the contact surgery discussed in the foregoing sections becomes a (-1)–surgery. Recall in particular the 'intermediate' point of view at the end of Section 6.2, which gives the 'cut and paste' interpretation of contact surgery that we are using here.

Definition 6.4.1 *Contact surgery in the sense of Theorem 6.2.5 along a Legendrian knot in a contact 3–manifold is called* **contact (-1)–surgery**.

Remark 6.4.2 With this notion, we can now phrase Remark 6.3.4 more succinctly as follows: any contact 3–manifold obtained from (S^3, ξ_{st}) by contact (-1)–surgeries admits a Stein filling. In fact, the handlebody corresponding to these surgeries carries such a Stein structure. (The same is true if there are additional 1–handles.)

In [50] it was shown how to define a more general notion of contact r–

surgery for any $r \in \mathbb{Q} \cup \{\infty\}$. Recall from Example 2.5.10 that a model for the contact structure in the neighbourhood of a Legendrian $K = S^1$ is provided by

$$\cos\theta\, dx - \sin\theta\, dy = 0.$$

We saw on page 95 that the tubular neighbourhood $\{x^2 + y^2 \leq \varepsilon\}$ has a convex boundary in the sense of Definition 4.6.16. When we cut out this tubular neighbourhood νK and glue back a solid torus (corresponding to some surgery coefficient r relative to the contact framing), we should like to extend the contact structure ξ on $\overline{M \setminus \nu K}$ over the solid torus we glue back. This amounts to finding contact structures on the solid torus $S^1 \times D^2$ with a certain convex boundary condition determined by r.

When $r = 0$, the meridional disc of $S^1 \times D^2$ gives rise to an overtwisted disc. When $r \neq 0$, then by results of Giroux [105] and Honda [137] there are finitely many tight contact structures (and at least one such) on $S^1 \times D^2$ with the corresponding convex boundary condition.† This allows one to speak of **contact r–surgery**, with the caveat that the resulting contact structure on the surgered manifold is not, in general, uniquely determined.

For $r = 1/k$ with $k \in \mathbb{Z}$, however, the results of Giroux and Honda say that there is indeed a unique tight contact structure on $S^1 \times D^2$ satisfying the relevant convex boundary condition. Thus, contact $(1/k)$–surgery is a well-defined process also as regards the resulting contact structure. In particular, this extension procedure can be taken as the definition of contact (-1)–surgery, in which case the above definition becomes a theorem – proved by combining Example 6.2.7 and the fact that the contact structure on the upper boundary $\partial^+ H = D^2 \times \partial D^2$ of a 4–dimensional symplectic 2–handle is tight. This last statement is a consequence of the fact that attaching such a handle to the symplectic 4–ball gives rise to a fillable and hence‡ tight contact 3–manifold.

Thus, whereas integer Dehn surgeries have a special topological meaning, the $(1/k)$–surgeries (with respect to the contact framing) are the ones a contact geometer would prefer. The contact topologist, therefore, should be most interested in contact (± 1)–surgeries.

In the arguments of Section 6.2 it is feasible to exchange the roles of the upper and lower boundary of the symplectic handle, and thus to consider handles that can be attached to concave ends of symplectic cobordisms. (This is reasonable only if we take $k \geq n$, for the attaching sphere is now

† N.B. This is not saying that the contact structure on the surgered manifold will be tight, even if ξ was tight.

‡ See Theorem 6.5.6.

an S^{2n-k-1}, and we can only appeal to an h–principle if this is an isotropic sphere.)

Proposition 6.4.3 *The attaching of a 4–dimensional symplectic 2–handle along the concave boundary of a 4–dimensional symplectic cobordism corresponds to a contact $(+1)$–surgery on that concave boundary.*

Proof This follows from a computation completely analogous to that in Example 6.2.7. □

Unfortunately, as soon as we start to mix contact $(+1)$– and contact (-1)–surgeries on a given contact 3–manifold, we lose any correspondence with symplectic cobordisms. Even so, the following theorem from [51] has proved rather useful in applications, see e.g. [164] and [200].

Theorem 6.4.4 (Ding–Geiges) *Let (M, ξ) be a closed, connected contact 3–manifold. Then (M, ξ) can be obtained by contact (± 1)–surgery along a Legendrian link in (S^3, ξ_{st}).*

We need a few preparations for the proof of this theorem. Recall from page 95 the definition of a Legendrian push-off K' of a Legendrian knot K. By construction, it is a knot parallel to K that determines the contact framing of K. Observe that K may in turn be regarded as a Legendrian push-off of K'. It is not clear, *a priori*, whether two Legendrian push-offs of K in opposite directions transverse to ξ are Legendrian isotopic in $M \setminus K$. This, however, has no impact on the validity of the following statement.

Proposition 6.4.5 (Cancellation Lemma) *Let (M', ξ') be the contact manifold obtained from (M, ξ) by contact (-1)–surgery along a Legendrian knot K and contact $(+1)$–surgery along a Legendrian push-off K' of K. Then (M', ξ') is contactomorphic to (M, ξ).*

Proof As a warm-up, we deal with the topological part of this proposition, i.e. we prove that M' is diffeomorphic to M, even though the contact geometric argument below also proves that part.

Write T for a solid torus $S^1 \times D^2$ and νK for a closed tubular neighbourhood of K. The meridian on $\partial(\nu K)$ will be denoted by μ_K. We may assume that K' lies on $\partial(\nu K)$. Since K' represents the contact framing of K, contact (-1)–surgery along K means that topologically we form the manifold

$$M'' := \overline{M \setminus \nu K} \cup_g T,$$

where the gluing map g sends the meridian $\mu_T := * \times \partial D^2$ to $\mu_K - K'$ (and the longitude $\lambda_T := S^1 \times *$ to K', say). The gluing of λ_T with K' implies that in the surgered manifold M'' the knot K' is isotopic to $S^1 \times \{\mathbf{0}\} \subset T$. Hence, the second surgery along K' amounts to cutting out T and regluing it a second time.

Any parallel copy of K' on $\partial(\nu K) = \partial T$ represents the contact framing of K', so in the second surgery along $S^1 \times \{\mathbf{0}\} \subset T$, the surgery coefficient has to be read with respect to the longitude K' on ∂T. The meridian of $S^1 \times \{\mathbf{0}\}$ in M'' is $\mu_K - K'$. This means that for the second surgery we glue in T by sending μ_T to $(\mu_K - K') + K' = \mu_K$, which amounts to a trivial surgery.

Now we introduce contact structures into the picture by considering the symplectic model handle with $k = n = 2$. As in Example 6.2.7 we see that the contact framing of the Legendrian attaching sphere $S^1_H \subset \partial^- H$ is given by the vector field $q_1\,\partial_{p_1} + q_2\,\partial_{p_2}$ tangential to the hypersurface $g^{-1}(-1)$ along S^1_H, and transverse to the contact structure $\ker \alpha_0$. It is now easy to check that in terms of the coordinates (q_1, q_2, p_1, p_2), the Legendrian push-offs of S^1_H are given by

$$(\sqrt{2\varepsilon^2+2}\,\cos\theta, \sqrt{2\varepsilon^2+2}\,\sin\theta, \pm\varepsilon\cos\theta, \pm\varepsilon\sin\theta),\quad \theta \in S^1.$$

Likewise, the Legendrian push-offs of the belt sphere

$$\{q_1 = q_2 = 0,\ p_1^2 + p_2^2 = 1\} \subset \partial^+ H \subset g^{-1}(1)$$

are given by

$$(\pm\delta\cos\theta, \pm\delta\sin\theta, \sqrt{1+\delta^2/2}\,\cos\theta, \sqrt{1+\delta^2/2}\,\sin\theta),\quad \theta \in S^1.$$

The flow of the Liouville vector field $Y = -q_1\,\partial_{q_1} - q_2\,\partial_{q_2} + 2p_1\,\partial_{p_1} + 2p_2\,\partial_{p_2}$ is given by

$$(q_1, q_2, p_1, p_2) \longmapsto (\mathrm{e}^{-t}q_1, \mathrm{e}^{-t}q_2, \mathrm{e}^{2t}p_1, \mathrm{e}^{2t}p_2).$$

Hence this flow moves any Legendrian push-off of the attaching sphere to one of the belt sphere. We conclude that in the contact manifold obtained by contact (-1)–surgery along K, the Legendrian push-off K' of K is Legendrian isotopic to the belt sphere corresponding to this first surgery.

From the description of contact surgery on page 302 and in Figure 6.4 it is then immediate (with Proposition 6.4.3) that by a contact $(+1)$–surgery along this belt sphere we recover the original contact manifold. □

Proof of Theorem 6.4.4 A proof of this theorem pretty much from first principles was given in [51]. Here we give a much shorter proof.†

† The importance of the proof given in [51] lies in the fact that it also discusses equivalences of contact Dehn surgeries, see [55].

In [56] it was shown that any Lutz twist can be effected by two contact $(+1)$–surgeries. The proof given there is by exhibiting explicit surgery curves and calculations very similar to those in our proof of the Lutz–Martinet theorem in Chapter 4.

Thus, by performing two such contact $(+1)$–surgeries on the given contact manifold (M, ξ), we obtain an overtwisted contact structure ξ' on M. Thanks to the Lickorish–Wallace theorem (Thm. 1.7.5), we can recover S^3 from M by integer surgery. According to Remarks 6.3.15, ξ' being overtwisted, we can perform these surgeries as contact $(+1)$–surgeries, say. This produces some contact structure ξ'' on S^3.

By Eliashberg's classification of contact structures on S^3 (Thm 4.10.3) and our proof of the Lutz–Martinet theorem, one may obtain ξ'' from ξ_{st} by Lutz twists or, as indicated, by contact $(+1)$–surgeries. Hence, conversely, we can obtain ξ_{st} from ξ'' by contact (-1)–surgeries thanks to the cancellation lemma.

A further application of the cancellation lemma tells us that we can reverse all these surgeries so as to obtain (M, ξ) from (S^3, ξ_{st}) by a sequence of contact (± 1)–surgeries. Since each surgery curve may be assumed to miss the belt spheres of the preceding surgeries after a Legendrian isotopy, this is the same (because of isotopy extension, Thm. 2.6.2) as saying that (M, ξ) can be obtained from (S^3, ξ_{st}) by contact (± 1)–surgery along a Legendrian link. □

Prima facie, this theorem appears to be weaker than the Lutz–Martinet theorem based on surgery along transverse knots, because it does not assert the existence of a contact structure on every 3–manifold (or indeed in every homotopy class of tangent 2–plane fields). However, one can actually give a proof of the Lutz–Martinet theorem using contact Dehn surgery, see [56].

In one respect Theorem 6.4.4 is considerably stronger than the Lutz–Martinet theorem, because it says that *every* contact structure on a 3–manifold M can be obtained by contact (± 1)–surgery on (S^3, ξ_{st}).

6.5 Symplectic fillings

We are now going to use the surgery presentation of Theorem 6.4.4 to give a proof of Theorem 1.7.15, saying that any weak symplectic filling of a contact 3–manifold (M, ξ) embeds into a closed symplectic 4–manifold. In condensed form, this argument has appeared previously in [98] and is essentially due to Özbağcı and Stipsicz [200]. In fact, what the proof shows is that any boundary component of a weak filling can be capped off with a concave filling. This entails that a contact manifold realisable as one of several boundary components of a (weak or strong) filling – such a contact manifold used to be called *semi-fillable* – is in fact fillable in the sense of Definition 5.1.1; simply cap off all the other boundary components.

Lemma 6.5.1 *Strong fillings can be capped off.*

Proof Let (W, ω) be a strong symplectic filling of the contact 3–manifold (M, ξ). By Theorem 6.4.4, there is a Legendrian link $\mathbb{L} = \mathbb{L}^- \sqcup \mathbb{L}^+$ in (S^3, ξ_{st}) such that contact (-1)–surgery along the components of $\mathbb{L}^-$ and contact $(+1)$–surgery along those of $\mathbb{L}^+$ produces (M, ξ). By the cancellation lemma (Prop. 6.4.5), we can now attach symplectic 2–handles to the boundary (M, ξ) of (W, ω) corresponding to contact (-1)–surgeries that undo the contact $(+1)$–surgeries along $\mathbb{L}^+$. The result will be a symplectic manifold (W', ω') strongly filling a contact manifold (M', ξ'), and the latter can be obtained from $(S^3, \xi_{st}) = \partial(D^4, \omega_{st})$ by performing contact (-1)–surgeries (along $\mathbb{L}^-$) only.

A handlebody obtained from (D^4, ω_{st}) by attaching symplectic handles in this way is in fact, by Remark 6.4.2, a Stein filling of its boundary contact manifold, and for those a symplectic cap had been found earlier by Akbulut–Özbağcı [8] and Lisca–Matić [163]. By Proposition 5.2.5, the cap that fits on the Stein filling can also be glued to the strong filling (W', ω'). □

The following lemma allows us to attach symplectic handles also to weak fillings of contact 3–manifolds, see [87, Lemma 2.4].

Lemma 6.5.2 *Let (W, ω) be a weak filling of the contact 3–manifold (M, ξ), and L a Legendrian knot in (M, ξ). There is an arbitrarily C^∞–small perturbation of ξ to a new contact structure ξ', the perturbation being stationary outside an arbitrarily small neighbourhood of L in M, so that (W, ω) is still a weak filling of (M, ξ') and a strong filling near L. In other words, there is a Liouville vector field Y for ω, defined in a neighbourhood of L in M, such that $\xi' = \ker(i_Y\omega|_{TM})$ near L.*

Proof The idea of the proof is to take a Legendrian knot in the boundary of a strong filling as a model, and then to adjust ξ in a neighbourhood of L so that it looks like the contact structure in that 'strong' model. In fact, we only need to have a contact type boundary in the model; compactness of the filling is not essential.

Thus, let (W_0, ω_0) be any symplectic 4–manifold with contact type boundary $(M_0, \xi_0 = \ker \alpha_0)$, and L_0 a Legendrian knot in (M_0, ξ_0). For instance, we may take $M_0 = M$, $\alpha_0 = \alpha$ (where $\xi = \ker \alpha$), and $W_0 = (-\infty, 0] \times M$

with symplectic form $\omega_0 = d(e^t\alpha)$. We want to define a diffeomorphism ψ from a neighbourhood $\mathcal{N}(L)$ of L in M to a neighbourhood $\mathcal{N}(L_0)$ of L_0 in M_0 such that

- $\psi(L) = L_0$,
- $\psi^*(\alpha_0|_{L_0}) = \alpha|_L$, so that in particular $T\psi(\xi|_L) = \xi_0|_{L_0}$,
- $\psi^*(\omega_0|_{\mathcal{N}(L_0)}) = \omega|_{\mathcal{N}(L)}$.

Here by restriction of a differential form to a subset U of M we mean the restriction of that differential form to $TM|_U$, likewise for subsets of M_0.

Assuming the existence of such a diffeomorphism ψ, the proof concludes as follows. The third condition on ψ allows us, by Lemma 5.1.4, to extend ψ to a symplectomorphism between a neighbourhood of L in (W,ω) and a neighbourhood of L_0 in (W_0,ω_0). With ψ also denoting that extended diffeomorphism, the vector field $T\psi^{-1}(\partial_t)$ is a Liouville vector field for ω inducing the contact form $\psi^*\alpha_0$ on the boundary M near $L \subset M$. So our task becomes to perturb ξ near L so that it coincides with $\ker(\psi^*\alpha_0)$.

Since, along L, the symplectic bundle structure

$$(\psi^*(d\alpha_0))|_\xi = (\psi^*\omega_0)|_\xi = \omega|_\xi$$

is in the (positive) conformal class of $d\alpha|_\xi$, and the 1–forms $\psi^*\alpha_0$ and α coincide on L, the 1–form

$$\alpha_t := (1-t)\alpha + t\psi^*\alpha_0$$

is a contact form in a neighbourhood of L for all $t \in [0,1]$.

Now apply the Gray stability theorem (Thm. 2.2.2). Define the time-dependent vector field X_t as in the proof of that theorem. At points of L we have $\dot{\alpha}_t \equiv 0$, so there $X_t \equiv \mathbf{0}$ by Remark 2.2.3. We may multiply X_t by a bump function that is identically 1 near L and supported in an arbitrarily small neighbourhood of L, so that the norm of this new vector field becomes arbitrarily small. The flow of such a vector field defines an isotopy between ξ and a contact structure C^∞–close to ξ that coincides with $\ker(\psi^*\alpha_0)$ near L.

It remains to construct the diffeomorphism ψ with the described properties. This is done in three steps.

(i) Choose a diffeomorphism $\psi\colon L \to L_0$. Of course, with our specific choice of L_0, we may simply take ψ to be the identity on L.

(ii) In M choose a transversal $\Sigma \supset L$ to the line field $\ker(\omega|_{TM})$ with $T\Sigma|_L = \xi|_L$. This Σ may be taken to be an open annulus $S^1 \times (-1,1)$, where L is identified with $S^1 \times \{0\}$. Define $\Sigma_0 \supset L_0$ in M_0 analogously.

Given two area forms ω_1 and ω_2 on $S^1 \times (-1,1)$, one can define, for some small $\varepsilon > 0$, a diffeomorphism ψ of $S^1 \times (-\varepsilon, \varepsilon)$ onto its image in $S^1 \times (-1,1)$ of the form

$$\begin{array}{ccccccc} S^1 & \times & (-\varepsilon,\varepsilon) & \longrightarrow & S^1 & \times & (-1,1) \\ (\theta & , & t) & \longmapsto & (\theta & , & g(\theta,t)), \end{array}$$

with $g(\theta,0) = 0$ for all $\theta \in S^1$ and $\psi^*\omega_2 = \omega_1$. Indeed, if we write $\omega_i = a_i(\theta,t)\, d\theta \wedge dt$, $i = 1,2$, then g is found as solution of the differential equation

$$\frac{\partial g}{\partial t}(\theta,t) = \frac{a_1(\theta,t)}{a_2(\theta, g(\theta,t))},$$

with initial condition $g(\theta,0) = 0$.

In our situation this translates into saying that we can extend the map ψ – after shrinking Σ and Σ_0 appropriately – to a diffeomorphism $\psi\colon \Sigma \to \Sigma_0$ such that $\psi^*(\omega_0|_{T\Sigma_0}) = \omega|_{T\Sigma}$.

(iii) Let R be a vector field spanning $\ker(\omega|_{TM}) = \ker(j^*\omega)$, with j denoting the inclusion of M in W, and R_0 an analogous vector field on M_0. For R_0 we may take the Reeb vector field of α_0; for R we may require that $\alpha(R) = 1$ along L. Then

$$\mathcal{L}_R(j^*\omega) = d(i_R j^*\omega) + i_R\big(d(j^*\omega)\big) \equiv 0,$$

i.e. the flow of R preserves $j^*\omega$. Similarly, the flow of R_0 preserves $j_0^*\omega_0$. Hence, by mapping the flow lines of R to those of R_0, the map ψ from step (ii) can be extended to the desired diffeomorphism. Notice that the condition $\psi^*(\alpha_0|_{L_0}) = \alpha|_L$ is guaranteed by our normalisation of R and R_0 along L and L_0, respectively. □

We now use this lemma to reduce the proof of Theorem 1.7.15 to the consideration of homology spheres only, that is, 3–manifolds whose integral homology is the same as that of S^3. First we need a topological lemma. This is a special case of a more general procedure for computing the homology of a manifold obtained by Dehn surgery, see [111, Prop. 5.3.11].

Lemma 6.5.3 *Let K, L be a pair of knots in a homology 3–sphere M_0 with linking number $\mathtt{lk}(K,L) = 1$. Then the manifold M_1 obtained from M_0 by Dehn surgery along L with surgery coefficient $p/q \neq \infty$ and Dehn surgery along K with surgery coefficient 0 is again a homology sphere.*

Proof Write C for the complement in M_0 of tubular neighbourhoods of K and L. By a simple Mayer–Vietoris argument as in Section 1.7.2, we see that $H_1(C) \cong \mathbb{Z} \oplus \mathbb{Z}$, generated by the meridians μ_K and μ_L. The surgered manifold M_1 is obtained from C by attaching two 2–cells along the curves

λ_K and $p\mu_L + q\lambda_L$, respectively, and then attaching two 3–cells. Hence

$$H_1(M_1) \cong (\mathbb{Z}^{\mu_K} \oplus \mathbb{Z}^{\mu_L})/\langle \lambda_K, p\mu_L + q\lambda_L \rangle.$$

Here the longitudes are the ones determined by the surface framing. So the longitude λ_K is the boundary of a Seifert surface $\Sigma_K \subset M_0$ for K. Cutting out a tubular neighbourhood of L amounts to removing a single (counting with signs) meridional disc of L from Σ_K, since $\mathtt{lk}(K, L) = 1$. This punctured Seifert surface gives the relation $\lambda_K = \mu_L$. Likewise, a Seifert surface for L gives the relation $\lambda_L = \mu_K$. It follows that $H_1(M_1) = 0$, so M_1 is a homology sphere by Poincaré duality. □

Lemma 6.5.4 *Let (W, ω) be a weak filling of (M, ξ). Then (W, ω) embeds into a symplectic manifold weakly filling a homology sphere.*

Proof Start from a contact surgery presentation of (M, ξ) as in the proof of Lemma 6.5.1. For each component L_i of $\mathbb{L}$ we choose a Legendrian knot K_i in (S^3, ξ_{st}) linked only with that component, with $\mathtt{lk}(K_i, L_i) = 1$. Moreover, we may assume that $\mathtt{tb}(K_i) = 1$; for instance, take the right-handed Legendrian trefoil knot from Figure 3.3. Then contact (-1)–surgery along K_i amounts to a surgery with coefficient 0 relative to the surface framing. By the preceding lemma, the result of the surgeries on M along the K_i will be a homology sphere. By Lemma 6.5.2 we can effect these surgeries by attaching symplectic 2–handles, so the result will be a weak filling of that homology sphere, containing the initial filling (W, ω). □

Lemma 6.5.5 *Any weak filling of a homology sphere can be deformed to a strong filling.*

Proof Let (W, ω) be a weak filling of (M, ξ), with M a homology sphere. We want to modify ω in a collar neighbourhood $[0, 1] \times M$ of the boundary $M \equiv \{1\} \times M$ such that the resulting symplectic manifold is a strong filling of the new induced contact structure on the boundary.

Since $H^2(M) = 0$, we can write $\omega = d\eta$ with some 1–form η in a collar neighbourhood as described. (We see that it would be enough to have M a rational homology sphere.) Choose a 1–form α on M with $\xi = \ker \alpha$ and $\alpha \wedge \omega|_{TM} > 0$, which is possible for a weak filling. Then set

$$\widetilde{\omega} = d(f\eta) + d(g\alpha)$$

on $[0, 1] \times M$, where the smooth functions $f(t)$ and $g(t)$, $t \in [0, 1]$, are chosen as follows. Fix a small $\varepsilon > 0$. Choose $f\colon [0, 1] \to [0, 1]$ identically 1 on $[0, \varepsilon]$

and identically 0 near $t = 1$. Choose $g\colon [0,1] \to \mathbb{R}_0^+$ identically 0 near $t = 0$ and with $g'(t) > 0$ for $t > \varepsilon/2$.

We compute

$$\widetilde{\omega} = f'\, dt \wedge \eta + f\omega + g'\, dt \wedge \alpha + g\, d\alpha,$$

whence

$$\begin{aligned} \widetilde{\omega}^2 &= 2ff'\, dt \wedge \eta \wedge \omega + 2f'g\, dt \wedge \eta \wedge d\alpha + f^2\omega^2 \\ &\quad + 2fg'\, \omega \wedge dt \wedge \alpha + 2fg\, \omega \wedge d\alpha + 2gg'\, dt \wedge \alpha \wedge d\alpha. \end{aligned}$$

The terms appearing with the factors f^2, fg' and gg' are positive volume forms. By choosing g small on $[0,\varepsilon]$ and g' large compared with $|f'|$ and g on $[\varepsilon, 1]$, one can ensure that these positive terms dominate the three terms we cannot control. Then $\widetilde{\omega}$ is a symplectic form on the collar and, in terms of the coordinate $s := \log g(t)$, the symplectic form looks like $d(e^s \alpha)$ near the boundary, with Liouville vector field ∂_s. □

Theorem 1.7.15 is now an immediate consequence of Lemmata 6.5.1, 6.5.4 and 6.5.5.

Here, as an application of this theorem, is a tightness criterion for contact structures. This is originally due to Gromov [118, 2.4.D$'_2$.b] and Eliashberg [63, Thm. 3.2.1]. A detailed proof along the lines of the original argument can be found in [246]. The proof given here is from [200].

Theorem 6.5.6 (Eliashberg, Gromov) *If a 3–dimensional contact manifold (M, ξ) is weakly symplectically fillable, then ξ is tight.*

Put the other way round, a 3–manifold with an overtwisted contact structure is not weakly symplectically fillable. A related phenomenon in higher dimensions is described in [199]. For the proof of Theorem 6.5.6 we need one simple and one hard topological fact. First, let us look at the simple one.

Lemma 6.5.7 *Let K be a knot bounding an embedded disc Δ in the 3–manifold $M = \partial W$. Let W' be the 4–manifold with boundary obtained from W by attaching a 2–handle along M with attaching 1–sphere K, corresponding to a surgery along K with surgery coefficient $n \in \mathbb{Z}$ with respect to the surface framing of K. Then Δ and the core disc of the 2–handle glue together to a 2–sphere Σ in W' with self-intersection number n.*

Proof Push Δ 'vertically' into W, and think of Σ as being obtained by gluing the core disc, a vertical cylinder over $\partial\Delta$, and Δ, see Figure 6.11.

In order to compute the self-intersection number of Σ, we need to construct

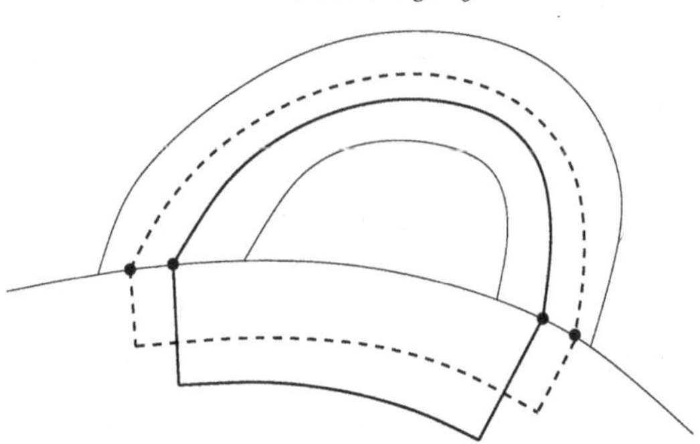

Fig. 6.11. Self-intersection of sphere in handle surgery.

a transverse push-off of Σ. In the 2–handle $D^2 \times D^2$, the core disc $D^2 \times \{\mathbf{0}\}$ can be pushed to a disc $D^2 \times *$. The boundary of this 2–disc will be a knot K' parallel to K with $\mathtt{lk}(K, K') = n$ in M. Complete this disc to a 2–sphere Σ' as before, where we push the disc Δ' bounded by K' to a 'depth' in W different from that of Δ. Then every intersection point of K with the Seifert surface Δ' of K' in M translates into an intersection point of Σ' with Σ (of the same sign) in W', so the result follows from Proposition 3.4.11. □

The following auxiliary result, however, is well beyond the scope of the present text. It is a special case of what is known as the adjunction inequality and can be proved via Seiberg–Witten theory. For the relevant references see [200, Chapter 13].

Theorem 6.5.8 *Let Σ be a smoothly embedded 2–sphere in a closed symplectic 4–manifold with $b_2^+ > 1$, i.e. with the intersection product on the second (real) homology being positive definite on an at least 2–dimensional subspace. If the class of Σ in homology is non-trivial, then the self-intersection number of Σ is negative.* □

Proof of Theorem 6.5.6 Let (W, ω) be a weak symplectic filling of (M, ξ). Arguing by contradiction, we assume that ξ is overtwisted. By the discussion in Section 6.3.3 we can find an embedded disc $\Delta \subset M$ with Legendrian boundary $\partial\Delta$ satisfying $\mathtt{tb}(\partial\Delta) = 2$. Use Lemma 6.5.2 to attach a symplectic 2–handle to (W, ω) along M with attaching 1–sphere $\partial\Delta$; call the resulting symplectic manifold (W', ω'). This is then a weak filling of $M' := \partial W'$. As we saw in the preceding section, M' is the result of a contact surgery on M with surgery coefficient -1 with respect to the contact framing, and hence with surgery coefficient $+1$ with respect to the surface framing of $\partial\Delta$. It follows that the 2–sphere Σ in W', constructed as in Lemma 6.5.7, has self-intersection number +1. By adding another such handle, and then applying Theorem 1.7.15, we obtain a closed symplectic 4–manifold containing a sphere Σ that contradicts Theorem 6.5.8. □

Remark 6.5.9 The converse of Theorem 6.5.6 does not hold. Etnyre and Honda [87] found the first examples of tight contact structures on certain Seifert fibred 3–manifolds that do not admit any symplectic filling.

Corollary 6.5.10 *The following contact structures are tight:*

(i) the standard contact structures ξ_{st} on S^3 and $\mathbb{R}^3$;

(ii) the standard contact structure on $S^2 \times S^1 \subset \mathbb{R}^3 \times S^1$, given by

$$z\,d\theta + x\,dy - y\,dx = 0;$$

(iii) the contact structures ξ_n on T^3, $n \in \mathbb{N}$, given by

$$\cos(n\theta)\,dx - \sin(n\theta)\,dy = 0.$$

Proof (i) The tightness of ξ_{st} on S^3 follows directly from Theorem 6.5.6, since (S^3, ξ_{st}) admits the strong filling $(D^4, \omega_{\mathrm{st}})$.

Any open subset of a tight contact 3–manifold is obviously again tight, so the tightness of ξ_{st} on $\mathbb{R}^3$ follows from Proposition 2.1.8.

(ii) The standard contact structure on $S^2 \times S^1$ may, by Gray stability, alternatively be described as

$$\alpha := z\,d\theta + \frac{1}{2}x\,dy - \frac{1}{2}y\,dx = 0.$$

Then $D^3 \times S^1$ with symplectic form $\omega := dx \wedge dy + dz \wedge d\theta$ is a strong filling, since the Liouville vector field $Y := (x\,\partial_x + y\,\partial_y)/2 + z\,\partial_z$ for ω is transverse to $\partial(D^3 \times S^1) = S^2 \times S^1$ and gives $\alpha = i_Y\omega$.

(iii) If $(M, \xi) \to (M', \xi')$ is a covering of contact 3–manifolds and ξ is tight, then so is ξ', for any overtwisted disc in (M', ξ') would lift to an overtwisted disc in (M, ξ). Under the covering $\mathbb{R}^3 \to T^3$ given by

$$(u, v, w) \mapsto (x, y, \theta) = (w\ \cos v + u\ \sin v, -w\ \sin v + u\ \cos v, v)\ \bmod (2\pi\mathbb{Z})^3,$$

the contact structure ξ_1 on T^3 lifts to $\ker(dw + u\,dv) = \xi_{\mathrm{st}}$ on $\mathbb{R}^3$. Hence ξ_1 is tight. Finally, ξ_n is the lift of ξ_1 under the covering $T^3 \to T^3$ given by $\theta \mapsto n\theta$, hence all the ξ_n are tight. □

7
Further constructions of contact manifolds

'Maybe I'm an open book...'

Sheena Easton,
For Your Eyes Only

Contact surgery, as we have just seen, is a powerful method for constructing contact manifolds, but by far not the only one. Concerning 3–manifolds, we saw in Section 4.4 that it is also possible to find contact structures with the help of open books and branched coverings. In Sections 7.3 and 7.5, respectively, we extend these two methods to contact manifolds of arbitrary dimension. The deep relation between open books and contact structures was discovered by Giroux. One of the remarkable applications of open books is the construction, due to Bourgeois, of contact structures on all odd-dimensional tori. The compatibility of contact structures with branched coverings in arbitrary dimensions was first observed by Gromov.

Before the advent of topological techniques for the construction of higher-dimensional contact manifolds, two methods of a more geometric nature had been used extensively: the explicit description of a contact form on Brieskorn manifolds, which we shall discuss in Section 7.1, and the contact form given by the connection 1–form on suitable principal S^1–bundles (Boothby–Wang construction), as described in Section 7.2.

In Section 7.4 I present the fibre connected sum of contact manifolds. This construction was originally suggested by Gromov, both for symplectic and contact manifolds. In symplectic topology, it was put on a sound footing and used prominently by Gompf [109].

Section 7.6 deals with the plumbing of cotangent disc bundles. This can be used to construct contact structures, notably on spheres, with strong symplectic fillings. Special properties of these fillings were one of the first means to detect 'exotic' contact structures, i.e. contact structures that are different from the standard contact structure (on S^{2n+1}, say) but cannot be distinguished from this standard structure by homotopical methods alone.

Finally, Section 7.7 treats contact manifolds with an action by a compact Lie group. In contrast with the symplectic situation, such actions always have a momentum map and give rise to a 'reduced' contact manifold. I describe some of the basic aspects of this contact reduction.

7.1 Brieskorn manifolds

Let $\mathbf{a} = (a_0, \ldots, a_n)$ be an $(n+1)$–tupel of integers $a_j > 1$, and set

$$V(\mathbf{a}) := \{\mathbf{z} := (z_0, \ldots, z_n) \in \mathbb{C}^{n+1}\colon\ f(\mathbf{z}) := z_0^{a_0} + \cdots + z_n^{a_n} = 0\}.$$

Further, with S^{2n+1} denoting the unit sphere in $\mathbb{C}^{n+1}$, we define

$$\Sigma(\mathbf{a}) := V(\mathbf{a}) \cap S^{2n+1}.$$

Lemma 7.1.1 $\Sigma(\mathbf{a})$ *is a smooth manifold of dimension* $2n-1$.

Proof Set

$$\rho_0(\mathbf{z}) := z_0\overline{z}_0 + \cdots + z_n\overline{z}_n,$$

and consider the map $(f, \rho_0)\colon \mathbb{C}^{n+1} \to \mathbb{C} \times \mathbb{R}$. Then $\Sigma(\mathbf{a})$ is the preimage under this map of the value $(0,1) \in \mathbb{C}\times\mathbb{R}$. By a standard result in differential topology [38, Lemma 5.9], it suffices to show that $(0,1)$ is a regular value. With a little Wirtinger calculus (and using the fact that f is holomorphic) one finds that this regularity condition is equivalent to the condition that the Jacobian matrix

$$\begin{pmatrix} a_0 z_0^{a_0-1} & \ldots & a_n z_n^{a_n-1} & 0 & \ldots & 0 \\ 0 & \ldots & 0 & a_0\overline{z}_0^{a_0-1} & \ldots & a_n\overline{z}_n^{a_n-1} \\ \overline{z}_0 & \ldots & \overline{z}_n & z_0 & \ldots & z_n \end{pmatrix}$$

have rank 3 over $\mathbb{C}$ in all points of $\Sigma(a)$.

If $\mathbf{z}$ is a point where this matrix has rank smaller than 3, there exists a non-zero complex number λ such that $\overline{z}_j = \lambda a_j z_j^{a_j-1}$, $j = 0, \ldots, n$, and hence

$$\sum_{j=0}^{n} \frac{z_j\overline{z}_j}{a_j} = \lambda \sum_{j=0}^{n} z_j^{a_j} = \lambda f(\mathbf{z}).$$

This equality is incompatible with the conditions $\rho_0(\mathbf{z}) = 1$ and $f(\mathbf{z}) = 0$ for a point to lie on $\Sigma(\mathbf{a})$. $\square$

The manifolds $\Sigma(\mathbf{a})$ are called **Brieskorn manifolds** after the seminal paper [36] by Brieskorn, whose main theorem gives a necessary and sufficient condition on $\mathbf{a}$ for $\Sigma(\mathbf{a})$ to be a topological sphere, and means to show that those $\Sigma(\mathbf{a})$ yield, in general, exotic differentiable structures on the topological $(2n-1)$–sphere. For a more comprehensive account see [133].

Theorem 7.1.2 *The real* 1*–form*

$$\alpha := \frac{\mathrm{i}}{4}\sum_{j=0}^{n}(z_j\, d\overline{z}_j - \overline{z}_j\, dz_j)$$

induces a Stein fillable contact structure on any Brieskorn manifold $\Sigma(\mathbf{a})$.

Proof The function $\rho(\mathbf{z}) := (\|\mathbf{z}\|^2 - 1)/4$ is strictly plurisubharmonic on $\mathbb{C}^{n+1}$, and the same is true for its restriction to the complex submanifold $V(\mathbf{a}) \setminus \{\mathbf{0}\}$. Notice that $\alpha = -d\rho \circ J$, that is, α and ρ are related to each other as in Section 5.4. From Proposition 5.4.2 and the definitions in Section 5.4 it is therefore immediate that $\ker\alpha$ defines the complex tangencies of the strictly Levi pseudoconvex boundary of $V(\mathbf{a}) \cap D^{2n+2}$, and thus a contact structure.

Unfortunately, $V(\mathbf{a})$ is not a complex manifold – it has a singular point at the origin $\mathbf{0}$. In order to get a Stein filling, we replace $V(\mathbf{a})$ by $V_s(\mathbf{a})$, defined by the equation

$$z_0^{a_0} + \cdots + z_n^{a_n} = s,$$

and $\Sigma(\mathbf{a})$ by $\Sigma_s(\mathbf{a}) := V_s(\mathbf{a}) \cap S^{2n+1}$, where s is a positive real number. Now $V_s(\mathbf{a})$ is indeed a non-singular complex manifold. As shown in [133, Satz 14.3], there is an $\varepsilon > 0$ such that $\Sigma_s(\mathbf{a})$ is diffeomorphic to $\Sigma(\mathbf{a}) = \Sigma_0(\mathbf{a})$ for all $s \in [0, \varepsilon]$. In fact, the proof given there shows that the $\Sigma_s(\mathbf{a})$ form the fibres of a differentiable fibre bundle.

Our previous argument shows that the 1–form α induces a contact structure on each $\Sigma_s(\mathbf{a})$, with a Stein filling given by $V_s(\mathbf{a}) \cap D^{2n+2}$ in case $s > 0$. With the help of a connection on the fibre bundle formed by the $\Sigma_s(\mathbf{a})$, we may regard these contact structures as a smooth family ξ_s, $s \in [0, \varepsilon]$, of contact structures on $\Sigma(\mathbf{a})$. By Gray stability (Thm. 2.2.2), the ξ_s are isotopic, and since ξ_s is Stein fillable for $s > 0$, so is ξ_0. □

In this argument, we did not use any special features of the complex submanifold $V_s(\mathbf{a})$. Also, notice that α is the standard contact form on S^{2n+1}. So the same argument proves the following statement, which was first observed by Abe and Erbacher [2] (without the part about Stein fillings).

Theorem 7.1.3 *Let V be a complex submanifold of $\mathbb{C}^{n+1}$. If V intersects S^{2n+1} transversely, then the standard contact form on S^{2n+1} restricts to a contact form on the differentiable manifold $V \cap S^{2n+1}$. If, moreover, V is a proper submanifold, the induced contact structure is Stein fillable.* □

The fact that α induces a contact structure on $\Sigma(\mathbf{a})$ was also established by Sasaki and Hsü [217], using a direct computation.

The contact form $\alpha_0 := \alpha$ on $\Sigma(\mathbf{a})$ has one essential drawback: the dynamics of its associated Reeb vector field is rather complicated. From that viewpoint, it is more convenient to work with the form

$$\alpha_1 := \frac{\mathrm{i}}{4}\sum_{j=0}^{n} a_j (z_j\, d\overline{z}_j - \overline{z}_j\, dz_j).$$

The disadvantage here is that the verification of the contact condition for α_1 is a lot more involved.

Assuming, for the time being, that α_1 does indeed induce a contact form on $\Sigma(\mathbf{a})$, we can easily check that its associated Reeb vector field is given by

$$R_{\alpha_1} = 2\mathrm{i}\sum_{j=0}^{n} \frac{1}{a_j}(z_j\, \partial_{z_j} - \overline{z}_j\, \partial_{\overline{z}_j}).$$

Indeed, one computes

$$df(R_{\alpha_1}) = f(\mathbf{z}) \quad \text{and} \quad d\rho_0(R_{\alpha_1}) = 0,$$

which shows that R_{α_1} is tangent to $\Sigma(\mathbf{a})$. Furthermore, the defining equations for the Reeb vector field are satisfied, since

$$\alpha_1(R_{\alpha_1}) \equiv 1 \quad \text{and} \quad i_{R_{\alpha_1}} d\alpha_1 = -d\rho_0,$$

the latter form being zero on the tangent spaces to $\Sigma(\mathbf{a})$.

The flow of R_{α_1} is given by

$$\psi_t(z_0, \ldots, z_n) = (\mathrm{e}^{2\mathrm{i}t/a_0} z_0, \ldots, \mathrm{e}^{2\mathrm{i}t/a_n} z_n).$$

All the orbits of this flow are closed, and the flow defines an effective S^1–action on $\Sigma(\mathbf{a})$.

It remains to check the contact condition for α_1. In fact, the following proposition tells us, together with Gray stability, that the contact structures on $\Sigma(\mathbf{a})$ induced by α_0 and α_1 are isotopic. The proof is based on ideas from [171], where it was first shown that α_1 defines a contact form on $\Sigma(\mathbf{a})$. The isotopy between α_0 and α_1 was observed by Ustilovsky in his thesis [234].

Proposition 7.1.4 *Set* $c_j(t) := (1-t+t/a_j)^{-1}$, $t \in [0,1]$. *Then the* 1*–form*

$$\alpha_t := \frac{\mathrm{i}}{4}\sum_{j=0}^{n} c_j(t)(z_j\, d\overline{z}_j - \overline{z}_j\, dz_j)$$

defines a contact form on $\Sigma(\mathbf{a})$ *for each* $t \in [0,1]$.

Proof The restriction of the skew-symmetric form $d\alpha_t$ to each tangent space of $\Sigma(\mathbf{a})$ has at least a 1–dimensional kernel. In order to verify that the restriction of α_t to $T\Sigma(\mathbf{a})$ satisfies the contact condition, we need to show that α_t does not vanish on that kernel (which forces that kernel to be one-dimensional). In other words, we need to show that there is *no* non-zero vector $\mathbf{v} \in T\mathbb{C}^{n+1}|_{\Sigma(\mathbf{a})}$ – with $\mathbb{C}^{n+1}$ thought of as the real manifold $\mathbb{R}^{2n+2}$ – that satisfies the following system of equations, where λ can be any complex number and μ any real number:

$$\begin{aligned} df(\mathbf{v}) &= 0,\\ d\rho_0(\mathbf{v}) &= 0,\\ \alpha_t(\mathbf{v}) &= 0,\\ i_{\mathbf{v}} d\alpha_t &= \lambda\, df + \overline{\lambda}\,\overline{df} + \mu\, d\rho_0. \end{aligned}$$

This system of equations can be written more explicitly as

$$\sum_{j=0}^{n} \frac{\partial f}{\partial z_j} dz_j(\mathbf{v}) = 0, \tag{7.1}$$

$$\sum_{j=0}^{n} \big(z_j\, d\overline{z}_j(\mathbf{v}) + \overline{z}_j\, dz_j(\mathbf{v})\big) = 0, \tag{7.2}$$

$$\sum_{j=0}^{n} c_j(t)\big(z_j\, d\overline{z}_j(\mathbf{v}) - \overline{z}_j\, dz_j(\mathbf{v})\big) = 0, \tag{7.3}$$

and

$$\frac{\mathrm{i}}{2}\sum_{j=0}^{n} c_j(t)\big(dz_j(\mathbf{v})\, d\overline{z}_j - d\overline{z}_j(\mathbf{v})\, dz_j\big) = \lambda \sum_{j=0}^{n} \frac{\partial f}{\partial z_j} dz_j + \overline{\lambda}\sum_{j=0}^{n} \overline{\frac{\partial f}{\partial z_j}}\, d\overline{z}_j + \mu \sum_{j=0}^{n} (z_j\, d\overline{z}_j + \overline{z}_j\, dz_j). \tag{7.4}$$

Here and in the sequel it is understood that all functions are evaluated at the footpoint of the tangent vector $\mathbf{v}$. By comparing coefficients in (7.4) we find

$$\frac{\mathrm{i}}{2} c_j(t)\, dz_j(\mathbf{v}) = \overline{\lambda} \overline{\frac{\partial f}{\partial z_j}} + \mu z_j, \tag{7.5}$$

$$-\frac{\mathrm{i}}{2} c_j(t)\, d\overline{z}_j(\mathbf{v}) = \lambda \frac{\partial f}{\partial z_j} + \mu \overline{z}_j. \tag{7.6}$$

Multiply (7.5) by $\overline{z}_j$, (7.6) by z_j, and take the sum over j. With (7.3) and $\rho_0(\mathbf{z}) = 1$ along $\Sigma(\mathbf{a})$ this gives

$$\mu = -\mathrm{Re}\Big(\lambda \sum_j z_j \frac{\partial f}{\partial z_j} \Big). \tag{7.7}$$

Multiply (7.5) by $c_j(t)^{-1}(\partial f/\partial z_j)$ and sum over j. With (7.1) we obtain

$$0 = \overline{\lambda} \sum_j c_j(t)^{-1} \left| \frac{\partial f}{\partial z_j} \right|^2 - \mathrm{Re}\Big(\lambda \sum_j z_j \frac{\partial f}{\partial z_j} \Big) \cdot \Big(\sum_j \frac{z_j}{c_j(t)} \frac{\partial f}{\partial z_j} \Big). \tag{7.8}$$

I claim that from this equation it follows that $\lambda = 0$. Assuming this claim for the moment, we then see from (7.7) that $\mu = 0$, and from (7.5) and (7.6) we then find $\mathbf{v} = \mathbf{0}$.

It remains to prove the claim. Here we distinguish two cases. If we have $\mathrm{Re}\big(\lambda \sum z_j (\partial f/\partial z_j)\big) = 0$, then $\lambda = 0$ is immediate from (7.8), for the $\partial f/\partial z_j$ do not vanish simultaneously along $\Sigma(\mathbf{a})$.

If $\mathrm{Re}\big(\lambda \sum z_j (\partial f/\partial z_j)\big) \neq 0$, then from (7.8) we see that the complex number

$$\sum_j c_j(t)^{-1} z_j \frac{\partial f}{\partial z_j}$$

must be a real multiple of $\overline{\lambda}$, and this allows us to rewrite (7.8) as

$$0 = \overline{\lambda} \left\{ \sum_j c_j(t)^{-1} \left| \frac{\partial f}{\partial z_j} \right|^2 - \mathrm{Re}\left[\Big(\sum_j \overline{z}_j \overline{\frac{\partial f}{\partial z_j}} \Big) \Big(\sum_j \frac{z_j}{c_j(t)} \frac{\partial f}{\partial z_j} \Big) \right] \right\}.$$

Thus, we are done if we can show that the expression in braces is never zero along $\Sigma(\mathbf{a})$. For then this last equation will again force $\lambda = 0$.

Observe that our argument immediately shows that α_1 is a contact form, since $c_j(1) = a_j$ and

$$\sum_j \frac{z_j}{a_j} \frac{\partial f}{\partial z_j} = f,$$

so the second summand within the braces vanishes along $\Sigma(\mathbf{a})$, while the first does not. Similarly, α_t is seen to be a contact form whenever f is any homogeneous polynomial (use the Euler identity). Now, however, we consider the polynomial $f(\mathbf{z}) = \sum z_j^{a_j}$ as before, and compute

$$\begin{aligned}
\sum_j c_j(t)^{-1} a_j^2 |z_j|^{2(a_j-1)} &\geq (1-t)\sum_j a_j^2 |z_j|^{2(a_j-1)} \\
&\geq (1-t)\Big|\sum_j a_j z_j^{a_j-1} z_j\Big|^2 \\
&= (1-t)\sum_j a_j \overline{z}_j^{a_j} \sum_j a_j z_j^{a_j} \\
&= \mathrm{Re}\left[\left(\sum_j \overline{z}_j \overline{\frac{\partial f}{\partial z_j}}\right)\sum_j\left(1-t+\frac{t}{a_j}\right)a_j z_j^{a_j}\right],
\end{aligned}$$

where the last equality follows from $f(\mathbf{z}) = 0$. The first inequality is strict for $t \in (0,1]$. The second inequality (Cauchy–Schwarz with $\sum |z_j|^2 = 1$) is strict for $t \in [0,1)$ since, as shown at the beginning of this section, the vector $(\overline{z}_0, \ldots, \overline{z}_1)$ is never proportional to the vector $(a_0 z_0^{a_0-1}, \ldots, a_n z_n^{a_n-1})$ along $\Sigma(\mathbf{a})$. □

In his thesis and in [235], Ustilovsky computes the contact homology of these contact structures on Brieskorn manifolds diffeomorphic to S^{4m+1}, which allows him to detect infinitely many non-diffeomorphic but homotopically standard contact structures on these spheres. Here 'homotopically standard' means that the contact structure induces the same almost contact structure (see Section 8.1 below) as the standard contact structure ξ_{st}, and thus cannot be distinguished from ξ_{st} by homotopical methods alone. For the computation of the contact homology of Brieskorn manifolds see also [33] and [150].

The fact that a (homotopically standard) contact structure on an odd-dimensional sphere is exotic, i.e. not diffeomorphic to the standard contact structure, can be detected by classical† methods. On spheres of dimension $4n+1$, such exotic structures were first found by Eliashberg [67], on S^7 and spheres of dimension $8n+3$, by this author, using Brieskorn manifolds in an essential way. Exotic contact structures on the remaining spheres of dimension $8n+7$ were finally found by Ding–Geiges [52].

For other interesting applications of Brieskorn manifolds in contact geometry see [34] and [152].

† According to the usage of C. T. C. Wall, as related to me by Charles Thomas, any result or method prior to one's thesis may be referred to as 'classical'. The classical method here is Gromov's theory of pseudoholomorphic curves.

7.2 The Boothby–Wang construction

In this section I shall assume a certain familiarity with the notions of connection and curvature on principal bundles. A good reference is [58, §25], which treats in detail the case of principal S^1–bundles. The Boothby–Wang construction is concerned with contact forms on exactly such principal S^1–bundles, where the fibres are the orbits of the Reeb vector field. Another very useful reference, dealing exclusively with principal S^1–bundles, is the paper [148] by Kobayashi.

As we shall see, the essential point in this construction is a certain regularity property of the Reeb vector field. We begin by describing this property for arbitrary nowhere vanishing vector fields. Thus, let R be such a vector field on an $(m+1)$–dimensional manifold M. According to the tubular flow theorem [203, Thm. 2.1.1], given any point $p \in M$, we can find an $\varepsilon > 0$ and local coordinates $(x_0, \ldots, x_m)$ around p such that in the 'flow box'

$$\{(x_0, \ldots, x_m) \in \mathbb{R}^{m+1} \colon |x_i| < \varepsilon,\ i = 0, \ldots, m\}$$

the point $p \in M$ corresponds to the origin $\mathbf{0} \in \mathbb{R}^{m+1}$, and the vector field R is given by ∂_{x_0}.†

Definition 7.2.1 A nowhere vanishing vector field R on M is called **regular** if around each point of M there is a flow box that is pierced at most once by any integral curve of R. A flow box with this property will be referred to as a **regular flow box**.

Observe that if R is a regular vector field on a closed manifold M, then all its orbits are embedded circles. (If R had an immersed copy of $\mathbb{R}$ as a flow line, we could find a sequence of points $(p_\nu)_{\nu\in\mathbb{N}}$ on this flow line with $p_\nu \to \infty$ in $\mathbb{R}$ and $p_\nu \to p \in M$ for $\nu \to \infty$ and some $p \in M$, contradicting the existence of a regular flow box around p.)

Definition 7.2.2 A contact form α is called **regular** if its Reeb vector field R_α is a regular vector field.

A simple example is given by the standard contact form on S^{2n+1} from Example 2.1.7. Here the orbits of the Reeb vector field are the fibres of the Hopf fibration $S^{2n+1} \to \mathbb{C}P^n$.

The easy half of the theorem of Boothby and Wang [30] claims the existence of a regular contact form on suitable circle bundles over symplectic manifolds. In order to state it succinctly, we introduce some further terminology.

† The tubular flow theorem holds for arbitrary vector fields R in the neighbourhood of any point $p \in M$ where $R_p \neq 0$.

Given a symplectic manifold (B,ω), the symplectic form ω, being a closed 2–form, represents a cohomology class $[\omega]$ in the second de Rham cohomology group of B, which we may identify with $H^2(B;\mathbb{R})$, thc singular cohomology group with real coefficients. There is a natural map $H^2(B;\mathbb{Z}) \to H^2(B;\mathbb{R})$ induced by the inclusion $\mathbb{Z} \subset \mathbb{R}$. If $[\omega]$ lies in the image of this map, we call ω an **integral symplectic form**.

Strictly speaking, a connection 1–form on a principal S^1–bundle is 1–form on the total space with values in the Lie algebra $i\mathbb{R}$ of S^1. But there is little harm in regarding it as an $\mathbb{R}$–valued 1–form, i.e. a differential form in the usual sense. Then the normalisation and invariance property of a connection on a principal bundle can, in this particular case, be summarised as follows.

Definition 7.2.3 Let $\pi\colon M \to B$ be a principal S^1–bundle, with the S^1–action on M generated by the flow of a vector field R (with period 2π). A **connection 1–form** on this bundle is a differential 1–form α on M with the following properties:

- invariance: $\mathcal{L}_R\alpha \equiv 0$,
- normalisation: $\alpha(R) \equiv 1$.

From these conditions it follows with the Cartan formula that

$$\mathcal{L}_R(d\alpha) \equiv 0 \text{ and } i_R d\alpha \equiv 0.$$

This means that $d\alpha$ induces a well-defined 2–form ω on the quotient B under the S^1–action, i.e. there is a 2–form ω with $\pi^*\omega = d\alpha$. We call ω the **curvature form** of the connection form α.

Since the differential $T\pi$ is surjective, the pull-back map π^* is injective. So from $\pi^* d\omega = d^2\alpha = 0$ we get $d\omega = 0$. This means that ω defines a de Rham cohomology class. The class $-[\omega/2\pi]$ turns out to be integral; it is the **Euler class** of the S^1–bundle. This class determines the bundle up to isomorphism. Useful references for these facts – besides [58] – are [177, Thm. 2.72, Example 3.48] and [31, §6], notably for the sign conventions in this construction. A related discussion (about the fact that the Euler class determines the bundle) can be found in Section 8.1 below.

Theorem 7.2.4 (Boothby–Wang – First Part) *Let (B,ω) be a closed symplectic manifold (of dimension $2n$) with integral symplectic form $\omega/2\pi$. Let $\pi\colon M \to B$ be the principal S^1–bundle with Euler class $-[\omega/2\pi] \in H^2(B;\mathbb{Z})$. Then there is a connection 1–form α on M with the following properties:*

- *α is a regular contact form,*
- *the curvature form of α is ω,*

- *the vector field R defining the principal S^1–action on M coincides with the Reeb vector field of α.*

Proof Choose any connection 1–form α' on M. Then $d\alpha' = \pi^*\omega'$, with ω' a closed 2–form on B satisfying $[\omega'] = [\omega]$. It follows that $\omega - \omega' = d\beta$ for some 1–form β on B. Set $\alpha = \alpha' + \pi^*\beta$. This is also a connection 1–form for the given S^1–bundle, and it satisfies $d\alpha = \pi^*\omega$, i.e. its curvature form is ω.

Since ω is a symplectic form on B, we have

$$\alpha \wedge (d\alpha)^n = \alpha \wedge \pi^*\omega^n \neq 0,$$

so α is a contact form.

As we saw above, the invariance and normalisation condition on α imply that the vector field R defining the S^1–action also satisfies $i_R d\alpha \equiv 0$, which means that R is the Reeb vector field of α. In particular, the orbits of the Reeb vector field are the fibres of the S^1–bundle, which – by the local triviality of the bundle – ensures that α is a regular contact form. □

In [30] this theorem was used to construct contact manifolds – starting from an algebraic variety $B \subset \mathbb{C}P^N$ with symplectic form given by the Kähler form induced from $\mathbb{C}P^N$ – that are different from spheres and cotangent sphere bundles, which were essentially the only known examples of contact manifolds (see Lemma 1.2.3) at that time.

The other half of the theorem of Boothby and Wang states that the construction we just described is in fact the only way (up to scaling by constants) in which regular contact forms can arise. The proof of this part is slightly more involved – in fact, the proof given in [30] is incomplete in one important respect.

Theorem 7.2.5 (Boothby–Wang – Second Part) *Let α be a regular contact form on a closed manifold M. Then, after rescaling by a suitable constant, α is the connection 1–form on a principal S^1–bundle $\pi\colon M \to B$ over a symplectic manifold (B, ω), with ω the curvature form of α.*

The key point in proving this theorem is to show that the function $p \mapsto \lambda(p)$, assigning to each point $p \in M$ the period $\lambda(p)$ of the flow line of the Reeb vector field R_α through p, is constant.† The argument that follows is a variant of that given in [198], see also [229].

† In [30] it is shown that λ is a smooth function, but not that it is constant. Rescaling by a non-constant function, however, might dramatically alter the dynamics of the Reeb flow, as we saw in Example 2.2.5.

Lemma 7.2.6 *Let R be a regular vector field on a closed manifold M. Then the function $M \to \mathbb{R}^+$, $p \mapsto \lambda(p)$ is smooth.*

Proof First of all we notice that the function λ is bounded from below. This follows from the fact that the compact manifold M can be covered by finitely many regular flow boxes, and for points in a flow box $(-\varepsilon, \varepsilon)^{m+1}$ the function λ is bounded from below by 2ε.

Choose any point $p_0 \in M$. We want to show that λ is differentiable at this point. Set $t_0 = \lambda(p_0)$ and consider the time-t_0-map ψ_{t_0} of the flow of R. On an open neighbourhood of $p_0 \equiv \mathbf{0}$ contained in a regular flow box around p_0, this map is given by

$$\psi_{t_0}(x_0, x_1, \ldots, x_m) = (\widetilde{x}_0, x_1, \ldots, x_m),$$

with some smooth function $\widetilde{x}_0 = \widetilde{x}_0(x_0, x_1, \ldots, x_m)$ with $\widetilde{x}_0(\mathbf{0}) = 0$. For small $|t|$ and $(x_0, x_1, \ldots, x_m)$ sufficiently close to $\mathbf{0}$ we have

$$\psi_t(x_0, x_1, \ldots, x_m) = (x_0 + t, x_1, \ldots, x_m).$$

Then the equation $\psi_{t_0} \circ \psi_{x_0} = \psi_{x_0} \circ \psi_{t_0}$ (with $|x_0|$ small), applied at the point $(0, x_1, \ldots, x_m)$, translates into

$$\widetilde{x}_0(x_0, x_1, \ldots, x_m) = \widetilde{x}_0(0, x_1, \ldots, x_m) + x_0.$$

Define a smooth function λ^* in a neighbourhood of $\mathbf{0}$ by

$$\lambda^*(x_0, x_1, \ldots, x_m) := t_0 - \widetilde{x}_0(0, x_1, \ldots, x_m).$$

This function also associates with each point a return time:

$$\begin{aligned} & \psi_{\lambda^*(x_0, x_1, \ldots, x_m)}(x_0, x_1, \ldots, x_m) = \\ &= \psi_{-\widetilde{x}_0(0, x_1, \ldots, x_m)} \circ \psi_{t_0}(x_0, x_1, \ldots, x_m) \\ &= \psi_{-\widetilde{x}_0(0, x_1, \ldots, x_m)}(\widetilde{x}_0(0, x_1, \ldots, x_m) + x_0, x_1, \ldots, x_m) \\ &= (x_0, x_1, \ldots, x_m). \end{aligned}$$

Moreover, $\lambda^*(p_0) = t_0 = \lambda(p_0)$. We now want to show that λ^* coincides with λ near p_0; this will prove in particular that λ is smooth.

For p close to p_0, the return time $\lambda^*(p)$ is close to t_0, and therefore positive. It follows that

$$\lambda^*(p) = k(p)\lambda(p) \text{ with } k(p) \in \mathbb{N}.$$

We need to show that $k \equiv 1$ near p_0.

Arguing by contradiction, we assume that there is a sequence of points $(p_\nu)_{\nu \in \mathbb{N}}$ with $\lim_{\nu \to \infty} p_\nu = p_0$ and $k(p_\nu) \geq 2$ for all $\nu \in \mathbb{N}$. Since λ is bounded from above by the smooth function λ^* and from below by some positive constant, we may choose this sequence in such a way that $\lambda_0 := \lim_{\nu\to\infty} \lambda(p_\nu)$ exists (and is positive). Then, using the fact that the map $(p,t) \mapsto \psi_t(p)$ is continuous, we have

$$p_0 = \lim_{\nu\to\infty} p_\nu = \lim_{\nu\to\infty} \psi_{\lambda(p_\nu)}(p_\nu) = \psi_{\lambda_0}(p_0).$$

It follows that $\lambda_0 = lt_0$ for some $l \in \mathbb{N}$. Then

$$\begin{aligned} t_0 &= \lambda^*(p_0) = \lim_{\nu\to\infty} \lambda^*(p_\nu) = \lim_{\nu\to\infty} \big(k(p_\nu)\lambda(p_\nu)\big) \\ &\geq 2 \lim_{\nu\to\infty} \lambda(p_\nu) = 2\lambda_0 = 2lt_0 \\ &\geq 2t_0. \end{aligned}$$

This contradiction proves the lemma. □

Lemma 7.2.7 *Let R_α be the Reeb vector field of a regular contact form α on a closed manifold M. Then the function $M \to \mathbb{R}^+$, $p \mapsto \lambda(p)$ is constant.*

Proof The smooth vector field $R := \lambda R_\alpha / 2\pi$ is regular and has all orbits of period 2π, so the flow of this vector field generates a free S^1–action on M. This defines on M the structure of a principal S^1–bundle, see [142].

In local coordinates $(\theta, x_1, \ldots, x_{2n})$ defined by a bundle chart around any given point $p \in M$ – where θ ranges over S^1, and $(x_1, \ldots, x_{2n})$, over some open subset of $\mathbb{R}^{2n}$ – we can write α as

$$\alpha = f(\theta, x_1, \ldots, x_{2n})\, d\theta + \sum_{j=1}^{2n} g_j(\theta, x_1, \ldots, x_{2n})\, dx_j.$$

The condition $\alpha(R_\alpha) \equiv 1$, and the fact that $R_\alpha = 2\pi\partial_\theta/\lambda$ in this bundle chart, gives $f = \lambda/2\pi$. In particular, f does not depend on θ.

The condition $i_{R_\alpha} d\alpha \equiv 0$ translates into

$$0 \equiv i_{\partial_\theta} d\alpha = \sum_{j=1}^{2n} \left(\frac{\partial g_j}{\partial \theta} - \frac{\partial f}{\partial x_j} \right) dx_j,$$

that is, $\partial g_j / \partial\theta = \partial f / \partial x_j$ for $j = 1, \ldots, 2n$. Hence

$$\frac{\partial^2 g_j}{\partial \theta^2} = \frac{\partial^2 f}{\partial x_j \partial \theta} = 0,$$

since $\partial f / \partial \theta = 0$. So the functions g_j are both linear and 2π–periodic in θ, hence constant in θ. Thus $\partial f / \partial x_j = \partial g_j / \partial \theta = 0$ for $j = 1, \ldots, 2n$. This proves that $\lambda = 2\pi f$ is a constant function. □

Proof of Theorem 7.2.5 By the preceding lemma, we may assume – after rescaling α by a constant – that the flow of the Reeb vector field R_α (which is simply rescaled by the inverse of this constant) generates a free S^1–action on M. This gives M the structure of a principal S^1–bundle $\pi\colon M \to B$ over some manifold B, and the defining equations for the Reeb vector field mean that (the rescaled) α is a connection 1–form for this bundle.

The curvature 2–form ω on B is defined by $d\alpha = \pi^*\omega$. We saw above that ω is closed, and from $\pi^*\omega^n = (d\alpha)^n \neq 0$ it follows that ω is a symplectic form. □

The notion of a regular contact form has been generalised to that of a so-called **almost regular** contact form in [229]. Here the Reeb vector field is allowed to pierce small flow boxes a finite number of times, with a global upper bound on this number. This corresponds to S^1–actions without fixed points. Examples are provided by the contact form α_1 on Brieskorn manifolds from Section 7.1.

As observed by Gromov [119, p. 343], the Boothby–Wang construction can be used to define a notion of blow-up for contact manifolds. Suppose M' is a submanifold of a (strict) contact manifold (M, α) with the property that $\alpha|_{TM'}$ is a regular contact form. Assume further that the corresponding S^1–action on M' lifts to a symplectic S^1–action on the symplectic normal bundle $\mathrm{SN}_M(M')$. Then one can get a new contact manifold by blowing up M along M', essentially by an S^1–invariant extension of the symplectic blow-up [177, 7.1]. For an application of this observation see [210, Cor. 8.3].

7.3 Open books

The aim of the present section is to extend the construction of contact structures on open books from the 3–dimensional setting in Section 4.4.2 to arbitrary (odd) dimensions. Most of the ideas in this section belong to Giroux [106], [107], partially in collaboration with Mohsen [108].

Some features of the 3–dimensional construction in Section 4.4.2 do not readily generalise to higher dimensions. We therefore need to begin with two simple observations concerning the differential topology of open books. Our notation is as in Alexander's theorem (Thm. 4.4.3) and Definition 4.4.4, except that M is now of dimension $2n+1$ and Σ of dimension $2n$.

Lemma 7.3.1 *Let ϕ_t, $t \in [0,1]$, be a technical isotopy of diffeomorphisms of Σ, i.e. $\phi_t \equiv \phi_0$ for t near 0, and $\phi_t \equiv \phi_1$ for t near 1, with ϕ_t equal to the identity map near $\partial\Sigma$ for all t. Then $M(\phi_0)$ is diffeomorphic to $M(\phi_1)$.*

Proof It is enough to show that there is a diffeomorphism from the mapping torus $\Sigma(\phi_1)$ to $\Sigma(\phi_0)$, equal to the identity near the boundary. Define a diffeomorphism of $\Sigma \times [0, 2\pi]$ by

$$(x, \varphi) \longmapsto (\phi_{\varphi/2\pi} \circ \phi_1^{-1}(x), \varphi).$$

This diffeomorphism gives rise to the commutative diagram

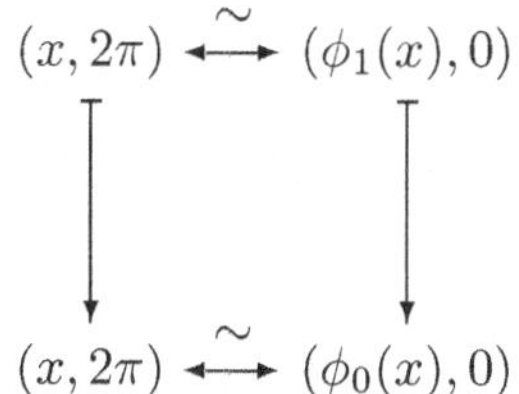

and hence it descends to the desired diffeomorphism. □

The second observation concerns a less restrictive definition of the notion of mapping torus. Let $\overline{\varphi}\colon \Sigma \to \mathbb{R}^+$ be a smooth function, constant near $\partial\Sigma$. Define the **(generalised) mapping torus** $\Sigma_{\overline{\varphi}}(\phi)$ as the quotient space of

$$\{(x, \varphi) \in \Sigma \times \mathbb{R} \colon 0 \leq \varphi \leq \overline{\varphi}(x)\},$$

by identifying $(x, \overline{\varphi}(x))$ with $(\phi(x), 0)$. With c denoting the value of $\overline{\varphi}$ near $\partial\Sigma$,† we form $M_{\overline{\varphi}}(\phi)$ as the quotient space

$$\left(\Sigma_{\overline{\varphi}}(\phi) + \partial\Sigma \times D^2\right)/\sim,$$

where $(x, e^{i\varphi}) \subset \partial(\partial\Sigma \times D^2)$ is identified with $[x, c\varphi/2\pi] \in \partial(\Sigma_{\overline{\varphi}}(\phi))$.

A diffeomorphism from $M(\phi)$ to $M_{\overline{\varphi}}(\phi)$ is then given by the identity map on $\partial\Sigma \times D^2$, and by

$$[x, \varphi] \longmapsto \left[x, \frac{\varphi}{2\pi}\,\overline{\varphi}(x)\right] \quad \text{for } [x, t] \in \Sigma(\phi).$$

Before we turn to the construction of contact structures on higher-dimensional open books, we revisit the 3–dimensional case. The next lemma shows that we may without loss of generality impose the condition that the monodromy map ϕ be a symplectomorphism. In the higher-dimensional situation, this condition will have to figure as an additional assumption.

Let β be a 1–form on a surface Σ as in Lemma 4.4.6, and denote the symplectic form $d\beta$ by ω.

Lemma 7.3.2 *The diffeomorphism ϕ of Σ is isotopic, via diffeomorphisms equal to the identity near $\partial\Sigma$, to a symplectomorphism of (Σ, ω).*

† For notational convenience only I pretend that $\overline{\varphi}$ takes the *same* constant value c near all boundary components of Σ.

Proof Set $\omega_t = (1-t)\omega + t\phi^*\omega$, $t \in [0,1]$, so that $\omega_0 = \omega$ and $\omega_1 = \phi^*\omega$. Notice that ω_t is an area form for all $t \in [0,1]$. Our aim will be to construct an isotopy ψ_t of Σ (as always equal to the identity near $\partial\Sigma$) satisfying $\psi_t^*\omega_t = \omega_0$. Given such an isotopy, the map $\phi \circ \psi_1$ is the desired symplectomorphism isotopic to ϕ.

We apply the Moser trick to the equation $\psi_t^*\omega_t = \omega_0$. Suppose ψ_t is the flow of the time-dependent vector field X_t. Then differentiation of the equation $\psi_t^*\omega_t = \omega_0$ with respect to t yields

$$\psi_t^*\left(\dot{\omega}_t + \mathcal{L}_{X_t}\omega_t\right) = 0.$$

Since $\dot{\omega}_t = \phi^*\omega - \omega = d(\phi^*\beta - \beta)$ and $\mathcal{L}_{X_t}\omega_t = d(i_{X_t}\omega_t)$, that equation will be satisfied if

$$\phi^*\beta - \beta + i_{X_t}\omega_t = 0.$$

That last equation can be solved uniquely for X_t, because ω_t is an area form. Moreover, we have $\phi^*\beta - \beta = 0$ near $\partial\Sigma$, which entails $X_t \equiv \mathbf{0}$ near the boundary. This guarantees both the global existence of the flow ψ_t, and ψ_t to be the identity map near the boundary. □

This lemma (together with Lemma 7.3.1) allows us to assume – still in the 3–dimensional situation – that ϕ is a symplectomorphism of (Σ, ω). Observe that with $\omega = d\beta$ the condition $\phi^*\omega = \omega$ translates into saying that $\phi^*\beta - \beta$ is closed.

In order for the construction from Section 4.4.2 to work in higher dimensions, this assumption has to be imposed from the start. The following will be the main result of this section.

Theorem 7.3.3 *Let M a closed odd-dimensional manifold such that $M = M(\phi)$ (in the notation of Theorem 4.4.3), with the following properties:*

- *Σ is a compact manifold admitting an exact symplectic form $\omega = d\beta$,*
- *the Liouville vector field Y for ω defined by $i_Y\omega = \beta$ is transverse to $\partial\Sigma$, pointing outwards,*
- *the monodromy map ϕ is a symplectomorphism of (Σ, ω), equal to the identity near the boundary $\partial\Sigma$.*

Then M admits a contact structure supported by the open book decomposition (B, p) corresponding to the abstract open book (Σ, ϕ).

Write β_∂ for the restriction of β to $T(\partial\Sigma)$. This is a contact form on $\partial\Sigma$ by Lemma 1.4.5. The conditions in the theorem imply that $\omega = d(e^s\beta_\partial)$ on a collar neighbourhood $(-\varepsilon, 0] \times \partial\Sigma$ of $\partial\Sigma$ in Σ, just as in Lemma 4.4.6. In higher dimensions, however, there is no analogue of the convexity statement in that lemma. Instead one needs a suitable modification of the monodromy map.

Lemma 7.3.4 *The symplectomorphism ϕ of $(\Sigma, \omega = d\beta)$ is isotopic, via symplectomorphisms equal to the identity near $\partial\Sigma$, to an exact symplectomorphism ϕ_1, i.e. a symplectomorphism such that $\phi_1^*\beta - \beta$ is exact.*

Proof We mimic the argument of Lemma 7.3.2 in the attempt to 'improve' the given symplectomorphism ϕ. Thus, define a vector field X on Σ by the equation

$$\phi^*\beta - \beta + i_X\omega = 0,$$

and let ψ_t denote its flow. As before, we have $X \equiv \mathbf{0}$ near $\partial\Sigma$, so ψ_t is globally defined and equal to the identity map near the boundary.

We claim that $\phi \circ \psi_1$ is an exact symplectomorphism as desired. Set

$$\gamma := \phi^*\beta - \beta = -i_X\omega.$$

Then $d\gamma = 0$ and

$$\mathcal{L}_X\gamma = d(\gamma(X)) + i_X d\gamma = 0,$$

so we have $\psi_t^*\gamma = \gamma$ for all $t \in \mathbb{R}$. Likewise, we have

$$\mathcal{L}_X\omega = d(i_X\omega) = -d\gamma = 0,$$

so $\psi_t^*\omega = \omega$ for all t. This implies that $\phi \circ \psi_1$ is isotopic to ϕ via symplectomorphisms that equal the identity map near $\partial\Sigma$. Observe further that

$$\mathcal{L}_X\beta = d(\beta(X)) + i_X d\beta = d(\beta(X)) - \gamma.$$

Finally, we compute

$$\begin{aligned}
(\phi \circ \psi_1)^*\beta - \beta &= \psi_1^*\phi^*\beta - \beta \;=\; \psi_1^*(\beta + \gamma) - \beta \;=\; \psi_1^*\beta + \gamma - \beta \\
&= \gamma + \int_0^1 \frac{d}{dt}\psi_t^*\beta \;=\; \gamma + \int_0^1 \psi_t^*(\mathcal{L}_X\beta) \\
&= d\int_0^1 \psi_t^*(\beta(X));
\end{aligned}$$

this shows that $\phi \circ \psi_1$ is exact symplectic. □

Proof of Theorem 7.3.3 By Lemmata 7.3.1 and 7.3.4 we may assume that ϕ is an exact symplectomorphism of $(\Sigma, d\beta)$, equal to the identity near $\partial\Sigma$. The equation

$$\phi^*\beta - \beta =: d\varphi$$

defines a function $\overline{\varphi}$ on Σ, unique up to adding a constant. Thanks to the compactness of Σ we may assume that $\overline{\varphi}$ takes positive values only.

The 1–form

$$\alpha := \beta + d\varphi$$

is a contact form on $\Sigma \times \mathbb{R}$ invariant under the transformation

$$(x, \varphi) \longmapsto (\phi(x), \varphi - \overline{\varphi}(x)).$$

It follows that α descends to a contact form on $\Sigma_{\overline{\varphi}}(\phi)$.

Near the boundary $\partial\Sigma$ the map ϕ is the identity, and $\overline{\varphi}$ is a locally constant function. So the contact form α looks near the boundary $\partial(\Sigma_{\overline{\varphi}}(\phi))$ as in the construction in Section 4.4.2, and we can define the extension over $\partial\Sigma \times D^2$ as we did there in order to obtain a contact structure on $M_{\overline{\varphi}}(\phi) \cong M(\phi)$ adapted to the open book decomposition. Indeed, we now have to work with an ansatz $\alpha = h_1(r)\, \beta_\partial + h_2(r)\, d\varphi$ on $\partial\Sigma \times D^2_{1+\varepsilon}$, with boundary conditions (1) and (2) as on page 153. With the functions h_1, h_2 described there, we compute

$$\begin{aligned} \alpha \wedge (d\alpha)^n &= n h_1(r)^{n-1} \big(h_1(r) h_2'(r) - h_1'(r) h_2(r)\big) \cdot \\ &\quad\; \beta_\partial \wedge (d\beta_\partial)^{n-1} \wedge dr \wedge d\varphi \\ &> 0. \end{aligned}$$

I leave it to the reader to check that all the conditions of Definition 4.4.7, notably the orientation conventions, are satisfied. □

For examples of higher-dimensional manifolds with contact structures supported by open books see [151] and [152].

Rather remarkably, there is the following converse to Theorem 7.3.3. Unfortunately, complete details of its proof have not yet appeared, but see [106].

Theorem 7.3.5 (Giroux–Mohsen) *To any contact structure ξ on a closed manifold M of dimension at least equal to 3 one can find an open book decomposition (B, p) of M supporting ξ.* □

In 1979 Lutz discovered a contact structure on the 5–torus T^5. Ever since, it had been a prominent question whether all odd-dimensional tori T^{2n+1} admit a contact structure. This question was finally settled by Bourgeois [32] in 2002. Using the theorem of Giroux–Mohsen he was able to generalise the construction by Lutz considerably.

Theorem 7.3.6 (Bourgeois) *Let (M, ξ) be a closed contact manifold. Then $M \times T^2$ admits a contact structure $\tilde{\xi}$ invariant under the obvious T^2–action, and with $M \times \{q\}$ a contact submanifold of $(M \times T^2, \tilde{\xi})$ for each $q \in T^2$.*

Proof Needless to say, we may assume that M is connected. Write $2n-1$ for the dimension of M. If $n=1$, then $M=S^1$. A contact structure on $S^1\times T^2$ of the desired kind is given by

$$\tilde{\xi}=\ker\big(d\theta+\cos\theta\,d\theta_1-\sin\theta\,d\theta_2\big);$$

here θ denotes the S^1–coordinate, while θ_1,θ_2 are the circle-valued coordinates on T^2.

Now suppose $n\geq 2$. Write $\xi=\ker\alpha$ with α satisfying the conditions (o), (i) and (ii) of Definition 4.4.7 for some open book decomposition (B,p) of M. Let (r,φ) be polar coordinates on the D^2–factor of a neighbourhood $B\times D^2$ of the binding B, such that $p\colon M\setminus B\rightarrow S^1$ is given by φ in that neighbourhood.

Choose a smooth function ρ of the variable r on $B\times D^2$ satisfying the requirements that

- $\rho(r)=r$ near $B\equiv B\times\{\mathbf{0}\}$,
- $\rho'(r)\geq 0$,
- $\rho\equiv 1$ near $B\times\partial D^2$.

Extend ρ to a smooth function on M by setting it equal to 1 outside $B\times D^2$. Then $x_1:=\rho\cos\varphi$ and $x_2:=\rho\sin\varphi$ are smooth functions on M that coincide with the Cartesian coordinate functions on the D^2–factor near $B\times\{\mathbf{0}\}\subset B\times D^2$. The usual identities

$$x_1\,dx_2-x_2\,dx_1=\rho^2\,d\varphi \text{ and } dx_1\wedge dx_2=\rho\,d\rho\wedge d\varphi$$

continue to hold in this situation (on all of M).

Now define a 1–form $\widetilde{\alpha}$ on $M\times T^2$ by

$$\widetilde{\alpha}=x_1\,d\theta_1-x_2\,d\theta_2+\alpha.$$

Then

$$d\widetilde{\alpha}=dx_1\wedge d\theta_1-dx_2\wedge d\theta_2+d\alpha,$$

hence

$$\begin{aligned}(d\widetilde{\alpha})^n &= n(d\alpha)^{n-1}\wedge(dx_1\wedge d\theta_1-dx_2\wedge d\theta_2)\\ &\quad -n(n-1)(d\alpha)^{n-2}\wedge dx_1\wedge d\theta_1\wedge dx_2\wedge d\theta_2.\end{aligned}$$

In order to verify the contact condition, we now compute

$$\begin{aligned}\widetilde{\alpha}\wedge(d\widetilde{\alpha})^n &= n(d\alpha)^{n-1}\wedge(x_1\,dx_2-x_2\,dx_1)\wedge d\theta_1\wedge d\theta_2\\ &\quad +n(n-1)\alpha\wedge(d\alpha)^{n-2}\wedge dx_1\wedge dx_2\wedge d\theta_1\wedge d\theta_2\\ &= n(d\alpha)^{n-1}\wedge(\rho^2\,d\varphi)\wedge d\theta_1\wedge d\theta_2\\ &\quad +n(n-1)\alpha\wedge(d\alpha)^{n-2}\wedge(\rho\,d\rho\wedge d\varphi)\wedge d\theta_1\wedge d\theta_2.\end{aligned}$$

Condition (i) of Definition 4.4.7 guarantees that the first summand is a positive volume form away from $B \times T^2$; along $B \times T^2$ this summand vanishes. The second summand is non-negative on all of $M \times T^2$ as well, and it is positive near $B \times T^2$ by condition (ii).

So $\tilde{\xi} := \ker \tilde{\alpha}$ is a contact structure on $M \times T^2$. The other properties of $\tilde{\xi}$ stated in the theorem are easily verified. □

7.4 Fibre connected sum

The fibre connected sum is a method for constructing manifolds that is closely related to surgery: the connected sum of manifolds is an example for either method and, as I shall explain, any fibre connected sum can be effected as a sequence of surgeries.

Let M, M' be two closed, oriented manifolds with $\dim M' < \dim M$. No assumption is made on these manifolds being connected. Suppose we are given two embeddings $j_0, j_1 \colon M' \to M$ with disjoint images. Furthermore, it is assumed that the normal bundles N_i of the submanifolds $j_i(M') \subset M$, $i = 0, 1$, are isomorphic under a fibre-orientation-reversing bundle isomorphism Φ covering $j_1 \circ j_0^{-1}|_{j_0(M')}$. Choose a bundle metric on N_0, and give N_1 the induced metric that turns Φ into a bundle isometry. Write $\|\mathbf{v}\|$ for the corresponding norm of an element $\mathbf{v} \in N_i$. Identify N_i with an open tubular neighbourhood of $j_i(M')$ in M, with $N_0, N_1 \subset M$ still disjoint. For any interval $I \subset \mathbb{R}_0^+$, denote by N_i^I the subset of N_i given by $\{\mathbf{v} \in N_i \colon \|\mathbf{v}\| \in I\}$.

Definition 7.4.1 The **fibre connected sum** $\#_\Phi M$ is the closed, oriented manifold given as the quotient space

$$\big(M \setminus (j_0(M') \cup j_1(M'))\big)/\sim,$$

where we identify along open subsets with an *orientation-preserving* diffeomorphism as follows:

$$N_0^{(0,\varepsilon)} \ni \mathbf{v} \stackrel{\sim}{\longleftrightarrow} \frac{\sqrt{\varepsilon^2 - \|\mathbf{v}\|^2}}{\|\mathbf{v}\|} \cdot \Phi(\mathbf{v}) \in N_1^{(0,\varepsilon)}.$$

Remarks 7.4.2 (1) As in the handle construction on page 292, one only needs to check the Hausdorff property of the quotient space in order to guarantee that the identification via a diffeomorphism of open subsets yields a differential manifold.

(2) The fibre connected sum $\#_\Phi M$ can also be written as the quotient space

$$\Big(M \setminus \big(N_0^{[0,\varepsilon/2]} \cup N_1^{[0,\varepsilon/2]}\big)\Big)/\sim,$$

where the identification is given by

$$N_0^{(\varepsilon/2,\sqrt{3}\varepsilon/2)} \ni \mathbf{v} \stackrel{\sim}{\longleftrightarrow} \frac{\sqrt{\varepsilon^2 - \|\mathbf{v}\|^2}}{\|\mathbf{v}\|} \cdot \Phi(\mathbf{v}) \in N_1^{(\varepsilon/2,\sqrt{3}\varepsilon/2)}. \tag{7.9}$$

(3) If M is a disjoint union $M_0 \sqcup M_1$ of two connected manifolds, and if j_i is an embedding of M' into M_i, we also write $M_0 \#_\Phi M_1$ for $\#_\Phi M$. In the special case of M' being a single point, the fibre connected sum $M_0 \#_\Phi M_1$ is the same as the ordinary connected sum $M_0 \# M_1$.

(4) If the M_i as in (3) are product manifolds $M_i = M' \times M_i'$, and if the embeddings j_i are inclusion maps $M' \equiv M' \times \{q_i\} \subset M' \times M_i'$, then $M_0 \#_\Phi M_1$ is diffeomorphic to $M' \times (M_0' \# M_1')$.

(5) There is a cobordism from M to $\#_\Phi M$, obtained from the trivial cobordism $[0,1] \times M$ by identifying *closed* tubular neighbourhoods of the submanifolds $\{1\} \times j_0(M')$ and $\{1\} \times j_1(M')$ in $\{1\} \times M$ via Φ and smoothing the resulting corners. As is well known, any cobordism translates with the help of Morse theory into a sequence of surgeries, see [153, Thm. VII.1.1].

We now want to show that, under suitable assumptions, one can form the fibre connected sum of a contact manifold along codimension 2 contact submanifolds. The following theorem was stated as an exercise in [119, p. 343]. A proof was published in [93]; full details, however, were only given for the case of trivial normal bundles. The argument below covers the general case and is simpler than the one suggested in [93].

Theorem 7.4.3 *Let (M,ξ) and (M',ξ') be contact manifolds of dimension* $\dim M' = \dim M - 2$, *where the contact structures ξ, ξ' are assumed to be cooriented; these cooriented contact structures induce orientations of M and M'. Let $j_0, j_1\colon (M',\xi') \to (M,\xi)$ be disjoint contact embeddings that respect the coorientations, and such that there exists a fibre-orientation-reversing bundle isomorphism $\Phi\colon N_0 \to N_1$ of the normal bundles of $j_0(M')$ and $j_1(M')$. Then the fibre connected sum $\#_\Phi M$ admits a contact structure that coincides with ξ outside tubular neighbourhoods of the submanifolds $j_0(M')$ and $j_1(M')$.*

Proof Write $\pi_i\colon N_i \to j_i(M')$ for the bundle projections. Let γ_1 be a connection 1–form on the unit circle bundle $\{\mathbf{v} \in N_1\colon \|\mathbf{v}\| = 1\}$ associated with N_1. With the help of the natural retraction $\mathbb{R}^2 \setminus \{\mathbf{0}\} \to S^1$ we may pull back γ_1 to a 1–form on $N_1 \setminus j_1(M')$.† Set $\gamma_0 := -\Phi^*\gamma_1$. Denote by r the radial coordinate in the fibres of N_i. Observe that $r^2\gamma_i$ defines a smooth

† Up to a scaling factor 2π, this 1–form is the global angular form of [31, p. 73].

1–form on N_i whose exterior derivative restricts to an area form $2r\,dr\wedge\gamma_i$ on each fibre.

Let α' be a contact form defining $\xi'=\ker\alpha'$ on $M'\equiv j_i(M')$. Under the identification of N_i with a tubular neighbourhood of $j_i(M')$ in M, Theorem 2.5.15 and its proof allow us to write $\xi=\ker(\pi_i^*\alpha'+r^2\gamma_i)$ on a neighbourhood $N_i^{[0,2\varepsilon)}$ of the zero section $j_i(M')$ of N_i.

We now use the description of the fibre connected sum from Remark 7.4.2 (2). Choose a smooth function $f\colon(\varepsilon/2,2\varepsilon)\to\mathbb{R}$ with the following properties:

- $f'(r)>0$,
- $f(r)=r^2$ on an open interval containing $[\varepsilon,2\varepsilon)$,
- $f(r)=r^2-\varepsilon^2/2$ on the interval $(\varepsilon/2,\sqrt{3}\varepsilon/2)$.

By the first two conditions, the modified 1–form $\pi_i^*\alpha'+f(r)\gamma_i$ is a contact form on $N_i^{(\varepsilon/2,2\varepsilon)}$ (provided that $\varepsilon>0$ has been chosen sufficiently small) that coincides with the previously given one on a neighbourhood of $N_i^{[\varepsilon,2\varepsilon)}$ in $N_i^{(\varepsilon/2,2\varepsilon)}$. The third condition implies that $f(\sqrt{\varepsilon^2-r^2})=-f(r)$ on $(\varepsilon/2,\sqrt{3}\varepsilon/2)$. Since $\gamma_0=-\Phi^*\gamma_1$, this behaviour of f on $(\varepsilon/2,\sqrt{3}\varepsilon/2)$ ensures that the identification map (7.9) is actually a contactomorphism. So the contact structure on M descends to a contact structure on $\#_\Phi M$. □

Remark 7.4.4 Given an embedding $j\colon M'\to M$, the normal bundle $N:=TM/T(j(M'))$ of $j(M')$ is isomorphic to the bundle $j^*N=j^*(TM)/TM'$ over M'. One may take the latter as the definition of the normal bundle; this definition makes sense also for immersions. The condition on the existence of a fibre-orientation-reversing bundle isomorphism Φ in the theorem is equivalent to stipulating that the bundles $j_0^*N_0$ and $j_1^*N_1$ over M' have Euler classes of the opposite sign; see page 370 *et seq.* for a proof that orientable 2–plane bundles are classified by their Euler class (or first Chern class), and [31, p. 286] for the simple observation that changing the fibre orientation amounts to reversing the sign of the Euler class.

Example 7.4.5 Let M be a closed, orientable 3–manifold. By combining the theorems of Martinet (Thm. 4.1.1) and Bourgeois (Thm. 7.3.6), we obtain a contact structure on $M\times T^2$ for which each submanifold $M\times\{q\}$ is a contact submanifold. An application of Theorem 7.4.3 to several copies of such a 5–manifold as described in Remark 7.4.2 (4) produces a contact structure on $M\times\Sigma$ for all closed, orientable surfaces Σ.

7.5 Branched covers

Branched covers were encountered previously in Section 4.4.1, for the special case of 3–manifolds. There we built a contact structure on the branched cover from one on the base (there: S^3) by using an explicit description of the covering map near the upstairs branching set.

In the present section we want to extend this result to branched coverings in arbitrary dimensions. Beware, though, that the branching set, by definition, is always of codimension 2. Like the fibre connected sum, the branched covering construction for contact manifolds was posed as an exercise in [119, p. 343], and a solution of this exercise first appeared in [93].

Definition 7.5.1 A differentiable map $p\colon M \to M_0$ between m–dimensional manifolds is a **branched covering** if every point $x_0 \in M_0$ has a neighbourhood U_0 with the following property. For each component U of $p^{-1}(U_0)$ there is a number $k \in \mathbb{N}$ and a commutative diagram

$$\begin{array}{ccc} D^2 \times [-1,1]^{m-2} & \xrightarrow{\;h\;} & U \\ {\scriptstyle p_k}\big\downarrow & & \big\downarrow{\scriptstyle p|_U} \\ D^2 \times [-1,1]^{m-2} & \xrightarrow[\;h_0\;]{} & U_0 \end{array}$$

where h and h_0 are diffeomorphisms, $h_0(\mathbf{0}) = x_0$, and with D^2 regarded as a subset of $\mathbb{C}$ the map p_k is described by

$$p_k(z, \mathbf{t}) = (z^k, \mathbf{t}).$$

The natural number k is called the **branching index** of the point $p^{-1}(x_0) \cap U$. Write $M' \subset M$ for the codimension 2 submanifold made up of points with branching index greater than 1. The submanifold $M_0' := p(M') \subset M_0$ is called the **branch set**.

Remark 7.5.2 The branch set M_0' is sometimes called the **downstairs** branch set. There is some confusion in the literature, however, what to call the **upstairs** branch set. Some authors use this term to refer to M', others use it for the set $p^{-1}(M_0')$. As we have seen in Theorem 4.4.1, M' is, in general, a strict subset of $p^{-1}(M_0')$, i.e. there are points in the preimage of the branch set where the map p is unbranched.

Notice that both the maps $p|_{M\setminus p^{-1}(M_0')}\colon M \setminus p^{-1}(M_0') \to M_0 \setminus M_0'$ and $p|_{M'}\colon M' \to M_0'$ are unbranched coverings. Likewise, p is an unbranched covering when restricted to a neighbourhood of the codimension 2 submanifold $p^{-1}(M_0') \setminus M'$.

Example 7.5.3 The map $S^1 \to S^1$, $z \mapsto z^k$ is an unbranched covering of degree k. The $(m-1)$–fold suspension of this map gives a branched covering $S^m \to S^m$, branched along an unknotted $S^{m-2} \subset S^m$. For $m = 2$ and under the identification of S^2 with the Riemann sphere $\mathbb{C} \cup \{\infty\}$, this map is still described by $z \mapsto z^k$, with branch points 0 and ∞.

Here is the branched cover theorem for contact manifolds. The notation is as in the definition, and it is understood that M and M_0 are *closed* manifolds.

Theorem 7.5.4 *Suppose that M_0 admits a contact structure $\xi_0 = \ker \alpha_0$ such that $\xi_0' := TM_0' \cap \xi_0$ is a contact structure on M_0'.*† *Then M admits a contact structure $\xi = \ker \alpha$, with α equal to $p^*\alpha_0$ outside a neighbourhood of M'.*

Proof Set $\widetilde{\alpha} := p^*\alpha_0$. By the remark above, this defines a contact form on $M \setminus M'$, and $\widetilde{\alpha}|_{TM'}$ is likewise a contact form. Using a tubular map, we may identify a neighbourhood of $M' \subset M$ with the normal bundle $NM' \to M'$. Choose a Riemannian metric on NM'. As in Section 7.4 we write r for the corresponding radial coordinate on the fibres of NM'. Take a connection 1–form γ on the unit circle bundle associated with NM'. As in the proof of Theorem 7.4.3, we may regard $r^2\gamma$ as a smooth 1–form on NM'.

Choose a bump function $g\colon \mathbb{R}_0^+ \to [0,1]$ with $g(r) \equiv 1$ near $r = 0$ and $g(r) \equiv 0$ for $r \geq 1$. (For convenience we also assume $g'(r) \equiv 0$ for $r \geq 1$.) With C some positive real number, we set

$$\alpha := C\widetilde{\alpha} + g(r)r^2\gamma;$$

this may be regarded as a 1–form defined on all of M.

We compute

$$d\alpha = C\,d\widetilde{\alpha} + \big(g'(r)\cdot r + 2g(r)\big)\cdot r\,dr \wedge \gamma + g(r)\cdot r^2\,d\gamma.$$

With the dimension of M, M_0 equal to $2n+1$, we need to study the form

$$\begin{aligned}\alpha \wedge (d\alpha)^n &= C^{n+1}\widetilde{\alpha} \wedge (d\widetilde{\alpha})^n \\ &\quad + nC^n\big(g'(r)\cdot r + 2g(r)\big)\widetilde{\alpha} \wedge (d\widetilde{\alpha})^{n-1} \wedge r\,dr \wedge \gamma \\ &\quad + A(r^2, C^n);\end{aligned}$$

here $A(r^2, C^n)$ is a term in which each summand contains a factor r^2 or higher powers of r, and at most an nth power of C.

The first summand in this expression is a positive volume form on $M \setminus M'$, and it vanishes along M' (this follows from Theorem 2.5.15, again see the

† Then (M_0', ξ_0') is a contact submanifold of (M_0, ξ_0). Equivalently, $\alpha_0|_{TM_0'}$ is a contact form.

proof of Theorem 7.4.3). The second summand is positive on $TM|_{M'}$; the remainder term $A(r^2, C^n)$ vanishes along M'.

We conclude that there is a $\delta \in (0,1)$, independent of $C \geq 1$, such that the $(2n+1)$–form $\alpha \wedge (d\alpha)^n$ is positive for $r \leq \delta$. The same is true on the complement of the tubular neighbourhood $\{r < 1\} \subset M$, because here all summands but the first vanish. On the compact set $\{\delta \leq r \leq 1\}$ the first summand dominates for $C \gg 1$. So for C sufficiently large, the 1–form α does indeed define a contact form of the desired type on M. □

Example 7.5.5 Here is an alternative argument for finding a contact structure on $M \times \Sigma_g$, with M a closed, orientable 3–manifold, and Σ_g a surface of genus g, see Example 7.4.5. The basis for this example is the observation that Σ_g is a branched cover of S^2, branched along $2g+2$ points: take a model of $\Sigma_g \subset \mathbb{R}^3$ with an axis of symmetry intersecting Σ_g in $2g+2$ points; then form the quotient under the involution defined by this symmetry; see [209, p. 129]. So $M \times \Sigma_g$ is a branched cover of $M \times S^2$, branched along $2g+2$ copies of M of the form $M \times \{q_i\} \subset M \times S^2$.

Take a contact form α_1 on M with $\ker \alpha_1$ trivial as a 2–plane bundle (Prop. 4.3.2). Equip the cotangent bundle T^*M with a bundle metric, and normalise α_1 (or the metric) so that this 1–form may be regarded as a section of the unit cotangent bundle $ST^*M \rightarrow M$. The fact that $\ker \alpha_1$ is trivial translates into saying that we may choose a further section β of the unit cotangent bundle orthogonal to α_1. For $\varepsilon > 0$ sufficiently small, each of the 1–forms

$$\alpha_i := \frac{\alpha_1 + (i-1)\varepsilon\beta}{\sqrt{1+(i-1)^2\varepsilon^2}}, \quad i = 1, \ldots, 2g+2,$$

is a section of the unit cotangent bundle of M and a contact structure on M. In terms of the trivialisation $ST^*M \cong M \times S^2$ defined by α_1 and β (and a third 1–form orthonormal to either of them), the α_i are constant sections of the trivial bundle $M \times S^2 \rightarrow M$.

Let λ be the Liouville form on ST^*M. If we regard the α_i as maps $M \rightarrow ST^*M$, then $\alpha_i = \alpha_i^*\lambda$, as shown in Lemma 1.4.1. This means that the $\alpha_i(M)$ are contact submanifolds of $(ST^*M, \ker \lambda)$. Now apply the branched cover theorem to $M_0 := ST^*M$ and $M_0' = \sqcup_i \alpha_i(M)$.

7.6 Plumbing

Let B_i be a closed n–dimensional manifold, and let $\pi_i \colon E_i \rightarrow B_i$ be a D^n–bundle over B_i, $i = 1, 2$. Let x_i be a point of B_i, and choose a disc neighbourhood D_i^n of x_i in B_i. Then there is a fibre-preserving diffeomorphism $\pi_i^{-1}(D_i^n) \rightarrow D_i^n \times D^n$.

Definition 7.6.1 The **plumbing** of E_1 with E_2 at x_1, x_2 is the manifold with boundary obtained by identifying $D_1^n \times D^n$ with $D_2^n \times D^n$, using the map

$$D_1^n \times D^n \ni (\mathbf{q}, \mathbf{p}) \longmapsto (-\mathbf{p}, \mathbf{q}) \in D_2^n \times D^n,$$

and smoothing corners; see Figure 7.1.

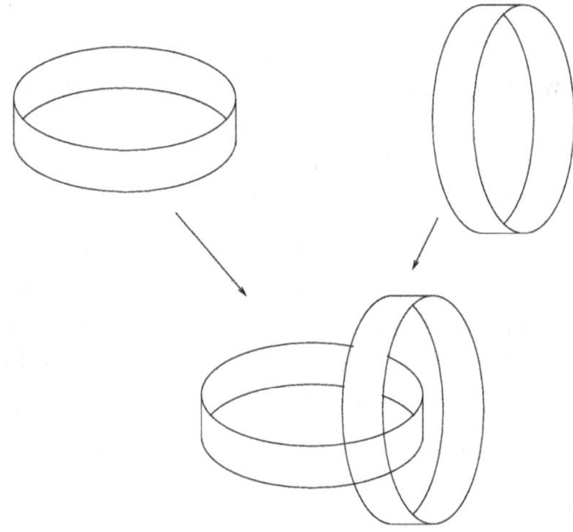

Fig. 7.1. Plumbing of disc bundles.

This procedure can be carried out with any finite number of disc bundles. For plumbing in the context of Kirby diagrams of 4–manifolds see [111, *passim*]; for the role of plumbing in higher-dimensional surgery theory, in particular the construction of exotic spheres, see [39, Chapter V] and [133, §7 *et seq.*].

Many of the constructions of $(2n-1)$–dimensional (exotic) spheres via plumbing use (co-)tangent disc bundles of S^n. So the following plumbing theorem is a powerful tool for the construction of contact manifolds. This theorem is, for most practical purposes, equivalent to a result of Eliashberg [65, Thm. 1.3.5], saying that the image of a 'nice' totally real immersion into a complex manifold has a neighbourhood with strictly pseudoconvex boundary. In [67], Eliashberg rephrased this result in the language of Lagrangian immersions into symplectic manifolds and used it for the construction of exotic, but homotopically standard, contact structures on spheres as described on page 338. While the extension of Eliashberg's result in [92] employed Brieskorn manifolds, the construction in [52], dealing with the spheres of dimension $8n+7$, was again based on plumbing.

Theorem 7.6.2 *Let W be the result of plumbing cotangent disc bundles. Then there is a symplectic form ω on W, and a contact structure ξ on ∂W, such that (W, ω) is a strong filling of $(\partial W, \xi)$.*

Proof It suffices to deal with the plumbing of two cotangent disc bundles $\pi_i\colon T^*B_i \to B_i$, $i=1,2$. Choose local coordinates $\mathbf{q}^i = (q_1^i, \ldots, q_n^i)$ on a disc neighbourhood D_i^n of $x_i \in B_i$, and let $\mathbf{p}^i$ be the dual coordinates on the fibres of $T^*B_i|_{D_i^n}$. The Liouville form λ_i on T^*B_i can be written as $\lambda_i = \mathbf{p}^i\, d\mathbf{q}^i$ in these local coordinates (Lemma 1.4.1). The coordinates $(\mathbf{q}^i, \mathbf{p}^i)$ define a trivialisation $\pi_i^{-1}(D_i^n) \cong D_i^n \times D^n$, so these coordinates are adapted to the plumbing construction. Under the map in Definition 7.6.1, the Liouville form λ_2 pulls back to $-\mathbf{q}^1\, d\mathbf{p}^1$ on $D_1^n \times D^n$. Observe that the symplectic form $d\lambda_2$ pulls back to the symplectic form $d\lambda_1$, so we get an induced symplectic form on the plumbed manifold (at least before we smooth corners).

Therefore, rather than straightening the angle in the strict sense of [38, (13.12)], we use an alternative procedure where we replace the initial non-smooth boundary by a smooth one. The symplectic form will not be affected by this procedure.

In order to create 'room' for this alternative procedure, we glue 'unit' disc bundles† as before, but we form the actual plumbed manifold by smoothing the corners on the manifold obtained by gluing ε–disc bundles for some $\varepsilon \in (0, 1/3)$. The identification of $D_{1,\varepsilon}^n \times D_\varepsilon^n$ with $D_{2,\varepsilon}^n \times D_\varepsilon^n$ yields a manifold W_0 with corner along $\{\|\mathbf{q}\| = \|\mathbf{p}\| = \varepsilon\}$. Figure 7.2 gives a first idea how we are going to smooth the corner. As coordinates $(\mathbf{q}, \mathbf{p})$ in this local description we use $(\mathbf{q}^1, \mathbf{p}^1)$. So W_0 carries a symplectic form ω; on $W_0 \cap \{\|\mathbf{q}\|, \|\mathbf{p}\| \leq 1\}$ this form is given by $\omega = d\mathbf{p} \wedge d\mathbf{q}$. Away from the corner, ∂W_0 is a convex boundary. On $\partial W_0 \cap \{\varepsilon < \|\mathbf{q}\| \leq 3\varepsilon\}$, the contact form is given by $\mathbf{p}\, d\mathbf{q}$, induced by the Liouville vector field $\mathbf{p}\,\partial_{\mathbf{p}}$. On $\partial W_0 \cap \{\varepsilon < \|\mathbf{p}\| \leq 3\varepsilon\}$, the contact form is given by $-\mathbf{q}\, d\mathbf{p}$, induced by the Liouville vector field $\mathbf{q}\,\partial_{\mathbf{q}}$.

To obtain the plumbed manifold W, we replace

$$\partial W_0 \cap \{\|\mathbf{q}\|, \|\mathbf{p}\| \leq 1\} = \{\varepsilon \leq \|\mathbf{q}\| \leq 1,\ \|\mathbf{p}\| = \varepsilon\} \cup \{\|\mathbf{q}\| = \varepsilon,\ \varepsilon \leq \|\mathbf{p}\| \leq 1\}$$

† Here 'unit' is tautological: we first choose local coordinates, and then take the standard metric with respect to these coordinates.

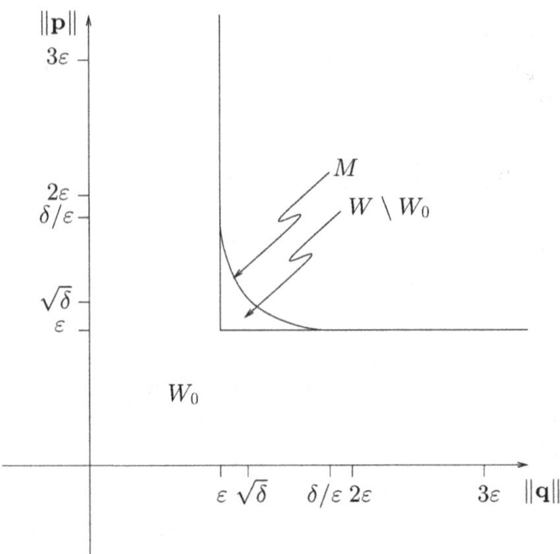

Fig. 7.2. Smoothing corners in a plumbing.

by a smooth hypersurface M of the form $\{F(\|\mathbf{q}\|, \|\mathbf{p}\|) = 0\}$ with the following properties:

- M is transverse to the Liouville vector field $(\mathbf{q}\,\partial_{\mathbf{q}} + \mathbf{p}\,\partial_{\mathbf{p}})/2$,
- M coincides with ∂W_0 for $\|\mathbf{q}\| \geq 2\varepsilon$ and for $\|\mathbf{p}\| \geq 2\varepsilon$.

It is geometrically clear that such a hypersurface can be found; a formal construction of M will be given in the remark following this proof. The desired manifold W will be the manifold that is bounded – in a neighbourhood of the plumbing region – by the hypersurface M. Since the 2–form $d\mathbf{p} \wedge d\mathbf{q}$ is defined on $\{\|\mathbf{q}\|, \|\mathbf{p}\| \leq 1\}$, the symplectic form ω on W_0 extends in the obvious way to a symplectic form on W.

To prove that W has convex boundary ∂W we need only find an interpolation between $(\mathbf{q}\,\partial_{\mathbf{q}} + \mathbf{p}\,\partial_{\mathbf{p}})/2$ and $\mathbf{p}\,\partial_{\mathbf{p}}$ as Liouville vector fields for $d\mathbf{p}\wedge d\mathbf{q}$ transverse to the part of ∂W in our coordinate neighbourhood where $\|\mathbf{q}\| \in (\varepsilon, 3\varepsilon]$; for the complementary part of ∂W in this coordinate neighbourhood the argument is analogous.

Choose a smooth function $\mathbb{R} \to \mathbb{R}$, $x \mapsto f(x)$ with the following properties:

(1) $f \equiv 0$ for $x \leq (2\varepsilon)^2$,
(2) $f \equiv 1/2$ for $x \geq (3\varepsilon)^2$,
(3) $f' \geq 0$.

I claim that the vector field

$$Y := \left(\frac{1}{2} - f(\|\mathbf{q}\|^2)\right)\mathbf{q}\,\partial_{\mathbf{q}} + \left(\left(\frac{1}{2} + f(\|\mathbf{q}\|^2)\right)\mathbf{p} + 2(\mathbf{qp})f'(\|\mathbf{q}\|^2)\cdot\mathbf{q}\right)\partial_{\mathbf{p}}$$

is a smooth interpolation with all the desired properties.

For $\|\mathbf{q}\| \leq 2\varepsilon$, the vector field Y coincides with $(\mathbf{q}\,\partial_{\mathbf{q}} + \mathbf{p}\,\partial_{\mathbf{p}})/2$, which is transverse to the hypersurface M (which constitutes ∂W in the region under consideration) by construction of M (which I still owe you).

For $\|\mathbf{q}\| \geq 2\varepsilon$, the vector field Y is transverse to ∂W because of

$$\mathbf{p}\,d\mathbf{p}(Y) = \left(\frac{1}{2} + f(\|\mathbf{q}\|^2)\right)\|\mathbf{p}\|^2 + 2(\mathbf{qp})^2 f'(\|\mathbf{q}\|^2) > 0.$$

It only remains to check that Y is a Liouville vector field for $d\mathbf{p}\wedge d\mathbf{q}$. But this is immediate from

$$\begin{aligned} i_Y(d\mathbf{p}\wedge d\mathbf{q}) &= \frac{1}{2}(\mathbf{p}\,d\mathbf{q} - \mathbf{q}\,d\mathbf{p}) \\ &\quad + f(\|\mathbf{q}\|^2)\,(\mathbf{p}\,d\mathbf{q} + \mathbf{q}\,d\mathbf{p}) + (\mathbf{qp})f'(\|\mathbf{q}\|^2)\cdot 2\mathbf{q}\,d\mathbf{q} \\ &= \frac{1}{2}(\mathbf{p}\,d\mathbf{q} - \mathbf{q}\,d\mathbf{p}) + d\big(f(\|\mathbf{q}\|^2)\,\mathbf{qp}\big). \end{aligned}$$

This concludes the proof. □

Remark 7.6.3 Here, as promised, is the construction of the smooth function $\mathbb{R}^+ \times \mathbb{R}^+ \to \mathbb{R}$, $(x,y) \mapsto F(x,y)$ that allows us to define the hypersurface $M := \{F(\|\mathbf{q}\|, \|\mathbf{p}\|) = 0\}$.

Choose $\delta \in (\varepsilon^2, 2\varepsilon^2)$. Then

$$\varepsilon < \sqrt{\delta} < \frac{\delta}{\varepsilon} < 2\varepsilon.$$

Let $h\colon \mathbb{R}^+ \to \mathbb{R}^+$ be a smooth function with

(1) $h(x) = x$ in a neighbourhood of the interval $(0, \sqrt{\delta}]$,
(2) $h \equiv \delta/\varepsilon$ in a neighbourhood of the interval $[\delta/\varepsilon, \infty)$,
(3) $h' \geq 0$.

The construction of M will be symmetric in $\|\mathbf{q}\|$ and $\|\mathbf{p}\|$ about the set $\{\|\mathbf{q}\| = \|\mathbf{p}\| = \sqrt{\delta}\}$, so we consider the part with $\|\mathbf{q}\| \geq \sqrt{\delta}$ only. Set $F(x,y) := h(x)\cdot y - \delta$. Then, near $\|\mathbf{q}\| = \sqrt{\delta}$, the hypersurface M is given by $\{\|\mathbf{q}\|\cdot\|\mathbf{p}\| = \delta\}$. For $\|\mathbf{q}\|$ in a neighbourhood of $[\delta/\varepsilon, \infty)$, the hypersurface is given by $\{\|\mathbf{p}\| = \varepsilon\}$.

So it remains to verify that $\mathbf{q}\,\partial_{\mathbf{q}} + \mathbf{p}\,\partial_{\mathbf{p}}$ is transverse to the hypersurface $\{F(\|\mathbf{q}\|, \|\mathbf{p}\|) = 0\}$. With $\widetilde{F}(\mathbf{q},\mathbf{p}) := F(\|\mathbf{q}\|, \|\mathbf{p}\|)$ we compute

$$d\widetilde{F} = h'(\|\mathbf{q}\|)\cdot\frac{\mathbf{q}\,d\mathbf{q}}{\|\mathbf{q}\|}\cdot\|\mathbf{p}\| + h(\|\mathbf{q}\|)\cdot\frac{\mathbf{p}\,d\mathbf{p}}{\|\mathbf{p}\|},$$

hence

$$d\widetilde{F}(\mathbf{q}\,\partial_{\mathbf{q}} + \mathbf{p}\,\partial_{\mathbf{p}}) = h'(\|\mathbf{q}\|)\cdot\|\mathbf{q}\|\cdot\|\mathbf{p}\| + h(\|\mathbf{q}\|)\cdot\|\mathbf{p}\| > 0,$$

as desired.

7.7 Contact reduction

Let U be a coisotropic subspace of a symplectic vector space (V,ω). Then the symplectic orthogonal complement $U^{\perp}$ is an isotropic subspace of (V,ω) contained in U, and the symplectic linear form ω on V induces a symplectic structure on the quotient $U/U^{\perp}$. This quotient is the **linear symplectic reduction** of U [177, p. 40].

This construction carries over to a coisotropic submanifold Q of a symplectic manifold W. It can be shown that the distribution $(TQ)^{\perp}$ is integrable, so Q is foliated by isotropic leaves. If each leaf is a compact connected manifold, then the leaf space turns out to be a symplectic manifold. The most important special case of this construction is the Marsden–Weinstein (or Marsden–Weinstein–Meyer) symplectic quotient [173], [179].

Here one considers a symplectic manifold (W,ω) with a **Hamiltonian action** (from the left) by a compact Lie group G, whose Lie algebra we shall denote by $\mathfrak{g}$ (and its dual by $\mathfrak{g}^*$). This means that one requires the existence of a **momentum map** μ, that is, a smooth map

$$\mu\colon W \longrightarrow \mathfrak{g}^*$$

with the following properties.

(1) For each $X \in \mathfrak{g}$, the function $H_{\underline{X}}\colon W \to \mathbb{R}$, defined by

$$H_{\underline{X}}(p) = \langle \mu(p), X\rangle,$$

is a Hamiltonian function for the vector field $\underline{X}$ on M generated by X. (Here $\langle -,-\rangle$ denotes the duality pairing between $\mathfrak{g}^*$ and $\mathfrak{g}$.)

(2) The map μ is equivariant with respect to the given G–action on M and the coadjoint action of G from the left on $\mathfrak{g}^*$.

All the terms in this definition that may be unfamiliar at this point will be explained when we discuss the contact analogue of the momentum map.

We make the following simplifying assumptions.

(i) The value $\mathbf{0}$ is a regular value of μ, so that $\mu^{-1}(\mathbf{0})$ is a submanifold.

(ii) The group G acts freely† on $\mu^{-1}(\mathbf{0})$, so that the quotient $\mu^{-1}(\mathbf{0})/G$ is also a manifold.

† In the case of non-compact, finite-dimensional Lie groups one needs to require in addition that the action be proper, see [3, Prop. 4.1.23].

One can then prove that $\mu^{-1}(\mathbf{0})$ is a coisotropic submanifold, and its isotropic foliation is given by the orbits of G contained in $\mu^{-1}(\mathbf{0})$. The symplectic manifold $\mu^{-1}(\mathbf{0})/G$ is called the **Marsden–Weinstein quotient** or **symplectic quotient**. Observe that

$$\dim\bigl(\mu^{-1}(\mathbf{0})/G\bigr) = \dim M - 2\dim G.$$

For the details of this construction see [40, Chapter IX], [177, Section 5.4], or [3, Section 4.3]; the latter also explains the role of this reduction procedure in classical mechanics. A beautiful survey article on the history of symplectic reduction is [174].

Here is the analogue of this construction in contact geometry. I describe only the simplest situation, as in [93]. There is now a vast literature on this subject. Some papers where the fundamentals of contact reduction are described in various degrees of generality are [9], [166], and [244].

Thus, we now consider a strict contact manifold (M,α) and a *strict* contact action (from the left) by a compact Lie group G, that is, an action preserving the contact form α.† The G–action is a smooth map

$$\begin{array}{ccccc} G & \times & M & \longrightarrow & M \\ (g & , & p) & \longmapsto & g\cdot p \end{array}$$

satisfying $g_1\cdot(g_2\cdot p) = (g_1g_2)\cdot p$ and $e\cdot p = p$, for all $g_1, g_2 \in G$, $p \in M$, and the neutral element $e \in G$. By slight abuse of notation I write g for the map from M to itself sending p to $g\cdot p$. Then the condition for a strict contact action is $g^*\alpha = \alpha$ for all $g \in G$.

Before we turn to the general situation, let me describe the **contact quotient** in the case of an S^1–action, generated by the flow of some vector field $\underline{X}$ on M. The condition for this to be a strict contact action is that $\mathcal{L}_{\underline{X}}\alpha \equiv 0$. Define the **momentum map** $\mu\colon M \to \mathbb{R}$ by $\mu = \alpha(\underline{X})$, or more precisely, $\mu(p) = \alpha_p(\underline{X}_p)$.

Proposition 7.7.1 *Let (M,α) be a contact manifold with a strict contact S^1–action, generated by the flow of a vector field $\underline{X}$ on M. Then $\underline{X}$ is tangent to the level sets of the momentum map μ. The value 0 is a regular value of μ if and only if $\underline{X}$ is nowhere zero on the level set $\mu^{-1}(0)$. Hence, in this case the S^1–action on M restricts to a locally free‡ action on $\mu^{-1}(0)$. If this restricted action is free, α induces a contact form on the quotient manifold $\mu^{-1}(0)/S^1$.§*

† Given a contact *structure* preserved by a G–action, one can find a G–invariant contact *form* by averaging.

‡ That is, with all isotropy groups $G_p := \{g \in G\colon\ g\cdot p = p\}$ finite.

§ In [93] I suggested wrongly that the restricted action is always free (for 0 a regular

Proof We compute

$$d\mu = d(\alpha(\underline{X})) = \mathcal{L}_{\underline{X}}\alpha - i_{\underline{X}}d\alpha = -i_{\underline{X}}d\alpha. \tag{7.10}$$

Thus, $d\mu(\underline{X}) \equiv 0$, which proves the first statement.

To have $p \in \mu^{-1}(0)$ is the same as $\underline{X}_p \in \ker \alpha_p$. Hence, along the 0–level of μ, the contact condition for α entails that $d\mu_p$ vanishes if and only if $\underline{X}_p$ does.

Now assume that 0 is indeed a regular value of μ. The conditions $\mathcal{L}_{\underline{X}}\alpha \equiv 0$ and $\alpha(\underline{X}) \equiv 0$ along $\mu^{-1}(0)$ are precisely what is needed for α to descend to a well-defined 1–form on the quotient manifold $\mu^{-1}(0)/S^1$.

The restriction of the 2–form $d\alpha$ to $T_p(\mu^{-1}(0)) \cap \ker \alpha_p$ can have at most a 1–dimensional kernel, and since we know from Equation (7.10) that $\underline{X}_p \neq \mathbf{0}$ lies in this kernel, it is indeed 1–dimensional. When we pass to the quotient $\mu^{-1}(0)/S^1$, the 1–form induced by α is given by restricting α to hyperplanes in $T_p(\mu^{-1}(0))$ complementary to $\underline{X}_p$. Similarly, the differential of the induced 1–form is given by restricting $d\alpha$ to such hyperplanes. It follows, as claimed, that α induces a contact form on $\mu^{-1}(0)/S^1$. □

Now to the general case of a strict contact action by any compact Lie group G.

Definition 7.7.2 *Let $\underline{X}$ be the vector field on M generated by $X \in \mathfrak{g}$. This means that with* exp: $\mathfrak{g} \to G$ *denoting the exponential map, we set*

$$\underline{X}_p := \frac{d}{dt}(\exp(tX) \cdot p)|_{t=0}.$$

The **momentum map**

$$\mu\colon M \longrightarrow \mathfrak{g}^*$$

of the contact G–action is defined by

$$\langle \mu(p), X \rangle = \alpha_p(\underline{X}_p).$$

The analogue of condition (1) above is immediate from this definition. Indeed, our definitions entail that $H_{\underline{X}} = \alpha(\underline{X})$, and since G acts by strict contact transformations we have $\mathcal{L}_{\underline{X}}\alpha \equiv 0$. It now follows from Theorem 2.3.1 that $H_{\underline{X}}$ is the Hamiltonian function corresponding to $\underline{X}$.

Before stating the equivariance property (2) formally, recall the basics of the (co-)adjoint representation of a Lie group, see [238, Section 3.46].

value of μ). However, a locally free action is sufficient to allow one to give the quotient in a natural way the structure of a *contact orbifold.*

The Lie group G acts on itself from the left by conjugation:

$$\begin{array}{ccccc} G & \times & G & \longrightarrow & G \\ (g & , & h) & \longmapsto & c_g(h) := ghg^{-1}. \end{array}$$

Since $c_g(e) = e$, we can use the identification of the tangent space T_eG with the Lie algebra $\mathfrak{g}$ to define the **adjoint representation**

$$\begin{array}{rccc} \mathrm{Ad}: & G & \longrightarrow & \mathrm{Aut}\,(\mathfrak{g}) \\ & g & \longmapsto & T_e c_g. \end{array}$$

Given any homomorphism $\varphi\colon H \to G$ of Lie groups, and with exp denoting the two exponential maps $\mathfrak{g} \to G$ and $\mathfrak{h} \to H$, it is straightforward to check that

$$\varphi(\exp(tY)) = \exp(tT_e\varphi(Y))$$

for any $Y \in \mathfrak{h} = T_eH$ and $t \in \mathbb{R}$. Applied to $H = G$ and $\varphi = c_g$ this gives

$$g\exp(tX)g^{-1} = \exp(t\mathrm{Ad}\,(g)(X)). \tag{7.11}$$

The **coadjoint action** of G from the left on $\mathfrak{g}^*$ is given by

$$\begin{array}{ccccc} G & \times & \mathfrak{g}^* & \longrightarrow & \mathfrak{g}^* \\ (g & , & u) & \longmapsto & g(u) := u \circ \mathrm{Ad}\,(g^{-1}). \end{array}$$

Lemma 7.7.3 *The momentum map μ is equivariant with respect to the given G-action on M and the coadjoint action of G on $\mathfrak{g}^*$, i.e.*

$$\mu(g \cdot p) = g(\mu(p)) \text{ for all } g \in G,\ p \in M.$$

Proof We have

$$\begin{aligned} \underline{X}_{g\cdot p} &= \frac{d}{dt}(\exp(tX)\cdot(g\cdot p))|_{t=0} \\ &= \frac{d}{dt}((gg^{-1}\exp(tX)g)\cdot p)|_{t=0} \\ &= T_pg\Big(\frac{d}{dt}(g^{-1}\exp(tX)g)\cdot p)|_{t=0}\Big) \\ &= T_pg\big(\underline{\mathrm{Ad}\,(g^{-1})(X)}_p\big), \end{aligned}$$

where for the last equality we used Equation (7.11). Then, for any $X \in \mathfrak{g}$,

$$\begin{aligned}
\mu(g \cdot p)(X) &= \alpha_{g \cdot p}(\underline{X}_{g \cdot p}) \\
&= \alpha_{g \cdot p}\Big(T_p g\big(\underline{\mathrm{Ad}\,(g^{-1})(X)}_p\big)\Big) \\
&= (g^* \alpha)_p \big(\underline{\mathrm{Ad}\,(g^{-1})(X)}_p\big) \\
&= \alpha_p \big(\underline{\mathrm{Ad}\,(g^{-1})(X)}_p\big) \\
&= \mu(p)(\mathrm{Ad}\,(g^{-1})(X)) \\
&= g(\mu(p))(X).
\end{aligned}$$

This proves the lemma. □

Thus, in contrast with the symplectic case, a strict contact action always comes with a momentum map having the desired properties.

Notice that the G–equivariance of μ implies that the G–action on M restricts to an action on $\mu^{-1}(\mathbf{0})$.

Lemma 7.7.4 *(a) For all $p \in M$, $\mathbf{v} \in T_p M$, and $X \in \mathfrak{g}$, we have*

$$\langle T_p \mu(\mathbf{v}), X \rangle = d\alpha(\mathbf{v}, \underline{X}_p);$$

here we identify $T_{\mu(p)}\mathfrak{g}^*$ with $\mathfrak{g}^*$.

(b) The flow of the Reeb vector field R_α preserves the level sets of μ.

(c) If $\mu(p) = \mathbf{0}$, then $T_p(G \cdot p)$, the tangent space to the orbit through p, is an isotropic subspace of the symplectic vector space $(\ker \alpha_p, d\alpha_p)$.

(d) If $\mathbf{0}$ is a regular value of μ, then the isotropic subspace in (c) is of the same dimension as G, and it equals the symplectic orthogonal complement of $\ker \alpha_p \cap T_p(\mu^{-1}(\mathbf{0}))$.

Proof (a) This follows from

$$\langle T_p \mu(\mathbf{v}), X \rangle = d(\alpha(\underline{X}))(\mathbf{v})$$

and $\mathcal{L}_{\underline{X}} \alpha \equiv 0$ (together with the Cartan formula).

(b) From (a) we have

$$\langle T_p \mu(R_\alpha), X \rangle = 0 \text{ for all } X \in \mathfrak{g},$$

hence $T_p \mu(R_\alpha) = \mathbf{0}$.

(c) The tangent space $T_p(G \cdot p)$ is spanned by vectors of the form $\underline{X}_p$, with X ranging over $\mathfrak{g}$. In particular, it is a subspace of $\ker \alpha_p$, since $\alpha_p(\underline{X}_p) =$

$\mu(p)(X) = 0$. With $\mathbf{v} = \underline{Y}_p$ for some $Y \in \mathfrak{g}$ we find with (a) that

$$\begin{aligned} d\alpha_p(\underline{Y}_p, \underline{X}_p) &= \langle T_p\mu(\underline{Y}_p), X\rangle \\ &= \left\langle \frac{d}{dt}(\mu(\exp(tY)\cdot p))|_{t=0}, X\right\rangle \\ &= \left\langle \frac{d}{dt}(\exp(tY)(\mu(p)))|_{t=0}, X\right\rangle \\ &= 0, \end{aligned}$$

where we have used the equivariance of μ.

(d) In order to prove that $\dim T_p(G\cdot p) = \dim G$, we need to show that $\underline{X}_p \neq \mathbf{0}$ for any non-zero $X \in \mathfrak{g}$. Given such an X, the fact that $\mathbf{0}$ is a regular value of μ allows us to choose a tangent vector $\mathbf{v} \in T_pM$ such that $\langle T_p\mu(\mathbf{v}), X\rangle \neq 0$. Then $\underline{X}_p \neq 0$ follows from (a).

Since $\mathbf{0}$ is a regular value of μ, and the intersection of the hyperplane $\ker\alpha_p$ with $T_p(\mu^{-1}(\mathbf{0}))$ is transverse by (b), the vector subspaces

$$T_p(G\cdot p) \text{ and } \ker\alpha_p \cap T_p(\mu^{-1}(\mathbf{0}))$$

are of complementary dimension in $\ker\alpha_p$. From (a), where we take $\mathbf{v} \in \ker\alpha_p \cap T_p(\mu^{-1}(\mathbf{0}))$, we see that

$$\left(T_p(G\cdot p)\right)^{\perp} \supset \ker\alpha_p \cap T_p(\mu^{-1}(\mathbf{0})).$$

For dimensional reasons, this inclusion must be an equality. □

Theorem 7.7.5 *Let G be a compact Lie group acting by strict contact transformations on the contact manifold (M,α). If $\mathbf{0} \in \mathfrak{g}^*$ is a regular value of the momentum map μ of this action, then G acts locally freely on the level set $\mu^{-1}(\mathbf{0})$. If that action is free, α induces a contact form on the quotient manifold $\mu^{-1}(\mathbf{0})/G$.*

Proof The fact that the action of G is locally free on $\mu^{-1}(\mathbf{0})$ (provided that $\mathbf{0}$ is a regular value of μ), follows from the dimension statement in (d) of the preceding lemma.

The 1–form α descends to a well-defined 1–form on the quotient manifold by the same reasoning as in the proof of Proposition 7.7.1. Finally, in passing to this quotient, we take the quotient of the coisotropic subspace $\ker\alpha_p \cap T_p(\mu^{-1}(\mathbf{0})) \subset (\ker\alpha_p, d\alpha_p)$ by its symplectic orthogonal space $T_p(G\cdot p)$, that is, by the kernel of $d\alpha_p$ on that coisotropic subspace. So the induced 1–form on the quotient is a contact form. □

8
Contact structures on 5–manifolds

> 'One done to foure makyth the seconde odde nombre,
> that is the nombre of fiue and hyghte Quinarius.'
>
> Bartholomaeus Anglicus,
> *De proprietatibus rerum*†

In the present chapter we discuss the analogue of the Lutz–Martinet theorem for simply connected 5–manifolds. Throughout, we assume contact structures ξ to be cooriented, i.e. defined as $\xi = \ker \alpha$ by a global 1–form defining the coorientation of ξ. Moreover, if an orientation of the 5–manifold has been chosen, it is understood that the contact structure is positive, that is, $\alpha \wedge (d\alpha)^2$ is a positive volume form. As we saw in Section 2.4, a cooriented contact structure on an oriented manifold M induces an almost contact structure, that is, in the case of 5–manifolds, a reduction of the structure group of the tangent bundle TM from SO(5) to U(2) $\times$ 1.

Theorem 8.0.6 *Every closed, oriented, simply connected* 5*–manifold admits a contact structure in every homotopy class of almost contact structures.*

This theorem was (essentially) proved in [91]. In retrospect I regard my treatment of orientations in that paper as somewhat frivolous; in order to address this issue the present chapter includes a discussion of self-diffeomorphisms of simply connected 5–manifolds.

A word of caution is in order. The theorem does not claim that every simply connected 5–manifold admits a contact structure, for in contrast to the 3–dimensional case there is now a non-trivial topological obstruction to the existence of an almost contact structure, namely, the third integral Stiefel–Whitney class.

The proof of this theorem presented here is more or less the one I gave in [91]; the main differences concern the topological aspects of the proof.

† From the entry for 'five' in the *Oxford English Dictionary*.

Whereas in [91] I quoted Barden's classification of simply connected 5–manifolds, in Section 8.2 I give a direct proof of a structure theorem for those 5–manifolds which admit an almost contact structure. For the most part, these differential topological considerations only involve an embedding theorem due to Haefliger, and elementary surgery theory. But in one special case we cannot avoid quoting the h–cobordism theorem and Thom's result about the structure of the 5–dimensional oriented cobordism group.

There are two further deviations from the proof given in [91] that are of more contact geometric interest. First of all, in Section 8.1 I give a more elementary derivation of the homotopy classification of almost contact structures. Secondly, the only manifolds where one has to deal with more than one homotopy class of almost contact structures are the two S^3–bundles over S^2. While in [91] I treated the corresponding cases of Theorem 8.0.6 by appealing to certain existence results for symplectic structures on 4–manifolds (and then using the Boothby–Wang construction), in Section 8.3 I derive the result by analysing the possible framings for contact surgeries along an isotropic S^1 in S^5. An alternative proof of Theorem 8.0.6 via open book decompositions can be found in [149], [151]. A large part of the theorem can also be proved via explicit realisations of simply connected 5–manifolds as Brieskorn manifolds, see [92], [149], [150], [229].

The direct classification of simply connected 5–manifolds admitting an almost contact structure should not be regarded as a vain exercise. In view of the Eliashberg–Weinstein contact surgery and Giroux's results about the relation between contact structures and open books, structure theorems for manifolds admitting an almost contact structure constitute one possible route towards general existence statements for contact structures.

8.1 Almost contact structures

The aim of the present section is to identify the obstruction to the existence of an almost contact structure on a given closed, oriented 5–manifold M, and to classify almost contact structures (up to homotopy). We do not yet impose the condition that M be simply connected.

Recall that the third integral Stiefel–Whitney class $W_3(M)$ is defined as $W_3(M) := \beta w_2(M)$, where $w_2(M) \in H^2(M;\mathbb{Z}_2)$ is the usual second Stiefel–Whitney class and β the Bockstein homomorphism

$$\beta\colon\, H^2(M;\mathbb{Z}_2) \longrightarrow H^3(M;\mathbb{Z})$$

in the long exact sequence

$$\ldots \to H^2(M;\mathbb{Z}) \longrightarrow H^2(M;\mathbb{Z}) \xrightarrow{\rho_2} H^2(M;\mathbb{Z}_2) \xrightarrow{\beta} H^3(M;\mathbb{Z}) \to \ldots$$

induced by the coefficient sequence

$$0 \longrightarrow \mathbb{Z} \xrightarrow{\times 2} \mathbb{Z} \longrightarrow \mathbb{Z}_2 \longrightarrow 0.$$

Hence, the condition $W_3(M) = 0$ is equivalent to saying that $w_2(M) \in H^2(M;\mathbb{Z}_2)$ lifts to an integral class $c \in H^2(M;\mathbb{Z})$, that is, there exists a class c with $\rho_2 c = w_2(M)$.

Now, an almost contact structure is a $(\mathrm{U}(2) \times 1)$–structure† on the tangent bundle TM, so the first Chern class of the almost contact structure is an integral cohomology class that reduces modulo 2 to $w_2(M)$. This implies that $W_3(M) = 0$ is an obvious necessary condition for the existence of an almost contact structure. The following proposition shows that this condition is also sufficient. Notice that because of $\mathrm{U}(2) \times 1 \subset \mathrm{SO}(5)$, an almost contact structure induces an orientation of M. When we speak of an almost contact structure on an oriented manifold, it is understood that this structure induces the given orientation.

Proposition 8.1.1 *Let M be a closed, oriented 5–manifold with $W_3(M) = 0$. Then M admits an almost contact structure. Homotopy classes of almost contact structures are in one-to-one correspondence with integral lifts of $w_2(M)$. The correspondence is given by associating to an almost contact structure its first Chern class.*

This proposition is going to be proved by an obstruction-theoretic argument. I have tried to use as little general theory as possible, *viz.*:

(i) there is a sequence of obstructions in certain cohomology groups that allow one to decide whether an almost contact structure exists and whether two almost contact structures are homotopic;
(ii) the class $W_3(M)$ is the obstruction to finding a certain tangential frame.

All other obstructions are identified by 'elementary' means. I make no apology for the resulting length of the argument.

Write F_5 for the homogeneous space $\mathrm{SO}(5)/\mathrm{U}(2)$. (For the basic theory of homogeneous manifolds see [238].) In order to be specific, we regard $\mathrm{U}(2)$ as being embedded in $\mathrm{SO}(5)$ in lower diagonal position, likewise for the various subgroups considered below. Almost contact structures correspond to sections of the principal F_5–bundle associated with the tangent bundle, see [224, Thm. 9.4]. This will be implicit in the discussion below involving the classifying spaces $\mathrm{BSO}(5)$ and $\mathrm{BU}(2)$. More explicitly, this is a consequence of the contact analogue of Lemma 8.1.7 below.

† Often I am simply going to say 'U(2)–structure'.

It then follows from general obstruction theory that the obstructions to the *existence* of an almost contact structure live in the cohomology groups $H^i(M;\pi_{i-1}(F_5))$, the obstructions to *homotopy* of almost contact structures in $H^i(M;\pi_i(F_5))$. The computation of the homotopy groups of F_5 is based on the following lemma from [117].

Lemma 8.1.2 *(a)* $\mathrm{U}(n+1)\cap\mathrm{SO}(2n+1)=\mathrm{U}(n)$,
(b) $\mathrm{U}(n+1)\cdot\mathrm{SO}(2n+1)=\mathrm{SO}(2n+1)\cdot\mathrm{U}(n+1)=\mathrm{SO}(2n+2)$,
(c) $F_{2n+1}:=\mathrm{SO}(2n+1)/\mathrm{U}(n)=\mathrm{SO}(2n+2)/\mathrm{U}(n+1)=:F_{2n+2}$.

Proof The statement in (a) is obvious. As to (b), we have to show that every element $t\in\mathrm{SO}(2n+2)$ can be written as $t=us=s'u'$ with $u,u'\in\mathrm{U}(n+1)$ and $s,s'\in\mathrm{SO}(2n+1)$.

The natural action of $\mathrm{SO}(2n+2)$ on the sphere $S^{2n+1}\subset\mathbb{R}^{2n+2}$ induces a simply transitive action on the space of orthonormal $(2n+1)$–frames on that sphere. The action of $\mathrm{SO}(2n+1)\subset\mathrm{SO}(2n+2)$ fixes the point $p_0=(1,0,\ldots,0)\in S^{2n+1}$ and is simply transitive on the orthonormal $(2n+1)$–frames at that point. Likewise, the natural action of $\mathrm{U}(n+1)$ on $S^{2n+1}\subset\mathbb{C}^{n+1}$ is transitive. Thus, given $t\in\mathrm{SO}(2n+2)$, we can find an element $u\in\mathrm{U}(n+1)$ such that $u^{-1}t$ fixes the point p_0, and hence $u^{-1}t=s$ for a unique $s\in\mathrm{SO}(2n+1)$. The decomposition $t=s'u'$ is proved similarly. Observe that the commutativity statement of (b) implies that the product $\mathrm{SO}(2n+2)=\mathrm{SO}(2n+1)\cdot\mathrm{U}(n+1)$ is in fact the same as the group generated by $\mathrm{SO}(2n+1)$ and $\mathrm{U}(n+1)$.

Statement (c) now follows from (a) and (b). Indeed, every element of the homogeneous space $\mathrm{SO}(2n+2)/\mathrm{U}(n+1)=\mathrm{SO}(2n+1)\cdot\mathrm{U}(n+1)/\mathrm{U}(n+1)$ is of the form $s\mathrm{U}(n+1)$ with $s\in\mathrm{SO}(2n+1)$, so $\mathrm{SO}(2n+1)$ acts transitively on that space. The stabiliser of $e\mathrm{U}(n+1)$, with e denoting the trivial element of $\mathrm{SO}(2n+1)$, is the subgroup $\mathrm{U}(n+1)\cap\mathrm{SO}(2n+1)$, which by (a) equals $\mathrm{U}(n)$. This implies (c). □

The next proposition can be found, at least implicitly, in the work of Cartan [43], see also [127, p. 519]. We defer its proof to the end of this section.

Proposition 8.1.3 *The homogeneous space* $\mathrm{SO}(6)/\mathrm{U}(3)$ *is diffeomorphic to the complex projective space* $\mathbb{C}P^3$.

Proof of Proposition 8.1.1 Consider the generalised Hopf fibration

$$\begin{array}{ccc}\mathbb{C}^4\supset S^7 & \longrightarrow & \mathbb{C}P^3\\ (z_0,z_1,z_2,z_3) & \longmapsto & (z_0:z_1:z_2:z_3)\end{array}$$

with fibre S^1. From the homotopy exact sequence of this fibration and the preceding lemma and proposition we find that the homotopy groups of $F_5 = \mathrm{SO}(5)/\mathrm{U}(2)$ are given by $\pi_i(F_5) = 0$ for $i = 0, 1, 3, 4, 5$ and $\pi_2(F_5) \cong \mathbb{Z}$. Hence the only obstruction to the existence of an almost contact structure lies in the cohomology group $H^3(M; \pi_2(F_5))$.

Recall from [188, §12] that the Stiefel–Whitney class $W_3(M) \in H^3(M; \mathbb{Z})$ is the primary obstruction to the existence of a 3–frame (here $3 = 5 - 3 + 1$) on M. As in the discussion of the parallelisability of 3–manifolds in Chapter 4 we may work with orthonormal frames. Then the coefficient group $\mathbb{Z}$ of the homology group above is actually the homotopy group $\pi_2(V_3(\mathbb{R}^5))$ of the Stiefel variety $V_3(\mathbb{R}^5) = \mathrm{SO}(5)/\mathrm{SO}(2)$. The fibration

$$S^3 \cong \mathrm{U}(2)/\mathrm{U}(1) \hookrightarrow \mathrm{SO}(5)/\mathrm{SO}(2) \xrightarrow{p} \mathrm{SO}(5)/\mathrm{U}(2)$$

shows that the obvious projection map p induces an isomorphism of homotopy groups, $p_\#\colon \pi_2(V_3(\mathbb{R}^5)) \xrightarrow{\cong} \pi_2(F_5)$. It follows that $W_3(M)$ is also the primary (and, by the above, only) obstruction to the existence of an almost contact structure on M. (In fact, the analogous argument shows that W_3 is the primary obstruction to the existence of an almost contact structure on manifolds of arbitrary odd dimension.)

The only obstruction to homotopy of almost contact structures lies in the group $H^2(M; \pi_2(F_5))$. This obstruction we want to identify as the first Chern class. In order to do so, we interpret the question of finding and classifying almost contact structures as a lifting problem for classifying maps. A Riemannian metric on the oriented manifold M, or equivalently a reduction of the structure group of the tangent bundle TM to SO(5), allows us to describe TM in terms of a classifying map $f\colon M \to \mathrm{BSO}(5)$. The inclusion $\mathrm{U}(2) \to \mathrm{SO}(5)$ induces a fibration $\mathrm{BU}(2) \to \mathrm{BSO}(5)$ with fibre F_5, see [242, p. 680].† An almost contact structure may now be regarded as a lifting of f to a classifying map $\tilde{f}\colon M \to \mathrm{BU}(2)$, i.e. a map making the following diagram commutative.

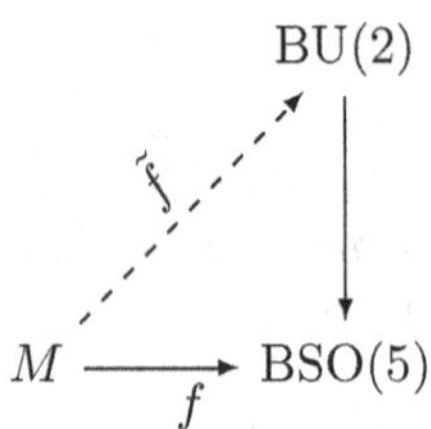

† In spite of the terrifying page number, this is actually a good reference, since the relevant discussion in Whitehead's book is contained in a very readable appendix about compact Lie groups.

The obstruction class in $H^2(M;\pi_2(F_5))$ is the obstruction to homotopy over the 2–skeleton $M^{(2)}$ of lifts $\tilde{f}$ of f. This being the only obstruction to homotopy of almost contact structures, the homotopy class of an almost contact structure is completely determined by the restriction $\tilde{f}|_{M^{(2)}}$.

From the homotopy exact sequence of the universal bundle

$$\mathrm{U}(2) \hookrightarrow \mathrm{EU}(2) \longrightarrow \mathrm{BU}(2),$$

where EU(2) is contractible, we have

$$\pi_1(\mathrm{BU}(2)) \cong \pi_0(\mathrm{U}(2)) = 0$$

and

$$\pi_2(\mathrm{BU}(2)) \cong \pi_1(\mathrm{U}(2)) \cong \mathbb{Z}.$$

Hence

$$[M^{(2)}, \mathrm{BU}(2)] \cong [M^{(2)}/M^{(1)}, \mathrm{BU}(2)],$$

and $M^{(2)}/M^{(1)}$ is a wedge of 2–spheres. We may therefore restrict attention to the case that $M^{(2)} = S^2$.

The first Chern class of the U(2)–bundle $\xi_{\tilde{f}}$ over S^2 corresponding to the map $\tilde{f}\colon S^2 \to \mathrm{BU}(2)$ is $c_1(\xi_{\tilde{f}}) = \tilde{f}^* c_1$, where c_1 is the first Chern class of the universal U(2)–bundle. This c_1 is a generator of $H^2(\mathrm{BU}(2);\mathbb{Z}) \cong \mathbb{Z}$, see [188]. Recall the Hurewicz isomorphism $\pi_2(\mathrm{BU}(2)) \to H_2(\mathrm{BU}(2);\mathbb{Z})$ given by $[\tilde{f}] \mapsto \tilde{f}_*[S^2]$. We also have an isomorphism $H_2(\mathrm{BU}(2);\mathbb{Z}) \to \mathbb{Z}$ given by $u \mapsto \langle c_1, u\rangle$, the Kronecker pairing with the generator c_1 of $H^2(\mathrm{BU}(2);\mathbb{Z})$. It follows that the isomorphism

$$[S^2, \mathrm{BU}(2)] = \pi_2(\mathrm{BU}(2)) \xrightarrow{\cong} \mathbb{Z}$$

can be described by

$$[\tilde{f}] \longmapsto \langle c_1, \tilde{f}_*[S^2]\rangle = \langle \tilde{f}^* c_1, [S^2]\rangle = \langle c_1(\xi_{\tilde{f}}), [S^2]\rangle.$$

We conclude that every element of $H^2(S^2;\mathbb{Z})$ arises as the first Chern class of a unique homotopy class of U(2)–bundles over S^2.

An entirely analogous argument, starting from the fact that $\pi_1(\mathrm{BSO}(5)) = 0$ and $\pi_2(\mathrm{BSO}(5)) = \mathbb{Z}_2$, shows that, up to homotopy, there are exactly two SO(5)–bundles over S^2, classified by the second Stiefel–Whitney class $w_2 \in H^2(S^2;\mathbb{Z}_2)$; see also the next section.

Write $i\colon M^{(2)} \to M$ for the inclusion of the 2–skeleton. The preceding arguments imply that for every integral lift $c \in H^2(M^{(2)};\mathbb{Z})$ of $i^* w_2(M) = w_2(i^*TM)$ there is a unique homotopy class of U(2)–structures on $i^*TM = TM|_{M^{(2)}}$ with first Chern class equal to c. It follows that almost contact structures on M with the same first Chern class are homotopic over $M^{(2)}$

and therefore homotopic over M. Obviously, almost contact structures on M with different first Chern class cannot be homotopic. This is consistent with the fact that the homomorphism $i^*\colon H^2(M;\mathbb{Z}) \to H^2(M^{(2)};\mathbb{Z})$ is injective, as follows from $H^2(M, M^{(2)};\mathbb{Z}) = 0$ and the cohomology long exact sequence of the pair $(M, M^{(2)})$.

It remains to be shown that every class $c \in H^2(M;\mathbb{Z})$ with $\rho_2 c = w_2(M)$ can be realised as the first Chern class of an almost contact structure on M. To this end, we relate almost contact structures on M to stable almost complex structures, i.e. almost complex structures on the stabilised tangent bundle. In Lemma 8.1.2 we had seen that $F_{2n+1} = F_{2n+2}$. From the homotopy exact sequence of the fibration

$$F_{2n+2} \hookrightarrow F_{2n+3} \longrightarrow F_{2n+3}/F_{2n+2} = \mathrm{SO}(2n+3)/\mathrm{SO}(2n+2) = S^{2n+2}$$

we see that the homotopy groups $\pi_q(F_{2n+1}) = \pi_q(F_{2n+2})$ are stable for $q \leq 2n$, i.e. in that range these groups only depend on q. This implies that an odd-dimensional manifold admits an almost contact structure if and only if it admits a stable almost complex structure. The group $\pi_{2n+1}(F_{2n+1})$ is not stable, and in general the homomorphism $\pi_{2n+1}(F_{2n+1}) \to \pi_{2n+1}(F_{2n+3})$ is surjective, but not injective. This means that non-homotopic almost contact structures may become homotopic when stabilised. In the case of interest to us, however, we have $\pi_5(F_m) = 0$ for all $m \geq 5$, so the homotopy classification of almost contact structures on our 5–manifold M actually coincides with the homotopy classification of stable almost complex structures.

Our obstruction-theoretic argument so far guarantees the existence of some almost contact structure η_0 on M, provided $W_3(M) = 0$. (This notation η constitutes a mild abuse of notation when compared with Definition 2.4.7.) To conclude the argument, for any $u \in H^2(M;\mathbb{Z})$ we need to find an almost contact structure (or stable almost complex structure) η_u with $c_1(\eta_u) = c_1(\eta_0) + 2u$. Complex line bundles over M are classified by maps $\tilde{f}\colon M \to \mathrm{BU}(1) = K(\mathbb{Z}, 2)$, and, as above, the first Chern class $c_1(L_{\tilde{f}})$ of the line bundle $L_{\tilde{f}}$ determined by $\tilde{f}$ is given by $c_1(L_{\tilde{f}}) = \tilde{f}^* c_1$. It then follows from our discussion in Section 3.4.2 that there is a one-to-one correspondence between complex line bundles on M and elements of $H^2(M;\mathbb{Z})$, the correspondence being given by the first Chern class.

Thus, given $u \in H^2(M;\mathbb{Z})$, let L_u be the complex line bundle over M with $c_1(L) = u$. Then the dual line bundle L_u^* satisfies $c_1(L_u^*) = -u$. The virtual bundle $L_u \ominus L_u^*$ (regarded as an element of the reduced K–group $\widetilde{K}(M)$) has $c_1(L_u \ominus L_u^*) = 2u$, and its underlying real bundle is trivial, since L_u and L_u^* are isomorphic as real bundles. Hence $\eta_u := \eta_0 \oplus L_u \ominus L_u^*$ is a stable almost complex structure on M with $c_1(\eta_u) = c_1(\eta_0) + 2u$. □

Remark 8.1.4 Here is a slightly more simple-minded formulation of that last step in the argument, for readers who feel uneasy about virtual bundles. Choose a complex vector bundle C over M such that $L_u^* \oplus C$ is trivial (as a complex vector bundle). It is not difficult to show that such a complementary bundle always exists, see [19, Cor. 1.4.14]. This forces $c_1(C) = u$. Then $c_1(L \oplus C) = 2u$, and since L and L^* are isomorphic as real bundles, $L \oplus C$ is trivial as real vector bundle. Then set $\eta_u = \eta_0 \oplus L \oplus C$.

It remains to supply a proof of the claimed identification of the manifold $\mathrm{SO}(6)/\mathrm{U}(3)$ (and hence $\mathrm{SO}(5)/\mathrm{U}(2)$) with $\mathbb{C}P^3$. The proof given here borrows ideas from [1] and [20].

We begin with some general linear algebra that explains in part why almost complex structures on a $2n$–dimensional oriented manifold W correspond to sections of the principal F_{2n}–bundle associated with the tangent bundle TW, and analogously almost contact structures on a $(2n+1)$–dimensional manifold to sections of the associated principal F_{2n+1}–bundle.

Notation 8.1.5 Write $\mathcal{J}^+(\mathbb{R}^{2n})$ for the space of complex structures on $\mathbb{R}^{2n}$ compatible with the standard metric and orientation, that is,

$$\mathcal{J}^+(\mathbb{R}^{2n}) = \{J \in \mathrm{SO}(2n) \colon\ J^2 = -\mathrm{id}_{\mathbb{R}^{2n}}\}.$$

Remark 8.1.6 Given $J \in \mathrm{O}(2n)$ with $J^2 = -\mathrm{id}_{\mathbb{R}^{2n}}$, one can always find an orthonormal basis $\mathbf{a}_1, \ldots, \mathbf{a}_{2n}$ for $\mathbb{R}^{2n}$ with $\mathbf{a}_{2j} = J\mathbf{a}_{2j-1}$, $j = 1, \ldots, n$. To see this, start with any unit vector $\mathbf{a}_1$, and set $\mathbf{a}_2 = J\mathbf{a}_1$. The condition $J^2 = -\mathrm{id}_{\mathbb{R}^{2n}}$ implies that the subspace U spanned by $\mathbf{a}_1$ and $\mathbf{a}_2$ is 2–dimensional. Now proceed by induction, observing that the orthogonal complement $U^\perp$ is J–invariant. The positivity condition $J \in \mathrm{SO}(2n)$ is equivalent to the existence of a positive basis of this form.

Lemma 8.1.7 *The manifold $\mathcal{J}^+(\mathbb{R}^{2n})$ is diffeomorphic to the homogeneous space $F_{2n} = \mathrm{SO}(2n)/\mathrm{U}(n)$.*

Proof Let $\mathbf{e}_1, \ldots, \mathbf{e}_{2n}$ be the standard basis for $\mathbb{R}^{2n}$, and J_0 the standard complex structure on $\mathbb{R}^{2n}$, defined by $J_0\mathbf{e}_{2j-1} = \mathbf{e}_{2j}$, $j = 1, \ldots, n$. Consider the map

$$\begin{array}{ccc} \mathrm{SO}(2n) & \longrightarrow & \mathcal{J}^+(\mathbb{R}^{2n}) \\ A & \longmapsto & AJ_0A^{-1}. \end{array}$$

In order to verify that $J := AJ_0A^{-1}$ does indeed lie in $\mathcal{J}^+(\mathbb{R}^{2n})$, we compute

$$JJ^t = AJ_0A^{-1}(A^{-1})^t J_0^t A^t = \mathrm{id}$$

and

$$J^2 = AJ_0A^{-1}AJ_0A^{-1} = -\mathrm{id},$$

and observe that if the columns of A are denoted by $\mathbf{a}_1, \ldots, \mathbf{a}_{2n}$, then $J = AJ_0A^{-1}$ is the complex structure on $\mathbb{R}^{2n}$ characterised by $J\mathbf{a}_{2j-1} = \mathbf{a}_{2j}$, $j = 1, \ldots, n$.

This last point, together with the remark preceding the lemma, implies that the map above is surjective, and we can define a transitive left-action of $\mathrm{SO}(2n)$ on $\mathcal{J}^+(\mathbb{R}^{2n})$ by

$$\begin{array}{ccccc} \mathrm{SO}(2n) & \times & \mathcal{J}^+(\mathbb{R}^{2n}) & \longrightarrow & \mathcal{J}^+(\mathbb{R}^{2n}) \\ (A & , & J) & \longmapsto & AJA^{-1}. \end{array}$$

The isotropy group of this action at $J_0 \in \mathcal{J}^+(\mathbb{R}^{2n})$ is the unitary group $\mathrm{U}(n)$, for the condition $AJ_0 = J_0A$ is equivalent to $A \in \mathrm{GL}(n, \mathbb{C})$, and $\mathrm{SO}(2n) \cap \mathrm{GL}(n, \mathbb{C}) = \mathrm{U}(n)$, see the remark below. The result then follows by the general theory of homogeneous spaces, see [238, Thm. 3.62]. □

Remark 8.1.8 In order to see that $\mathrm{SO}(2n) \cap \mathrm{GL}(n, \mathbb{C}) = \mathrm{U}(n)$, or equivalently, $\mathrm{O}(2n) \cap \mathrm{GL}(n, \mathbb{C}) = \mathrm{U}(n)$, it is convenient to arrange the basis of $\mathbb{R}^{2n}$ in the order $\mathbf{e}_1, \mathbf{e}_3, \ldots, \mathbf{e}_{2n-1}, \mathbf{e}_2, \mathbf{e}_4, \ldots, \mathbf{e}_{2n}$. Denote the $n \times n$ identity matrix by I_n and write $2n \times 2n$ matrices in block form

$$A = \begin{pmatrix} a & b \\ c & d \end{pmatrix}$$

(with $n \times n$ blocks). Then

$$J_0 = \begin{pmatrix} 0 & -I_n \\ I_n & 0 \end{pmatrix},$$

and the condition $AJ_0 = J_0A$ for A to be in $\mathrm{GL}(n, \mathbb{C})$ is equivalent to A being of the form

$$A = \begin{pmatrix} a & -b \\ b & a \end{pmatrix},$$

with the i–J_0–equivariant inclusion of $\mathrm{GL}(n, \mathbb{C})$ in $\mathrm{GL}(2n, \mathbb{R})$ given by

$$a + ib \longmapsto \begin{pmatrix} a & -b \\ b & a \end{pmatrix}.$$

The condition $A \in \mathrm{O}(2n)$ then translates into

$$(a + ib)(a - ib)^t = I_n,$$

which is exactly the condition for $a + ib$ to be unitary.

Our next goal is to identify $\mathcal{J}^+(\mathbb{R}^{2n})$ with a Grassmannian manifold whose points are certain complex subspaces of $\mathbb{C}^{2n}$. That latter manifold will in turn be proven, for $n = 3$, to be diffeomorphic to $\mathbb{C}P^3$.

Given $J \in \mathcal{J}^+(\mathbb{R}^{2n})$, extend it to a complex automorphism J^c of $\mathbb{C}^{2n} = \mathbb{R}^{2n} \otimes_{\mathbb{R}} \mathbb{C}$. This J^c is a normal automorphism satisfying $(J^c)^2 = -\mathrm{id}_{\mathbb{C}^{2n}}$, hence $\mathbb{C}^{2n}$ splits as a direct sum

$$\mathbb{C}^{2n} = (\mathbb{C}^{2n})_J^{(1,0)} \oplus (\mathbb{C}^{2n})_J^{(0,1)}$$

of eigenspaces for J^c corresponding to the eigenvalues $+\mathrm{i}$ and $-\mathrm{i}$, called the spaces of $(1,0)$–vectors and $(0,1)$–vectors, respectively. If $\mathbf{x} + \mathrm{i}\mathbf{y}$ is a $(1,0)$–vector, with $\mathbf{x}, \mathbf{y} \in \mathbb{R}^{2n}$, then

$$J\mathbf{x} + \mathrm{i}J\mathbf{y} = J^c(\mathbf{x} + \mathrm{i}\mathbf{y}) = \mathrm{i}(\mathbf{x} + \mathrm{i}\mathbf{y}) = -\mathbf{y} + \mathrm{i}\mathbf{x},$$

hence $\mathbf{x} + \mathrm{i}\mathbf{y} = \mathbf{x} - \mathrm{i}J\mathbf{x}$. Conversely, any vector of this form is a $(1,0)$–vector. Similarly, the space of $(0,1)$–vectors consists of vectors of the form $\mathbf{x} + \mathrm{i}J\mathbf{x}$ with $\mathbf{x} \in \mathbb{R}^{2n}$. In particular, both these eigenspaces are n–dimensional complex subspaces of $\mathbb{C}^{2n}$.

A complex subspace $V \subset \mathbb{C}^{2n}$ is called **isotropic** if $\langle \mathbf{v}, \mathbf{v} \rangle = 0$ for all $\mathbf{v} \in V$, where $\langle -, - \rangle$ denotes, *nota bene*, the standard symmetric, complex bilinear inner product on $\mathbb{C}^{2n}$ (rather than the Hermitian inner product) given by

$$\langle \mathbf{z}, \mathbf{w} \rangle := \sum_{j=1}^{2n} z_j w_j .$$

It follows that $\langle \mathbf{v}, \mathbf{v}' \rangle = 0$ for all elements $\mathbf{v}, \mathbf{v}'$ of an isotropic subspace V, i.e. V is contained in its $\langle -, - \rangle$–orthogonal complement $V^\perp$. By an argument completely analogous to the proof of Lemma 1.3.1, we have

$$\dim U + \dim U^\perp = 2n$$

for any complex subspace $U \subset \mathbb{C}^{2n}$, where dim denotes complex dimension. We conclude that $\dim V \leq n$ for any isotropic subspace $V \subset \mathbb{C}^{2n}$.

An isotropic subspace $V \subset \mathbb{C}^{2n}$ is called **maximal isotropic** if it has the maximal possible dimension, i.e. $\dim V = n$. With $\overline{V}$ denoting the complex conjugate of V, we clearly have $V \cap \overline{V} = \{\mathbf{0}\}$ for any isotropic subspace V, and hence $\mathbb{C}^{2n} = V \oplus \overline{V}$ if V is maximal isotropic.

Observe that the spaces of $(1,0)$– and $(0,1)$–vectors described above are examples of maximal isotropic subspaces of $\mathbb{C}^{2n}$. The former are **positive** in the sense that they possess a basis of the form

$$\{\mathbf{a}_1 - \mathrm{i}\mathbf{a}_2, \ldots, \mathbf{a}_{2n-1} - \mathrm{i}\mathbf{a}_{2n}\},$$

with $\mathbf{a}_1, \ldots, \mathbf{a}_{2n}$ a positive orthonormal basis for $\mathbb{R}^{2n}$.

Remark 8.1.9 Let V be the positive maximal isotropic subspace of $(1,0)$–vectors of a positive complex structure J on $\mathbb{R}^{2n}$ with $J\mathbf{a}_{2j-1} = \mathbf{a}_{2j}$, $j = 1, \ldots, n$. If n is even, then the space $\overline{V}$ of $(0,1)$–vectors is also positive maximal isotropic – it is the space of $(1,0)$–vectors of the positive complex structure defined by $J\mathbf{a}_{2j-1} = -\mathbf{a}_{2j}$. If n is odd, then $\overline{V}$ is negative maximal isotropic.

Notation 8.1.10 Write $\mathrm{Iso}^+(\mathbb{C}^{2n})$ for the manifold of positive maximal isotropic subspaces of $\mathbb{C}^{2n}$.

Lemma 8.1.11 *There is a diffeomorphism*

$$\mathcal{J}^+(\mathbb{R}^{2n}) \xrightarrow{\cong} \mathrm{Iso}^+(\mathbb{C}^{2n})$$

given by associating with $J \in \mathcal{J}^+(\mathbb{R}^{2n})$ *the space* $V_J \in \mathrm{Iso}^+(\mathbb{C}^{2n})$ *of* $(1,0)$–*vectors of* J^{c}.

Proof We define an inverse map as follows. Given $V \in \mathrm{Iso}^+(\mathbb{C}^{2n})$, define an automorphism $J^{\mathrm{c}}\colon \mathbb{C}^{2n} \to \mathbb{C}^{2n}$ by

$$J^{\mathrm{c}}(\mathbf{v}) = \mathrm{i}\mathbf{v} \;\text{ for }\; \mathbf{v} \in V, \quad J^{\mathrm{c}}(\overline{\mathbf{v}}) = -\mathrm{i}\overline{\mathbf{v}} \;\text{ for }\; \overline{\mathbf{v}} \in \overline{V}.$$

Observe that J^{c} is orthogonal with respect to the inner product $\langle -, - \rangle$.

Since $V \cap \overline{V} = \{\mathbf{0}\}$, the vector space V does not contain any real or purely imaginary vectors. It follows that $\mathbf{x} \in \mathbb{R}^{2n} \subset \mathbb{C}^{2n}$ can be written uniquely in the form

$$\mathbf{x} = \frac{1}{2}(\mathbf{x} - \mathrm{i}\mathbf{y}) + \frac{1}{2}(\mathbf{x} + \mathrm{i}\mathbf{y}) \in V \oplus \overline{V}$$

with $\mathbf{y} \in \mathbb{R}^{2n}$. Define $J\colon \mathbb{R}^{2n} \to \mathbb{R}^{2n}$ by $J\mathbf{x} = \mathbf{y}$. We have

$$\mathbf{y} = \frac{1}{2}(\mathbf{y} + \mathrm{i}\mathbf{x}) + \frac{1}{2}(\mathbf{y} - \mathrm{i}\mathbf{x})$$

with

$$\mathbf{y} + \mathrm{i}\mathbf{x} = \mathrm{i}(\mathbf{x} - \mathrm{i}\mathbf{y}) = J^{\mathrm{c}}(\mathbf{x} - \mathrm{i}\mathbf{y}) \in V$$

and

$$\mathbf{y} - \mathrm{i}\mathbf{x} = -\mathrm{i}(\mathbf{x} + \mathrm{i}\mathbf{y}) = J^{\mathrm{c}}(\mathbf{x} + \mathrm{i}\mathbf{y}) \in \overline{V}.$$

This shows $J\mathbf{y} = -\mathbf{x}$, that is, $J^2 = -\mathrm{id}_{\mathbb{R}^{2n}}$, and $J^{\mathrm{c}}|_{\mathbb{R}^{2n}} = J$, so that J^{c} is the complex linear extension of J.

The positivity of V translates into the positivity of J; thus $J \in \mathcal{J}^+(\mathbb{R}^{2n})$. Clearly $V_{J_V} = V$ and $J_{V_J} = J$ by construction, which means that the map $V \mapsto J_V$ is indeed the inverse of $J \mapsto V_J$.

Finally, the $(\pm\mathrm{i})$–eigenspaces of J^{c} depend smoothly on J; conversely, a smoothly varying V gives rise to a smoothly varying automorphism J^c and hence J. This proves that the map $J \mapsto V_J$ is a diffeomorphism. □

Remark 8.1.12 One may want to verify this last point in explicit local coordinates. Suffice it to say here that under the identification of $\mathcal{J}^+(\mathbb{R}^{2n})$ with $\mathrm{SO}(2n)/\mathrm{U}(n)$, the map $J \mapsto V_J$ is given by

$$[A] \longmapsto \mathrm{span}\{\mathbf{a}_1 - \mathrm{i}\mathbf{a}_2, \ldots, \mathbf{a}_{2n-1} - \mathrm{i}\mathbf{a}_{2n}\},$$

where $\mathbf{a}_1, \ldots, \mathbf{a}_{2n}$ are the columns of $A \in \mathrm{SO}(2n)$. This is clearly a smooth map. See also the end of the proof of Lemma 8.1.13 below.

We now specialise to the case $n = 3$.

Lemma 8.1.13 *The manifold* $\mathrm{Iso}^+(\mathbb{C}^6)$ *is diffeomorphic to* $\mathbb{C}P^3$.

This and the two preceding lemmata constitute a proof of Proposition 8.1.3.

Remark 8.1.14 For completeness, let me mention that the homogeneous space $\mathcal{J}^+(\mathbb{R}^4) \cong \mathrm{SO}(4)/\mathrm{U}(2)$ is diffeomorphic to S^2; obviously $\mathcal{J}^+(\mathbb{R}^2)$ equals $\{J_0\}$. Here is a hint. Regard S^3 as the space of unit quaternions. Then $\mathrm{SO}(4)$ can be identified with $(S^3 \times S^3)/\mathbb{Z}_2$, corresponding to left and right multiplication by unit quaternions, where $\mathbb{Z}_2$ is generated by $(-1, -1)$. Likewise, $\mathrm{U}(2)$ can be identified with $(S^1 \times S^3)/\mathbb{Z}_2$. For more details see [219].

Proof of Lemma 8.1.13 The idea is to realise $\mathbb{C}^6$ as the space $\bigwedge^2 \mathbb{C}^4$ of 2–vectors in $\mathbb{C}^4$. The key to the construction is that the antilinear Hodge $*$–operator is an antilinear involution on $\bigwedge^2 \mathbb{C}^4$, or what is called a **real structure** on that complex vector space. This allows us to identify a real subspace $\mathbb{R}^6 \subset \bigwedge^2 \mathbb{C}^4$ as the $*$–invariant subspace, to which the preceding constructions can be applied. Implicitly, this approach relies on the fact that $\mathrm{SU}(4)$ is a double covering of $\mathrm{SO}(6)$; such a double covering is described explicitly in [245].

Let $\mathbf{v}_0, \mathbf{v}_1, \mathbf{v}_2, \mathbf{v}_3$ be the standard basis for $\mathbb{C}^4$ and h the standard sesquilinear Hermitian inner product

$$h(\mathbf{z}, \mathbf{w}) = \sum_{k=0}^{3} z_k \overline{w}_k .$$

Fix the volume form

$$\mathrm{vol} := \mathbf{v}_0 \wedge \mathbf{v}_1 \wedge \mathbf{v}_2 \wedge \mathbf{v}_3 \in \textstyle\bigwedge^4 \mathbb{C}^4.$$

Write elements $\boldsymbol{\alpha}, \boldsymbol{\beta} \in \bigwedge^2 \mathbb{C}^4$ in the form

$$\boldsymbol{\alpha} = \sum_{k<l} a_{kl} \mathbf{v}_k \wedge \mathbf{v}_l, \quad \boldsymbol{\beta} = \sum_{k<l} b_{kl} \mathbf{v}_k \wedge \mathbf{v}_l,$$

and extend h to $\bigwedge^2 \mathbb{C}^4$ by setting

$$h(\boldsymbol{\alpha}, \boldsymbol{\beta}) = \sum_{k<l} a_{kl} \bar{b}_{kl}.$$

Define the *antilinear* $*$–operator

$$*\colon \bigwedge^2 \mathbb{C}^4 \longrightarrow \bigwedge^2 \mathbb{C}^4$$

by setting, for $k, l \in \{0, 1, 2, 3\}$, $k \neq l$,

$$*(\mathbf{v}_k \wedge \mathbf{v}_l) = \mathbf{v}_{k'} \wedge \mathbf{v}_{l'},$$

with (k, l, k', l') an even permutation of $(0, 1, 2, 3)$, and antilinear extension, so that

$$*(\lambda \boldsymbol{\alpha}) = \overline{\lambda}(*\boldsymbol{\alpha}) \text{ for } \lambda \in \mathbb{C}.$$

Then $*^2$ is the identity on $\bigwedge^2 \mathbb{C}^4$, so the $*$–operator defines a real structure, as claimed. Further, for arbitrary $\boldsymbol{\alpha}, \boldsymbol{\beta} \in \bigwedge^2 \mathbb{C}^4$, we have

$$\boldsymbol{\alpha} \wedge \boldsymbol{\beta} = h(\boldsymbol{\alpha}, *\boldsymbol{\beta}) \,\mathrm{vol}.$$

Remark 8.1.15 Let $\mathbf{v}_0', \mathbf{v}_1', \mathbf{v}_2', \mathbf{v}_3'$ be any other unitary basis of $\mathbb{C}^4$ with respect to the Hermitian inner product h, i.e.

$$h(\mathbf{v}_k', \mathbf{v}_l') = \delta_{kl} := \begin{cases} 1 & \text{for } k = l, \\ 0 & \text{for } k \neq l. \end{cases}$$

Assume further that

$$\mathbf{v}_0' \wedge \mathbf{v}_1' \wedge \mathbf{v}_2' \wedge \mathbf{v}_3' = \mathrm{vol}.$$

These conditions are equivalent to requiring that $\mathbf{v}_0', \mathbf{v}_1', \mathbf{v}_2', \mathbf{v}_3'$ be obtained from the standard basis $\mathbf{v}_0, \mathbf{v}_1, \mathbf{v}_2, \mathbf{v}_3$ by a transformation with an element in SU(4). The equation preceding this remark gives, by the non-degeneracy of h, a coordinate-free definition of the $*$–operator, which is seen to take the same explicit form as above in terms of any such basis $\mathbf{v}_0', \mathbf{v}_1', \mathbf{v}_2', \mathbf{v}_3'$, which is reasonably called a **special unitary basis**.

Now define

$$\begin{aligned}
\mathbf{e}_1 &:= \tfrac{1}{\sqrt{2}}(\mathbf{v}_0 \wedge \mathbf{v}_1 + \mathbf{v}_2 \wedge \mathbf{v}_3), & \mathbf{e}_2 &:= \tfrac{i}{\sqrt{2}}(\mathbf{v}_0 \wedge \mathbf{v}_1 - \mathbf{v}_2 \wedge \mathbf{v}_3), \\
\mathbf{e}_3 &:= \tfrac{1}{\sqrt{2}}(\mathbf{v}_0 \wedge \mathbf{v}_2 + \mathbf{v}_3 \wedge \mathbf{v}_1), & \mathbf{e}_4 &:= \tfrac{i}{\sqrt{2}}(\mathbf{v}_0 \wedge \mathbf{v}_2 - \mathbf{v}_3 \wedge \mathbf{v}_1), \\
\mathbf{e}_5 &:= \tfrac{1}{\sqrt{2}}(\mathbf{v}_0 \wedge \mathbf{v}_3 + \mathbf{v}_1 \wedge \mathbf{v}_2), & \mathbf{e}_6 &:= \tfrac{i}{\sqrt{2}}(\mathbf{v}_0 \wedge \mathbf{v}_3 - \mathbf{v}_1 \wedge \mathbf{v}_2).
\end{aligned}$$

These elements of $\bigwedge^2 \mathbb{C}^4$ satisfy $*\mathbf{e}_i = \mathbf{e}_i$, $i = 1, \ldots, 6$, and they form a unitary basis with respect to the Hermitian inner product h. We identify $\mathrm{span}_{\mathbb{R}}\{\mathbf{e}_1, \ldots, \mathbf{e}_6\}$ with $\mathbb{R}^6$ (with orientation defined by this basis) and regard $\bigwedge^2 \mathbb{C}^4$ as the complexification

$$\bigwedge^2 \mathbb{C}^4 = \mathbb{R}^6 \otimes_{\mathbb{R}} \mathbb{C} = \mathbb{R}^6 \oplus \mathrm{i}\mathbb{R}^6.$$

Observe that with respect to this splitting of elements of $\bigwedge^2 \mathbb{C}^4$ into a real and imaginary part, complex conjugation is given by the $*$–operator.

We claim that the standard symmetric, complex bilinear inner product on $\bigwedge^2 \mathbb{C}^4$ (relative to the basis $\mathbf{e}_1, \ldots, \mathbf{e}_6$) is given by

$$\langle \boldsymbol{\alpha}, \boldsymbol{\beta} \rangle = h(\boldsymbol{\alpha}, *\boldsymbol{\beta}).$$

This follows from the fact that $h(\boldsymbol{\alpha}, *\boldsymbol{\beta})$ is complex bilinear in $\boldsymbol{\alpha}$ and $\boldsymbol{\beta}$ and satisfies

$$h(\mathbf{e}_i, *\mathbf{e}_j) = h(\mathbf{e}_i, \mathbf{e}_j) = \delta_{ij}.$$

We then have

$$\langle \boldsymbol{\alpha}, \boldsymbol{\beta} \rangle \,\mathrm{vol} = \boldsymbol{\alpha} \wedge \boldsymbol{\beta},$$

so the condition for a complex subspace $V \subset \bigwedge^2 \mathbb{C}^4$ to be isotropic becomes

$$\boldsymbol{\alpha} \wedge \boldsymbol{\alpha} = 0 \;\text{ for all }\; \boldsymbol{\alpha} \in V,$$

which entails $\boldsymbol{\alpha} \wedge \boldsymbol{\beta} = 0$ for all $\boldsymbol{\alpha}, \boldsymbol{\beta} \in V$.

Now define a map

$$\begin{array}{rcl} \mathbb{C}P^3 & \longrightarrow & \mathrm{Iso}^+(\bigwedge^2 \mathbb{C}^4) \\ [\mathbf{z}] & \longmapsto & V_{\mathbf{z}} := \{\mathbf{z} \wedge \mathbf{w} \colon \mathbf{w} \in \mathbb{C}^4\} \subset \bigwedge^2 \mathbb{C}^4, \end{array}$$

where $\mathbf{z} \in \mathbb{C}^4$ is a representative of $[\mathbf{z}] \in \mathbb{C}P^3 = (\mathbb{C}^4 \setminus \{\mathbf{0}\})/\mathbb{C}^*$. The subspace $V_{\mathbf{z}}$ depends on the equivalence class $[\mathbf{z}]$ only and has complex dimension 3. Moreover, each $\boldsymbol{\alpha} \in V_{\mathbf{z}}$ satisfies $\boldsymbol{\alpha} \wedge \boldsymbol{\alpha} = 0$, so $V_{\mathbf{z}}$ is a maximal isotropic subspace of $\bigwedge^2 \mathbb{C}^4$. In order to verify that $V_{\mathbf{z}}$ is positive for all $\mathbf{z} \in \mathbb{C}^4 \setminus \{\mathbf{0}\}$, it suffices (by a continuity argument) to check this for one particular $\mathbf{z}$. We see that $V_{\mathbf{v}_0}$ is spanned by

$$\mathbf{v}_0 \wedge \mathbf{v}_1 = \frac{1}{\sqrt{2}}(\mathbf{e}_1 - \mathrm{i}\mathbf{e}_2), \quad \mathbf{v}_0 \wedge \mathbf{v}_2 = \frac{1}{\sqrt{2}}(\mathbf{e}_3 - \mathrm{i}\mathbf{e}_4), \quad \mathbf{v}_0 \wedge \mathbf{v}_3 = \frac{1}{\sqrt{2}}(\mathbf{e}_5 - \mathrm{i}\mathbf{e}_6),$$

which identifies $V_{\mathbf{v}_0}$ as the positive maximal isotropic subspace V_{J_0}.

Let $\mathbf{w} \in \mathbb{C}^4$. The condition $\mathbf{w} \wedge \boldsymbol{\alpha} = 0$ for all $\boldsymbol{\alpha} \in V_{\mathbf{z}}$ is equivalent to $\mathbf{w}$ being a multiple of $\mathbf{z}$. This implies that $[\mathbf{z}]$ is determined by $V_{\mathbf{z}}$, and hence that the map $[\mathbf{z}] \mapsto V_{\mathbf{z}}$ is injective.

We next prove that this map is also surjective. Thus, let $V \subset \bigwedge^2 \mathbb{C}^4$ be a positive maximal isotropic subspace. Choose a basis $\boldsymbol{\alpha}, \boldsymbol{\beta}, \boldsymbol{\gamma}$ for V. The isotropy condition implies

$$\boldsymbol{\alpha} \wedge \boldsymbol{\alpha} = \boldsymbol{\beta} \wedge \boldsymbol{\beta} = \boldsymbol{\gamma} \wedge \boldsymbol{\gamma} = 0.$$

By standard linear algebra this implies that $\boldsymbol{\alpha}$ can be written in the form $\mathbf{w}_0 \wedge \mathbf{w}_1$ for some $\mathbf{w}_0, \mathbf{w}_1 \in \mathbb{C}^4$. Here is a quick argument: Write $\boldsymbol{\alpha} \in \bigwedge^2 \mathbb{C}^4$ in terms of the basis $\mathbf{v}_0, \mathbf{v}_1, \mathbf{v}_2, \mathbf{v}_3$ for $\mathbb{C}^4$ as

$$\boldsymbol{\alpha} = \sum_{k<l} a_{kl} \mathbf{v}_k \wedge \mathbf{v}_l.$$

If $\boldsymbol{\alpha} \neq 0$, we may assume without loss of generality that $a_{01} \neq 0$. Then

$$\boldsymbol{\alpha} = (a_{01}\mathbf{v}_0 - a_{12}\mathbf{v}_2 - a_{13}\mathbf{v}_3) \wedge \left(\mathbf{v}_1 + \frac{a_{02}}{a_{01}}\mathbf{v}_2 + \frac{a_{03}}{a_{01}}\mathbf{v}_3\right) + \lambda \mathbf{v}_2 \wedge \mathbf{v}_3$$

for some $\lambda \in \mathbb{C}$. Define

$$\mathbf{w}_0 := a_{01}\mathbf{v}_0 - a_{12}\mathbf{v}_2 - a_{13}\mathbf{v}_3 \;\text{ and }\; \mathbf{w}_1 := \mathbf{v}_1 + \frac{a_{02}}{a_{01}}\mathbf{v}_2 + \frac{a_{03}}{a_{01}}\mathbf{v}_3.$$

Then $\mathbf{w}_0, \mathbf{w}_1, \mathbf{v}_2, \mathbf{v}_3$ form a basis for $\mathbb{C}^4$, and

$$\boldsymbol{\alpha} \wedge \boldsymbol{\alpha} = 2\lambda\, \mathbf{w}_0 \wedge \mathbf{w}_1 \wedge \mathbf{v}_2 \wedge \mathbf{v}_3.$$

Hence, if $\boldsymbol{\alpha} \wedge \boldsymbol{\alpha} = 0$, then $\lambda = 0$ and $\boldsymbol{\alpha} = \mathbf{w}_0 \wedge \mathbf{w}_1$.

By adding a suitable multiple of $\mathbf{w}_0$ to $\mathbf{w}_1$ and renormalising (which amounts to rescaling $\boldsymbol{\alpha}$), we may assume that $h(\mathbf{w}_k, \mathbf{w}_l) = \delta_{kl}$ for $k, l \in \{0, 1\}$. Choose vectors $\mathbf{w}_2, \mathbf{w}_3 \in \mathbb{C}^4$ that complete $\mathbf{w}_0, \mathbf{w}_1$ to a special unitary basis for $\mathbb{C}^4$ (see Remark 8.1.15). Write

$$\boldsymbol{\beta} = \sum_{k<l} b_{kl} \mathbf{w}_k \wedge \mathbf{w}_l.$$

Because of $\boldsymbol{\alpha} \wedge \boldsymbol{\beta} = 0$ we have $b_{23} = 0$, hence

$$0 = \boldsymbol{\beta} \wedge \boldsymbol{\beta} = (b_{03}b_{12} - b_{02}b_{13})\,\mathrm{vol}.$$

This means we can find a non-trivial solution of the equation

$$\begin{pmatrix} b_{12} & b_{02} \\ b_{13} & b_{03} \end{pmatrix} \begin{pmatrix} \lambda_0 \\ -\lambda_1 \end{pmatrix} = \begin{pmatrix} 0 \\ 0 \end{pmatrix},$$

and any such solution will satisfy

$$(\lambda_0 \mathbf{w}_0 + \lambda_1 \mathbf{w}_1) \wedge \boldsymbol{\beta} = 0.$$

By rescaling we may assume $|\lambda_0|^2 + |\lambda_1|^2 = 1$. Set

$$\mathbf{w}_0' = \lambda_0 \mathbf{w}_0 + \lambda_1 \mathbf{w}_1 \;\text{ and }\; \mathbf{w}_1' = -\overline{\lambda}_1 \mathbf{w}_0 + \overline{\lambda}_0 \mathbf{w}_1,$$

so that still $h(\mathbf{w}_k', \mathbf{w}_l') = \delta_{kl}$ for $k, l \in \{0, 1\}$. Then $\boldsymbol{\alpha} = \mathbf{w}_0' \wedge \mathbf{w}_1'$, and since

$\mathbf{w}_0' \wedge \boldsymbol{\beta} = 0$, we can complete $\mathbf{w}_0', \mathbf{w}_1'$ to a special unitary basis $\mathbf{w}_0', \mathbf{w}_1', \mathbf{w}_2', \mathbf{w}_3'$ for $\mathbb{C}^4$ such that (after rescaling $\boldsymbol{\beta}$)

$$\boldsymbol{\alpha} = \mathbf{w}_0' \wedge \mathbf{w}_1' \text{ and } \boldsymbol{\beta} = \mathbf{w}_0' \wedge \mathbf{w}_2'.$$

Because of $\boldsymbol{\alpha} \wedge \boldsymbol{\gamma} = \boldsymbol{\beta} \wedge \boldsymbol{\gamma} = 0$, we can write

$$\boldsymbol{\gamma} = c_{01}\mathbf{w}_0' \wedge \mathbf{w}_1' + c_{02}\mathbf{w}_0' \wedge \mathbf{w}_2' + c_{03}\mathbf{w}_0' \wedge \mathbf{w}_3' + c_{12}\mathbf{w}_1' \wedge \mathbf{w}_2'.$$

Then

$$0 = \boldsymbol{\gamma} \wedge \boldsymbol{\gamma} = c_{03}c_{12}\,\mathrm{vol}.$$

If $c_{12} = 0$, the isotropic subspace V is spanned by

$$\mathbf{w}_0' \wedge \mathbf{w}_1', \quad \mathbf{w}_0' \wedge \mathbf{w}_2' \text{ and } \mathbf{w}_0' \wedge \mathbf{w}_3',$$

and hence V equals the positive maximal isotropic subspace $V_{\mathbf{w}_0'}$. If $c_{03} = 0$, then V is spanned by

$$\mathbf{w}_0' \wedge \mathbf{w}_1', \quad \mathbf{w}_0' \wedge \mathbf{w}_2' \text{ and } \mathbf{w}_1' \wedge \mathbf{w}_2'.$$

This case, however, corresponds to V being a negative maximal isotropic space, for in this case the complex conjugate $*V$ is the positive maximal isotropic subspace $V_{\mathbf{w}_3'}$, see Remark 8.1.9.

It remains to check that the map $[\mathbf{z}] \mapsto V_{\mathbf{z}}$ is a local (and hence global) diffeomorphism. We do this in local coordinates around the points $[\mathbf{v}_0]$ and $V_{\mathbf{v}_0}$, respectively. Consider the splitting

$$\mathbb{C}^6 = \bigwedge^2\mathbb{C}^4 = V_{\mathbf{v}_0} \oplus *V_{\mathbf{v}_0}.$$

Points in a neighbourhood of $V_{\mathbf{v}_0}$ in $\mathrm{Iso}^+(\mathbb{C}^6)$ can be written as the graph of a linear map $L\colon V_{\mathbf{v}_0} \to *V_{\mathbf{v}_0}$. The condition for the graph of L to lie in $\mathrm{Iso}^+(\mathbb{C}^6)$ is that

$$(\boldsymbol{\alpha} + L(\boldsymbol{\alpha})) \wedge (\boldsymbol{\beta} + L(\boldsymbol{\beta})) = 0 \text{ for all } \boldsymbol{\alpha}, \boldsymbol{\beta} \in V_{\mathbf{v}_0}.$$

Since both $V_{\mathbf{v}_0}$ and $*V_{\mathbf{v}_0}$ are isotropic, this translates into

$$\boldsymbol{\alpha} \wedge L(\boldsymbol{\beta}) + L(\boldsymbol{\alpha}) \wedge \boldsymbol{\beta} = 0 \text{ for all } \boldsymbol{\alpha} \in V_{\mathbf{v}_0}.$$

Write $L = (L_{ij})$ in terms of the bases

$$\boldsymbol{\alpha}_1 := \mathbf{v}_0 \wedge \mathbf{v}_1, \quad \boldsymbol{\alpha}_2 := \mathbf{v}_0 \wedge \mathbf{v}_2, \quad \boldsymbol{\alpha}_3 := \mathbf{v}_0 \wedge \mathbf{v}_3$$

for $V_{\mathbf{v}_0}$ and

$$*\boldsymbol{\alpha}_1 = \mathbf{v}_2 \wedge \mathbf{v}_3, \quad *\boldsymbol{\alpha}_2 = \mathbf{v}_3 \wedge \mathbf{v}_1, \quad *\boldsymbol{\alpha}_3 = \mathbf{v}_1 \wedge \mathbf{v}_2$$

for $*V_{\mathbf{v}_0}$, so that $L(\boldsymbol{\alpha}_i) = \sum_j L_{ij}(*\boldsymbol{\alpha}_j)$. Then

$$\boldsymbol{\alpha}_i \wedge L(\boldsymbol{\alpha}_j) + L(\boldsymbol{\alpha}_i) \wedge \boldsymbol{\alpha}_j = (L_{ji} + L_{ij})\,\mathrm{vol},$$

so the condition for the graph of L to be isotropic is that L be skew-symmetric. Write L in the form

$$L(z_1, z_2, z_3) = \begin{pmatrix} 0 & -z_3 & z_2 \\ z_3 & 0 & -z_1 \\ -z_2 & z_1 & 0 \end{pmatrix}.$$

Then z_1, z_2, z_3 are local holomorphic coordinates for $\mathrm{Iso}^+(\mathbb{C}^6)$.

Write $\mathbf{z} = \mathbf{v}_0 + z_1\mathbf{v}_1 + z_2\mathbf{v}_2 + z_3\mathbf{v}_3 \in \mathbb{C}^4$, so that z_1, z_2, z_3 are also local holomorphic coordinates for $\mathbb{C}P^3$. Then the isotropic subspace $V(z_1, z_2, z_3)$ corresponding to $L(z_1, z_2, z_3)$ is spanned by

$$\begin{aligned} \mathbf{v}_0 \wedge \mathbf{v}_1 + z_3\,\mathbf{v}_3 \wedge \mathbf{v}_1 - z_2\,\mathbf{v}_1 \wedge \mathbf{v}_2 &= \mathbf{z} \wedge \mathbf{v}_1, \\ \mathbf{v}_0 \wedge \mathbf{v}_2 - z_3\,\mathbf{v}_2 \wedge \mathbf{v}_3 + z_1\,\mathbf{v}_1 \wedge \mathbf{v}_2 &= \mathbf{z} \wedge \mathbf{v}_2, \\ \mathbf{v}_0 \wedge \mathbf{v}_3 + z_2\,\mathbf{v}_2 \wedge \mathbf{v}_3 - z_1\,\mathbf{v}_3 \wedge \mathbf{v}_1 &= \mathbf{z} \wedge \mathbf{v}_3. \end{aligned}$$

This shows that $V(z_1, z_2, z_3) = V_{\mathbf{z}}$. In other words, in terms of the local coordinates z_1, z_2, z_3 for $\mathbb{C}P^3$ and $\mathrm{Iso}^+(\mathbb{C}^6)$, respectively, the map $[\mathbf{z}] \mapsto V_{\mathbf{z}}$ is actually the identity map, and so it is clearly a local diffeomorphism. □

8.2 On the structure of 5–manifolds

For the remainder of this chapter, M denotes a closed, simply connected, oriented 5–manifold with $W_3(M) = 0$, that is, one admitting an almost contact structure. Before we can formulate a structure theorem for such manifolds, we need to discuss two special examples, namely, the two S^3–bundles over S^2. That there are indeed precisely two such bundles follows from the fact that any such bundle is obtained by gluing together two trivial bundles $D^2_\pm \times S^3$ over the upper and lower hemisphere of S^2, respectively, along the equator S^1; for this fact implies that isomorphism classes of S^3–bundles over S^2, which have structure group $\mathrm{SO}(4)$, are in one-to-one correspondence with elements of the fundamental group $\pi_1(\mathrm{SO}(4))$, i.e. based homotopy classes of maps $S^1 \to \mathrm{SO}(4)$, see Example 6.1.3 (3). To the characteristic map $\alpha \in \pi_1(\mathrm{SO}(4))$ corresponds the bundle

$$D^2_+ \times S^3 \cup_{g_\alpha} D^2_- \times S^3,$$

where the gluing map $g_\alpha \colon \partial D^2_+ \times S^3 \to \partial D^2_- \times S^3$, under a fixed identification of $\partial D^2_\pm$ with S^1, is given by

$$g_\alpha(\theta, p) = \big(\theta, \alpha(\theta)(p)\big), \quad \theta \in S^1,\ p \in S^3;$$

see [224, Thm. 18.5 and §26] for a formal discussion of this issue. Alternatively, observe that $[S^2, \mathrm{BSO}(4)] \cong \pi_1(\mathrm{SO}(4))$. Similar elementary facts about fibre bundles, in particular over spheres, will often be used below without further comment; everything we use can be found in a standard text such as the one by Husemoller [140] or the classic monograph by Steenrod [224].

The following proposition gives a characterisation of the total spaces of these S^3–bundles over S^2. This, as well as the main result of the present section (Thm. 8.2.9), follows directly from Barden's classification of simply connected 5–manifolds [22]. However, our aim here is to provide proofs of these results from first principles, the purpose of this being twofold: firstly, to indicate how Barden's general arguments simplify under the additional assumption $W_3(M) = 0$, secondly, to elucidate the structure of the 5–manifolds of interest to us, which will be important later when discussing questions about contact surgery.

Proposition 8.2.1 *The total space of the trivial S^3–bundle over S^2, denoted by $M_\infty = S^2 \times S^3$, is characterised by $H_2(M_\infty) \cong \mathbb{Z}$ and $w_2(M_\infty) = 0$.*

The total space of the non-trivial S^3–bundle over S^2, denoted by X_∞, is characterised by $H_2(X_\infty) \cong \mathbb{Z}$ and $w_2(X_\infty) \neq 0$, and it satisfies the condition $W_3(X_\infty) = 0$.

Remark 8.2.2 By saying a manifold M (satisfying the general assumption made at the beginning of this section) is characterised by certain homological data I mean that any M' with the same data is diffeomorphic to M (questions of orientation will be discussed later). The notation M_∞ and X_∞ is that of Barden.

First we formulate a basic lemma about characteristic classes of sphere bundles.

Lemma 8.2.3 *Let $\eta = (S^k, E, \pi, B)$ be an oriented fibre bundle with fibre S^k, total space E, projection mapping π and orientable base B. Then*

$$w_2(E) = \pi^*(w_2(\eta) + w_2(B)).$$

If $k = 1$, then $w_2(E) = \pi^* w_2(B)$.

Remark 8.2.4 The proof will show that we only use the additivity of w_2, which is a consequence of $w_1(B) = 0$ (that is, the orientability of B). So one has similar statements for the first Pontrjagin class p_1 in the absence of 2–torsion, etc.

Proof Let $\widetilde{\eta} = (\mathbb{R}^{k+1}, \widetilde{E}, \widetilde{\pi}, B)$ be the vector bundle associated with η. After choosing some auxiliary Riemannian metric we can write

$$T\widetilde{E} \cong T_1\widetilde{E} \oplus T_2\widetilde{E},$$

where $T_1\widetilde{E} \cong \widetilde{\pi}^*\widetilde{\eta}$ is the bundle of vectors tangent to the fibre and $T_2\widetilde{E} \cong \widetilde{\pi}^*(TB)$ the bundle of vectors normal to the fibre. Let $i\colon E \to \widetilde{E}$ be the natural inclusion (in particular $\widetilde{\pi} \circ i = \pi$). Then $i^*(T\widetilde{E}) \cong TE \oplus NE$, where NE is the line bundle of vectors normal to E in $\widetilde{E}$. This bundle is orientable and hence trivial, so

$$\begin{aligned} w_2(TE) &= w_2(TE \oplus NE) = w_2(i^*(T\widetilde{E})) \\ &= i^* w_2(T_1\widetilde{E} \oplus T_2\widetilde{E}) \\ &= i^*\widetilde{\pi}^* w_2(\widetilde{\eta} \oplus TB) \\ &= \pi^*(w_2(\eta) + w_2(B)). \end{aligned}$$

For $k = 1$ the Gysin sequence takes the form

$$\ldots \to H^0(B) \xrightarrow{\cup e} H^2(B) \xrightarrow{\pi^*} H^2(E) \to \ldots,$$

where $e = e(\eta)$ is the Euler class of η. The exactness of the Gysin sequence implies $\pi^* e(\eta) = 0$, and *a fortiori* $\pi^* w_2(\eta) = 0$ since $w_2(\eta)$ is the mod 2 reduction of $e(\eta)$. □

Next we show that the two bundles over S^2 with fibre S^k or $\mathbb{R}^{k+1}$ and structure group $\mathrm{SO}(k+1)$, where $k \geq 2$, are distinguished by the second Stiefel–Whitney class. Notice that $\pi_1(\mathrm{SO}(k+1)) \cong \mathbb{Z}_2$ for $k \geq 2$.

Lemma 8.2.5 *The non-trivial $\mathbb{R}^{k+1}$–bundle η over S^2 ($k \geq 2$) satisfies $w_2(\eta) = 1 \in H^2(S^2; \mathbb{Z}_2)$.*

Proof Let η_0 be an $\mathbb{R}^{k+1}$–bundle over S^2 with $w_2(\eta_0) = 0$. The second Stiefel–Whitney class is the primary obstruction to finding a k–frame over the 2–skeleton of the base [188, p. 140]. Here the base is 2–dimensional, and the complementary 1–dimensional bundle is orientable (the base being simply connected) and hence trivial, so $w_2(\eta_0) = 0$ implies that η_0 is the trivial bundle. □

Proof of Proposition 8.2.1 – Part One First we show that M_∞ and X_∞ have the homological data as claimed. From the Gysin sequence of an S^3–bundle over S^2,

$$\ldots \to H^i(S^2) \xrightarrow{\cup e} H^{i+4}(S^2) \xrightarrow{\pi^*} H^{i+4}(E) \to H^{i+1}(S^2) \to \ldots,$$

it is clear that the total space E has $H^i(E) \cong \mathbb{Z}$ for $i = 0, 2, 3, 5$ and zero otherwise. Then also $H_2(E) \cong \mathbb{Z}$ by Poincaré duality. From the homotopy exact sequence of the fibration it follows that E is simply connected.

The fact that $w_2(M_\infty) = 0$ and $w_2(X_\infty) \neq 0$ is a consequence of the two preceding lemmata.

The Bockstein sequence for X_∞ takes the form

$$\dots \to H^2(X_\infty; \mathbb{Z}_2) \xrightarrow{\beta} H^3(X_\infty; \mathbb{Z}) \xrightarrow{m_*} H^3(X_\infty; \mathbb{Z}) \to \dots$$

Here the homomorphism $m_*\colon \mathbb{Z} \to \mathbb{Z}$ is multiplication by 2 and in particular injective, so β is necessarily the zero homomorphism. Hence $W_3(X_\infty) = \beta w_2(X_\infty) = 0$. □

For the converse, that is, to show that any M with the given data is diffeomorphic to M_∞ or X_∞, respectively, we need the following fundamental theorem of Haefliger [122].

Theorem 8.2.6 (Haefliger) *Let V be a closed, connected manifold of dimension s and W a manifold of dimension m. Let $f\colon V \to W$ be a (continuous) map such that the induced homomorphism $f_\#\colon \pi_i(V) \to \pi_i(W)$ is an isomorphism for $i \leq k$ and surjective for $i = k+1$. Then*

(i) f is homotopic to an embedding if $m \geq 2s - k$ and $s > 2k + 2$ (equivalently, $m \geq 2s - k$ and $2m \geq 3s + 3$);
(ii) two embeddings of V into W homotopic to f are isotopic if $m > 2s - k$ and $s \geq 2k + 2$ (or $m > 2s - k$ and $2m > 3s + 3$). □

Typically we use this theorem for $V = S^s$, and we formulate this case as a separate corollary.

Corollary 8.2.7 *Let W be an m–dimensional manifold.*

(i) If $2m \geq 3s + 3$ and W is $(2s - m + 1)$–connected, then any map $S^s \to W$ is homotopic to an embedding.
(ii) If $2m > 3s + 3$ and W is $(2s - m + 2)$–connected, then two homotopic embeddings $S^s \to W$ are isotopic. □

In conjunction with the Hurewicz isomorphism between the groups $\pi_k(W)$ and $H_k(W)$ of a $(k-1)$–connected manifold W, this corollary will allow us to represent homology classes in the lowest non-trivial dimension by embedded spheres, provided the inequalities of the corollary are satisfied. In the present chapter this will mean classes in $H_2(M)$. The following theorem of Thom shows that we can also represent homology classes in codimensions one and two by submanifolds, though not necessarily embedded spheres. This theorem and its proof are the obvious generalisations of our discussion on representing homology classes by submanifolds in Section 3.4.2, see in particular Remark 3.4.6. (For deeper results in this direction see [230].)

Theorem 8.2.8 (Thom) *Let W be a closed, oriented m–dimensional manifold. Then any homology class in $H_{m-1}(W)$ or $H_{m-2}(W)$ can be represented by the fundamental class of a smooth submanifold, possibly disconnected in the case of codimension one.* □

Proof of Proposition 8.2.1 – Part Two We are now ready to identify $M_\infty = S^2 \times S^3$ from its homological data. This very simple example serves as illustration for the more general arguments used later. The case of X_∞ is considerably more involved and will be deferred to the end of this section.

So let M be a (closed, simply connected) 5–manifold with $H_2(M) \cong \mathbb{Z}$ and $w_2(M) = 0$. By the Hurewicz isomorphism $\pi_2(M) \cong H_2(M)$ we can represent a generator of $H_2(M)$ by a map $f\colon\ S^2 \to M$. By Corollary 8.2.7 we may assume f to be an embedding (and the isotopy class of this embedding is uniquely defined). Let $N := N(f(S^2))$ be the normal bundle of $f(S^2)$ in M. Then

$$0 = f^* w_2(M) = w_2(TS^2 \oplus f^* N) = w_2(f^* N),$$

so by Lemma 8.2.5 the normal bundle of $f(S^2)$ in M is trivial. This allows us to perform surgery along $f(S^2)$. That means we remove a tubular neighbourhood $S^2 \times \mathrm{Int}\,(D^3)$ of $f(S^2)$ and glue back $D^3 \times S^2$ with standard identification along the boundary $S^2 \times S^2$. Notice that the identification of a (closed) tubular neighbourhood of $f(S^2)$ with $S^2 \times D^3$ is unique up to homotopy, since the choice of framing lies in the zero group $\pi_2(\mathrm{SO}(3))$, see page 135. By the work of Wall [236], which will be discussed in great detail below, this surgery has the effect of killing $\pi_2(M)$, so the resulting manifold is a homotopy 5–sphere and hence diffeomorphic to S^5 by [145], see also [153]. This shows that M is constructed from S^5 by cutting out $\mathrm{Int}\,(D^3) \times S^2$ and gluing back $S^2 \times D^3$, which again is a well-defined process because of the uniqueness up to isotopy of embeddings $S^2 \to S^5$ and $\pi_2(\mathrm{SO}(3)) = 0$. The complement of $\mathrm{Int}\,(D^3) \times S^2$ in S^5 is diffeomorphic to $S^2 \times D^3$, so M is the result of gluing together two copies of $S^2 \times D^3$ with the identity map along the boundary, which yields the manifold $S^2 \times S^3$. □

Now we can formulate the promised structure theorem.

Theorem 8.2.9 *Let M be a closed, simply connected 5–manifold satisfying*

$W_3(M) = 0$, and decompose the second homology group $H_2(M)$ as $H_2(M) \cong \mathbb{Z}^r \oplus G$, where G is a finite group.†

(i) *If $w_2(M) = 0$, then M is diffeomorphic to $\#_r(S^2 \times S^3)\#M_t$, where M_t satisfies $w_2(M_t) = 0$ and $H_2(M_t) \cong G$.*
(ii) *If $w_2(M) \neq 0$, then M is diffeomorphic to*

$$\#_{r-1}(S^2 \times S^3)\#X_\infty\#M_t$$

with M_t as before.

Moreover, M_t can be obtained from S^5 by surgery along a link of 2–spheres.

The proof of this theorem will be split into a sequence of lemmata. The assumptions on M remain the same throughout. The reader may, without loss of continuity, take this theorem for granted and skip ahead to the next section, where we discuss the contact geometric implications of this theorem.

We have $H^2(M;\mathbb{Z}_2) \cong \mathrm{Hom}(H_2(M), \mathbb{Z}_2)$ by the universal coefficient theorem, so we may regard w_2 as a homomorphism $H_2(M) \to \mathbb{Z}_2$.

Lemma 8.2.10 *If $w_2(M) \neq 0$, there is a set of generators $f_1, \ldots, f_r$ and $t_1, \ldots, t_s$ of $H_2(M)$ (where $f_1, \ldots, f_r$ generate the free part $\mathbb{Z}^r$ and $t_1, \ldots, t_s$ the torsion part G of $H_2(M)$) such that $w_2(f_r) = 1$ and $w_2 = 0$ on all other generators.*

Proof Once again, consider the Bockstein sequence for M,

$$\ldots \to H^2(M;\mathbb{Z}) \longrightarrow H^2(M;\mathbb{Z}_2) \stackrel{\beta}{\longrightarrow} H^3(M;\mathbb{Z}) \stackrel{m_*}{\longrightarrow} H^3(M;\mathbb{Z}) \to \ldots$$

The relevant cohomology groups can easily be computed using the universal coefficient theorem and Poincaré duality, and the sequence becomes

$$\mathrm{Hom}(\mathbb{Z}^r, \mathbb{Z}) \longrightarrow \mathrm{Hom}(\mathbb{Z}^r, \mathbb{Z}_2) \oplus \mathrm{Hom}(G, \mathbb{Z}_2) \stackrel{\beta}{\longrightarrow} \mathbb{Z}^r \oplus G \stackrel{m_*}{\longrightarrow} \mathbb{Z}^r \oplus G,$$

with the first homomorphism in this sequence being the obvious one. This implies that β is injective on the summand $\mathrm{Hom}(G, \mathbb{Z}_2)$ and identically zero on $\mathrm{Hom}(\mathbb{Z}^r, \mathbb{Z}_2)$, so we can only have $W_3(M) = 0$ if w_2 is zero on G. Choose a basis $f_1', \ldots, f_{r-1}', f_r$ for the free part of $H_2(M)$ such that $w_2(f_r) \neq 0$. Set $f_i = f_i' - w_2(f_i')f_r$, $i = 1, \ldots, r-1$. These are the desired generators. □

From now on we fix a generating set $\{f_1, \ldots, f_r, t_1, \ldots, t_s\}$ for $H_2(M)$ with w_2 equal to zero on all generators except possibly f_r.

† In this decomposition, the torsion part $G := TH_2(M)$ is uniquely determined, but the free part $FH_2(M) \cong \mathbb{Z}^r$ is not.

Lemma 8.2.11 *If there is a free generator on which w_2 is zero (that is, $w_2(M) = 0$ and $r \geq 1$ or $w_2(M) \neq 0$ and $r \geq 2$), then M splits off a summand $S^2 \times S^3$. In other words, M is diffeomorphic to a manifold of the form $S^2 \times S^3 \# M_1$, and the manifold M_1 satisfies $H_2(M_1) \cong \mathbb{Z}^{r-1} \oplus G$ and $W_3(M_1) = 0$.*

Proof Represent the generators $f_1, \ldots, f_r, t_1, \ldots, t_s$ of $H_2(M)$ by embedded 2–spheres. These embeddings may be assumed pairwise mutually transverse, i.e. disjoint. By Lemma 8.2.5 and the argument used in Part Two of the proof of Proposition 8.2.1, the condition of w_2 being zero or non-zero on one of these generators is equivalent to the normal bundle of the corresponding sphere being trivial or non-trivial. By assumption $w_2(f_1) = 0$, so we can perform surgery along the 2–sphere $S_1^2 \subset M$ representing f_1. The statement of the lemma translates into two claims.

(a) Surgery along S_1^2 has the effect of killing the generator f_1, i.e. the resulting manifold M_1 has homology as described.
(b) The reverse surgery to obtain M from M_1 is along a 2–sphere that is contractible in M_1.

Indeed, performing surgery along $S^2 \subset S^5$ yields an S^2–bundle over S^3 (for we glue together two copies of $D^3 \times S^2$), and because of $\pi_2(\mathrm{SO}(3)) = 0$ there is only the trivial bundle. Hence surgery along a contractible $S^2 \subset M_1$ is the same, by Corollary 8.2.7, as taking the connected sum with $S^2 \times S^3$.

By Poincaré duality, $H_3(M)$ is isomorphic to $H^2(M)$, which in turn is isomorphic to the free part $FH_2(M) \cong \mathbb{Z}^r$ of $H_2(M)$ by the universal coefficient theorem. We therefore find a class $g_1 \in H_3(M)$ which has intersection number 1 with f_1 and intersection number 0 with all other generators of $H_2(M)$. By Theorem 8.2.8, g_1 can be represented by a submanifold $Q \subset M$.

Our aim is to ensure that Q intersects S_1^2 in exactly one point, but none of the spheres representing the other generators, i.e. that we only have the geometric intersections forced on us by the algebraic conditions.† The usual way to achieve this is via the Whitney trick for removing intersection points of opposite sign (see [153], [187, Thm. 6.6]). Unfortunately, the codimension of Q is only 2 and the condition $\pi_1(M \setminus Q) = 0$, necessary for applying the Whitney lemma, may be violated. We are saved by the fact that we are free to change the submanifold Q, as long as we stay in the same homology class $[Q] \in H_3(M)$.

Thus we proceed as follows. Let P be a 2–sphere representing one of

† The condition that Q be disjoint from the spheres representing the generators other than f_1 is superfluous in the proof of the present lemma, but it will be used below (page 392).

the generators of $H_2(M)$. Suppose $p_-, p_+ \in P \cap Q$ are intersection points of opposite sign (of course all intersections may and will be assumed transverse, so we only have to deal with a finite number of isolated intersection points).

Join p_- and p_+ by an arc γ in P avoiding other intersection points with Q. A thin tubular neighbourhood $\mathcal{N}$ of P intersects Q in two 3–discs around p_- and p_+. Remove these 3–discs from Q and replace them by a tube $S^2 \times [-1,1]$ given by the part of $\partial\mathcal{N}$ lying over γ. By smoothing the corners we obtain a new embedded 3–manifold $Q' \subset M$ with $[Q]_M = [Q']_M$, for $Q' \sqcup \overline{Q}$ is the boundary of the 4–ball given by the part of $\mathcal{N}$ lying over γ (see Figure 8.1, illustrating the situation with reduced dimensions). Iterating this construction has the effect of removing all superfluous intersection points at the cost of adding 1–handles to Q.

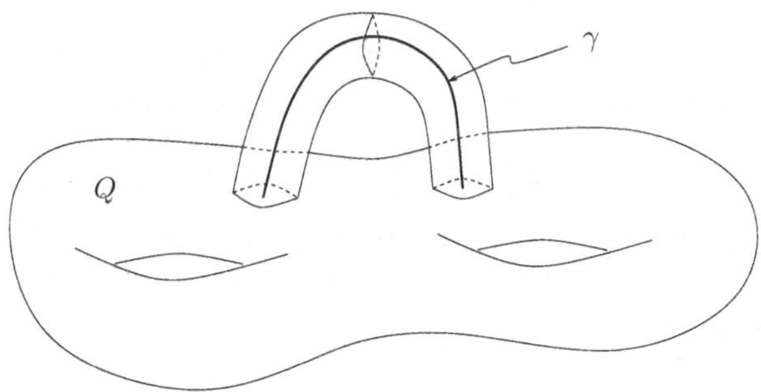

Fig. 8.1. Removing pairs of intersection points.

So we may assume that Q has one point of intersection with S_1^2 and none with the other spheres representing generators of $H_2(M)$.

Now we consider the effect of surgery along S_1^2. Write $\mathcal{N}_1 := S_1^2 \times \mathrm{Int}\,(D^3)$ for a tubular neighbourhood of S_1^2, let $M_0 := M \setminus \mathcal{N}_1$ be its complement, and denote by

$$M_1 := (M \setminus \mathcal{N}_1) \cup_{S^2 \times S^2} (D^3 \times S^2)$$

the manifold obtained by surgery along S_1^2. Observe that $Q \setminus \mathcal{N}_1$ is a submanifold of M_0 with boundary $S_0^2 := \partial(Q \setminus \mathcal{N}_1) \subset \partial M_0$ a 2–sphere, which may be thought of as a fibre of the normal sphere bundle of S_1^2 in M.

This surgery clearly kills the generator f_1. The trouble is that we might have created a new element in $H_2(M_1)$ represented by $* \times S^2$. However, by choosing as a representative of that class the one given by S_0^2, we see that it is zero in $H_2(M_1)$, for it bounds the 3–chain $Q \setminus \mathcal{N}_1$ in M_1. This proves claim (a).

Because of the Hurewicz isomorphism $H_2(M_1) \cong \pi_2(M_1)$ (for clearly

$\pi_1(M_1) = 0$), the sphere S_0^2 is even homotopically trivial in M_1. Since M is obtained from M_1 by performing surgery along S_0^2, this proves claim (b).

The fact that $W_3(M_1) = 0$ follows immediately from the interpretation of W_3 as the obstruction to the existence of a 3–frame over the 3–skeleton, once we take into account that $W_3(M) = 0$ and $M = S^2 \times S^3 \# M_1$. □

Remark 8.2.12 This proof constitutes a refinement of an argument due to Wall [236].

Repeated application of this lemma proves – except for the claimed surgical description of M_t – case (i) of our structure theorem, and case (ii) has been reduced to $H_2(M) \cong \mathbb{Z} \oplus G$, generated by $f_1, t_1, \ldots, t_s$, and $w_2 \neq 0$ only on the free generator f_1.

We have seen in the proof of the preceding lemma that a single surgery suffices to kill each of the free generators of $H_2(M)$ on which w_2 vanishes. For the torsion generators the problem becomes more complicated and can only be handled with some additional algebraic information. The papers by Kervaire–Milnor [145] and Wall [236] offer two alternative approaches to this problem. Wall obtains more detailed information about the structure of M, but in our situation the ideas of Kervaire–Milnor lead most quickly to the desired structure results.

We begin with a couple of preparatory lemmata of an algebraic nature. Let W be a compact, orientable 6–manifold with boundary ∂W. Denote by $\chi(W)$ the Euler characteristic of W and by

$$\chi^*(\partial W) = \sum_{i=0}^{2} \dim H_i(\partial W; \mathbb{R})$$

the **semi-characteristic** of ∂W.

Lemma 8.2.13 *The sum $\chi(W) + \chi^*(\partial W)$ is even.*

Proof The intersection pairing

$$H_3(W; \mathbb{R}) \otimes H_3(W; \mathbb{R}) \longrightarrow \mathbb{R}$$

is skew-symmetric and hence has even rank r. So the lemma is proved if we can show that r is congruent modulo 2 to $\chi(W) + \chi^*(\partial W)$.

Consider the homology exact sequence of the pair $(W, \partial W)$,

$$H_3(W) \xrightarrow{j_*} H_3(W, \partial W) \xrightarrow{\partial_*} H_2(\partial W) \xrightarrow{i_*} \ldots \xrightarrow{j_*} H_0(W, \partial W) \to 0,$$

with real coefficients understood. By relative Poincaré duality, the intersection product

$$H_3(W;\mathbb{R}) \otimes H_3(W,\partial W;\mathbb{R}) \longrightarrow \mathbb{R}$$

constitutes a perfect pairing. For $u, v \in H_3(W;\mathbb{R})$ and with $\bullet$ denoting the intersection product, we have $u \bullet v = u \bullet j_* v$, so the kernel of j_* consists precisely of those elements v for which $u \bullet v = 0$ for all $u \in H_3(W;\mathbb{R})$. It follows that

$$\begin{aligned} r &= \dim H_3(W;\mathbb{R}) - \dim(\ker j_*) \\ &= \dim(\operatorname{im} j_*) \;=\; \dim(\ker \partial_*) \\ &= b_3(W,\partial W) - \dim(\operatorname{im} \partial_*), \end{aligned}$$

with $b_3(W,\partial W) := \dim H_3(W,\partial W;\mathbb{R})$ denoting the third Betti number of $(W,\partial W)$. Continuing to use the exactness of the above sequence in this way, i.e.

$$\dim(\operatorname{im} \partial_*) = \dim(\ker i_*) = b_2(\partial W) - \dim(\operatorname{im} i_*), \quad \text{etc.},$$

we find that r is the alternating sum of the dimensions of the vector spaces to the right of j_* in that sequence.

From relative Poincaré duality we have $b_i(W,\partial W) = b_{6-i}(W)$. Using this and reducing modulo 2, we obtain

$$\begin{aligned} r &\equiv \sum_{i=0}^{6} b_i(W) + \sum_{i=0}^{2} b_i(\partial W) \\ &\equiv \chi(W) + \chi^*(\partial W) \bmod 2, \end{aligned}$$

as was claimed. □

Lemma 8.2.14 *Let M be a closed, simply connected 5–manifold and $S^2 \times D^3 \hookrightarrow M$ an arbitrary embedding. Let M' be the manifold obtained from M by surgery along $S^2 \times \{\mathbf{0}\}$, that is,*

$$M' = \left(M \setminus S^2 \times \operatorname{Int}(D^3)\right) \cup_{S^2 \times S^2} (D^3 \times S^2).$$

Then the second Betti numbers $b_2(M)$ and $b_2(M')$ are different.

Proof The surgery operation defines a cobordism

$$W = ([0,1] \times M) \cup (D^3 \times D^3),$$

obtained from $[0,1] \times M$ by attaching a 3–handle $D^3 \times D^3$ to $\{1\} \times M$ via the given embedding $\partial D^3 \times D^3 \hookrightarrow \{1\} \times M$. Then $\partial W = M \sqcup M'$. We notice that W has the homotopy type of M with a 3–cell attached, hence

$$\chi(W) = \chi(M) - 1 = -1,$$

since the Euler characteristic of the odd-dimensional manifold M is equal to 0. The preceding lemma then asserts that

$$\chi^*(M) \not\equiv \chi^*(M') \mod 2.$$

Now the fact that $H_1(M) = H_1(M') = 0$ implies $b_2(M) \neq b_2(M')$, even modulo 2. □

Lemma 8.2.15 *Let M and M' be as in the preceding lemma. Write $M_0 = M \setminus S^2 \times \mathrm{Int}\,(D^3)$. Let u be the class in $H_2(M_0)$ represented by $S^2 \times *$ and u' the class represented by $* \times S^2$, where $*$ denotes a base point in S^2. Then*

$$H_2(M) \cong H_2(M_0)/\langle u' \rangle \quad \textit{and} \quad H_2(M') \cong H_2(M_0)/\langle u \rangle.$$

Proof We have the homology exact sequence of the pair (M, M_0),

$$H_3(M, M_0) \xrightarrow{\partial_*} H_2(M_0) \longrightarrow H_2(M) \longrightarrow H_2(M, M_0).$$

By excision, $H_i(M, M_0)$ is isomorphic to $H_i(S^2 \times D^3, S^2 \times S^2)$, which is the zero group unless $i = 3$, when it is isomorphic to $\mathbb{Z}$ with generator $[* \times D^3]$. Under the boundary homomorphism ∂_* this generator maps to $u' = [* \times S^2] \in H_2(M_0)$. This proves the statement about $H_2(M)$, and that for $H_2(M')$ is completely analogous. □

Now we return to the situation where $H_2(M)$ is isomorphic to $\mathbb{Z} \oplus G$ with generators $f_1, t_1, \ldots, t_s$, where $w_2(f_1) \neq 0$ and $w_2(t_i) = 0$. We want to prove the structure theorem for this case. Write $S^2_{f_1}, S^2_{t_i} \subset M$, $i = 1, \ldots, s$, for disjointly embedded spheres representing the generators of $H_2(M)$.

Now perform surgery along $S^2_{t_1}$. Write $M_0 = M \setminus S^2_{t_1} \times \mathrm{Int}\,(D^3)$ and $M' = M_0 \cup_{S^2 \times S^2} D^3 \times S^2$. Denote by t'_1 the class in $H_2(M_0)$ and also $H_2(M')$ represented by $S^2_{t'_1} := * \times S^2$. By the preceding lemma we know that $H_2(M_0)$ is generated by $f_1, t_1, \ldots, t_s, t'_1$, and $H_2(M')$ by $f_1, t_2, \ldots, t_s, t'_1$.

By the proof of Lemma 8.2.11 we find an embedded 3–manifold $Q \subset M$ representing a generator of $H_3(M) \cong \mathbb{Z}$, having one transverse intersection point with $S^2_{f_1}$ and being disjoint from the other 2–spheres. So we may regard Q and $S^2_{f_1}$ also as submanifolds of M'. Their having intersection number 1 implies that f_1 is still a free generator in $H_2(M')$. With Lemma 8.2.14 we deduce $b_2(M') \geq 2$, so among $t_2, \ldots, t_s, t'_1 \in H_2(M')$ there must be a free generator f'.

Remark 8.2.16 The work of Wall [236] shows that the torsion generators $t_1, \ldots, t_s$ can be chosen in such a way that t_2 becomes a free generator in $H_2(M')$ with $t_1' = mt_2$, $m \in \mathbb{Z}$, and $b_2(M') = 2$, but here we do not need this precise information.

The 2–sphere $S^2_{t_1'}$ clearly has trivial normal bundle in M', and so do the spheres representing $t_2, \ldots, t_s$. Hence $w_2(f') = 0$, and we can perform surgery along a 2–sphere $S^2_{f'}$ representing f'. By the proof of Lemma 8.2.11, this has the effect of killing the class f'. The second homology group of the resulting manifold thus has one generator less than $H_2(M)$, and we may continue in this fashion until we reach a manifold with $H_2 \cong \mathbb{Z}$ and $w_2 \neq 0$, which by Proposition 8.2.1 (which has yet to be proved) is diffeomorphic to X_∞.

Finally, consider the reverse surgeries that lead from X_∞ to M. Observe that all surgeries on M were done in the complement of Q, so we may regard Q as a submanifold of X_∞, where it represents a generator of $H_3(X_\infty) \cong \mathbb{Z}$, since Q has intersection number one with $S^2_{f_1}$ (which likewise survives from M to X_∞). In X_∞, the non-trivial S^3–bundle over S^2, consider a fibre S^3_0. From the Wang sequence of this fibration, see [223],

$$\ldots \to H_i(S^3_0) \to H_i(X_\infty) \to H_{i-2}(S^3_0) \to H_{i-1}(S^3_0) \to \ldots,$$

it is easy to see that S^3_0 represents a generator of $H_3(X_\infty)$, so it represents up to sign the same class as Q and therefore has *algebraic* intersection number zero with all the 2–spheres in X_∞ along which we want to perform surgery. Now $X_\infty \setminus S^3_0 \cong S^3 \times D^2$ is simply connected; this allows us to apply the Whitney trick and thus to assume that those 2–spheres and S^3_0 have *geometric* intersection number 0.

Hence M is obtained from X_∞ by surgery along a link of contractible 2–spheres, which is the same as taking the connected sum of X_∞ with some manifold M_t obtained from S^5 by surgery along a link of 2–spheres. (In fact, we may take $S^2_{f_1}$ to be a section of the bundle $X_\infty \to S^2$. Then the complement of $S^2_{f_1}$ and S^3_0 in X_∞ is a 5–disc.)

This concludes the proof of Theorem 8.2.9. □

Proof of Proposition 8.2.1 – Part Three It remains to be shown that a simply connected 5–manifold M with $H_2(M;\mathbb{Z}) \cong \mathbb{Z}$ and $w_2(M) \neq 0$ is diffeomorphic to X_∞. *For the remainder of this section, M denotes just such a manifold.* The arguments employed here are for the most part special cases of those in Sections 3 and 4 of [22]. In a step by step procedure we are going to construct a simply connected h–cobordism between M and X_∞.

The only potentially non-zero Stiefel–Whitney number of a simply connected 5–manifold is w_2w_3. From the Bockstein sequence we see that $W_3(M)$

is zero. The third Stiefel–Whitney class $w_3(M)$ being the mod 2 reduction of $W_3(M)$, it vanishes *a fortiori*. So both M and X_∞ have vanishing Stiefel–Whitney numbers. By a deep result of Thom [230], see also [188], this implies that M and X_∞ are oriented cobordant.† Let V_0 be such a cobordism, $\partial V_0 = M \sqcup \overline{X}_\infty$, and write i, i_∞ for the inclusion of M or X_∞, respectively, into V_0 (we shall keep that notation also for the modified cobordisms below). Then

$$i^* w_2(V_0) = w_2(TV_0|M) = w_2(TM \oplus \epsilon) = w_2(M) \neq 0,$$

so $w_2(V_0) \neq 0$.

Lemma 8.2.17 *There is a simply connected cobordism V_1 between M and X_∞ with $H_2(V_1) \cong \mathbb{Z}_2$.*

Proof First of all, we may represent the generators of $\pi_1(V_0)$ by embedded 1–spheres. Since V_0 is orientable, these 1–spheres have trivial normal bundle, and surgery along them will kill the fundamental group. Call the resulting simply connected cobordism V_0'.

The second Stiefel–Whitney class $w_2(V_0')$, considered as a homomorphism $H_2(V_0') \to \mathbb{Z}_2$, can only be non-zero on elements of $H_2(V_0')$ of order 2^k, with $k = \infty$ not excluded. If $e_1, \ldots, e_l$ is any generating set for $H_2(V_0')$ with $w_2(e_1) \neq 0$ and e_1 of minimal order among the generators on which w_2 does not vanish, then

$$e_1,\; e_2' := e_2 - w_2(e_2)e_1,\; \ldots,\; e_l' := e_l - w_2(e_l)e_1$$

is a new generating set for $H_2(V_0')$ such that w_2 is non-zero only on the generator e_1.

Now perform surgery on V_0' along 2–spheres representing the elements $2e_1, e_2', \ldots, e_l'$ of $H_2(V_0')$. All those spheres have trivial normal bundle because w_2 vanishes on these elements of $H_2(V_0')$. This being surgery below the middle dimension, it clearly has the effect of killing these elements, see [183], [153]. Thus we have constructed a simply connected cobordism V_1 with $H_2(V_1) \cong \mathbb{Z}_2$ (see the following remark). □

Remark 8.2.18 In Lemma 2 of [183] it is only shown that surgery along a 2–sphere representing $2e_1$ kills a subgroup containing that element, so we might inadvertently be killing e_1 as well. The computation in Proposition X.1.1 of [153] shows that indeed we kill $2e_1$, but not e_1. Alternatively, the fact that $H_2(V_1)$ equals $\mathbb{Z}_2$ rather than the zero group follows from $i^* w_2(V_1) =$

† In general, the Stiefel–Whitney numbers can only detect non-oriented cobordisms, but in dimension 5 both the non-oriented and the oriented cobordism group equals $\mathbb{Z}_2$.

$w_2(M) \neq 0$. From this we also conclude that $i_*\colon H_2(M) \to H_2(V_1), \mathbb{Z} \to \mathbb{Z}_2$, is reduction modulo 2 (rather than the zero homomorphism). The same is true for $i_{\infty *}\colon H_2(X_\infty) \to H_2(V_1)$.

Lemma 8.2.19 *There is a simply connected cobordism V_2 between M and X_∞ with $H_2(V_2) \cong \mathbb{Z}$ and i, i_∞ inducing isomorphisms on H_2.*

Proof Before turning to the formal argument, it may be instructive to give a geometric motivation for the following construction. Suppose we were already given the desired cobordism V_2. By performing surgery along a sphere representing twice the generator of $H_2(V_2)$ we would obtain a cobordism V_1 with $H_2(V_1) \cong \mathbb{Z}_2$. Here we have cut out $S^2 \times \operatorname{Int}(D^4)$, with $S^2 \times *$ representing twice the generator of $H_2(V_2)$, and replaced it with $D^3 \times S^3$. So the reverse surgery that yields V_2 from V_1 ought to be along a 3–sphere which has intersection number one with the 3–chain in V_1 that is bounded by twice the generator of $H_2(V_1) \cong \mathbb{Z}_2$.

Now to the details. Write x_0 for a generator of $H_2(M)$, and x_∞ for a generator of $H_2(X_\infty)$. We want to think of $H_2(\partial V_1)$ as a direct sum

$$H_2(\partial V_1) \cong \mathbb{Z}^{x_0} \oplus \mathbb{Z}^{x_\infty - x_0}.$$

Write $j = i \sqcup i_\infty$ for the inclusion of ∂V_1 into V_1. Then the induced homomorphism $j_*\colon H_2(\partial V_1) \to H_2(V_1)$ is given by $ax_0 + b(x_\infty - x_0) \mapsto a \bmod 2$, so $\ker j_*$ is generated by $2x_0$ and $x_\infty - x_0$.

From the homology exact sequence of the pair $(V_1, \partial V_1)$,

$$H_3(V_1, \partial V_1) \xrightarrow{\partial_*} H_2(\partial V_1) \xrightarrow{j_*} H_2(V_1),$$

we know that any element in $\ker j_*$ is the image under ∂_* of an element in $H_3(V_1, \partial V_1)$. Choose $g_0 \in H_3(V_1, \partial V_1)$ such that $\partial_* g_0 = 2x_0$. Since x_0 has infinite order, so does g_0. By forming the connected sum (in the interior) of V_1 with $S^3 \times S^3$ and adding to g_0 the class represented by the sphere $S^3 \times *$, say, we may assume in addition that g_0 is indivisible. This enables us to extend g_0 to a generating set $g_0, g_1, \ldots, g_p$ for $H_3(V_1, \partial V_1)$ such that $\partial_* g_i$ is a multiple of $x_\infty - x_0$ for each $i = 1, \ldots, p$, and $\partial_* g_1, \ldots, \partial_* g_p$ generate $\mathbb{Z}^{x_\infty - x_0}$.

This choice of g_0 also allows us to find (by relative Poincaré duality) a class h_0 in $H_3(V_1)$ with intersection numbers $h_0 \bullet g_0 = 1$ and $h_0 \bullet g_i = 0$ for $i = 1, \ldots, p$. By the generalised Hurewicz theorem [35, p. 488], the Hurewicz homomorphism $\pi_3(V_1) \to H_3(V_1)$ is surjective. So we see, using Haefliger's theorem again, that h_0 can be represented by an embedded 3–sphere $S^3_{h_0}$. Because of $\pi_2(\mathrm{SO}(3)) = 0$, this sphere has trivial normal bundle in V_1 and hence we find an embedding $S^3_{h_0} \times D^3 \hookrightarrow V_1$.

The classes $g_0, \dots, g_p$ are not necessarily represented by submanifolds, and so we cannot use the differential topological argument from the proof of Lemma 8.2.11 to remove superfluous intersection points. But we observe that $(S^3_{h_0} \setminus *) \times D^3$ is a 6–cell, so we may choose representative (simplicial) chains for $g_0, \dots, g_p$ which avoid this cell. (Moreover, we may assume that we are working with a simplicial decomposition of V_1 containing the 3–cell $* \times D^3$.) By removing copies of $* \times D^3$ of opposite sign from these chains, we obtain representative chains for $g_0, \dots, g_p$ whose geometric intersection number with $S^3_{h_0}$ equals the algebraic intersection number with h_0. Now set $V_1' = V_1 \setminus S^3_{h_0} \times \mathrm{Int}\,(D^3)$ and $V_2 = V_1' \cup (D^4 \times S^2)$. So V_2 is the cobordism obtained from V_1 by surgery along $S^3_{h_0}$. Clearly V_2 is simply connected (by Seifert–van Kampen). There is a choice of framing in $\pi_3(\mathrm{SO}(3)) \cong \mathbb{Z}$ for this surgery, but it will turn out that this choice does not affect the relevant properties of V_2.

By excision we have

$$H_i(V_2, V_1') \cong H_i(D^4 \times S^2, S^3 \times S^2) \cong \begin{cases} \mathbb{Z} & \text{for } i = 4, \\ 0 & \text{otherwise.} \end{cases}$$

In particular, from the homology exact sequence of (V_2, V_1') it follows that the inclusion $V_1' \hookrightarrow V_2$ induces an isomorphism $H_2(V_1') \xrightarrow{\cong} H_2(V_2)$. Since we are only interested in the second homology group, we may restrict our attention to V_1'.

Write $j' = i' \sqcup i'_\infty$ for the inclusion of ∂V_1 into V_1' and consider the homology exact sequence of the pair $(V_1', \partial V_1)$,

$$H_3(V_1', \partial V_1) \xrightarrow{\partial_*} H_2(\partial V_1) \xrightarrow{j'_*} H_2(V_1').$$

I claim that $\ker j'_*$ is generated by $x_\infty - x_0$. In order to see that $x_\infty - x_0$ is contained in $\ker j'_*$, observe that the class $x_\infty - x_0$ can be written as a $\mathbb{Z}$–linear combination of $\partial_* g_1, \dots, \partial_* g_p$. The classes g_i, $i = 1, \dots, p$, have, as shown above, representative chains that miss $S^3_{h_0} \times D^3$. So the $\partial_* g_i$ are homologous to zero in $H_2(V_1')$. Conversely, any element $z \in \ker j'_*$ is the image under ∂_* of an element in $H_3(V_1', \partial V_1)$, any representative chain of which misses $S^3_{h_0} \times D^3$, so it represents a linear combination of $g_1, \dots, g_p$ in $H_3(V_1, \partial V_1)$. Hence z lies in the group generated by $\partial_* g_1, \dots, \partial_* g_p$, which is $\mathbb{Z}^{x_\infty - x_0}$.

It follows immediately that $H_2(V_1')$ contains a copy of $\mathbb{Z}$ generated by $j'_* x_0$. We end the proof by showing that this constitutes all of $H_2(V_1')$. In the homology exact sequence of the pair (V_1, V_1'),

$$H_3(V_1, V_1') \xrightarrow{\partial_*} H_2(V_1') \to H_2(V_1) \to H_2(V_1, V_1'),$$

we have $H_3(V_1, V_1') \cong \mathbb{Z}$ and $H_2(V_1, V_1') = 0$ by excision. The generator of $H_3(V_1, V_1')$ is represented by a disc fibre $* \times D^3$ of $S^3_{h_0} \times D^3$, which under the boundary homomorphism ∂_* maps to the class $[* \times S^2]$. That latter class equals $2j'_*x_0$ (a potential sign here is irrelevant), since together they bound the 3–chain $g_0 \setminus [* \times D^3]$. So the sequence becomes

$$\begin{array}{ccccccc} \mathbb{Z} & \longrightarrow & H_2(V_1') & \longrightarrow & \mathbb{Z}_2 & \longrightarrow & 0 \\ 1 & \longmapsto & 2j'_*x_0. & & & & \end{array}$$

This forces $H_2(V_1') \cong \mathbb{Z}$ with generator $j'_*x_0 = i'_*x_0$. Observe finally that

$$(i'_\infty)_*x_\infty = j'_*x_\infty = j'_*x_0 + j'_*(x_\infty - x_0) = j'_*x_0$$

likewise generates $H_2(V_1')$. □

The following lemma, at last, completes the proof of Proposition 8.2.1.

Lemma 8.2.20 *There is a simply connected h–cobordism V_3 between M and X_∞, so M is diffeomorphic to X_∞.*

Proof We have $H_2(V_2, M) = H_2(V_2, X_\infty) = 0$ from the preceding lemma and the homology exact sequences of these pairs. Hence

$$H_3(V_2, M) \cong H^3(V_2, X_\infty) \cong FH_3(V_2, X_\infty)$$

by relative Poincaré duality and the universal coefficient theorem. For the same reason, $H_3(V_2, X_\infty)$ is a free abelian group, and the intersection product defines a non-singular pairing

$$H_3(V_2, M) \otimes H_3(V_2, X_\infty) \longrightarrow \mathbb{Z}.$$

If $H_3(V_2, M) \neq 0$, choose classes $m \in H_3(V_2, M)$ and $x \in H_3(V_2, X_\infty)$ with $m \bullet x = 1$. These elements lift to $H_3(V_2)$ by the homology exact sequence of these pairs, and may be represented in V_2 by embedded spheres S^3_m, S^3_x meeting transversely in a single point (once again we use the theorem of Haefliger and the Whitney trick). Because of $\pi_2(\mathrm{SO}(3)) = 0$, these spheres have trivial normal bundle, so the embeddings of S^3_m and S^3_x extend to an embedding of a neighbourhood of the 3–skeleton $S^3 \vee S^3$ of $S^3 \times S^3$ into V_2. This shows the existence of a cobordism V_2' between M and X_∞ such that $V_2 \cong V_2' \# S^3 \times S^3$. Now $H_3(V_2', M)$ has smaller rank than $H_3(V_2, M)$, and we may repeat the procedure until we obtain a simply connected cobordism V_3 between M and X_∞ with $H_i(V_3, M) = H_i(V_3, X_\infty) = 0$ for $i \leq 3$. By relative Poincaré duality and the universal coefficient theorem we find $H_*(V_3, M) = 0$. This shows that V_3 is an h–cobordism and hence – by the h–cobordism theorem [153] – a product $V_3 \cong [0,1] \times M \cong [0,1] \times X_\infty$. □

8.3 Existence of contact structures

We are now going to prove Theorem 8.0.6. By the discussion in Section 8.1 we can restrict our attention to (closed, oriented, simply connected) 5–manifolds M with $W_3(M) = 0$, as classified in Theorem 8.2.9. Our task, then, is to prove that for any integral lift $c \in H^2(M;\mathbb{Z})$ of $w_2(M)$ there is a contact structure, inducing the given orientation, with first Chern class equal to c.

By Theorem 6.2.5 (for connected sums), it suffices to prove Theorem 8.0.6 individually for the manifolds $S^2 \times S^3$, X_∞, and those of type M_t (in the notation of Theorem 8.2.9).

We begin with the latter. Since $H^2(M_t) = 0$, there is only one homotopy class of almost contact structures (that is, one for either orientation), so it suffices simply to construct some contact structure.

By the structure theorem, M_t is obtained from S^5 by surgery along a link of 2–spheres. The 5–sphere admits contact structures realising either orientation, e.g. $\xi_\pm := \ker(\pm\alpha_0)$ from Example 2.1.7.

Lemma 8.3.1 *Any embedding $j\colon S^2 \to (S^5, \xi_\pm)$ is covered by a fibrewise injective complex bundle map $TS^2 \otimes \mathbb{C} \to \xi_\pm$.*

Proof Let $j_0\colon S^2 \to S^5$ be a Legendrian embedding that sends S^2 to the submanifold of S^5 given by $\{y_1 = y_2 = y_3 = 0\}$. By Corollary 8.2.7, the given embedding j is isotopic to j_0. Hence, by the homotopy properties of vector bundles, $j^*\xi_\pm$ and $j_0^*\xi_\pm$ are isomorphic U(2)–bundles. There is a fibrewise injective complex bundle map (over id_{S^2})

$$\begin{array}{ccc} TS^2 \otimes \mathbb{C} & \longrightarrow & j_0^*\xi_\pm \\ T_xS^2 \otimes \mathbb{C} \ni \mathbf{u} + \mathrm{i}\mathbf{v} & \longmapsto & (x, Tj_0(\mathbf{u}) + J_\pm Tj_0(\mathbf{v})), \end{array}$$

where $J_\pm$ denotes the complex bundle structure on $\xi_\pm$. So such a bundle map exists likewise for $j^*\xi$, which is the same as saying that j is covered by a fibrewise injective complex bundle map $TS^2 \otimes \mathbb{C} \to \xi_\pm$. □

Remark 8.3.2 The discussion on page 371 showed that U(2)–bundles over S^2 are classified by the first Chern class. Moreover, see [188, §15], if $E \to B$ is a real vector bundle, the odd Chern classes of the complexified vector bundle $E \otimes \mathbb{C} \to B$ are cohomology elements of order 2. In particular, we have $c_1(TS^2 \otimes \mathbb{C}) = 0$. So the condition for an embedding $j\colon S^2 \to (M^5, \xi)$ to be covered by a fibrewise injective (and hence isomorphic) complex bundle map $TS^2 \otimes \mathbb{C} \to \xi$ can be formulated as $c_1(j^*\xi) = 0$.

Thus, by the h–principle for isotropic embeddings (Prop. 6.3.6), we may assume that M_t is obtained from $(S^5, \xi_\pm)$, as oriented manifold, via surgery along a link of Legendrian 2–spheres. The topological framing of a surgery along a 2–sphere in a 5–manifold is unique, since $\pi_2(\mathrm{SO}(3))$ is the trivial group. So when we perform contact surgery along this link of Legendrian 2–spheres, we do indeed obtain a contact structure (realising either of the desired orientations) on M_t.

Each of the two S^3–bundles over S^2 admits an orientation-reversing diffeomorphism that induces the identity on the cohomology group H^2. For instance, since the inclusion $\mathrm{SO}(3) \subset \mathrm{SO}(4)$ induces an isomorphism on π_1, we may assume that the characteristic map of the bundle is in $\pi_1(\mathrm{SO}(3))$, and hence equivariant with respect to a suitable reflection in the fibre. Such a reflection then induces the desired diffeomorphism. It therefore suffices to construct contact structures on $S^2 \times S^3$ and X_∞ realising, respectively, any even or odd integer in $H^2 \cong \mathbb{Z}$ as first Chern class, without any attention to orientations.

We have seen in Example 6.1.3 (3) that the two S^3–bundles over S^2 can be obtained from S^5 by surgery along a standard $S^1 \subset S^5$ by choosing the framing of the surgery appropriately. In the contact geometric setting, we start with the standard contact sphere (S^5, ξ_{st}) from Example 2.1.7 and the isotropic circle

$$S^1 = \{x_1^2 + x_2^2 = 1,\ x_3 = y_1 = y_2 = y_3 = 0\}.$$

The conformal symplectic normal bundle of this S^1 is spanned by the vector fields $\partial_{x_3}, \partial_{y_3}$. We are free to choose any symplectic trivialisation of this bundle, which amounts to a freedom in $\pi_1(\mathrm{U}(1)) \cong \mathbb{Z}$ for the choice of framing when performing contact surgery along S^1. Since the inclusion of $\mathrm{U}(1) = \mathrm{SO}(2)$ in $\mathrm{SO}(4)$ induces a surjection $\mathbb{Z} \to \mathbb{Z}_2$ on π_1, it is immediately clear that we can produce a contact structure on both $S^2 \times S^3$ and X_∞. Finally, we want to convince ourselves that the choice of integer in $\pi_1(\mathrm{U}(1))$ translates directly into the first Chern class of the resulting contact structure.

In Example 6.2.8 we saw that the handlebody corresponding to contact surgery along the standard Legendrian $S^1 \subset (S^3, \xi_{\mathrm{st}})$ may be regarded as the cotangent disc bundle $\pi\colon DT^*S^2 \to S^2$ of S^2 with its natural symplectic structure. Since this is a bundle with Lagrangian fibres, its tangent bundle $T(DT^*S^2)$ is isomorphic, as a complex vector bundle, to $\pi^*(T^*S^2 \otimes \mathbb{C})$. From Remark 8.3.2 it follows that the handlebody is an almost complex manifold with vanishing first Chern class.

Surgery along $S^1 \subset (S^5, \xi_{\mathrm{st}})$ corresponds to attaching a 6–dimensional 2–handle; the resulting handlebody is a D^4–bundle over S^2. When we think of D^4 as $D^2 \times D^2 \in i\mathbb{R}^2 \times \mathbb{C}$, the characteristic map of this bundle splits accordingly. On the first D^2–factor it is the one considered in the preceding paragraph. On the second D^2–factor the characteristic map is determined by the choice of trivialisation of the conformal symplectic normal bundle. These trivialisations are in one-to-one correspondence with $\pi_1(\mathrm{U}(1)) \cong \mathbb{Z}$; a correspondence with the integers can be fixed by identifying $0 \in \mathbb{Z}$ with the trivialisation given by $\partial_{x_3}, \partial_{y_3}$.

With this choice, the integer describing the framing is the Euler class in $H^2(S^2) \cong \mathbb{Z}$ of the resulting D^2–bundle, and hence the first Chern class of the total space of the D^4–bundle; see the arguments for proving Lemma 8.2.3.

The inclusion of the corresponding S^3–bundle into the D^4–bundle induces an isomorphism on H^2, and the outward normal defines a trivial complex line bundle complementary to the resulting contact structure on this S^3–bundle. It follows that the first Chern class of the (total space of the) D^4–bundle coincides with the first Chern class of the contact structure on the corresponding S^3–bundle.

This concludes the proof of Theorem 8.0.6.

Appendix A
The generalised Poincaré lemma

In Remark 2.5.12 we first alluded to the generalised Poincaré lemma. In Chapter 6 this was used to prove a version of the neighbourhood theorem for isotropic submanifolds of contact manifolds strong enough to be applicable to the construction of symplectic handlebodies. In this appendix we provide a proof of that lemma. Our argument is analogous to that used in the textbook [23] by Barden and Thomas for proving the classical Poincaré lemma.

We start with a differential manifold M and consider the maps

$$\begin{array}{rccc} j_t\colon & M & \longrightarrow & M \times \mathbb{R} \\ & x & \longmapsto & (x,t) \end{array}$$

and

$$\begin{array}{rccc} \phi_t\colon & M \times \mathbb{R} & \longrightarrow & M \times \mathbb{R} \\ & (x,s) & \longmapsto & (x,s+t). \end{array}$$

Observe that ϕ_t may be regarded as the flow of the vector field ∂_t on $M \times \mathbb{R}$, and $j_t = \phi_t \circ j_0$.

Write $\Omega^k(-)$ for the space of differential k–forms on a given manifold. We define an operator

$$I\colon\ \Omega^k(M \times \mathbb{R}) \longrightarrow \Omega^{k-1}(M),$$

called *integration along fibres*, by requiring that it send $\eta \in \Omega^k(M \times \mathbb{R})$ to

$$I(\eta) := \int_0^1 j_t^*(i_{\partial_t}\eta)\,dt.$$

Lemma A.1 *The operator I defines a cochain homotopy between the cochain maps j_0^* and j_1^* of the de Rham cochain complex, i.e.*

$$j_1^*\eta - j_0^*\eta = d(I(\eta)) + I(d\eta).$$

Proof We compute

$$\begin{aligned}
j_1^*\eta - j_0^*\eta &= \int_0^1 \frac{d}{dt}(j_t^*\eta)\,dt = \int_0^1 j_0^*\frac{d}{dt}(\phi_t^*\eta)\,dt \\
&= \int_0^1 j_0^*\phi_t^*(\mathcal{L}_{\partial_t}\eta)\,dt \\
&= \int_0^1 j_t^*\big(d(i_{\partial_t}\eta) + i_{\partial_t}(d\eta)\big)\,dt \\
&= d\int_0^1 j_t^*(i_{\partial_t}\eta)\,dt + \int_0^1 j_t^* i_{\partial_t}(d\eta)\,dt \\
&= d(I(\eta)) + I(d\eta),
\end{aligned}$$

which is the claimed identity. □

In the next proposition we apply this integration along fibres to the construction of a cochain homotopy between the pull-back maps of homotopic maps.

Proposition A.2 *If $f_0, f_1\colon M \to N$ are homotopic smooth maps between manifolds M, N, then f_0^* and f_1^* are cochain homotopic maps of the de Rham cochain complex.*

Proof Let $h\colon M \times \mathbb{R} \to N$ be a homotopy such that $h_t := h \circ j_t = f_t$ for $t = 0, 1$. We may assume without loss of generality that h is a technical homotopy, that is, $h_t \equiv f_0$ for $t \leq \varepsilon$ and $h_t \equiv f_1$ for $t \geq 1 - \varepsilon$. Furthermore, one may then assume h to be smooth. For a given k–form η on N we compute, using the foregoing lemma,

$$\begin{aligned}
f_1^*(\eta) - f_0^*(\eta) &= j_1^* \circ h^*(\eta) - j_0^* \circ h^*(\eta) \\
&= d(I(h^*\eta)) + I(dh^*\eta) \\
&= d\big((I \circ h^*)(\eta)\big) + (I \circ h^*)(d\eta),
\end{aligned}$$

i.e. a cochain homotopy is defined by $I \circ h^*$. □

We can now formulate the *generalised Poincaré lemma.* We say that a differential form η on a manifold E **vanishes on a submanifold** B if $\eta = 0$ on $TM|_B$. This is stronger, in general, than saying that the pull-back of η to B under the inclusion map vanishes, for the latter is equivalent to saying merely that $\eta = 0$ on TB.

Proposition A.3 *Let $\pi\colon E \to B$ be a smooth vector bundle. Identify B with a submanifold of E via the zero section σ_0. Let U be a neighbourhood of B in E such that $U \cap E_x \subset E_x$ is star-shaped for all fibres $E_x := \pi^{-1}(x)$,*

$x \in B$. Let η be a differential k-form on U that is closed, $d\eta = 0$, and vanishes on TB, i.e. $\sigma_0^*\eta = 0$. Then there is a $(k-1)$-form $\zeta \in \Omega^{k-1}(U)$ vanishing on B with $\eta = d\zeta$.

If, in addition, η vanishes on B, then ζ may be chosen to be vanishing to second order on B, i.e. such that the first-order partial derivatives of its components with respect to local coordinates (of U), in any chart, vanish at points of B.

Proof Let $h\colon U \times \mathbb{R} \to U$ be a technical homotopy between $h_0 = \mathrm{id}_U$ and $h_1 = \sigma_0 \circ \pi$, defined by contracting radially along the fibres. In particular, we have $h_t|_B = \mathrm{id}_B$ for all $t \in \mathbb{R}$. Let $H = I \circ h^*$ be the corresponding cochain homotopy found in the preceding proof. Then

$$\eta = \pi^*\sigma_0^*\eta - d(H(\eta)) - H(d\eta) = -d(H(\eta)).$$

We claim that $\zeta := -H(\eta)$ has the desired properties. To this end, we rewrite $H(\eta)$ as follows:

$$\begin{aligned} H(\eta) &= \int_0^1 j_t^*(i_{\partial_t}(h^*\eta))\,dt \\ &= \int_0^1 j_t^*h^*(i_{Th(\partial_t)}\eta)\,dt \\ &= \int_0^1 h_t^*(i_{Th(\partial_t)}\eta)\,dt. \end{aligned}$$

Observe that $Th(\partial_t)$ is the time-dependent vector field on U whose flow is h. Since $h_t|_B = \mathrm{id}_B$ for all $t \in \mathbb{R}$, this vector field vanishes on B. Those two conditions taken together yield $H(\eta) = 0$ on B.

If, in addition, η vanishes on B, then the coefficients of $i_{Th(\partial_t)}\eta$ at points of B (in any local coordinates) are sums of products of two differentiable functions, both of which vanish on B. This proves the proposition. □

Applied to the normal bundle of a submanifold $L \subset M$, this yields the following corollary, where the inclusion of L in M is denoted by j_L.

Corollary A.4 *Let L be a submanifold of a differential manifold M. Let $\eta \in \Omega^k(M)$ be a closed k-form with $j_L^*\eta = 0$. Then there is an open neighbourhood U of L in M and a $(k-1)$-form ζ on U vanishing on L and with $d\zeta = \eta$ on U.*

If, in addition, η vanishes on L, then ζ may be chosen to be vanishing to second order on L, i.e. such that the first-order partial derivatives of its components with respect to local coordinates (of M), in any chart, vanish at points of L. □

Appendix B
Time-dependent vector fields

In this appendix we collect some formulæ for time-dependent vector fields that were used in Chapter 2.

We begin with a situation where the time-dependent vector field is only defined along the image of a smooth family of maps. The following lemma entered the proof of Theorem 2.6.2.

Lemma B.1 *Let* $j_t \colon L \to M$, $t \in [0,1]$, *be a smooth family of smooth maps from a manifold* L *to a manifold* M. *Define a time-dependent vector field* X_t *along* $j_t(L)$ *by*

$$X_t \circ j_t = \frac{d}{dt} j_t.$$

Then, for any differential k*-form* α *on* M, *we have*

$$\frac{d}{dt}(j_t^* \alpha) = d(j_t^* i_{X_t} \alpha) + j_t^* (i_{X_t} d\alpha).$$

Proof (i) The formula holds for functions $f \in C^\infty(M)$:

$$\frac{d}{dt}(j_t^* f)(p) = \frac{d}{dt}(f \circ j_t)(p) = (i_{X_t} df)(j_t(p)) = j_t^* (i_{X_t} df)(p).$$

(ii) If the formula holds for a k–form α and an l–form β, then also for their wedge product $\alpha \wedge \beta$:

$$\begin{aligned} & d\big(j_t^* i_{X_t}(\alpha \wedge \beta)\big) + j_t^*\big(i_{X_t} d(\alpha \wedge \beta)\big) = \\ = \;& d\big(j_t^*(i_{X_t}\alpha \wedge \beta + (-1)^k \alpha \wedge i_{X_t}\beta)\big) \\ & + j_t^* i_{X_t}(d\alpha \wedge \beta + (-1)^k \alpha \wedge d\beta) \\ = \;& d(j_t^* i_{X_t}\alpha) \wedge j_t^*\beta + (-1)^{k-1} j_t^* i_{X_t}\alpha \wedge j_t^* d\beta \\ & + (-1)^k j_t^* d\alpha \wedge j_t^* i_{X_t}\beta + j_t^*\alpha \wedge d(j_t^* i_{X_t}\beta) \\ & + j_t^* i_{X_t} d\alpha \wedge j_t^*\beta + (-1)^{k+1} j_t^* d\alpha \wedge j_t^* i_{X_t}\beta \\ & + (-1)^k j_t^* i_{X_t}\alpha \wedge j_t^* d\beta + j_t^*\alpha \wedge j_t^* i_{X_t} d\beta \\ = \;& \big(d(j_t^* i_{X_t}\alpha) + j_t^* i_{X_t} d\alpha\big) \wedge j_t^*\beta \\ & + j_t^*\alpha \wedge \big(d(j_t^* i_{X_t}\beta) + j_t^* i_{X_t} d\beta\big) \\ = \;& \frac{d}{dt}(j_t^*\alpha) \wedge j_t^*\beta + j_t^*\alpha \wedge \frac{d}{dt}(j_t^*\beta) \\ = \;& \frac{d}{dt}(j_t^*\alpha \wedge j_t^*\beta) \\ = \;& \frac{d}{dt}\big(j_t^*(\alpha \wedge \beta)\big). \end{aligned}$$

(iii) If the formula holds for α, then also for $d\alpha$:

$$\begin{aligned} \frac{d}{dt}(j_t^* d\alpha) &= d\Big(\frac{d}{dt}(j_t^*\alpha)\Big) \\ &= d\big(j_t^* i_{X_t} d\alpha\big) \\ &= d\big(j_t^* i_{X_t} d\alpha\big) + j_t^*\big(i_{X_t} d^2\alpha\big). \end{aligned}$$

(iv) Locally, functions and differentials of functions generate the algebra of differential forms. So from (i)–(iii) the lemma follows. □

A time-dependent vector field X_t on a manifold M – where for simplicity we assume that X_t and its flow ψ_t, given by the equation

$$X_t \circ \psi_t = \dot{\psi}_t,$$

are defined for all $t \in \mathbb{R}$ – gives rise to a vector field $\widetilde{X}$ in the usual sense on $M \times \mathbb{R}$ by setting

$$\widetilde{X}(p,t) = X_t(p) + \partial_t.$$

Then the flow $\widetilde{\psi}_h$ of $\widetilde{X}$ takes the form

$$\widetilde{\psi}_h(p,0) = (\psi_h(p), h),$$

or more generally

$$\widetilde{\psi}_h(p,t) = (\psi_{t+h} \circ \psi_t^{-1}(p), t+h).$$

It is now sensible to define the Lie derivative of a differential form α on M with respect to X_t as

$$\mathcal{L}_{X_t}\alpha := \lim_{h\to 0}\frac{(\psi_{t+h}\circ\psi_t^{-1})^*\alpha-\alpha}{h}. \tag{B.1}$$

Then

$$\psi_t^*(\mathcal{L}_{X_t}\alpha) = \lim_{h\to 0}\frac{\psi_{t+h}^*\alpha-\psi_t^*\alpha}{h} = \frac{d}{dt}(\psi_t^*\alpha), \tag{B.2}$$

as was used in the proof of Lemma 2.2.1.

We have frequently alluded to the Cartan formula $\mathcal{L}_X = d\circ i_X + i_X\circ d$ for the Lie derivative of differential forms. This holds true also for time-dependent vector fields.

Lemma B.2 *Let X_t, $t\in\mathbb{R}$, be a time-dependent vector field on a manifold M. Then*

$$\mathcal{L}_{X_t} = d\circ i_{X_t} + i_{X_t}\circ d.$$

Proof Since X_t is globally defined on M, the formula from the preceding lemma – assuming that ψ_t is globally defined – gives

$$\begin{aligned}\psi_t^*(\mathcal{L}_{X_t}\alpha) &= \frac{d}{dt}(\psi_t^*\alpha)\\ &= d(\psi_t^* i_{X_t}\alpha)+\psi_t^*(i_{X_t}d\alpha)\\ &= \psi_t^*\big(d(i_{X_t}\alpha)+i_{X_t}d\alpha\big).\end{aligned}$$

Since ψ_t is a diffeomorphism, the Cartan formula follows.

Alternatively, one can give a direct proof that does not rely on Lemma B.1 and the assumption that the flow of X_t be globally defined. Write

$$\pi\colon M\times\mathbb{R}\longrightarrow M$$

for the projection and

$$\sigma_t\colon M\longrightarrow M\times\mathbb{R}$$

for the inclusion map defined by $\sigma_t(p)=(p,t)$. Let $\tilde{d}$ denote the exterior derivative of differential forms on $M \times \mathbb{R}$. Then

$$\begin{aligned}
\mathcal{L}_{X_t}\alpha &= \lim_{h\to 0} \frac{\left(\psi_{t+h}\circ\psi_t^{-1}\right)^*\alpha - \alpha}{h} \\
&= \lim_{h\to 0} \frac{\left(\pi\circ\widetilde{\psi}_h\circ\sigma_t\right)^*\alpha - (\pi\circ\sigma_t)^*\alpha}{h} \\
&= \sigma_t^*\left(\lim_{h\to 0}\frac{\widetilde{\psi}_h^*\pi^*\alpha - \pi^*\alpha}{h}\right) \\
&= \sigma_t^*\left(\mathcal{L}_{\widetilde{X}}(\pi^*\alpha)\right) \\
&= \sigma_t^*\left(\tilde{d}(i_{\widetilde{X}}(\pi^*\alpha)) + i_{\widetilde{X}}\tilde{d}(\pi^*\alpha)\right) \\
&= d(i_{X_t}\alpha) + i_{X_t}d\alpha.
\end{aligned}$$

This concludes the alternative proof. □

From this direct proof we see, conversely, that the statement of Lemma B.1 for an isotopy $\psi_t\colon M \to M$ is a straightforward consequence of the definitions. Moreover, we obtain $(\mathcal{L}_{X_t}\alpha)|_{t=t_0} = \mathcal{L}_{X_{t_0}}\alpha$.

Thus, with hindsight, one could take this last identity as the definition of $L_{X_t}\alpha$. This would render the Cartan formula tautological. Instead, one would now need the proof of Lemma B.2 for showing identity (B.2).

A reasonable generalisation of definition (B.1) to the Lie derivative $L_{X_t}\alpha_t$ of a *time-dependent* differential form α_t would seem to be

$$\lim_{h\to 0}\frac{(\psi_{t+h}\circ\psi_t^{-1})^*\alpha_{t+h} - \alpha_t}{h}.$$

The accepted convention in the literature, however, is to interpret the expression $L_{X_t}\alpha_t$ at $t = t_0$ as a conventional Lie derivative $L_{X_{t_0}}\alpha_{t_0}$.

This gives the identities

$$\mathcal{L}_{X_t}\alpha_t = \lim_{h\to 0}\frac{(\psi_{t+h}\circ\psi_t^{-1})^*\alpha_t - \alpha_t}{h}$$

and

$$\mathcal{L}_{X_t}\alpha_t = d(i_{X_t}\alpha_t) + i_{X_t}d\alpha_t.$$

Observe that this 'conventional' definition of $\mathcal{L}_{X_t}\alpha_t$ differs from the 'reasonable' one above by the term

$$\lim_{h\to 0}\frac{(\psi_{t+h}\circ\psi_t^{-1})^*\alpha_{t+h} - (\psi_{t+h}\circ\psi_t^{-1})^*\alpha_t}{h} = \dot{\alpha}_t.$$

References

[1] E. Abbena, S. Garbiero and S. Salamon, Hermitian geometry on the Iwasawa manifold, *Boll. Un. Mat. Ital. B (7)* **11** (1997), no. 2, suppl., 231–249.

[2] K. Abe and J. Erbacher, Nonregular contact structures on Brieskorn manifolds, *Bull. Amer. Math. Soc.* **81** (1975), 407–409.

[3] R. Abraham and J. E. Marsden, *Foundations of Mechanics*, Benjamin Cummings Publishing Co., Reading, Mass. (1978).

[4] N. A'Campo and D. Kotschick, Contact structures, foliations, and the fundamental group, *Bull. London Math. Soc.* **26** (1994), 102–106.

[5] J. Adachi, Liouville setup and contact cobordisms, *Hokkaido Math. J.* **25** (1996), 637–650.

[6] D. Adams, *The Hitch Hiker's Guide to the Galaxy*, Pan Books, London (1979).

[7] B. Aebischer, M. Borer, M. Kälin, Ch. Leuenberger and H. M. Reimann, *Symplectic Geometry*, Progr. Math. **124**, Birkhäuser Verlag, Basel (1994).

[8] S. Akbulut and B. Özbağci, On the topology of compact Stein surfaces, *Int. Math. Res. Not.* **2002**, 769–782.

[9] C. Albert, Le théorème de réduction de Marsden–Weinstein en géométrie cosymplectique et de contact, *J. Geom. Phys.* **6** (1989), 627–649.

[10] J. W. Alexander, A lemma on systems of knotted curves, *Proc. Nat. Acad. Sci. U.S.A.* **9** (1923), 93–95.

[11] A. Andreotti and T. Frankel, The Lefschetz theorem on hyperplane sections, *Ann. of Math. (2)* **69** (1959), 713–717.

[12] Apollonius of Perga, *Treatise on Conic Sections* (T. L. Heath, ed.), Cambridge University Press (1896).

[13] V. I. Arnold, *Mathematical Methods of Classical Mechanics*, Grad. Texts in Math. **60**, Springer-Verlag, Berlin (1978).

[14] V. I. Arnold, *Geometrical Methods in the Theory of Ordinary Differential Equations* (2nd edition), Grundlehren Math. Wiss. **250**, Springer-Verlag, Berlin (1988).

[15] V. I. Arnold, *Contact Geometry and Wave Propagation*, Monogr. Enseign. Math. **34**, L'Enseignement Mathématique, Genève (1989); also in *Enseign. Math. (2)* **36** (1990), 215–266.

[16] V. I. Arnold, Contact geometry: the geometrical method of Gibbs's thermodynamics, in: *Proceedings of the Gibbs Symposium* (New Haven, 1989), American Mathematical Society, Providence (1990), 163–179.

[17] V. I. Arnold, Symplectic geometry and topology, *J. Math. Phys.* **41** (2000), 3307–3343.

[18] V. I. Arnold, *Lectures on Partial Differential Equations*, Universitext, Springer-

Verlag, Berlin (2004).

[19] M. ATIYAH, *K-Theory*, Adv. Book Classics, Addison-Wesley, Redwood City (1989).

[20] P. BAIRD AND J. C. WOOD, Hermitian structures and harmonic morphisms in higher-dimensional Euclidean spaces, *Internat. J. Math.* **6** (1995), 161–192.

[21] A. BANYAGA, *The Structure of Classical Diffeomorphism Groups*, Math. Appl. **400**, Kluwer Academic Publishers Group, Dordrecht (1997).

[22] D. BARDEN, Simply connected five-manifolds, *Ann. of Math. (2)* **82** (1965), 365–385.

[23] D. BARDEN AND C. B. THOMAS, *An Introduction to Differential Manifolds*, Imperial College Press, London (2003).

[24] BARTHOLOMAEUS ANGLICUS, *De proprietatibus rerum* (1235), translated as *On the Properties of Things* by John de Trevisa (1398).

[25] E. BEDFORD AND B. GAVEAU, Envelopes of holomorphy of certain 2-spheres in $\mathbb{C}^2$, *Amer. J. Math.* **105** (1983), 975–1009.

[26] D. BENNEQUIN, Entrelacements et équations de Pfaff, in: *IIIe Rencontre de Géométrie du Schnepfenried, Vol. 1*, Astérisque **107–108** (1983), 87–161.

[27] R. H. BING AND J. M. MARTIN, Cubes with knotted holes, *Trans. Amer. Math. Soc.* **155** (1971), 217–231.

[28] F. A. BOGOMOLOV AND B. DE OLIVEIRA, Stein small deformations of strictly pseudoconvex surfaces, in: *Birational Algebraic Geometry* (Baltimore, 1996), Contemp. Math. **207**, American Mathematical Society, Providence (1997), 25–41.

[29] W. M. BOOTHBY, Transitivity of the automorphisms of certain geometric structures, *Trans. Amer. Math. Soc.* **137** (1969), 93–100.

[30] W. M. BOOTHBY AND H. C. WANG, On contact manifolds, *Ann. of Math. (2)* **68** (1958), 721–734.

[31] R. BOTT AND L. W. TU, *Differential Forms in Algebraic Topology*, Grad. Texts in Math. **82**, Springer-Verlag, Berlin (1982).

[32] F. BOURGEOIS, Odd dimensional tori are contact manifolds, *Int. Math. Res. Not.* **2002**, 1571–1574.

[33] F. BOURGEOIS, A Morse–Bott approach to contact homology, in: *Symplectic and Contact Topology: Interactions and Perspectives* (Toronto/Montréal, 2001), Fields Inst. Commun. **35**, American Mathematical Society, Providence (2003), 55–77.

[34] C. P. BOYER, K. GALICKI AND M. NAKAMAYE, Sasakian geometry, homotopy spheres and positive Ricci curvature, *Topology* **42** (2003), 981–1002.

[35] G. E. BREDON, *Topology and Geometry*, Grad. Texts in Math. **139**, Springer-Verlag, Berlin (1993).

[36] E. BRIESKORN, Beispiele zur Differentialtopologie von Singularitäten, *Invent. Math.* **2** (1966), 1–14.

[37] TH. BRÖCKER AND T. TOM DIECK, *Representations of Compact Lie Groups*, Grad. Texts in Math. **98**, Springer-Verlag, Berlin (1985).

[38] TH. BRÖCKER AND K. JÄNICH, *Einführung in die Differentialtopologie*, Springer-Verlag, Berlin (1973).

[39] W. BROWDER, *Surgery on Simply-Connected Manifolds*, Ergeb. Math. Grenzgeb. **65**, Springer-Verlag, Berlin (1972).

[40] A. CANNAS DA SILVA, *Lectures on Symplectic Geometry*, Lecture Notes in Math. **1764**, Springer-Verlag, Berlin (2001).

[41] H.-D. CAO AND X.-P. ZHU, A complete proof of the Poincaré and geometrization conjectures – application of the Hamilton–Perelman theory of the Ricci flow, *Asian J. Math.* **10** (2006), 169–492.

[42] M. P. do Carmo, *Riemannian geometry*, Math. Theory Appl., Birkhäuser Verlag, Basel (1992).

[43] E. Cartan, Les groupes réels simples finis et continus, *Ann. Sci. École Norm. Sup.* **31** (1914), 263–355.

[44] J. Cerf, *Sur les difféomorphismes de la sphère de dimension trois* ($\Gamma_4 = 0$), Lecture Notes in Math. **53**, Springer-Verlag, Berlin (1968).

[45] Yu. Chekanov, Differential algebra of Legendrian links, *Invent. Math.* **150** (2002), 441–483.

[46] Yu. Chekanov, Invariants of Legendrian knots, in: *Proceedings of the International Congress of Mathematicians, Vol. II* (Beijing, 2002), Higher Education Press, Beijing (2002), 385–394.

[47] K. Cieliebak and Ya. Eliashberg, *Symplectic Geometry of Stein Manifolds*, book in preparation.

[48] V. Colin, Chirurgies d'indice un et isotopies de sphères dans les variétés de contact tendues, *C. R. Acad. Sci. Paris Sér. I Math.* **324** (1997), 659–663.

[49] V. Colin, E. Giroux and K. Honda, On the coarse classification of tight contact structures, in: *Topology and Geometry of Manifolds* (Athens, GA, 2001), Proc. Sympos. Pure Math. **71**, American Mathematical Society, Providence (2003), 109–120.

[50] F. Ding and H. Geiges, Symplectic fillability of tight contact structures on torus bundles, *Algebr. Geom. Topol.* **1** (2001), 153–172.

[51] F. Ding and H. Geiges, A Legendrian surgery presentation of contact 3–manifolds, *Math. Proc. Cambridge Philos. Soc.* **136** (2004), 583–598.

[52] F. Ding and H. Geiges, E_8–plumbings and exotic contact structures on spheres, *Int. Math. Res. Not.* **2004**, 3825–3837.

[53] F. Ding and H. Geiges, Legendrian knots and links classified by classical invariants, *Commun. Contemp. Math.* **9** (2007), 135–162.

[54] F. Ding and H. Geiges, A unique decomposition theorem for tight contact 3–manifolds, *Enseign. Math. (2)*, to appear.

[55] F. Ding, H. Geiges and A. I. Stipsicz, Surgery diagrams for contact 3–manifolds, *Turkish J. Math.* **28** (2004), 41–74; also in: *Proceedings of the 10th Gökova Geometry-Topology Conference 2003*, 41–74.

[56] F. Ding, H. Geiges and A. I. Stipsicz, Lutz twist and contact surgery, *Asian J. Math.* **9** (2005), 57–64.

[57] A. C. Doyle, *The Boscombe Valley Mystery*, in: *The Adventures of Sherlock Holmes*, George Newnes Ltd., London (1892).

[58] B. A. Dubrovin, A. T. Fomenko and S. P. Novikov, *Modern Geometry – Methods and Applications: Part II. The Geometry and Topology of Manifolds*, Grad. Texts in Math. **104**, Springer-Verlag, Berlin (1985).

[59] T. Duchamp, The classification of Legendre immersions, preprint (1982, 1996), available at `http://www.math.washington.edu/~duchamp`.

[60] K. Dymara, Legendrian knots in overtwisted contact structures on S^3, *Ann. Global Anal. Geom.* **19** (2001), 293–305.

[61] C. H. Edwards, Jr., Concentricity in 3–manifolds, *Trans. Amer. Math. Soc.* **113** (1964), 406–423.

[62] T. Ekholm, J. B. Etnyre and M. Sullivan, Non-isotopic Legendrian submanifolds in $\mathbb{R}^{2n+1}$, *J. Differential Geom.* **71** (2005), 85–128.

[63] Ya. Eliashberg, Three lectures on symplectic topology in Cala Gonone – Basic notions, problems and some methods, in: *Conference on Differential Geometry and Topology* (Sardinia, 1988), Rend. Sem. Fac. Sci. Univ. Cagliari **58** (1988), suppl., 27–49.

[64] YA. ELIASHBERG, Classification of overtwisted contact structures on 3–manifolds, *Invent. Math.* **98** (1989), 623–637.

[65] YA. ELIASHBERG, Topological characterization of Stein manifolds of dimension > 2, *Internat. J. Math.* **1** (1990), 29–46.

[66] YA. ELIASHBERG, Filling by holomorphic discs and its applications, in: *Geometry of Low-Dimensional Manifolds, Vol. 2* (Durham, 1989), London Math. Soc. Lecture Note Ser. **151**, Cambridge University Press (1990), 45–67.

[67] YA. ELIASHBERG, On symplectic manifolds with some contact properties, *J. Differential Geom.* **33** (1991), 233–238.

[68] YA. ELIASHBERG, New invariants of open symplectic and contact manifolds, *J. Amer. Math. Soc.* **4** (1991), 513–520.

[69] YA. ELIASHBERG, Contact 3–manifolds twenty years since J. Martinet's work, *Ann. Inst. Fourier (Grenoble)* **42** (1992), 165–192.

[70] YA. ELIASHBERG, Legendrian and transversal knots in tight contact 3–manifolds, in: *Topological Methods in Modern Mathematics* (Stony Brook, 1991), Publish or Perish, Houston (1993), 171–193.

[71] YA. ELIASHBERG, Unique holomorphically fillable contact structure on the 3–torus, *Internat. Math. Res. Notices* **1996**, 77–82.

[72] YA. ELIASHBERG, Symplectic geometry of plurisubharmonic functions, in: *Gauge Theory and Symplectic Geometry* (Montréal, 1995), Nato Adv. Sci. Inst. Ser. C Math. Phys. Sci. **488**, Kluwer Academic Publishers, Dordrecht (1997), 49–67.

[73] YA. ELIASHBERG, A few remarks about symplectic filling, *Geom. Topol.* **8** (2004), 277–293.

[74] YA. ELIASHBERG AND M. FRASER, Classification of topologically trivial Legendrian knots, in: *Geometry, Topology and Dynamics* (Montréal, 1995), CRM Proc. Lecture Notes **15**, American Mathematical Society, Providence (1998), 17–51.

[75] YA. ELIASHBERG, A. GIVENTAL AND H. HOFER, Introduction to symplectic field theory, *Geom. Funct. Anal.* **2000**, Special Volume, Part II, 560–673.

[76] YA. ELIASHBERG AND M. GROMOV, Convex symplectic manifolds, in: *Several Complex Variables and Complex Geometry* (Santa Cruz, 1989), Proc. Sympos. Pure Math. **52**, Part 2, American Mathematical Society, Providence (1991), 135–162.

[77] YA. ELIASHBERG AND N. MISHACHEV, *Introduction to the h–Principle*, Grad. Stud. Math. **48**, American Mathematical Society, Providence (2002).

[78] T. ERLANDSSON, The standard contact structure on S^{2n+1} and its stereographic image in $\mathbb{R}^{2n+1}$, unpublished note.

[79] J. B. ETNYRE, Symplectic convexity in low-dimensional topology, *Topology Appl.* **88** (1998), 3–25.

[80] J. B. ETNYRE, Introductory lectures on contact geometry, in: *Topology and Geometry of Manifolds* (Athens, GA, 2001), Proc. Sympos. Pure Math. **71**, American Mathematical Society, Providence (2003), 81–107.

[81] J. B. ETNYRE, On symplectic fillings, *Algebr. Geom. Topol.* **4** (2004), 73–80.

[82] J. B. ETNYRE, Legendrian and transversal knots, in: *Handbook of Knot Theory* (W. Menasco and M. Thistlethwaite, eds.), Elsevier, Amsterdam (2005), 105–185.

[83] J. B. ETNYRE, Lectures on open book decompositions and contact structures, in: *Floer Homology, Gauge Theory, and Low-Dimensional Topology*, Clay Math. Proc. **5**, American Mathematical Society, Providence (2006), 103–141.

[84] J. B. ETNYRE, Convex surfaces in contact geometry: class notes (2004), available at `http://www.math.gatech.edu/~etnyre`.

[85] J. B. ETNYRE AND K. HONDA, Knots and contact geometry I. Torus knots and figure eight knots, *J. Symplectic Geom.* **1** (2001), 63–120.

[86] J. B. Etnyre and K. Honda, On the nonexistence of tight contact structures, *Ann. of Math. (2)* **153** (2001), 749–766.

[87] J. B. Etnyre and K. Honda, Tight contact structures with no symplectic fillings, *Invent. Math.* **148** (2002), 609–626.

[88] R. A. Fenn, *Techniques of Geometric Topology*, London Math. Soc. Lecture Note Ser. **57**, Cambridge University Press (1983).

[89] L. E. Fraenkel, *An Introduction to Maximum Principles and Symmetry in Elliptic Problems*, Cambridge Tracts in Math. **128**, Cambridge University Press (2000).

[90] H. Freudenthal, Zum Hopfschen Umkehrhomomorphismus, *Ann. of Math. (2)* **38** (1937), 847–853.

[91] H. Geiges, Contact structures on 1–connected 5–manifolds, *Mathematika* **38** (1991), 303–311.

[92] H. Geiges, Applications of contact surgery, *Topology* **36** (1997), 1193–1220.

[93] H. Geiges, Constructions of contact manifolds, *Math. Proc. Cambridge Philos. Soc.* **121** (1997), 455–464.

[94] H. Geiges, A brief history of contact geometry and topology, *Expo. Math.* **19** (2001), 25–53; Chinese translation in *Shuxue Yilin* **25** (2006), 133–148.

[95] H. Geiges, *h–Principles and Flexibility in Geometry*, Mem. Amer. Math. Soc. **164** (2003), no. 779.

[96] H. Geiges, Christiaan Huygens and contact geometry, *Nieuw Arch. Wiskd. (5)* **6** (2005), 117–123.

[97] H. Geiges, Contact geometry, in: *Handbook of Differential Geometry, Vol. 2* (F. J. E. Dillen and L. C. A. Verstraelen, eds.), North-Holland, Amsterdam (2006), 315–382.

[98] H. Geiges, Contact Dehn surgery, symplectic fillings, and Property P for knots, *Expo. Math.* **24** (2006), 273–280.

[99] H. Geiges, A contact geometric proof of the Whitney–Graustein theorem, preprint.

[100] P. Ghiggini, Strongly fillable contact 3–manifolds without Stein fillings, *Geom. Topol.* **9** (2005), 1677–1687.

[101] P. Ghiggini, Linear Legendrian curves in T^3, *Math. Proc. Cambridge Philos. Soc.* **140** (2006), 451–473.

[102] P. Ghiggini and S. Schönenberger, On the classification of tight contact structures, in: *Topology and Geometry of Manifolds* (Athens, GA, 2001), Proc. Sympos. Pure Math. **71**, American Mathematical Society, Providence (2003), 121–151.

[103] V. Ginzburg, On closed characteristics of 2–forms, Ph.D. Thesis, UC Berkeley (1990).

[104] E. Giroux, Convexité en topologie de contact, *Comment. Math. Helv.* **66** (1991), 637–677.

[105] E. Giroux, Structures de contact en dimension trois et bifurcations des feuilletages de surfaces, *Invent. Math.* **141** (2000), 615–689.

[106] E. Giroux, Géométrie de contact: de la dimension trois vers les dimensions supérieures, in: *Proceedings of the International Congress of Mathematicians, Vol. II* (Beijing, 2002), Higher Education Press, Beijing (2002), 405–414.

[107] E. Giroux, Contact structures and symplectic fibrations over the circle, transparencies of a seminar talk.

[108] E. Giroux and J.-P. Mohsen, Structures de contact et fibrations symplectiques au-dessus du cercle, in preparation.

[109] R. E. Gompf, A new construction of symplectic manifolds, *Ann. of Math. (2)* **142** (1995), 527–595.

[110] R. E. GOMPF, Handlebody constructions of Stein surfaces, *Ann. of Math. (2)* **148** (1998), 619–693.

[111] R. E. GOMPF AND A. I. STIPSICZ, *4-Manifolds and Kirby Calculus*, Grad. Stud. Math. **20**, American Mathematical Society, Providence (1999).

[112] J. GONZALO, Branched covers and contact structures, *Proc. Amer. Math. Soc.* **101** (1987), 347–352.

[113] F. GONZÁLEZ-ACUÑA, Dehn's construction on knots, *Bol. Soc. Mat. Mexicana (2)* **15** (1970), 58–79.

[114] C. McA. GORDON AND J. LUECKE, Knots are determined by their complements, *J. Amer. Math. Soc.* **2** (1989), 371–415.

[115] D. H. GOTTLIEB, Partial transfers, in: *Geometric Applications of Homotopy Theory I*, Lecture Notes in Math. **657**, Springer-Verlag, Berlin (1978), 255–266.

[116] H. GRAUERT, On Levi's problem and the imbedding of real-analytic manifolds, *Ann. of Math. (2)* **68** (1958), 460–472.

[117] J. W. GRAY, Some global properties of contact structures, *Ann. of Math. (2)* **69** (1959), 421–450.

[118] M. GROMOV, Pseudoholomorphic curves in symplectic manifolds, *Invent. Math.* **82** (1985), 307–347.

[119] M. GROMOV, *Partial Differential Relations*, Ergeb. Math. Grenzgeb. (3) **9**, Springer-Verlag, Berlin (1986).

[120] R. C. GUNNING AND H. ROSSI, *Analytic Functions of Several Complex Variables*, Prentice-Hall, Englewood Cliffs (1965).

[121] C. GUTIERREZ AND B. PIRES, On Peixoto's conjecture for flows on non-orientable 2–manifolds, *Proc. Amer. Math. Soc.* **133** (2005), 1063–1074.

[122] A. HAEFLIGER, Plongements différentiables de variétés dans variétés, *Comment. Math. Helv.* **36** (1961), 47–82.

[123] P. HARTMAN, On local homeomorphisms of Euclidean spaces, *Bol. Soc. Mat. Mexicana (2)* **5** (1960), 220–241.

[124] Y. HATAKEYAMA, Some notes on the group of automorphisms of contact and symplectic structures, *Tôhoku Math. J. (2)* **18** (1966), 338–347.

[125] A. HATCHER, *Algebraic Topology*, Cambridge University Press (2002).

[126] G. HECTOR AND U. HIRSCH, *Introduction to the Theory of Foliations – Part A*, Aspects Math. **1**, Friedr. Vieweg & Sohn, Braunschweig (1981).

[127] S. HELGASON, *Differential Geometry, Lie Groups, and Symmetric Spaces*, Grad. Stud. Math. **34**, American Mathematical Society, Providence (2001).

[128] J. HEMPEL, *3-Manifolds*, Ann. of Math. Stud. **86**, Princeton University Press (1976).

[129] H. M. HILDEN, Three-fold branched coverings of S^3, *Amer. J. Math.* **98** (1976), 989–997.

[130] H. M. HILDEN, J. M. MONTESINOS AND T. THICKSTUN, Closed oriented 3–manifolds as 3–fold branched coverings of S^3 of special type, *Pacific J. Math.* **65** (1976), 65–76.

[131] C. D. HILL AND M. NACINOVICH, Stein fillability and the realization of contact manifolds, *Proc. Amer. Math. Soc.* **133** (2005), 1843–1850.

[132] M. W. HIRSCH, *Differential Topology*, Grad. Texts in Math. **33**, Springer-Verlag, Berlin (1976).

[133] F. HIRZEBRUCH AND K. H. MAYER, O(n)*-Mannigfaltigkeiten, exotische Sphären und Singularitäten*, Lecture Notes in Math. **57**, Springer-Verlag, Berlin (1968).

[134] T. HOBBES, *Leviathan*, Andrew Crooke, London (1651).

[135] H. HOFER, Holomorphic curves and dynamics in dimension three, in: *Symplectic Geometry and Topology* (Park City, 1997), IAS/Park City Math. Ser. **7**, American Mathematical Society, Providence (1999).

[136] H. Hofer, K. Wysocki and E. Zehnder, Pseudoholomorphic curves and dynamics in three dimensions, in: *Handbook of Dynamical Systems, Vol. 1A* (B. Hasselblatt and A. Katok, eds.), North-Holland, Amsterdam (2002), 1129–1188.

[137] K. Honda, On the classification of tight contact structures I, *Geom. Topol.* **4** (2000), 309–368; erratum: Factoring nonrotative $T^2 \times I$ layers, *Geom. Topol.* **5** (2001), 925–938.

[138] K. Honda, Gluing tight contact structures, *Duke Math. J.* **115** (2002), 435–478.

[139] H. Hopf, Zur Algebra der Abbildungen von Mannigfaltigkeiten, *J. Reine Angew. Math.* **105** (1930), 71–88.

[140] D. Husemoller, *Fibre Bundles* (3rd edition), Springer-Verlag, Berlin (1994).

[141] C. Isherwood, *Goodbye to Berlin*, The Hogarth Press, London (1939).

[142] K. Jänich, *Differenzierbare G–Mannigfaltigkeiten*, Lecture Notes in Math. **59**, Springer-Verlag, Berlin (1968).

[143] J. Jost and X. Li-Jost, *Calculus of Variations*, Cambridge Stud. Adv. Math. **64**, Cambridge University Press (1998).

[144] Y. Kanda, The classification of tight contact structures on the 3–torus, *Comm. Anal. Geom.* **5** (1997), 413–438.

[145] M. A. Kervaire and J. W. Milnor, Groups of homotopy spheres I, *Ann. of Math. (2)* **77** (1963), 504–537.

[146] R. C. Kirby, *The Topology of 4–Manifolds*, Lecture Notes in Math. **1374**, Springer-Verlag, Berlin (1989).

[147] H. Kneser, Geschlossene Flächen in dreidimensionalen Mannigfaltigkeiten, *Jahresber. Deutsch. Math.-Verein.* **38** (1929), 248–260.

[148] S. Kobayashi, Principal fibre bundles with the 1–dimensional toroidal group, *Tôhoku Math. J. (2)* **8** (1956), 29–45.

[149] O. van Koert, Open books for contact five-manifolds and applications of contact homology, Inaugural-Dissertation, Universität zu Köln (2005).

[150] O. van Koert, Contact homology of Brieskorn manifolds, *Forum Math.*, to appear.

[151] O. van Koert, Open books on contact 5–manifolds, *Ann. Inst. Fourier (Grenoble)*, to appear.

[152] O. van Koert and K. Niederkrüger, Open book decompositions for contact structures on Brieskorn manifolds, *Proc. Amer. Math. Soc.* **133** (2005), 3679–3686.

[153] A. A. Kosinski, *Differential Manifolds*, Academic Press, Boston (1993).

[154] P. B. Kronheimer and T. S. Mrowka, Witten's conjecture and Property P, *Geom. Topol.* **8** (2004), 295–310.

[155] J. Lanchester, *The Debt to Pleasure*, Picador, London (1996).

[156] S. Lefschetz, *L'Analysis Situs et la Géométrie Algébrique*, Gauthier-Villars, Paris (1950).

[157] W. Lewis, *Letters* (W. K. Rose, ed.), W. W. Norton, New York (1963).

[158] W. B. R. Lickorish, A representation of orientable combinatorial 3–manifolds, *Ann. of Math. (2)* **76** (1962), 531–540.

[159] S. Lie, Zur Theorie partieller Differentialgleichungen, *Göttinger Nachrichten* (1872), pp. 480 ff.

[160] S. Lie, *Geometrie der Berührungstransformationen* (dargestellt von S. Lie und G. Scheffers), B. G. Teubner, Leipzig (1896).

[161] S. Lie and F. Engel, *Theorie der Transformationsgruppen*, three volumes, B. G. Teubner, Leipzig (1888–1893).

[162] P. Lisca, On fillable contact structures up to homotopy, *Proc. Amer. Math. Soc.* **129** (2001), 3437–3444.

[163] P. Lisca and G. Matić, Tight contact structures and Seiberg–Witten invariants, *Invent. Math.* **129** (1997), 509–525.

[164] P. Lisca and A. I. Stipsicz, Tight, not semi-fillable contact circle bundles, *Math. Ann.* **328** (2004), 285–298.

[165] P. Lisca and A. I. Stipsicz, Ozsváth–Szabó invariants and contact surgery, in: *Floer Homology, Gauge Theory, and Low-Dimensional Topology*, Clay Math. Proc. **5**, American Mathematical Society, Providence (2006), 171–180.

[166] F. Loose, Reduction in contact geometry, *J. Lie Theory* **11** (2001), 9–22.

[167] R. Lutz, Sur l'existence de certaines formes différentielles remarquables sur la sphère S^3, *C. R. Acad. Sci. Paris Sér. A–B* **270** (1970), A1597–A1599.

[168] R. Lutz, Sur quelques propriétés des formes differentielles en dimension trois, Thèse, Strasbourg (1971).

[169] R. Lutz, Structures de contact sur les fibrés principaux en cercles de dimension trois, *Ann. Inst. Fourier (Grenoble)* **27** (1977), no. 3, 1–15.

[170] R. Lutz, Quelques remarques historiques et prospectives sur la géométrie de contact, in: *Conference on Differential Geometry and Topology* (Sardinia, 1988), Rend. Sem. Fac. Sci. Univ. Cagliari **58** (1988), suppl., 361–393.

[171] R. Lutz and C. Meckert, Structures de contact sur certaines sphères exotiques, *C. R. Acad. Sci. Paris Sér. A–B* **282** (1976), no. 11, A591–A593.

[172] H. C. Lyon, Torus knots in the complements of links and surfaces, *Michigan Math. J.* **27** (1980), 39–46.

[173] J. E. Marsden and A. Weinstein, Reduction of symplectic manifolds with symmetry, *Rep. Mathematical Phys.* **5** (1974), 121–130.

[174] J. E. Marsden and A. Weinstein, Comments on the history, theory and applications of symplectic reduction, in: *Quantization of Singular Symplectic Quotients*, Progr. Math. **198**, Birkhäuser Verlag, Basel (2001), 1–19.

[175] J. Martinet, Formes de contact sur les variétés de dimension 3, in: *Proc. Liverpool Singularities Sympos. II*, Lecture Notes in Math. **209**, Springer-Verlag, Berlin (1971), 142–163.

[176] D. McDuff, Symplectic manifolds with contact type boundaries, *Invent. Math.* **103** (1991), 651–671.

[177] D. McDuff and D. Salamon, *Introduction to Symplectic Topology* (2nd edition), Oxford Math. Monogr., The Clarendon Press, Oxford (1998).

[178] C. Meckert, Forme de contact sur la somme connexe de deux variétés de contact de dimension impaire, *Ann. Inst. Fourier (Grenoble)* **32** (1982), no. 3, 251–260.

[179] K. R. Meyer, Symmetries and integrals in mechanics, in: *Dynamical Systems* (Salvador, 1971), Academic Press, New York (1973), 259–272.

[180] S. Milligan, The Starlings, in: *The Goon Show*, Series 4, Special Programme, broadcast 31st August 1954.

[181] R. S. Millman and G. D. Parker, *Elements of Differential Geometry*, Prentice-Hall, Englewood Cliffs (1977).

[182] J. W. Milnor, On spaces having the homotopy type of a *CW*–complex, *Trans. Amer. Math. Soc.* **90** (1959), 272–280.

[183] J. W. Milnor, A procedure for killing homotopy groups of differentiable manifolds, in: *Differential Geometry*, Proc. Sympos. Pure Math. **3** (1961), 39–55.

[184] J. W. Milnor, A unique decomposition theorem for 3–manifolds, *Amer. J. Math.* **84** (1962), 1–7.

[185] J. W. Milnor, *Morse Theory*, Ann. of Math. Stud. **51**, Princeton University Press (1963).

[186] J. W. Milnor, *Topology from the Differentiable Viewpoint*, The University Press of Virginia, Charlottesville (1965).

[187] J. W. Milnor, *Lectures on the h–Cobordism Theorem*, Princeton University Press (1965).

[188] J. W. Milnor and J. D. Stasheff, *Characteristic Classes*, Ann. of Math. Stud. **76**, Princeton University Press (1974).

[189] E. E. Moise, Affine structures on 3–manifolds V: The triangulation theorem and Hauptvermutung, *Ann. of Math. (2)* **56** (1952), 96–114.

[190] E. E. Moise, *Geometric Topology in Dimensions* 2 *and* 3, Grad. Texts in Math. **47**, Springer-Verlag, Berlin (1977).

[191] J. M. Montesinos, Three-manifolds as 3–fold branched covers of S^3, *Quart. J. Math. Oxford Ser. (2)* **27** (1976), 85–94.

[192] J. C. Moore, On a theorem of Borsuk, *Fund. Math.* **43** (1956), 195–201.

[193] J. Moser, On the volume elements on a manifold, *Trans. Amer. Math. Soc.* **120** (1965), 286–294.

[194] J. Munkres, Obstructions to the smoothing of piecewise differentiable homeomorphisms, *Ann. of Math. (2)* **72** (1960), 521–554.

[195] J. Munkres, Differentiable isotopies on the 2–sphere, *Michigan Math. J.* **7** (1960), 193–197.

[196] J. Munkres, Obstructions to imposing differentiable structures, *Illinois J. Math.* **8** (1964), 361–376.

[197] R. Myers, Open book decompositions of 3–manifolds, *Proc. Amer. Math. Soc.* **72** (1978), 397–402.

[198] K. Niederkrüger, Compact Lie group actions on contact manifolds, Inaugural-Dissertation, Universität zu Köln (2005).

[199] K. Niederkrüger, A generalization of the overtwisted disk to higher dimensions, *Algebr. Geom. Topol.* **6** (2006), 2473–2508.

[200] B. Özbağci and A. I. Stipsicz, *Surgery on Contact* 3*–Manifolds and Stein Surfaces*, Bolyai Soc. Math. Stud. 13, Springer-Verlag, Berlin (2004).

[201] P. Ozsváth and Z. Szabó, An introduction to Heegaard Floer homology, in: *Floer Homology, Gauge Theory, and Low-Dimensional Topology*, Clay Math. Proc. **5**, American Mathematical Society, Providence (2006), 3–27.

[202] P. Ozsváth and Z. Szabó, Lectures on Heegaard Floer homology, in: *Floer Homology, Gauge Theory, and Low-Dimensional Topology*, Clay Math. Proc. **5**, American Mathematical Society, Providence (2006), 29–70.

[203] J. Palis, Jr. and W. de Melo, *Geometric Theory of Dynamical Systems – An Introduction*, Springer-Verlag, Berlin (1982).

[204] G. P. Paternain, *Geodesic Flows*, Progr. Math. **180**, Birkhäuser Verlag, Basel (1999).

[205] M. M. Peixoto, Structural stability on two-dimensional manifolds, *Topology* **1** (1962), 101–120.

[206] M. M. Peixoto, Structural stability on two-dimensional manifolds – a further remark, *Topology* **2** (1963), 179–180.

[207] M. M. Peixoto, On the classification of flows on 2–manifolds, in: *Dynamical Systems* (Salvador, 1971), Academic Press, New York (1973), 389–419.

[208] G. Perelman, The entropy formula for the Ricci flow and its geometric applications, `ArXiv: math.DG/0211159`.

[209] V. V. Prasolov and A. B. Sossinsky, *Knots, Links, Braids and 3–Manifolds*, Transl. Math. Monogr. **154**, American Mathematical Society, Providence (1997).

[210] F. Presas, Lefschetz type pencils on contact manifolds, *Asian J. Math.* **6** (2002), 277–301.

[211] R. M. Range, *Holomorphic Functions and Integral Representations in Several Complex Variables*, Grad. Texts in Math. **108**, Springer-Verlag, Berlin (1986).

[212] A. RANICKI, *High-Dimensional Knot Theory – Algebraic Surgery in Codimension 2*, with an appendix by E. Winkelnkemper, Springer Monogr. Math., Springer-Verlag, Berlin (1998).

[213] A. RANICKI, *Algebraic and Geometric Surgery*, Oxford Math. Monogr., The Clarendon Press, Oxford (2002).

[214] C. ROBINSON, *Dynamical Systems – Stability, Symbolic Dynamics, and Chaos* (2nd edition), Stud. Adv. Math., CRC Press, Boca Raton (1999).

[215] D. ROLFSEN, *Knots and Links*, Publish or Perish, Houston (1976).

[216] J. ROSENBERG, Surgery theory today – what it is and where it's going, in: *Surveys on Surgery Theory, Vol. 2* (S. Cappell, A. Ranicki, and J. Rosenberg, eds.), Ann. of Math. Stud. **149**, Princeton University Press (2001), 3–47.

[217] S. SASAKI AND C. J. HSÜ, On a property of Brieskorn manifolds, *Tôhoku Math. J. (2)* **28** (1976), 67–78.

[218] A. SCHOPENHAUER, *Gesammelte Werke* (L. Lütkehaus, Hrsg.), Haffmans Verlag, Zürich (1988).

[219] P. SCOTT, The geometries of 3–manifolds, *Bull. London Math. Soc.* **15** (1983), 401–487.

[220] W. SHAKESPEARE, *Hamlet*, The Clarendon Press, Oxford (1987).

[221] C. L. SIEGEL, *Vorlesungen über Himmelsmechanik*, Grundlehren Math. Wiss. **85**, Springer-Verlag, Berlin (1956).

[222] S. SMALE, Diffeomorphisms of the 2–sphere, *Proc. Amer. Math. Soc.* **10** (1959), 621–626.

[223] E. H. SPANIER, *Algebraic Topology*, McGraw Hill Book Co., New York (1966).

[224] N. STEENROD, *The Topology of Fibre Bundles*, Princeton University Press (1951).

[225] J. STILLWELL, *Classical Topology and Combinatorial Group Theory* (2nd edition), Grad. Texts in Math. **72**, Springer-Verlag, Berlin (1993).

[226] A. I. STIPSICZ, Contact surgery and Heegaard Floer theory, in: *Floer Homology, Gauge Theory, and Low-Dimensional Topology*, Clay Math. Proc. **5**, American Mathematical Society, Providence (2006), 143–170.

[227] I. TAMURA, *Topology of Foliations – An Introduction*, Transl. Math. Monogr. **97**, American Mathematical Society, Providence (1992).

[228] C. H. TAUBES, The Seiberg–Witten equations and the Weinstein conjecture, `ArXiv: math.SG/0611007`.

[229] C. B. THOMAS, Almost regular contact manifolds, *J. Differential Geometry* **11** (1976), 521–533.

[230] R. THOM, Quelques propriétés globales des variétés différentiables, *Comment. Math. Helv.* **28** (1954), 17–86.

[231] W. P. THURSTON, A norm for the homology of 3–manifolds, *Mem. Amer. Math. Soc.* **59** (1986), no. 339, 99–130.

[232] W. P. THURSTON, *Three-Dimensional Geometry and Topology, Vol. 1* (S. Levy, ed.), Princeton Math. Ser. **35**, Princeton University Press (1997).

[233] W. P. THURSTON AND H. E. WINKELNKEMPER, On the existence of contact forms, *Proc. Amer. Math. Soc.* **52** (1975), 345–347.

[234] I. USTILOVSKY, Contact homology and contact structures on S^{4m+1}, Ph.D. Thesis, Stanford University (1999).

[235] I. USTILOVSKY, Infinitely many contact structures on S^{4m+1}, *Internat. Math. Res. Notices* **1999**, 781–791.

[236] C. T. C. WALL, Killing the middle homotopy groups of odd dimensional manifolds, *Trans. Amer. Math. Soc.* **103** (1962), 421–433.

[237] A. H. WALLACE, Modifications and cobounding manifolds, *Canad. J. Math.* **12** (1960), 503–528.

[238] F. W. WARNER, *Foundations of Differentiable Manifolds and Lie Groups*, Grad. Texts in Math. **94**, Springer-Verlag, Berlin (1983).

[239] A. Weinstein, Symplectic manifolds and their Lagrangian submanifolds, *Advances in Math.* **6** (1971), 329–346.

[240] A. Weinstein, On the hypotheses of Rabinowitz' periodic orbit theorems, *J. Differential Equations* **33** (1979), 353–358.

[241] A. Weinstein, Contact surgery and symplectic handlebodies, *Hokkaido Math. J.* **20** (1991), 241–251.

[242] G. W. Whitehead, *Elements of Homotopy Theory*, Grad. Texts in Math. **61**, Springer-Verlag, Berlin (1978).

[243] H. Whitney, On regular closed curves on the plane, *Compositio Math.* **4** (1937), 276–284.

[244] C. Willett, Contact reduction, *Trans. Amer. Math. Soc.* **354** (2002), 4245–4260.

[245] I. Yokota, Explicit isomorphism between SU(4) and Spin(6), *J. Fac. Sci. Shinshu Univ.* **14** (1979), 29–34.

[246] K. Zehmisch, The Eliashberg–Gromov tightness theorem, Diplomarbeit, Universität Leipzig (2003).

Notation index

Symbol	Meaning
TM	tangent bundle of the manifold M 2
$\ker \alpha$	kernel of the differential 1–form α 2
$-/-$	quotient of bundle by sub-bundle 2
	quotient of vector space by subspace 15
$(-)^{\perp}$	orthogonal complement of sub-bundle 2
	symplectic orthogonal complement of subspace 14
	symplectic orthogonal complement of sub-bundle 68
$\oplus$	Whitney sum of vector bundles 2
$\cong$	isomorphism of vector bundles 2
	isomorphism of groups 43
	homeo- or diffeomorphism of manifolds 45
$(-)\vert_U$	restriction of vector bundle to the subset U of the base manifold 2
	restriction of map or linear form to the subspace U 15
C^{∞}	smooth or (infinitely often) differentiable 2
$\equiv$	identically equal to 2
	canonically identified with 58
T^*M	cotangent bundle of the manifold M 3
$(-)^*$	dual bundle 3
	dual vector space, dual Lie algebra 14, 360
	adjoint of matrix 18
T_pM	tangent space of the manifold M at the point p 3
ξ_p	subspace defined by the distribution ξ at the point p 3
$\wedge$	wedge or exterior product 3
$[X, Y]$	Lie bracket of the vector fields X and Y 3
	homotopy classes of maps from the space X to the space Y 107
X_p, $X(p)$	tangent vector at the point p 3, 166
$d\alpha$	exterior derivative of the differential form α 3

(M,ξ)	contact manifold 4
$\mathbb{R}$	field of real numbers 4
$\mathbb{R}^n$	Euclidean n–space 4
$B \setminus A$	complement of the subset $A \subset B$ in B 4
∂_x	velocity vector of the x–curves; vector field given by regarding the partial derivative $\frac{\partial}{\partial x}$ as a derivation 5
R_α	Reeb vector field of the contact form α 5
$\alpha\|_{T_pM}$, α_p	linear form on T_pM defined by the differential form α 5, 164
$\langle X \rangle$	line field spanned by the vector field X 5
$\mathbb{P}(-)$	projectivisation of a vector space or vector bundle 6
Tf	differential of a map f between manifolds 7
$\times$	Cartesian product 7
	vector product in $\mathbb{R}^3$ 39
$\mathbb{R}P^n$	real projective n–space 7
$(p_0 : \ldots : p_n)$	homogeneous coordinates on projective n–space 7
S^n	n–sphere 8
T^n	n–torus $S^1 \times \cdots \times S^1$ (n factors) 8
$:=$, $=:$	term on the left is defined by term on the right (respectively, the other way round) 8, 20
z', z''	first or second derivative, respectively, of the function $x \mapsto z(x)$ w.r.t. the variable x 9, 10
f^*	induced map on differential forms 9
	induced map on cohomology 106
$\mathbf{pq}$	$\sum_j p_j q_j$ 11
A^t	transpose of the matrix A 11
id	identity map 12
$(-,\omega)$	symplectic vector space 14
	symplectic manifold 20
	symplectic vector bundle 64
im, ker	image or kernel, respectively, of map or homomorphism 15
$\mathbf{0}$	zero element or origin of vector space 15
$[-]$	equivalence class in quotient space 16
	homology class 106
	homotopy class 107
δ_{ij}	Kronecker symbol (1 for $i = j$; 0 for $i \neq j$) 16
J	complex (bundle) structure 17, 64
$\mathbb{C}$	field of complex numbers 17
i	$\sqrt{-1} \in \mathbb{C}$ 17
$\mathcal{J}(\omega)$	space of ω–compatible complex structures 18

Author index

Subject index